COASTAL-MARINE CONSERVATION: SCIENCE AND POLICY

To Sally Lyons Brown
for her vision
and support

Slow down. . . .
You may be going the wrong way!

Library & Media Ctr.
Carroll Community College
1601 Washington Rd.
Westminster, MD 21157
WITHDRAWN

Coastal-Marine Conservation:

Science and Policy

G. Carleton Ray *and* **Jerry McCormick-Ray**

Both from
Department of Environmental Sciences
University of Virginia
Charlottesville
Virginia
USA

Illustrations by Robert L. Smith

Blackwell
Publishing

© 2004 by Blackwell Science Ltd
a Blackwell Publishing company

350 Main Street, Malden, MA 02148-5020, USA
108 Cowley Road, Oxford OX4 1JF, UK
550 Swanston Street, Carlton, Victoria 3053, Australia

The rights of G. Carleton Ray and Jerry McCormick-Ray to be identified
as the Authors of this Work have been asserted in accordance with the UK Copyright,
Designs, and Patents Act 1988.

All rights reserved. No part of this publication may be reproduced, stored in a retrieval system,
or transmitted, in any form or by any means, electronic, mechanical, photocopying, recording or
otherwise, except as permitted by the UK Copyright, Designs, and Patents Act 1988,
without the prior permission of the publisher.

First published 2004

Library of Congress Cataloging-in-Publication Data

Coastal-marine conservation : science and policy / [edited by] G. Carleton Ray and
Jerry McCormick-Ray ; illustrations by Robert L. Smith.
p. cm.
Includes bibliographical references (p.).
ISBN 0-632-05537-5 (pbk. : alk. paper)
1. Coastal ecology. 2. Nature conservation. 3. Endangered ecosystems.
I. Ray, G. Carleton. II. McCormick-Ray, Jerry.
QH541.5.C65 C5916 2003
333.95′616—dc21
2002038282

A catalogue record for this title is available from the British Library.

Set in 9½/11½pt Bembo
by Graphicraft Limited, Hong Kong
Printed and bound in the United Kingdom
by TJ International Ltd, Padstow, Cornwall

For further information on
Blackwell Publishing, visit our website:
http://www.blackwellpublishing.com

Contents

Authors of boxes

John E. Anderson, Ph.D., Assistant Research Professor, Department of Biology, Virginia Commonwealth University, Richmond, Virginia, and Research Biologist, Engineer Research and Development Center, Topographic Engineering Center, U.S. Army, Alexandria, Virginia, USA: Box 5.2 Created wetland mitigation: successes or failures?

John M. Baxter, Ph.D., Head, Marine and Coastal Section, Scottish Natural Heritage, Edinburgh, Scotland, UK: Box 2.2 The European Community Habitats Directive

Alan B. Bolten, Ph.D., Research Assistant Professor, and Karen A. Bjorndal, Ph.D., Professor of Zoology and Director, Archie Carr Center for Sea Turtle Research, University of Florida, Gainesville, Florida, USA: Box 7.5 Green turtles in the Caribbean: a shared resource

Michael R. Erwin, Ph.D., Wildlife Biologist, U.S. Geological Survey, and Research Professor, Department of Environmental Sciences, University of Virginia, Charlottesville, Virginia, USA: Box 5.3 Waterbirds: changing populations and changing habitats

James A. Estes, Ph.D., Wildlife Biologist, U.S. Geological Survey, and Adjunct Professor, University of California, Santa Cruz, California, USA: Box 9.2 How to design a sea otter reserve

Michael Garstang, Ph.D., Distinguished and Emeritus Professor, and Amber J. Soja, doctoral student, Department of Environmental Sciences, University of Virginia, Charlottesville, Virginia, USA: Box 3.1 Dust-to-dust: wind-blown material

Mark A. Hixon, Ph.D., Professor, Department of Zoology, Oregon State University, Corvallis, Oregon, USA: Box 7.4 How do so many kinds of coral-reef fishes co-exist?

Gary L. Hufford, Ph.D., Oceanographer, NOAA National Weather Service, Anchorage, Alaska, USA: Contributions to Introduction to case studies and to Chapter 6 Bering Sea: marine mammals in a regional sea; Box 9.6 Human factors: conflicts over the Kenai River, Alaska

Igor I. Krupnik, Ph.D., Ethologist, Arctic Studies Center, National Museum of Natural History, Smithsonian Institution, Washington, D.C., USA: Box 6.1 The subsistence era: early prehistory to Euro-American contacts; Box 6.2 The walrus in Native marine economies

Romuald N. Lipcius, Ph.D., Professor, and William T. Stockhausen, Ph.D., Senior Marine Scientist, Department of Fisheries Science, Virginia Institute of Marine Science, The College of William and Mary, Gloucester Point, Virginia, USA, and David B. Eggleston, Ph.D., Associate Professor, Department of Marine, Earth and Atmospheric Sciences, North Carolina State University, Raleigh, North Carolina, USA: Box 7.7 Metapopulation dynamics and marine reserves: Caribbean spiny lobster in Exuma Sound, Bahamas

Thomas R. Loughlin, Ph.D., Wildlife Biologist (Research), NOAA National Marine Mammal Laboratory, Seattle, Washington, USA: Contributions to Chapter 6 Bering Sea: marine mammals in a regional sea (Steller sea lion)

James G. Mead, Ph.D., Curator of Marine Mammals, Department of Systematic Biology, National Museum of Natural History, Smithsonian Institution, Washington, D.C., USA: Box 1.2 Cetacean strandings

Robert V. Miller, Ph.D. (retired), Snohomish, Washington, USA. Formerly Deputy Director, NOAA National Marine Mammal Laboratory, Seattle, Washington, USA: Box 2.4 U.S. legislation relating to marine mammals; Box 6.3 U.S.–Russia agreement on Cooperation in the Field of Protection of the Environment and Natural Resources

John C. Ogden, Ph.D., Director, Florida Institute of Oceanography, St. Petersburg, Florida, USA: Box 7.3 People and coral reefs

James E. Perry, Ph.D., Associate Professor, Department of Coastal and Ocean Policy, Virginia Institute of Marine Science, The College of William and Mary, Gloucester Point, Virginia, USA: Box 1.1 A perspective on an overabundant invasive species: common reed, *Phragmites australis*; Box 5.1 Temperate salt-marsh types of the U.S. east coast; Box 9.3 The Paraguay–Paraná hidrovía: protecting the Pantanal with lessons from the past

James H. Pipkin, Ll.D. (retired), Bethesda, Maryland, USA. Formerly U.S. Department of State Special Negotiator for Pacific Salmon (1994–2001); U.S. Federal Commissioner on the bilateral Pacific Salmon Commission (1994–2002); Counselor to the U.S. Secretary of the Interior (1993–8); and Director of the Interior Department's Office of Policy Analysis (1998–2001): Box 9.4 Pacific salmon: science policy

Frank M. Potter, J.D. (retired), Portland, Oregon, USA. Formerly Counsel, Subcommittee on Fisheries and Wildlife Conservation and the Environment, House Committee on Merchant Marine and Fisheries, and Chief Counsel, House Committee on Energy and Commerce, House of Representatives, Washington, D.C., USA: Box 9.7 Inventing foresight

Robert Prescott-Allen, PA Data, Victoria, British Columbia, Canada. Consultant on sustainable development and sustainability assessment; author of *The Wellbeing of Nations*; co-author of *Blueprint for Survival*, *World Conservation Strategy*, *The First Resource*, and *Caring for the Earth: A Strategy for Sustainable Living*: Box 9.1 Wellbeing assessment

Sam Ridgway, DVM, Ph.D., Senior Scientist, U.S. Navy Marine Mammal Program, and Professor of Comparative Pathology, Veterinary Medical Center, University of California, San Diego, California, USA: Box 1.3 Noise pollution: a threat to dolphins?

C. Richard Robins, Ph.D., Curator Emeritus, Natural History Museum and Biodiversity Research Center, University of Kansas, Lawrence, Kansas, USA: Box 7.2 Regional diversity among Caribbean fish species

K.O. Winemiller, Ph.D., Professor, Department of Wildlife and Fisheries Sciences, Texas A&M University, College Station, Texas, USA: Box 4.2 Life-history strategies of fishes

Preface

We have reached a point in history when biological knowledge is the sine qua non for a viable human future . . . A critical subset of society will have to understand the nature of life, the interaction of living creatures with their environment, and the strengths and limitations of the data and procedures of science itself. The acquisition of biological knowledge, so long a luxury except for those concerned with agriculture and the health sciences, has now become a necessity for all.

John A. Moore (1993)

Science is built up with facts, as a house is with stones. But a collection of facts is no more a science than a heap of stones is a house.

Jules Henri Poincaré, *La Science et L'Hypothèse* (1908)

We are at a time in history when science allows us better to understand our global environment, and when human societies clearly recognize the urgency of coastal-marine conservation and the need for sustainable use of resources. During the past century, humans have entered the sea in an era of the "Marine Revolution," acquiring the ability to intrude, exploit, and understand the last, previously unavailable portion of Earth – the oceans (Ray 1970). The rates and magnitude of change brought on by this Revolution follows 5–10 000 years of the Agricultural Revolution and two centuries of the Industrial Revolution. Observation of the quickening pace of change and the way that humans behave and manage themselves, and increasing knowledge of the way coastal and marine ecosystems function, have made apparent major ecosystem instabilities and incongruencies. Confronting these lies at the heart of conservation.

"Conservation," as defined in *Webster's Third New International Dictionary*, is the "deliberate, planned, or thoughtful preserving, guarding, or protecting." It is "planned management of a natural resource to prevent exploitation, destruction, or neglect." It is also "the wise utilization of a natural product," or "a field of knowledge concerned with coordination and plans for the practical application of data from ecology, limnology, pedology, or other sciences that are significant to preservation of natural resources." These definitions presume a working understanding of natural-resource science and illustrate that conservation is an issue-directed activity towards which science can provide guides for informed decision-making at all levels.

This book calls attention to the coastal realm as an ecosystem of global significance, where conservation science plays a fundamental role. This realm is a heterogeneous, extraordinarily complex, biologically diverse portion of Earth where the majority of humanity lives and conservation challenges are among the most urgent (Box P.1). This realm is also distinguished by fragmented jurisdictions, wherein no single agency has sole power to manage, conserve, or protect wide-ranging species or region-scale ecosystems. Many coastal-realm ecosystems are highly perturbed and have been altered into regimes not easily reversed. Often, environmental degradation is perpetuated by attitudes entrained into

Box P.1 Coastal-realm attributes

- Occupies 18% of Earth's surface, 8% of ocean surface, and less than 0.5% of ocean volume
- Provides up to 50% of global denitrification, 80% of global organic matter burial, 90% of global sedimentary mineralization, 75–90% of the global sink of suspended river load and associated elements and pollutants, and in excess of 50% of present-day global carbonate deposition
- Supplies approximately a quarter of global primary production, around 14% of global ocean production, and approximately 90% of world fish catch
- Hosts 60% of the world's people and two-thirds of the world's cities of more than 1.6 million people

Sources: From Holligan & Reiners (1992); Pernetta & Milliman (1995).

social practices and public policies, which assume the following: depleted resources and disturbed ecosystems will automatically recover if left alone: the size of the conservation budget is directly related to conservation effectiveness; conservation is a social science in which the people are managed, not ecosystems; legal mandates for resource protection result automatically, almost by fiat, from scientific information; and science is too uncertain and contradictory to guide conservation – that is, "action" is urgent and information can wait. These assumptions and the accelerating impact of human activities on the coastal realm force consideration of new ways of thinking and of applying conservation tools and practices, which demands an attempt to integrate human life styles and ecosystem behavior, and *vice versa*. History shows that unguided action lacking a scientific basis often leads to undesirable outcomes, and that without scientific information conservation action can lose direction. Conservation thus seeks better-informed, science-based and socially understood policies that can bring necessary changes into force before coastal ecosystems are propelled further into environmental debts from which recovery is difficult and expensive.

This book attempts to draw the reader into thinking about coastal-realm issues and conservation tools, and of linkages between conservation science and policy. Section I reviews the issues and tools for conservation; two chapters document the phenomena that concern conservationists directly, and the plethora of laws, regulations, agreements, organizations, and procedures that have been adopted to halt or minimize many of the effects. Section II provides the science background; its two chapters consider the coastal realm as a global ecosystem and how species' natural history affects and is affected by coastal-realm environments. Section III consists of three case studies: three regional examples, the temperate Chesapeake Bay, the sub-arctic Bering Sea, and the tropical Bahamas. A conservation issue is highlighted for each case study in the context of the region's history, its physical, chemical, and biological attributes, and its present social environment. These "real-world" coastal-realm conservation cases broadly reflect scientific and social-political challenges that are worldwide in scope. Section IV focuses on ecosystem health and conservation challenges; two chapters concern conservation issues in the broader contexts of ecosystem health and change, the environmental debt, and challenges facing coastal-realm conservation. Boxes by contributing authors provide detailed accounts of particular topics in order to amplify the text.

Our approach is not to provide specific answers or solutions to coastal-realm conservation problems. Rather, we attempt to raise discussion, with expectations that readers will explore and debate the many issues, approaches, and solutions that now exist – or that may be deployed in the future. While science can provide guide-posts for better informed decision-making at all levels, society as a whole needs to recognize ecosystem functions and the roles biota play, which together perpetuate ecosystem health, resource sustainability, and human wellbeing. As a result, we hope that future agents of conservation – students, conservationists, managers, policy-makers, legislators, and the concerned public – will be better prepared to tackle the difficult field of coastal-realm conservation.

References, scientific terms, Latin names, and units. This book provides readers with a window into a massive literature on conservation science and policy, including historical literature that provides a context for understanding the present state of knowledge of the coastal realm, its life, and its conservation and management. In-text citations are minimal because most subject material requires extensive references that would make the text cumbersome to read. The reference section at the end of the book is limited to these in-text citations and some suggested readings that can provide a guide to source material. We leave to readers the task of literature research to advance their own interest in particular topics, many of which are continually in the forefront of change. A complete list of references used during the writing of this book is given on the Blackwell Publishing website: <http://www.blackwellpublishing.com/ray/references.pdf>

The language of science is enormous and similar terms may have different, even contradictory, meanings among disciplines. We have attempted to minimize scientific terms by defining some of them in the text. Readers are referred to science dictionaries for other terminology. We use the International System of Units (SI units) and metric measurements (e.g., mt for tonnes) throughout the text, and refer readers to summary reference works on these units for definitions, measurements, and conversions. Exceptions occur when we wish to maintain the original works (e.g. Table 1.6).

Species are referred to by their vernacular ("common") names (herring, blue crab, etc.), with Latin names for identification. Most vernacular names are not standardized (birds and some fishes are exceptions). For example, "cod" is a common name for a valuable Atlantic fish of the cod family (Gadidae), but "cod" in Australia refers to groupers of the sea bass family (Serranidae); similarly, "rockfish" may refer to a number of fishes from a half dozen families of fishes. Therefore, scientific names are essential for identification, and are given with the vernacular the first time the species is mentioned in each chapter, or if far separated.

G. Carleton Ray and Jerry McCormick-Ray

Acknowledgments and permissions

We thank Sally Lyons Brown and the W.L. Lyons Brown Foundation for encouragement and support, without which the writing of this book would not have been possible. We also especially wish to thank all authors of boxes, listed in the previous section, who took time to write boxes included in the text, intended to give greater breadth to the coverage of this book. Many colleagues also responded generously to calls for information, imagery, advice, and text review. These include Charles Birkeland, Michael Brayneu, Douglas Burn, Jon Day, Jacques Descloitres, J. Frederick Grassle, John S. Gray, Samuel H. Gruber, Bruce P. Hayden, William J. Hargis, Christopher O. Justice, Victor S. Kennedy, Björn Kjerfve, Susan Larson, Ian G. Macintyre, Karen McGlathery, Ray Moore, Joseph S. Nelson, Roger Newell, Tim Parsons, David Pollard, Ken Sherman, Herman H. Shugart, Douglas Wartzok, and many others. We also thank reviewers of selected chapters for providing sound suggestions and advice: Richard Condrey, C. Richard Robins, and Kirk Winemiller, for Chapters 3 and 4; William Hargis, Victor S. Kennedy, Roger Newell, and James Perry, for Chapter 5; Robert Ginsburg for the geology portions and Mark Hixon and C. Richard Robins for the entirety of Chapter 7; F. Herbert Bormann and Chris Bormann, for Chapter 9. We are especially indebted to persons who have influenced our thinking during past decades: Frederick B. Bang, Raymond F. Dasmann, Francis H. Fay, J. Frederick Grassle, John A. Moore, Kenneth S. Norris, John C. Ogden, Fairfield Osborn, C. Richard Robins, William E. Schevill, John Twiss, William A. Watkins, Sir Peter Scott, and historical writers from whom sprang modern conservation. Special thanks also to two reviewers of the entire draft manuscript, Geoff Wigham of the University of Plymouth and Michael Rex of the University of Massachusetts, who offered helpful suggestions. We also thank our department colleagues, students, and librarians of Johns Hopkins University and University of Virginia, and Maryland Dept of National Resources, Virginia Marine Resources Commission, and the office of the Chesapeake Bay Program.

We are also indebted to the staff and contractors of Blackwell Publishing for their efficiency and positive encouragement, including: Ian Sherman, Sarah Shannon, Katrina McCallum, Rosie Hayden, Janey Fisher, and Erica Schwarz.

Finally, we thank funding organizations and agencies which have granted us support or for which we have served an advisory or consultancy role, most particularly: the Arctic Institute of North America, Bahamas National Trust, Henry Foundation, International Union for the Conservation of Nature and Natural Resources, Munson Foundation, National Aeronautics and Space Administration, National Geographic Society, National Oceanic and Atmospheric Administration, National Science Foundation, New York Aquarium (New York Zoological Society), Office of Naval Research, UNESCO, U.S. Man and the Biosphere Program, U.S. Marine Mammal Commission, U.S. National Park Service, and private donors to the University of Virginia Global Biodiversity Fund.

We also wish to thank the authors, illustrators, journals, and publishers who have given permission to use their work. References used for figures and tables are given in the captions. In the figure captions, "with permission" indicates that the publisher, and/or author, has permitted us to use the figure as originally published in the reference cited. In a few cases, we have made full reference to the figure used, in accord with the publisher's instructions. "Adapted from", "modified from", or "compiled from" indicate that significant modifications to original works have been made, requiring no permission. Nor in the case of government publications are permissions required. However, in all cases, reference is made to works from which information or concepts have been acquired. Permissions to use original works have been obtained from the following publishers: Academic Press (London, UK), Alliance Communications Group for Allen Press (Lawrence, Kansas), American Institute of Biological Sciences (Washington, D.C.), Blackwell Scientific Publications (Oxford, UK, and Boston, Massachusetts), Cambridge University Press (Cambridge, UK), Dorset House Publishing Co., Inc. (New York), Earthscan Publications Ltd. (London, UK) for World Commission on Dams, Ecological Society of America (Washington, D.C.), Elsevier Science (Oxford, UK), Geological Society of America, Inc. (Boulder, Colorado), Kluwer Academic Publishers (The Netherlands), National Academy of Sciences Press (Washington, D.C.), National Geographic Society (Washington, D.C.), National Museum of Natural History (Smithsonian Institution, Washington, D.C.), Open University (Milton Keynes, UK), Sigma Xi (The Scientific Research Society, Research Triangle Park, North Carolina), Society for Marine Mammalogy (Lawrence, Kansas), Society for the Study of Evolution (Lawrence, Kansas), Springer-Verlag (New York, and Heidelberg, Germany), and the University of Chicago Press (Chicago, Illinois).

Section I

Issues and mechanisms

A football-sized jellyfish (*Periphylla* spp.) under antarctic sea ice. McMurdo Sound, Southern Ocean. Photograph by the authors.

The two chapters in this Section document the many issues facing conservation and the many tools for dealing with them. Conservation issues relate specifically to biological issues (i.e., species depletions or habitat change), to issues that involve human activities (i.e., resource extractions and pollution), or to less apparent issues that affect ecosystem structure, function, and resiliency. Together these issues result in environmental change that is increasing in rate, magnitude, and duration, and from which recovery is difficult. Collectively, these issues contribute to an environmental debt, which is the total adjustment required to recover from change. The many current conservation and management mechanisms now in use overwhelmingly direct methods towards resolving specific issues, one by one. Although many successes are apparent, the environmental debt looms ever greater. Therefore, how can conservation mechanisms be made more effective to deal with the critical individual issues while confronting the mounting environmental debt?

Top: A healthy, high-diversity reef with corals, sponges, and gorgonians, Exuma Cays, Bahamas. Bottom: Turtlegrass (*Thalassia testudinum*) bed with a common sea star (*Oreaster reticulatus*) Cay Sal Bank, Bahamas. Photographs by the authors.

Chapter 1

Conservation issues

The materials of wealth are in the earth, on the seas, and in their natural and unaided productions.
Daniel Webster

It is time to understand "the environment" for what it is: the *national-security issue of the early twenty-first century.*
Robert D. Kaplan

1.1 Introduction

Three sets of issues compel recognition of forces that are global in scope, deepen the environmental debt, and raise ethical concerns about sustainability. Primary issues focus on species and their habitats and have long been the major focus of conservation. Secondary issues, conversely, direct attention toward human activities as causes for change. These activities have only relatively recently received conservation attention. However, the accelerating rates and momentum of human activities, and consequent ecosystem changes, have become high conservation priorities.

Efforts to address primary and secondary issues have slowed the depletion of life and its habitats, but have mostly failed to reverse trends of depletion and loss. This situation requires attention to systemic tertiary issues, involving emergent environmental phenomena, altered ecosystem states, and rates and dimensions of environmental and social change. Tertiary issues cause coastal-marine conservation to face complex, chronic problems that confound simple solutions.

1.2 Primary issues

Historically, most emphasis has been placed on depletion and extinction of species and protection of habitat. Yet, primary issues expand these domains to include species overabundance, ill-health, abnormal behaviors, and deteriorating habitats. Together, primary issues bring public attention to the need for conservation action.

1.2.1 Species extinction and depletion

Extinctions and depletions of coastal-marine species are documented worldwide and across most taxa (Table 1.1). Of the more than 120 species of marine mammals, at least a quarter is presently depleted and several are extinct. The Atlantic gray whale probably became extinct during the earliest days of Native American and European whaling. The Steller sea-cow population had probably already been decimated by subsistence hunters when it was discovered in 1741, and was hunted to extinction only 27 years later. The four remaining sea cows (one dugong and three manatees) are all endangered. The Caribbean monk seal was last reliably sighted in the 1950s near Jamaica. The Mediterranean monk seal population hovers around 300–400 individuals and the Hawaiian monk seal is also endangered. Large baleen whales seemed to be on the road to extinction until almost all whaling ceased in the 1980s; the North Atlantic right whale remains at risk. The Chinese river dolphin is critically endangered, as is the vaquita, a small porpoise endemic to the northeastern Gulf of California; both are among the world's rarest mammals. Sea birds and sea turtles have suffered similar fates. Most sea turtles are endangered. The Labrador duck and the flightless great auk became extinct in the 1800s, due to decimation by hunters for food, and many large sea birds, particularly albatrosses, are still being depleted.

Invertebrates and fishes are difficult to observe. The first documented extinction of a marine invertebrate was the North Atlantic eelgrass limpet, which disappeared in the early 1930s when a disease exterminated eelgrass beds, its sole habitat. The California white abalone existed in the thousands in the 1960s; now, the few remaining individuals are so widely separated that they may not be able to reproduce. Lately, fishes have elicited great public concern. Once abundant populations of cod, haddock, swordfish, salmons, tunas, sharks, and others have become severely depleted. The barndoor skate of the northwest Atlantic has been

Table 1.1 Selected extinct, endangered, and depleted species, illustrating worldwide depletion and extinction among a wide range of species groups.

Common name	Latin name	Range
Invertebrates		
Eelgrass limpet	*Lottia alveus*	North Atlantic
Rocky, mid-intertidal limpet	*"Collissella" edmitchell*	California (USA)
White abalone	*Haliotis sorenseni*	California (USA)
Fish		
Shortnose sturgeon	*Acipenser brevirostrum*	North Atlantic
Chinook salmon	*Oncorhynchus tshawytscha*	North Pacific
Totoaba (sea trout)	*Cynoscion macdonaldi*	Gulf of California
Barndoor skate	*Diptura laevis*	North Atlantic
Reptiles		
Green turtle	*Chelonia mydas*	Worldwide
Kemp's Ridley turtle	*Lepidochelys kempii*	Gulf of Mexico
Birds		
Auckland Island merganser	*Mergus australis*	New Zealand
Tahitian sandpiper	*Prosobonia leucoptera*	French Polynesia
New Providence hummingbird	*Chlorostilbon bracei*	Bahamas
Guadalupe storm petrel	*Oceanodroma macrodactyla*	Mexico
Bonin night heron	*Nycticorax caledonicus*	Japan
Steller's spectacled cormorant	*Phalacrocorax persipicillatus*	Komandorskiye Is. (Russia)
Canarian black oyster catcher	*Haematopus meadewaldoi*	Canary Islands (Spain)
Javanese wattled lapwing	*Vanellus macropterus*	Indonesia
Labrador duck	*Camptorhynchus labradorius*	North Atlantic
Great auk	*Pinguinus impennis*	North Atlantic
Short-tailed albatross	*Diomedea albatrus*	North Pacific
Cahow (Bermuda petrel)	*Pterodroma cahow*	North Atlantic
Mammals		
Steller sea cow	*Hydrodamalis gigas*	Komandorskiye Is. (Russia)
Caribbean monk seal	*Monachus tropicalis*	Caribbean Sea
Mediterranean monk seal	*Monachus monachus*	Mediterranean
Hawaiian monk seal	*Monachus schauinslandi*	Hawaiian Islands
Atlantic gray whale	*Eschrichtius robustus*	North Atlantic
Northern right whale	*Balaena glacialis*	North Atlantic
Chinese white flag dolphin	*Lipotes vexillifer*	Yangtze River
Vaquita porpoise	*Phocoena sinus*	Gulf of California
Marine otter	*Lutra felina*	South Pacific
Sea mink	*Mustela macrodon*	Nova Scotia to New England

Compiled from Carlton (1993); Norse (1993); Upton (1992); WCMC (1992).

drastically reduced due to by-catch in nets intended for other species. In the northern Gulf of California, the estuarine totoaba has become threatened by fishing, damming, and massive extraction of water from the Colorado River.

Coastal fishes that ascend or descend rivers and estuaries to spawn are especially vulnerable. Sturgeons, salmons, shad, menhaden, and others are widely depleted. Most sturgeons that inhabit fresh and coastal waters of Europe, Siberia, and North America are listed as threatened or endangered. The European sturgeon is one of the largest and most valuable of all fishes, and was common in the 1800s from the Baltic to the Black Sea. By the 1940s, spawning apparently occurred in only two rivers of western Europe. North American sturgeons have suffered similar fates. Many fishes that return to natal rivers to spawn, notably the "king" of salmons, the chinook, have been depleted or extirpated due to combinations of habitat alteration, pollution, and overfishing.

These few examples exemplify an issue of unknown dimensions. Many species remain to be discovered and many species, especially of smaller size and lesser economic importance, have no doubt disappeared without documentation of their existence. Data on population trends are especially sparse because original population numbers are not known.

1.2.2 Overabundance

Conversely to depletion, many species have recently flourished and become a concern when they dominate their communities, threaten human livelihood, depress densities of other species, deplete their own habitats, or cause a change in ecosystem function (Table 1.2). Overabundance is of greatest concern when native or introduced (exotic) species become invasive and do harm to other species or to their ecosystem. In areas of the Indo-Pacific, the native crown-of-thorns starfish recently became so numerous that it decimated coral reefs. The introduced American comb jelly, *Mnemiopsis*, has reduced plankton biomass and altered food webs of the Black and Azov seas. A fast-growing exotic alga, *Caulerpa*, is transforming portions of the Mediterranean seafloor into a dense, single-species cover, and has recently appeared in southern California and Australia. The invasive, native, common reed now dominates many U.S. wetlands (Box 1.1).

Box 1.1 A perspective on an overabundant invasive species: common reed, *Phragmites australis*

James E. Perry

An "invasive species" is defined as an alien species (i.e., any plant or animal species that is not native to that ecosystem) whose introduction causes, or is likely to cause, economic or environmental harm or harm to human health (U.S. Presidential Executive Order 13112). While the term "native" implies that invasive species come from a foreign country or continent, it is important to note that this is not always the case. Given the dynamic nature of ecosystems and the propensity for humans to modify these habitats (such as filling or dredging of wetlands), we now know that some of our native species will become invasive if given the opportunity. The common reed, *Phragmites australis*, is one of these species. This vascular plant occurs in wetlands throughout the world. It is an aggressive colonizer of disturbed sites and exhibits rapid vegetative propagation (1–2 m yr^{-1}), and is capable of suppressing competitors by shading and litter mat formation, which gives the plant a distinct advantage over other species. Once established in a wetland, it is extremely difficult and expensive to eradicate.

Common reed has become a distinct problem in restored and/or created wetlands in the mid-Atlantic region of the United States. Because of the disturbance nature of wetland restoration and/or construction (earth removal and movement), newly constructed marshes are highly susceptible to common reed invasion. This species is considered undesirable by resource managers of this region partly because of its ability to replace the dominant species of numerous tidal and non-tidal wetland plant communities and, therefore, to reduce habitat diversity. On highway I-95 in New York–New Jersey just south of New York City, hundreds of acres of common reed have replaced a mixed community of salt, salt meadow hay, and tall cordgrass marshes. Common reed has become invasive through its ability to rapidly colonize disturbed, particularly human-disturbed, habitats. Thus, for every road crossing, every dredge-spoil sidecasting, and every parking lot next to a marsh, common reed has invaded into other marsh types, and has lowered the biodiversity of these important systems.

Attempts to limit or eradicate common reed have met with variable success. Small populations can be removed by pulling up plants, but this needs to be done before flowering to avoid dispersing seeds during removal. Application of herbicides can be effective, but care must be taken as these broad-spectrum herbicides can destroy adjacent desirable species. Most current use of herbicides is in conjunction with multiple burnings of the marsh. The financial cost of eradication is high. Herbicides themselves are expensive and spraying of large areas is done by helicopter. Some managers question these costs, particularly when one considers that eradication is usually temporary: in most cases the common reed will return from adjacent wetlands to repopulate the treated area. Salinity and flooding have also been shown to have adverse effects on common reed, and die-back has been reported at sites where soil salinity was higher than 15‰.

The questions common reed poses to wetland scientists, regulators, and managers are complex. Since wetland restoration and/or creation will continue in the mid-Atlantic United States, what role should common reed be allowed to play in future wetland plant communities? Can wetland restoration and/or creation be considered successful if these sites are quickly invaded by common reed?

Table 1.2 Examples of overabundance. Some are natives; others are exotics (E).

Organism/Origin	Location affected	Characteristic	Impact
Dinoflagellates (microflagellate) Worldwide	Coastal seas, worldwide	Red and brown blooms	Discolors water; may kill variety of species; produces shellfish toxins
Chrysophytes (micro-golden algae) Worldwide	Coastal seas, worldwide	Golden-brown blooms	Discolors water; shades aquatic plants; disrupts food webs; causes hypoxia
Diatoms Worldwide	Coastal seas, worldwide	Algal blooms	Sometimes lethal; mucus clogs fishing nets
Cyanobacteria (blue-green algae) Worldwide	Global ocean, coastal	Blue-green algal bloom	Range from harmless to toxic
Green alga *Caulerpa taxifolia* Aquaria	Mediterranean, California, Australia	Invasive macro-alga (E)	Threatens benthic community
Australian "pine" *Casuarina equisetifolia* India to Australia	Tropical islands and coasts, worldwide	Invasive tree on impoverished or disturbed soils (E)	Invasive; alters ecosystems; depletes biodiversity; causes erosion
Jelly *Aurelia aurita* Northwest Atlantic	Black Sea	Pelagic, plankton-feeder (E)	Nuisance; biomass ~450 million tons
American comb jelly *Mnemiopsis leidyi* Northwest Atlantic	Mediterranean, Black, Azov seas	Pelagic carnivore of plankton (E)	Threatens plankton biomass and fisheries
European green crab *Carcinus maenas* Northeast Atlantic	Northern California, Oregon	Voracious benthic invertebrate (E)	Threatens small shore crabs, native clams, near-shore invertebrates
Chinese clam *Potamocorbula* sp. Northwest Pacific	San Francisco, Northern California	Invasive benthic invertebrate (E)	Threatens benthic community
Indo-Pacific mussel *Perna perna* Tropical Pacific	Gulf of Mexico	Invasive filter-feeder (E)	Nuisance; beds extend for kilometers
Zebra mussel *Dreissena polymorpha* Europe	Baltic Sea, North America, Europe	Invasive freshwater filter-feeder (E)	Clogs pipes; eliminates benthic life; reduces plankton abundance
Crown-of-thorns starfish *Acanthaster planci* Indo-Pacific	Indo-Pacific	May suddenly increase in numbers	Consumes coral polyps; decimates reefs
Lionfish *Pterois volitans* Indo-Pacific	Northwest Atlantic	Introduced from aquaria (E)	Highly toxic, voracious carnivore
Snow goose *Chen caerulescens* North America	North American Arctic	Explosive overabundance	Degrades arctic nesting area
Gray seal *Halichoerus grypus* North Atlantic	North Sea	Abundant and protected	Said to deplete cod fisheries
California sea lion *Zalophus californianus* Northeast Pacific	Northeast Pacific	Increasing and protected	Consumes salmons; has become a "pest" on docks

Compiled from Carlton (1985); Duxbury & Duxbury (1997); Kenney et al. (1996); Pollard & Hutchings (1990a,b); NAS (1995); Raloff (1998); Schneider & Heinemann (1996); Sherman et al. (1996); Stephens et al. (1988); Steneck & Carlton (2001); Whitfield et al. (2002); Woodham (1997).

Fig. 1.1 Satellite image of the Bering Sea, September 1998. A plankton (coccolithophore) bloom is the light-shaded area, on right. Causes unknown; no deleterious effects discovered. Courtesy of Jacques Descloitres, NASA Goddard Space Flight Center.

Planktonic species are especially notable invaders, and often "bloom" in such massive numbers that they discolor the water, as "red tides" (dinoflagellates), "green films" (cyanobacteria), and "brown tides" (chrysophytes). Some blooms are localized in bays or estuaries; others cover thousands of square kilometers (Fig. 1.1). Blooms may last for weeks. Some occur at the same time and place each year; others occur unpredictably. Some are harmless; others kill marine life and produce noxious gases. Causes are often obscure; some occur in response to nutrient inputs.

Overabundance also occurs in aquatic birds and mammals. The snow goose has exploded in numbers and is causing extensive habitat damage in coastal breeding areas. Fishermen claim that overabundant seals deplete valuable fisheries: for example, gray seals that consume cod in the North Sea and California sea lions that consume salmons on North America's Pacific coast.

1.2.3 Ill-health

Ill-health focuses on abnormal physiological states and compromised immune systems, expressed as lesions, diseases, and deformities. These have been recorded for a wide variety of taxa, in epidemic proportions, regionwide, associated with pollution, as massive die-offs, or as individual strandings (Table 1.3). Ill-health brings into question what constitutes normalcy. Little is known about the "normal state" of health of most marine species. For example, in corals and their relatives, ill-health is described as "white line," "black line," and fungus diseases of uncertain etiology and taxonomy.

Many diseases are recently discovered. Farmed Atlantic salmon (*Salmo salar*) are susceptible to contagious infectious anemia. A tumor mass (fibropapillomatosis) caused by a virus is commonly observed in sea turtles. A brain disease (avian vacuolar myelinopathy) has inflicted coastal bald eagles and waterfowl in the southeastern United States. And in Scotland's northeast waters, more than 90% of a population of adult bottlenose dolphins (*Tursiops truncatus*) had epidermal lesions.

In polluted harbors and contaminated waters, ill-health takes many forms, including skeletal deformities, tumors, sores, fin rot, pathogenic viruses and bacteria, fungi, protozoans, and various invertebrate parasites. Pollution has also been shown to induce malformation and mortality, associated with abnormal embryonic chromosome division, for example in planktonic eggs

Table 1.3 Examples of ill-health.

Condition	Group(s) affected (Disease)	Location
Orange disease	Coralline algae: *Porolithon*	South Pacific
Abnormalities: shell/spine deformities, abnormal scales	Invertebrates; fishes	North Atlantic, Japan, South California
Moribund Withering Syndrome	Red abalone: *Haliotis*	South California
Lesions: skin, skull, brain, other organs. Pathogens mostly unidentified	Fishes (ulcerations, fin rot), marine mammals, birds	Baltic, North Sea, Southwest Atlantic
Fungal infections	Sea fans *Gorgonia ventalina*; corals, several; seagrass *Zostera marina* (wasting disease); fishes, herrings	North Atlantic, Florida, Central and South America, Caribbean
Tumors	Fishes; sea turtles (fibropapillomatosis)	Bays, harbors, coasts, banks, sounds, estuaries
Viral infections	Fish (lymphocystis) Birds (infectious bursal disease) Seals (phocine distemper) Sea otter *Enhydra lutris* (herpesvirus) Marine mammals (morbillivirus)	Worldwide; Northeast Pacific
Bacterial infections	Corals *Acropora palmata*: "white pox" (*Serratia marcescens*) Fish (hemorrhagic disease, fin rot, ulcerations, red sore)	Industrial-urban bays, harbors, coasts, reefs; Wider Caribbean
Protozoan infections	Oysters (dermo, MSX) Dolphins (fatal hepatic sarcocystosis)	Worldwide
Reproductive disorders	Fishes (defective eggs, larvae) Marine mammals (reproductive failure, aborted fetuses)	Baltic, North Sea, North Atlantic, North Pacific, Arctic

Compiled from Aguirre (1998); Cervino et al. (1998); Geiser et al. (1998); Littler & Littler (1995); Nagelkerken et al. (1997); Patterson et al. (2002); Raloff (1998); Reimer & Lipscomb (1998); Resendes et al. (2002); Sindermann (1996); Wilson et al. (2000).

of Atlantic mackerel (*Scomber scombrus*). Furthermore, some marine diseases are of human origin: a bacterial species of the human gut has recently been shown to be a pathogen of corals.

1.2.4 Abnormal behavior

Changes in distributions and behavior, such as altered times and places of breeding, have been observed in coastal-marine species in response to climate or to human-caused alterations of the land- or seascape. Most altered behaviors have been recorded for birds and mammals that are relatively easy to observe. Some waterfowl that normally feed on shallow-water vegetation now consume crop residues left on farms and may not migrate. Gulls have become nuisances around garbage dumps and fishing vessels. Gulls and sea lions follow fishing boats to feed on discarded offal. California sea lions become noisy nuisances when they haul out on docks. Florida manatees often avoid winter's cold by congregating in warm-water effluents of power plants. Some cetaceans hybridize, but whether this is abnormal is unknown.

Marine mammal strandings are of particular interest (Box 1.2). One cause may be avoidance of unfavorable water conditions, such as toxic algal blooms. Another cause may be noise generated by ships: commercial, recreational, and military uses of the seas all produce significant noise and have rapidly increased in recent years. Strandings almost always arouse great public attention and, nowadays, intense efforts are made to direct the animals back to sea or save them under veterinary care.

Attacks on humans may be abnormal, and causes for shark attack are speculative. Sharks have decreased in numbers recently; yet, shark attacks do not appear to be decreasing, possibly because of increased numbers of swimmers in nearshore waters. Occasional attacks by marine mammals on swimmers suggest responses to human harassment. On the other hand, some marine mammal interactions with humans are mutualistic, positive, and learned. The immortalized killer whale, "Old Tom" of Twofold Bay, Australia, guided Australian whalers to humpback whales in the early 1900s. Tom was rewarded with the tongue of the whale. Dolphins are also known to aid coastal fishermen in pursuit of

Box 1.2 Cetacean strandings

James G. Mead

The word "stranding" is derived from "strand" the shore of the sea. The word usually applies to a denizen of the marine environment that has come ashore, whether under its own volition or not. When whales and dolphins die they normally sink to the bottom. If the water is shallow, decomposition gases form in the tissues and the carcass floats and drifts ashore. In waters that are sufficiently deep, the gases remain in solution and the body decomposes on the bottom.

Single strandings represent individual mortality and are of interest only when they occur in unusual numbers. Mass strandings are events that involve a number of animals. Sometimes stranded animals do not have to be close together to constitute a mass stranding. A case in point is the stranding of short-finned pilot whales (*Globicephala macrorhynchus*) along the east and gulf coasts of the United States. From 1970 to 1980 the number of individuals averaged 1.8 per year (range 0–3). Suddenly, four strandings occurred in May 1973, scattered from Ocean City, New Jersey, to Bodie Island, North Carolina. This may be categorized as one mass stranding, extended in space. Any one of those would be classed as a single stranding, without information on the others. Baleen whales (mysticetes) have not been shown to be subject to the same mass strandings as toothed whales (odontocetes). A 1987–88 mortality of humpbacks, involving 14 individuals during 5 weeks and over about 300 km distance, could have been interpreted as a mass stranding. This event was, however, shown to be the result of paralytic shellfish poisoning.

Mass strandings have been occurring for millennia and have been subjects of speculation for almost as long. However, whether some events referred to in the literature are mass strandings or drive fisheries is problematic. Some of the largest strandings were false killer (*Pseudorca crassidens*) and long-finned pilot (*Globicephala melas*) whales. These strandings occurred in areas and at times when there was an active drive fishery (Japan and Massachusetts, respectively). The highest mortality recorded to date is 310 long-finned pilot whales that stranded in New Zealand in 1987.

Explanations of the causes of mass strandings are: confusion induced by coastal topography along shallow, sloping, sandy shorelines; errors in geomagnetic navigation; reversion to ancient migratory routes; pollution; ingestion of debris; chase or harassment; unusual underwater noises; sudden stress; shallow-water feeding; and diseases, parasites, neurological disorders, suicide, overpopulation, or getting blown ashore by storms. In 1987 and 1988 there was a spectacular increase in strandings of bottlenose dolphins (*Tursiops truncatus*) along the Atlantic coast of the United States. The mortality lasted from June 1987 to February 1988 and extended from New Jersey to Florida. Painstaking work by an army of investigators on those stranded animals determined that the mortality was due to morbillovirus infection.

Strandings provide scientists with an avenue for investigating cetaceans. The most basic information of all, the existence of a species, is frequently brought to light by strandings. Original descriptions of 20 out of 21 species of beaked whales (Family Ziphiidae) have been based on strandings. Strandings also contribute to the gradual assembly of life-history information, including feeding habits, reproduction, age, growth, and pathology. Ideas about the functional mechanisms that cetaceans have evolved may depend on comparative studies – for example, insights into thermoregulation of reproductive tissues. Recognizing pathologies depends on knowledge of the anatomical structure of "normal" animals. Strandings provide a way of monitoring adverse human interactions with cetaceans. Increased mortality due to fisheries interactions is frequently observed on stranded animals, for example subtle net marks. Stranded animals also provide a source of material for monitoring levels of anthropogenic contaminants (pollution) in marine systems.

fish, and reports of dolphins associating with people have become common.

Materials introduced by humans into the sea profoundly alter feeding behavior, for example when objects unfamiliar to sea life are ingested. Sea turtles, sea birds, and marine mammals are especially likely to ingest foreign objects, and each year many thousands may be injured, receive inadequate nutrition, or die because of this propensity. Additionally, sea birds, sea turtles, and large, pelagic fishes take longline fish hooks and many thousands die.

1.2.5 Deteriorating habitats

Interest in habitat extends to valued species, biodiversity, habitats created by living organisms (e.g., living reefs), and ecosystem functions. For thousands of years, coastal wetlands in China have been converted to rice production. In the conterminous United States, more than half of all coastal wetlands are degraded or have been converted. Globally, beds of aquatic vegetation have been decreasing, affecting the abundance of invertebrates and waterfowl. Since the 1960s, more than half of all tropical mangroves have been logged and replaced by croplands, shrimp ponds, and tourist resorts. Also, coral reefs are deteriorating globally as a result of bleaching, disease, and human disturbances. Temperate oyster reefs have been depleted even more intensively and for a longer period; those that remain are now vestiges of their former extents.

On a larger scale, bays, sounds, and estuaries have been dramatically affected by alteration of watershed flows, overfishing, pollution, and human occupation. Many productive coastal habitats are in critical states of deterioration and contamination, particularly in industrialized regions. A case in point is San Francisco Bay: during the past century and a half, approximately 60% of its area has disappeared under a massive sediment load released from upland mining and other forms of development. The Mississippi River has been physically changed by levées since the early 1800s; today, its delta region and the northern Gulf of Mexico are subject to nutrient enrichment, oxygen deficiencies, and depleted food webs. Similar changes have occurred in most of the world's major bays, rivers, and estuarine systems.

Islands and beaches are of special concern. On land, islands have experienced the world's highest extinction rates. Oceanic islands inspired Charles Darwin and Alfred Russel Wallace to formulate the Theory of Evolution in the mid-1800s. Islands are hotbeds of evolution and host a high degree of endemism; 12 m sunflowers, 250 kg tortoises, peculiar creatures such as marine iguanas, and many others occur nowhere else. The world's longest barrier-island/beach system occupies United States coasts from Maine to Texas. Many of its approximately 300 barrier islands are undergoing rapid change, affecting beach-nesting birds and sea turtles and the barriers' ability to protect nearshore environments from ocean forces. Islands are also notorious for the ease with which natural communities can be upset by exotic species and other disturbances. However, very little is known of extinction off island shores.

1.3 Secondary issues

Secondary issues focus on regulation of human activities and resolution of conflicts that are generated as people mass into the coastal realm. These issues draw attention to changes in which humans have, for centuries if not millennia, been foremost agents of alteration.

1.3.1 Extractions: removing natural resources

Human societies have long benefited from seemingly endless supplies of food, minerals, chemicals, industrial products, building materials, and energy extracted from coastal land and oceans. Even seawater is a significant resource, from which freshwater, salt, and power can be extracted. Extractions usually result in ecosystem adjustments.

1.3.1.1 Fisheries

Modern fishing is an analog of the ancient hunter–gatherer. Its history resembles a "slash-and-burn" activity, moving from resource to resource and from ocean to ocean, taking first the largest, most valuable, and easiest species to hunt, then the smallest, most abundant, and logistically more difficult. Commercial fishing is dominated by only a few nations, with Peru, Japan, Chile, China, and the United States being the leaders in terms of tons landed. Today, the number of fishermen worldwide approaches 20 million; 90% are small-scale fishers who may account for 25% of the global catch. Furthermore, the fisheries industry employs almost ten times more persons than fishermen, including processors, shippers, and marketers.

World marine and inland fisheries production trebled from 18 million metric tons (mts) in 1950 to more than 100 million mts by the turn of the century (Fig. 1.2), by which time nearly 70% of the world's

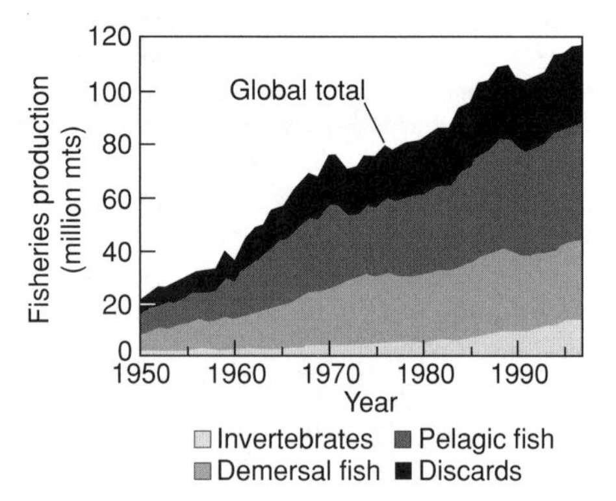

Fig. 1.2 Global fish catch has increased during the past half-century. Increasing catch of pelagic species is due mostly to increased catch of small species, while catch of demersal fish species has stagnated, and by-catch (mostly discarded) has increased. From Pauly et al. (2000), with permission from Sigma Xi.

marine commercial fish populations had become depleted. Today, few areas of the world remain unexploited. About 40% of the total catch enters an expanding market for human consumption. At least 20% is discarded as by-catch of non-target species. The remainder is divided among commercial and industrial uses, including pet food and fertilizer. As demand has increased, competition among fishers and the effort needed to catch a given weight of fish have increased, resulting in collapses of formerly flourishing fish populations. Technology also plays a significant role. Even small fishing vessels are equipped to pinpoint fishing grounds and to find fish. Therefore, despite collapses, new resources have been exploited and total landings have increased.

Whaling illustrates many aspects of commercial fishing. The most abundant and largest species of easiest access in coastal waters were hunted first, then depleted, and discovery of new populations followed. Intensive European whaling began in the 18th century for the right whale (*Eubalaena glacialis*) in the northeast Atlantic, where this species was soon extirpated. Later, New Englanders pursued right whales in the northwest Atlantic. Again, the whales were depleted; only a remnant population now remains, with dubious chances for recovery. In the 1820s, New Englanders took up whaling for sperm whales (*Physeter catodon*) for their valuable oil. At about the same time, bowheads (*Balaena mysticetus*) were pursued in the Bering Sea. Both species soon became depleted. Development of larger ships and better technology allowed whaling to expand to remote areas. The most lucrative whaling followed the late 19th-century development of fast ships equipped with harpoon guns that could navigate high-latitude, nutrient-rich seas where whales aggregated in summer to feed. Whaling soon became a major industry in the Southern Ocean surrounding Antarctica where whalers caught the speedy rorquals (*Balaenoptera* species) and humpbacks (*Megaptera novaeangliae*). Meanwhile, lucrative hunts were underway in other oceans. During World War II, global whaling was interrupted. The post-War period saw the most intensive whaling of all. The result has been that most populations of large whales have been reduced to remnant ones (Fig. 1.3). Today, only the little 6–10 m minke whale (*Balaenoptera acutorostrata*) is legally taken in the Southern Ocean and the North Atlantic.

Fishing reflects a similar history. A typical pattern is observed in the exploitation of temperate, schooling fishes in the North Atlantic. Herring (*Clupea harengus*),

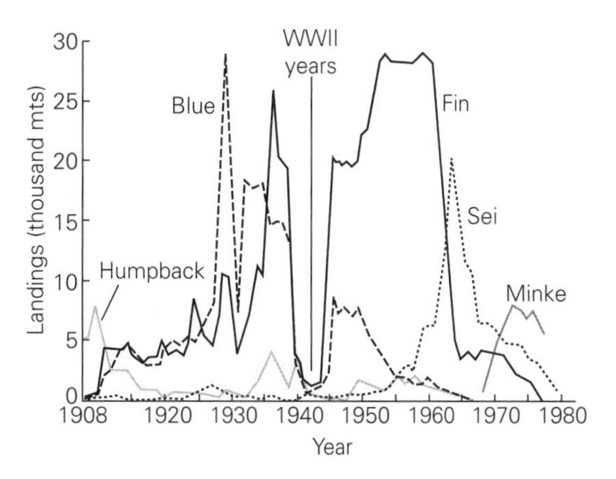

Fig. 1.3 Catch of baleen whales in the Southern Ocean from 1908 to the mid-1980s' moratorium established by the International Whaling Commission. Humpbacks were depleted first, then rorquals, in sequence from largest to smallest. From Jennings et al. (2001), with permission.

Atlantic cod (*Gadus morhua*), and others have supported European societies for many centuries (Fig. 1.4). Widespread depletions began in the 1800s. After World War II, increasing demand drove fisheries industries to exploit fishes more intensively. Depletions occurred first in the east, then in the west, and finally northward.

Today, fishing vessels are larger and more numerous than ever: in 1998, some 1124 fishing vessels greater than 100 tons were added to Lloyd's records. This increase in the fleet has been accompanied by major improvements in technology. Improvements in fish-finding devices and navigational aids have increased efficiency to the point that few fish schools go undetected. Trawl nets and seines have vastly increased in size, and towing two or more trawls can increase catch efficiency by 50–100%. Longlines with thousands of hooks may extend for tens of kilometers, and high seas drift nets can be more than 2 km long. All of these developments are highly efficient, but are non-selective and can result in greatly increased, unintended by-catch. Purse seines catch mammals and juvenile fish, and longlines and gillnets catch sharks, seabirds, and sea turtles. Furthermore, "ghost fishing" – the capture or entanglement of untargeted fish caught by discarded traps or nets at sea – results in considerable waste. Added to this are growing problems of illegal, unreported, and unregulated fishing that seem to be increasing worldwide. Finally, sport fishing is expanding and goes largely unrecorded, which substantially increases fisheries impacts and introduces uncertainty into management.

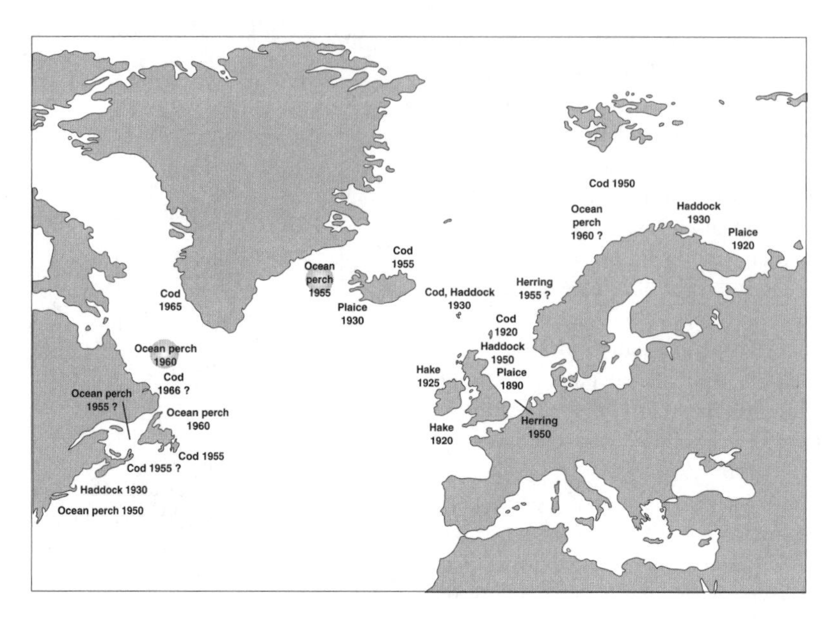

Fig. 1.4 Overfishing in the North Atlantic from before the 20th century. Plaice (*Pleuronectes platessa*) were depleted first in the North Sea; subsequently, fishing shifted to other areas, where the species was also depleted. The same sequence has occurred for other North Atlantic fisheries. Dates indicate approximately when increases in fishing effort no longer produced an increase in catch. From Holt (1969), with permission from Mr. N.H. Prentiss.

Table 1.4 Types of marine mineral resources.

	Unconsolidated (within water)		Consolidated (packed, hardened)	
Dissolved	**Surficial**	**In place**	**Surficial**	**In place**
Metals, salts in fresh- and seawater: Magnesium, potassium, sodium, calcium, bromine, sulfur, strontium, boron, uranium, and many other elements	*Shallow beaches; offshore placers*: Sand, gravel; heavy minerals: iron, silica, lime *Deep ocean-floor deposits*: Red clays, calcareous ooze, siliceous ooze, metalliferous ooze *Authigenic deposits*: Manganese nodules (Co, Ni, Cu, Mn), phosphorite nodules, phosphorite sands, glauconite sands	*Buried and in river placers*: Diamonds, gold, platinum, tin *Heavy minerals*: Magnetite, ilmenite, rutile, zircon, leucoxene, monazite, chromite, scheelite, wolframite	*Exposed stratified deposits*: Coal, iron ore, limestone *Authigenic coatings*: Manganese oxide, associated Co, Ni, Cu, phosphorite	*Disseminated massive, vein or tabular deposits*: Coal, iron, tin, gold, sulfur, metallic sulfides, metallic salts

Modified from Mangone (1991).

1.3.1.2 Minerals

The coastal realm is a distinct geological province, where mineral resources of geological and biological origin occur in solution, on the sea-bed surface, or buried (Table 1.4). Salt, magnesium, and bromine are recovered from seawater. Rock, coral, calcareous marls, shells, sand, gravel, and lime are commonly removed from coastal areas. Mining of phosphorite for fertilizer often results in stripping salt marshes. Tin is taken from alluvial deposits on shores or beaches in Malaysia, Indonesia, Thailand, Australia, Nigeria, China, and other nations. Coal is removed from mines that have been tunneled out under the sea. And many high-value, low-volume minerals including platinum, gold, silver, titanium, zirconium, chromium, and rare-earth minerals are removed from shores. The extraction of almost all minerals produces considerable amounts of waste, much of it toxic.

The most valuable nonrenewable resource is petroleum. About a third of the global supply comes from the coastal realm. Exploration suggests that considerable amounts remain in offshore deposits, where risks of extraction are often great. Reserves in high-latitude seas are subject to sea ice, harsh storms in winter, and a propensity toward severe earthquakes, such as off Russia's eastern Sakhalin Island and in the Sea of Okhotsk where exploration has recently begun. In the tropics, hurricanes are a constant threat.

Most offshore oil exploration has, until recently, been in shallow, continental-shelf waters. Given the continuing demand for petroleum, drilling is occurring farther out to sea and under more hazardous conditions. Drilling is now able to penetrate as deep as 8500 m or more. The extraction of hydrocarbons has many costly consequences beyond simply affecting sea life: for example, economic risks, land subsidence, erosion, and saltwater intrusion into freshwater aquifers.

1.3.1.3 Carving up coastal substrates

Substrate disturbance includes mineral mining, wetland ditching for public-health concerns and conversion for development, bottom trawling for fisheries, and dredging canals and ports for shipping. All of these activities are accompanied by resuspension and deposition of sediment elsewhere. Also, dynamiting and cyanide poisoning of tropical reefs are common practices for extracting fish; thus, reefs and fisheries are destroyed and the natural protection that reefs provide to shores against erosion is removed.

Use of mobile fishing gear for dredging and trawling produces among the greatest of impacts on coastal-ocean ecosystems. The physical effects are comparable to forest clear-cutting, but are estimated to affect approximately 150 times more area globally (Table 1.5). The frequency of dredging is also much greater than forest clear-cutting: some productive fishing areas are completely dredged up to three or four times per year. Dredging and trawling increase turbidity, alter benthic habitat, crush, bury, and smother non-target, sessile species, and expose infauna to predation. Hydraulic dredging to extract shellfish is less extensive than trawling for fish, but can be even more damaging as it leaves a highly disturbed benthos and creates a sediment wake that trails as a plume for some distance before sediment settles to smother benthic plants and animals far downstream. Of possibly even greater significance is release and reburial of nutrients and toxic substances, and oxygen depletion.

1.3.2 Introductions and additives

Humans continually add innovative products to watersheds and to the ocean's chemical soup that can affect biogeochemistry and the biota. Introductions include synthetic chemicals, toxic metals, trash, radioactivity, pathogens, exotic species, and excessive heat, noise, and artificial light. These enter directly from pipelines or offshore dumping, and indirectly when storms drench the land to wash solids, nutrients, metals, pesticides, pathogens, and other contaminants from streets, pavements, lawns, and farmlands. Anthropogenic introductions may be added slowly and chronically or suddenly and in concentrated forms, but rarely in accord with natural rhythms. Long-lasting carcinogenic, toxic, or lethal chemicals and metals collect into a contaminant profile when they accumulate within sediments.

Introductions originate mostly from the land. Many countries use the seas as dumps for land-based wastes and many nations and cities continue to dispose of untreated sewage directly into coastal waters. Even in developed, industrial nations, municipal sewage and toxic industrial wastes have been widely discharged directly into rivers and estuaries. Introductions also originate at sea. Ships increasingly collide with sea life, and substances that affect sea life are added through accidental (oil spills), and deliberate and routine operations (ocean dumping, bilge cleaning, antifouling paints). Equipment used in diverse ocean operations (fishing, monitoring instruments, submarines, submersibles, cables, military hardware, munitions, etc.) is often lost or discarded at sea. Occasionally, these introductions provide habitat opportunities for sea life (wrecks, artificial reefs, and floating materials that attract a variety of species).

Finally, introductions enter from the sky. Storms, winds, and rain transfer dust, plastics, debris, trash, nutrients, and microbes globally. Wind transports aerosols containing pollutants and nutrients, which fall with precipitation. Incomplete combustion from fossil fuel industries and automobiles releases metals and toxic chemicals to the air. Industrial plants introduce nitrogen and sulfur compounds that affect atmospheric chemistry, as is implicated in "acid rain". Radioactive materials released into the atmosphere from nuclear

Table 1.5 Comparison of forest clearcutting and trawling impacts.

Impact on:	Forest clear-cutting	Bottom trawling/dredging by fishing gear
Substrate	Exposes soils to erosion; compresses soils	Overturns, moves, and buries boulders and cobbles; homogenizes sediments; eliminates existing microtopography; leaves long-lasting grooves
Roots and infauna	Saprotrophs (that decay roots) are stimulated then eliminated	Infauna crushed and buried; others become susceptible to scavenging
Biogenic structures	Removes above-ground logs; buries structure-forming species	Removes, damages, or displaces structure-forming species
Associated species	Eliminates most late-succession species; encourages pioneer species	Eliminates most late-succession species; encourages pioneer species
Biogeochemistry	Releases large pulse of carbon to atmosphere by removing and oxidizing accumulated organic material; eliminates arboreal lichens that fix nitrogen	Releases large pulse of carbon to water column and atmosphere by removing and oxidizing accumulated organic material; increases oxygen demand
Recovery time to original structure	Decades to centuries	Years to centuries
Typical return time	40–200 years	40 days to 10 years
Global area affected per year	~ 0.1 million km^2 of net forest and woodland loss	~ 14.8 million km^2
Latitudinal range	Subpolar to tropical	Subpolar to tropical
Ownership	Private and public	Public
Scientific documentation (publications)	Many	Few
Public awareness	Substantial	Very little
Legal status	Modification of activity to lessen impacts and to prohibit or favor alternative logging methods and preservation	Activity restricted in only a few areas

From Watling & Norse (1998), with permission.

power plants and nuclear weapons testing, especially during the 1940s to 1960s, persist in some areas today.

1.3.2.1 Petroleum and related by-products

"Petroleum" is a broad term that describes naturally occurring and refined compounds and natural gases. Petroleum enters the coastal ocean from natural and anthropogenic sources, including submarine seeps, tanker accidents, deballasting operations, tank washing, refinery effluents, municipal and industrial discharges, losses from pipelines, offshore production, and industrial, municipal, urban, and river runoff. The United States National Research Council (NRC 2002) estimated that approximately 1.4 billion liters of petroleum and related hydrocarbons enter the oceans annually, chronically in low doses or catastrophically in high doses. A city of five million people might annually release roughly the equivalent of oil spilled when the tanker *Exxon Valdez* struck Bligh Reef in Prince William Sound, Alaska in 1989 and released approximately 37 000 tons of crude oil that spread more than 900 km from the spill site. This spill left an estimated 12% of its oil on subtidal sediments, affecting many species, some of which have apparently not yet recovered. The difference is that a city's input is chronic and an oil spill at sea is acute.

Oil spills can occur unpredictably at any time (Table 1.6). The effects are as varied as the contaminants themselves. Spills are most harmful in shallow, low-energy aquatic environments, sediments, wetlands, tidal flats, and sites with abundant wildlife. Relatively non-toxic petroleum tars that wash onto coastlines and beaches can smother biota and cripple recreation and tourism. Spills also result in oiling of birds, marine mammals, shorelines, and sediment, and release toxic substances that can affect marine life. The effects can be long-lasting or fleeting, depending on environmental conditions and the nature of the spill itself. The effects of spills are becoming fairly well known. However, little is known about how diffuse, chronic releases affect marine systems.

Table 1.6 Examples of major oil spills.

Location	Source	Amount	Cause	Date
Persian Gulf	Iraq bombing	~ 130 000 000 gal.	War	1991
Gulf of Mexico	Ixtoc I oil well	~ 600 000 mt	Blowout	1983
Nantucket, Massachusetts	*Argo Merchant*	7 700 000 gal.	Grounding	1976
Brittany, France	*Amoco Cadiz*	223 000 mt	Grounding	1978
Scilly Isles, U.K.	*Torrey Canyon*	119 000 mt	Grounding	1972
Prince William Sound, Alaska	*Exxon Valdez*	37 000 mt	Grounding	1989
Angola, 700 mi offshore	*ATB Summer*	260 000 mt	Fire/explosion	1991
West Delta, Louisiana	Pipeline break	6 720 000 gal.	Dragging anchor	1967
Russian Arctic	*Usinsk*	4 300 000 gal.	Pipeline rupture	1994
Galapagos Islands	*Jessica*	240 000 gal.	Fueling luxury ecotourism liner	2001
Brazil	Oil platform	~ 400 000 gal.	Explosion	2001
Spain, Galicia coast	Bahamian oil tanker *The Prestige*	77 000 mt	Crack in single hull of aging fuel tanker	2002

Compiled from Button (2003); Clark (1997); Irwin et al. (1997); Montevecchi & Kerry (2001); World Almanac Books (1998).

1.3.2.2 Industrial chemicals and metals

Industrial chemicals and metals are especially serious contaminants because of their toxicity, ubiquity, and varied effects (Table 1.7). Tens of thousands of chemical compounds are used as pesticides, defoliants, chlorinated solvents, and other industrial products; many result unintentionally in serious side-effects for life. Methyl mercury causes severe neurological effects and developmental problems; it is widespread and quickly enters aquatic food webs. Antifouling paints are used on ships and docks to inhibit fouling by marine organisms. Tributyltin (TBT), in particular, has saved ship owners and dock operators hundreds of millions of dollars, but is lethal to marine life, especially shellfish larvae; its use is now widely prohibited.

Some components of refined petroleum are highly toxic. Polynuclear aromatic hydrocarbons (PAHs) are ubiquitous; they are formed from incomplete combustion, released to the air, transported in particulates, and precipitated to the sea surface. Sources include creosote on dock pilings, industrial effluents, domestic sewage, oil spills, and bilge water. High-molecular-weight PAHs pose potential hazards to humans and sea life. Chronic exposure of bottom-dwelling fish to PAHs can produce lesions and deformities. The more persistent, heavier (4-, 5-, and 6-ring as opposed to lighter 2- and 3-ring) PAHs tend to have greater carcinogenic potential; benzo(a)pyrene, a high-molecular-weight, 5-ring PAH, is a carcinogen that can concentrate in organisms, especially shellfish.

Many synthetic compounds and pesticides used in households, on farms, and in industrial and military operations are hazardous wastes that persist in aquatic environments. Halogenated aromatic hydrocarbons, including organochlorines, are among the most important. DDT (dichloro-diphenyl-trichloroethane) and its derivatives, DDE (dichloro-diphenyl-ethane) and DDD (dichloro-diphenyl-dichloroethane), are highly toxic and can affect immune-system function. Organophosphates and carbamates are less persistent, but highly toxic, especially to aquatic life. PCBs (polychlorinated biphenyls) are derived from plastics and electrical equipment, can be stored in sediments, and can cause tumors, fetal death, and birth defects. Dioxins comprise a group of extremely toxic and persistent synthetic organic chemicals. Currently, their major source is incineration. Dioxins can cause abnormalities and tumors. These and a host of other chemicals and heavy metals enter the oceans, get incorporated into benthic sediments and fat tissues of animals, and are passed up food chains.

National legislation and international agreements now prohibit dumping hazardous wastes at sea. However, these wastes may persist in "hot spots," especially in sediments of industrial ports. The impact of dredging ever-deeper channels to accommodate shipping is a paramount cause for release of toxic chemicals from sediment. Millions of tons of contaminated dredged material have been dumped into harbors and seas, where sediments still contain mixtures of toxic contaminants that can be released by dredging and bottom stirring. For emerging industrial nations, these problems are especially serious, as these nations may not have the capacity to deal with contaminated sediments.

Table 1.7 Some major, ubiquitous, toxic industrial contaminants, their source and their impact.

Pollutant type	Industrial source	Biological impact
Heavy metals		Interact with biomolecules; impact varies with species and organ; liver is great accumulator
Cadmium	Mines, rivers, atmosphere, dredging	Collects in inshore mudflats, bacterial films, and organic matter, which modifies toxic exposure to benthic organisms
Copper	Electrical industry alloys and electrical wiring, algicides, acid mine drainage	Essential in biochemical processes; catalyst in hemoglobin formation; highly toxic to invertebrates
Mercury	Pulp and paper mills, fungicides, fossil fuel combustion, mercurial catalysts, weathering	Natural bacteria convert mercury to methylmercury that concentrates in fish, which can be toxic if eaten
Tin	Organotin production for pesticides, PVC stabilizer, biocide, etc. increased from 5000 tons in 1985 to 35 000 tons in 1995; now ~50 000 tons	Damages marine life worldwide. Tributyltin is extremely toxic to oysters. Sediment microorganisms convert metallic tin into methyltin
Zinc	Alloys, paints, cosmetics, etc.: used to coat steel against corrosion	Catalytic activity in essential biochemical processes
Synthetic organic compounds		Many are lipophilic. Can be carcinogenic, mutagenic, endocrine disrupters
Organochlorines: DDT, DDE, DDD, chlordane, Kepone®	Pesticides extensively used since 1940s	Remain decades in sediment: lipophilic, concentrate in liver, bioaccumulate, hormonally active
Organophosphates: diazinon, malathion	Pesticides	Fish liver neurotoxin after intracell biotransformation; repeated exposure may damage neuroendocrine function
Carbamates: atrazine, sevin	Agriculture, crop herbicides	Highly toxic; rapidly detoxified from animal tissues and eliminated
Polynuclear aromatic hydrocarbons (PAHs)	Crude oil, refined oil, related aerosols	Low solubility in water, remain in sediment, biomagnify: benzo(a)pyrene is a carcinogen and endocrine disrupter
Polychlorinated biphenyls (PCBs)	Dielectric fluids in electrical equipment (transformers, hydraulic systems)	Highly persistent; sequester in sediments; concentrate in liver and gonad tissue
Dioxins	Combustion and incineration, pulp and paper production	Among the most toxic: occur in seafood; affect food webs, carcinogenic, endocrine disrupters

Compiled from Calmano & Förstner (1996); Kline (1998); NRC (1999).

1.3.2.3 Litter and plastics

Most litter is accidentally or carelessly released. However, many commercial, fishing, and military disposal operations purposefully introduce tens of thousands of tons of litter annually into the seas. Plastics are a dominant component of litter and their durability has made them a major environmental problem. Plastic litter ranges in size from particles a few millimeters in diameter to large objects. The amount of plastic entering the sea from all sources is not known, but casual observation of littered beaches indicates that it is massive (Fig. 7.14B, page 224). Even in isolated areas such as the deep sea, trawls retrieve plastics and other debris. Plastics resist degradation, and many can persist for decades to centuries. When they accumulate on shores, at sea, or on the ocean floor they become a public disgrace, especially when mixed with medical wastes.

The harm that plastics cause marine life is often obvious, but difficult to quantify (Table 1.8). Each year, an unknown number of sea turtles, birds, and marine mammals suffer malnourishment or die after swallowing plastics. Also, thousands are crippled or killed by entanglement, entrapment, smothering, and strangulation due to encounters with foreign objects, especially discarded fishing gear (Fig. 1.5). Floating plastic sheets may also cling to plants, corals, and tidal animals and smother them. And tens of thousands of fish traps made entirely or partially of plastic are lost at sea annually, many of which continue to "ghost fish" for long periods of time.

Table 1.8 The impact of plastics on biota.

Species group	Total number of species worldwide	Number (%) of species with entanglement records	Number (%) of species with ingestion records
Sea turtles	7	6 (86%)	6 (86%)
Sea birds	312	51 (16%)	111 (36%)
Sphenisciformes (penguins)	16	6 (38%)	1 (6%)
Podicipediformes (grebes)	19	2 (10%)	0 (0%)
Procellarfiformes (albatrosses, petrels, shearwaters)	99	10 (10%)	62 (63%)
Pelicaniformes (pelicans, boobies, gannets, cormorants, frigate birds, tropic birds)	51	11 (22%)	8 (16%)
Charadriiformes (shorebirds, skuas, gulls, terns, auks)	122	22 (18%)	40 (33%)
Other birds	–	5	0
Marine mammals	115	32 (28%)	26 (23%)
Mysticeti (baleen whales)	10	6 (60%)	2 (20%)
Odontoceti (toothed whales)	65	5 (8%)	21 (32%)
Otariidae (fur seals, sea lions)	14	11 (79%)	1 (7%)
Phocidae (true seals)	19	8 (42%)	1 (5%)
Sirenia (manatees, dugongs)	4	1 (25%)	1 (25%
Mustellidae (sea otter)	1	1 (100%)	0 (0%)
Fish	–	34	33
Crustaceans	–	8	0
Squid	–	0	1
Species total	–	**136**	**177**

Compiled from Laist (1996); Marine Mammal Commission (1995).

Fig. 1.5 Dead Steller sea lion (*Eumetopias jubatus*) entangled in discarded fishing gear, Amak Island, southeastern Bering Sea, 1974. The skeleton shows signs of having been consumed by a predator, probably a grizzly bear. Photograph by the authors.

1.3.2.4 Biological pollution and exotic species

Marine biological pollution includes pathogens and microbes, and exotic, transgenic, and invasive species. Introduction of exotic species into new locations now occurs on a regular basis and some invasive species have become extremely abundant. A serious problem concerns ship ballast water, which is responsible for many problem species and their often-severe economic consequences: for example, a Japanese toxic dinoflagellate (*Gymnodimium catenatum*) introduced into Australian waters, the zebra mussel and European shore crab to North America, and the comb jelly to the Black Sea (Table 1.2). Pathogens, such as the cholera bacterium (*Vibrio cholerae*), may enter marine environments accidentally when ships pump their bilges. Microorganisms are also introduced into marine waters with wastewater from septic systems, sewer overflows, and untreated sewage. To avoid pathogenic introductions, bilge water can be pumped into holding tanks at ports for treatment. However, treatment is expensive, monitoring is difficult, violations of the few regulations that do exist are frequent, and most ports have few, if any, facilities.

Introductions often bring changes to the biological communities that they have invaded. A recent example is the discovery of a population of Indo-Pacific lionfish in subtropical waters of the U.S. east coast, probably

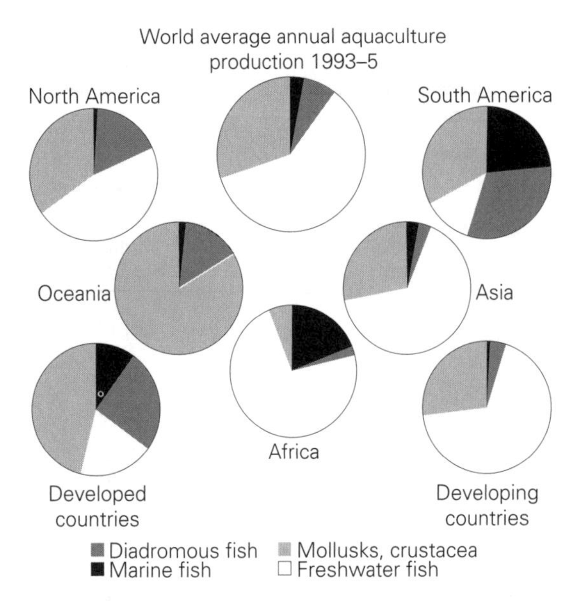

Fig. 1.6 Global aquaculture. Relative production by region and resource. Percentages are total average annual aquaculture production. Adapted from FAO (1995, 2001); WRI et al. (1998).

from an aquarium holding facility. This is the first known case of a successful introduction of a tropical Pacific fish into the northwest Atlantic. This attractive fish is a voracious predator that also bears highly venomous spines. In both appearance and predatory behavior, it is unique to the Atlantic and thus could be especially damaging to native prey species.

Aquaculture is a rapidly expanding source of biotic pollution. This activity is the raising of fishes, mollusks, crustaceans, and plants under confined, artificial conditions. Aquaculture is a very old practice that supplies much-needed food for many people, and has become a major global industry (Fig. 1.6). Brackish and saltwater aquaculture centers on a few species that are important export products and that attract high prices: for example, Atlantic (*Salmo salar*) and Pacific (*Onchorhynchus* species) salmons, and the marine fish called "dolphin" (or "mahi-mahi," *Coryphaena hippurus*). Aquaculture production systems differ widely, but most have four aspects in common. First, aquaculture depends heavily on natural sources of fishmeal and larvae for culture. Second, aquaculture usually requires conversion of natural habitat, such as estuaries, mangroves, and lagoons, into large ponds. Third, escaped individuals from aquacultural facilities may affect natural populations and communities through interbreeding, competition, and introduced diseases. The use of genetically engineered (transgenic) organisms is of special concern due to unknown interactions between farm-raised and naturally occurring populations. Farmed Atlantic salmon have escaped into Norwegian and North Pacific waters, with uncertain consequences for native fish populations. There is a long history of transfers of oysters into foreign waters. The Japanese oyster (*Crassostrea gigas*) is now farmed in North America, Europe, Australia, and elsewhere. To prevent interbreeding and replacement of native species, this species has been genetically altered so that it is presumably unable to reproduce; however, evidence indicates it can revert to breeding condition. Finally, aquaculture can become a source of pollution and disease. The crowding of individuals in restricted facilities creates ideal conditions for epidemics; antibiotics are used as preventatives. Thus, aquaculture generates large quantities of nutrients, pesticides, and antibiotics that can enter coastal waters, raising concerns about contamination and alteration of biochemical regimes.

1.3.2.5 Noise, heat, and light

Supertankers, huge fish-factory ships, cruise ships, and submersibles that explore the deepest parts of oceans create a noisy presence. Low-frequency sound is used for exploration of the ocean bottom for minerals and scientific measurement of water mass characteristics, and high- and low-frequency sounds are used in antisubmarine acoustic operations. All of these noises potentially can affect crustaceans, fishes, and marine mammals that use sound for underwater communication, food-finding, or navigation. Underwater noise has considerable potential to alter behavior or even to induce tissue damage if used intensively (Box 1.3).

Large amounts of heated water, often more than 10°C above ambient sea temperature, are introduced from power plants and other industrial activities into local waters. Heat can differentially affect local distributions and reproductive behavior of plants and animals. Elevated temperatures reduce oxygen supply and can exacerbate hypoxia. At levels above the physiological tolerance of a species, heated water can be lethal, notably in tropical corals that are sensitive to temperatures greater than 30°C.

Finally, artificial light from seaside development is a concern, especially for sea turtles, which return from the sea at night to nest on sandy shores – often where people prefer to recreate and live. Where urban lights add glow to dark shorelines, sea turtle hatchlings may

Box 1.3 Noise pollution: a threat to dolphins?

Sam Ridgway

The significance of human-made sound in disturbing or injuring cetaceans has been considered only recently. Earlier studies of dolphin hearing were motivated by the discovery of the animal's sonar. Audiograms, plots of hearing threshold at different sound frequencies (in Hz, cycles per second, or kHz; 1 kHz is equal to 1000 cycles of a sound wave), have been done on several species of the cetacean superfamily Delphinoidea (narwhals, white whales, all dolphins, and the porpoises). These audiograms showed sensitivity to sound frequencies up to about 60 to 150 kHz, almost eight times the frequency span of human hearing (humans are slightly more sensitive to sound pressure in air, but their frequency range is limited to about 20 kHz). In the frequency range of 40–80 kHz the bottlenose dolphin ear can sense a sound wave with a pressure of only 100 μPa (the pascal [Pa] is the standard measure of pressure, and a micropascal [μPa] is one millionth of a pascal). The ear is a phenomenal detector of pressure change. As a dolphin dives to a depth of 100 m the surrounding pressure is 1 trillion μPa yet the ear senses a pressure change one billionth the sea pressure. The dolphin ear can sense and respond to sounds as faint as 100 μPa and as loud as 10 billion μPa. Since these numbers are so large they become cumbersome to quantify, hearing is normally measured on a logarithmic decibel scale and related to a base value. For example, a dolphin threshold of 100 μPa is given as 40 dB re 1 μPa. A very loud sound that might overload the ear if it were continued for a second or more might be given as 200 dB re 1 μPa.

Sensitive ears connected to a massive auditory central nervous system are fundamental to the dolphin's echolocation and communication. It is reasonable to ask how an animal, with such excellent hearing, avoids damaging its own ears with the loud sounds it produces during echolocation. The dolphin ear, anatomically only a few centimeters away from its sound production mechanism, processes high-frequency echolocation pulses up to 230 dB re 1 μPa in peak to peak amplitude. Using intense pulses and sensitive ears, dolphins can detect echoes (as quiet as a human whisper) from small objects at 100 m and more. Because the dolphin's pulses are very brief, on the order of 40 μs (25 000 of these would equal a second of sound), the total energy within each pulse is miniscule. Anatomical structures, including highly reflective air sinuses that attenuate sound, probably help the animal avoid damaging its own ears.

On a comparative basis, the baleen whale auditory system does not appear as specialized as that of dolphins. The acoustic centers of the baleen whale brain are smaller than those of dolphins, for whom the auditory nerve is the largest cranial nerve; in baleen whales, the trigeminal nerve is larger. Unlike dolphins, whose sense of hearing predominates, baleen whales appear to rely most on the sense of touch. Although we have made no audiograms, observations show that baleen whales usually produce low-frequency sounds, often as low as 15 Hz.

If, as anatomical study suggests, the baleen whale ear is specialized for low frequencies, then the inference is that the animal's hearing is adapted to protection from considerable acoustic interference such as that which occurs from natural ocean background noise in the part of the acoustic spectrum below 1000 Hz. It is unlikely that baleen whales will be captured and trained for audiograms as have dolphins; nonetheless, physiological methods could be used to obtain audiograms on beached or entrapped whales.

The question arises: Can baleen whales detect calls of other whales by means other than auditory ones? The arrays of vibrissae about their heads suggest that baleen whales may use these adaptations to sense low-frequency vibrations including the calls of other baleen whales. Uses of tactile detection like these may explain the large trigeminal nerve in baleen whales. Until audiograms can be measured on baleen whales, we are left to speculate about their hearing thresholds and frequency sensitivities. The absence of definitive audiograms compounds the problem of determining what levels of human-generated sound may damage baleen whale hearing.

Dolphins have evolved robust mechanisms to protect their ears and body tissues from loud natural sounds such as lightning strikes, earthquakes, pounding surf, volcanic eruptions, whale calls, and even their own echolocation pulses. Year after year, these adaptations are eroded as oceanic shipping raises the ambient background noise in the oceans. These animals should not be continuously exposed to the equivalent of a boiler factory or even a loud discotheque. Intensified technology introduces loud noise for purposes of improved sonar, oil exploration, and acoustic communication modems. Increasing production of intrusive noise in the sea poses a serious threat to marine life. Science and technology must take action together in order to protect marine mammals such as dolphins and whales from dangerous noise.

Source: Popper et al. (2000).

become disoriented when they emerge from nests in the sand to scurry to the sea to avoid predators.

1.3.3 Physical alterations

Humans constitute a massive geophysical force as they transform the coastal and ocean horizon. Impervious substrates are replacing marshes and forests, urban centers are being joined into coastal conurbations, and a large portion of the land- and seascape now exists as patches of nature among areas of intense human use. Thus, the coastal land–seascape is being increasingly urbanized, compromising the natural flows and interchanges of coastal ecosystems (Table 1.9).

1.3.3.1 Reclamation and offshore development

Reclamation of coastal lands, water, and wetlands probably originated with the advent of the Agricultural Revolution 5000–10 000 years ago, and was greatly enhanced by establishment of coastal city-states. Salt marshes of England's Wash began to be reclaimed around AD 900 and only half remain intact today. In the Netherlands, great portions of the Zuidersea have been closed off and drained for agriculture. High dikes now prevent saltwater from intruding into agricultural land, and brackish lakes (polders) dominate large portions of the country; without this line of defense, the sea would threaten more than half of the Dutch population. Along Japan's coasts, where land is in short supply, man-made islands are being created to extend human occupation, industrial growth, and farming. A similar example is Hong Kong, where a large percent of coastal land is altered and former mangrove and saltmarsh environments have been converted for food production. In southeast Asia and Central and South America, large areas of coastal lowlands, particularly mangroves, have been converted to shrimp farms. Use of fertilizers and chemicals by farms and aquaculture increase pollution and eutrophy of coastal waters.

Table 1.9 Percents of coastline bordered by artificial structures.

Region	%	Region	%
Belgium	85	Scotland	8
Lake Erie, USA	61	Sweden	7
Japan	51	Portugal	7
South Carolina, USA	39	Oregon, USA	6
England	38	Ireland	<5
Kuwait	29	Victoria, Canada	4
Barbados	22	Montserrat	3
Lake Erie, Canada	21	Finland	<2
South Korea	21	Iceland	<2
Italy	13	Brazil	1
California, USA	9	Sierra Leone	<1
South Africa	9	Alaska, USA	<1

Modified from Walker (1990).

Reclamation is most obvious in estuaries and lagoons with prime port locations and where many of the world's largest cities are located (Table 1.10). Gigantic petrochemical complexes, steel mills, power plants, condominiums, and ship-building facilities occur on or adjacent to some of the richest farmlands and fishery areas on Earth. Offshore, structures are built to accommodate oil and gas extraction from fixed platforms, particularly where navigational requirements for free passage of ships can be met and where undersea pipelines can be constructed. To support these facilities, onshore facilities must be constructed, such as platform fabrication yards with boats to carry supplies and equipment to rigs and platforms, helicopters to ferry personnel to and from offshore sites, and barges to install pipelines between production and processing facilities. Conflicts arise where fisheries also exist. The full impact of such development on future fisheries is difficult to predict.

Military activities contribute directly and indirectly to coastal transformations. In many areas, military reservations that are "off-limits" may host some of the best-preserved natural habitats. However, areas of armed conflict can severely disrupt coastal systems. The use of defoliants by U.S. forces in Vietnam destroyed 14% of Vietnam's forests and severely damaged economically important mangrove swamp ecosystems, while also adding hazardous toxics to both land and sea. During the 1990 Persian–Arabian Gulf conflict, an estimated 3–6 million barrels of oil per day were deliberately burned, and oil spilled into the waters of the Gulf severely damaged coastal land and marine ecosystems.

1.3.3.2 Obstructing watershed flows

Freshwater is required for every aspect of life. However, freshwater distribution is uneven, a situation that has initiated extensive dam and channel construction to regulate and divert water supply to places where it may be used. The majority of large watersheds worldwide have been modified by dam construction. Great dams have been constructed across major rivers to produce energy, to mitigate against destructive floods and droughts, and to provide water to farms and cities.

Table 1.10 Some of the world's largest cities and major ports (•).

City/country	Population, × 1000	Population, × 1000 (projected)	Annual growth rate (%)	Coastal water	Port, maximum draft (m)
	1995	2015	1995–99		
Tokyo, Japan•	26 959	28 887	1.45	Tokyo Bay	10.5
Mexico City, Mexico	16 562	19 180	1.81		
São Paulo, Brazil	16 533	20 320	1.84		
New York City, USA •	16 332	17 602	0.34	New York Bight	13.5
Bombay (Mumbai), India	15 138	26 218	4.24	Arabian Sea	
Shanghai, China •	13 584	17 969	0.36	East China Sea	10.0
Los Angeles, USA •	12 410	14 217	1.60	Complex units	21.4
Calcutta, India •	11 923	17 305	1.81	Bay of Bengal	
Buenos Aires, Argentina •	11 802	13 856	1.15	Rio de la Plata	
Seoul, South Korea	11 609	12 980	1.92	Yellow Sea	
Beijing, China	11 299	15 572	0.87	Bohai Sea	
Osaka, Japan •	10 609	10 609	0.23	Osaka Bay	11.5
Lagos, Nigeria	10 287	24 640	5.68	Bight of Benin	
Rio de Janeiro, Brazil •	10 181	11 860	1.00	Guanabara Bay	10.1
Delhi, India	9 948	16 860	3.85		

Modified from World Almanac (1999).

Reservoirs behind dams also serve as valuable recreation assets. During the major dam-building period between 1950 and the mid-1980s, the number of large dams increased seven-fold worldwide. In 1997, an estimated 800 000 large dams existed, of which 45 000 are higher than a five-story building (Fig. 1.7).

Dams can affect very large portions of watersheds and landscapes. There are six dams on the highly industrialized Rhine watershed, which crosses seven countries and encompasses 68 large cities before it drains into the North Sea. The longest river in the world, the Nile, embraces more than 3 million km² of watershed in nine countries. The Nile's freshwater flow into the southeastern Mediterranean Sea is interrupted by Egypt's Aswan High Dam, and seven other large dams on the Nile have reduced flow to the sea to miniscule proportions, which has fundamentally affected the ecology and fisheries of the eastern Mediterranean. In southeastern South America, 29 large dams, with several more being planned, are intended to control flow and provide electric power for economic growth of the Paraná River, which drains the Pantanal, a 2.5 million km² wetland that straddles four countries (Box 9.3, page 276).

The world's largest dam is under construction in China. The Yangtze River drains a 1.7 million km² watershed and empties into the East China Sea. This river already has 45 600 dams, large and small, and the

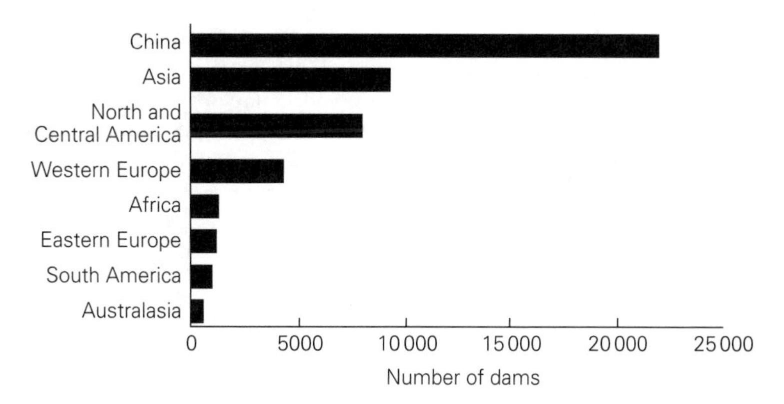

Fig. 1.7 Number and height of large dams worldwide. A large dam is one whose height is 15 m or higher, or whose height is between 10 and 15 m if it meets at least one of the following conditions: (i) a crest length of not less than 500 m; (ii) a spillway discharge potential of at least 2000 m³ s⁻¹; or (iii) a reservoir volume of not less than 1 million m³. From World Commission on Dams (2000), with permission.

Three Gorges Dam is to be the world's largest. The dam will be backed by a 690 km^2 reservoir with a maximum depth of 175 m. One major purpose is flood control, motivated by the death of 30 000 people after one of the worst Yangtze River floods, in 1954. The Three Gorges Dam is also intended for economic development and power generation. However, construction will displace more than a million people, and inundate many hectares of farmland. This dam will also affect valuable sea fisheries, habitats for rare fish, and the endangered white flag dolphin.

1.3.3.3 Urban land–seascapes

Productive coastal-marine areas are rapidly being physically transformed into commercialized urban complexes. A major contributor to this trend is that coastal environments satisfy cosmopolitan life-styles. The esthetic and recreational appeal of the ocean invites construction of hotels, marinas, residences, and condominiums on scenic coasts. Infrastructure and industry follow. Marina construction has rapidly increased worldwide to accommodate recreational boating and tourism, often with adverse effects on water quality. Canals and channels built to facilitate private boat uses cut into the land and create *cul de sacs* that alter hydrological patterns. In such areas, shores must be stabilized against erosive boat wakes, and canals must continually be dredged as a result of sedimentation. As tourism rapidly expands, huge docks are built to accommodate super cruise ships, some of which have been implicated in illegal dumping of wastes at sea. In many tropical nations, development has caused a general deterioration of local water quality, where sewage and construction wastes damage coral reefs and other productive environments. In the northern Mediterranean Sea, tourism and recreation may more than treble in the next few decades to further deplete already scarce resources and living space.

Tourism development is essential for many developing nations, but tourism places strains on limited water, land, and fisheries. Furthermore, tourism and leisure infrastructure often block local access to beaches and require constant maintenance against the forces of sea and storms. For more than a century, coastal engineers have attempted to fortify coastlines against erosion and storm damage and to protect human life and property by constructing groins, jetties, seawalls, and revetments, and by adding fill to eroding beaches. In some areas, bulldozers are continually deployed to reconstitute beaches after major storms (Fig. 1.8). Sophisticated engineering procedures can create a false sense of security that encourages overdevelopment and invites greater risks and costly shoreline defenses.

Fig. 1.8 Bulldozer restoring a beach at Scheveningen, the Netherlands, 1989. Photograph by the authors.

1.4 Tertiary issues

Tertiary issues are systemic and chronic, and are most difficult to address, as they relate directly to rates of change, irreversible conditions, and social values and norms, and thereby to difficult and controversial policies. These are poorly understood issues that focus attentions on complex interactions and difficult solutions.

1.4.1 Emergent phenomena

Emergent phenomena are not in history books, or publications, or recalled from memory or understanding. Presently, they are being widely reported, suggesting changes in environmental states or conditions that have occurred as a result of human interventions. Examples are many. Here, we consider only a few of the most notable.

1.4.1.1 Carcinogens and endocrine disrupters

Synthetic pollutants, once thought to be harmlessly dispersed and diluted in the environment, are causing new physiological disorders. Several synthetic compounds and pesticides are known for their persistence and for being endocrine disrupters, that is, chemicals that mimic hormones and alter biological functions such as reproduction. DDT, Kepone®, dioxins, PCBs, and others accumulate, persist, strongly associate with organic matter and sediments, and resist bacterial detoxification. DDT, long prohibited in many parts of the world, still occurs in sea life. PCBs also remain widespread in the environment, from tropical to polar regions. These and other compounds can bioaccumulate through food chains to pose potential health hazards to coastal-marine and human life. Many are known carcinogens that concentrate in organisms and can be metabolized into more toxic forms.

Many studies indicate effects of endocrine disrupters on reproduction, for example: juvenile male otters (*Lutra canadensis*) in the Columbia River with reduced penis size; male Western gulls (*Larus occidentalis*) with feminized behavior; male Florida panthers (*Felis concolor*) whose testicles failed to descend; male American alligators (*Alligator mississippiensis*) in Florida with non-functional testes; and masculinized female mosquito fish (*Gambusia affinis*). In the Baltic Sea, populations of gray seals (*Halichoerus grypus*) and ringed seals (*Phoca hispida*) have exhibited low fertility rates correlated with PCBs in their tissues. In the polluted portion of the Baltic, cod, salmon, and sea trout have shown poor recruitment. There is also evidence of endocrine disruption in common carp (*Cyprinus carpio*) and largemouth bass (*Micropterus salmoides*) in freshwaters that contain synthetic organic compounds. These effects have the potential to occur in humans.

1.4.1.2 Harmful algal blooms

Many species of single-celled algae live in the sea. Most are beneficial, serving as energy producers at the base of the food web. Occasionally they "bloom" and accumulate into dense patches. Such blooms have been recorded throughout history. What may be new is their proliferation and new forms. Harmful algal blooms that contain toxins or cause other negative effects are becoming common occurrences that affect natural communities by replacing species, shading vegetation, disrupting food-web dynamics, and causing oxygen depletion. Some blooms concern public health when toxins enter food webs to concentrate in species consumed by humans. Globally, several thousand cases of human poisoning are reported each year from consumption of fish or shellfish that have consumed toxic algae.

A microscopic benthic dinoflagellate is known to cause human ciguatera poisoning when fish are consumed that contain its toxin. Captain Cook suffered from ciguatera when visiting New Caledonia in 1774. Today, ciguatera is responsible for more cases of human illness than any other kind of seafood toxicity. In 1960–84, more than 24 000 cases of human poisoning were reported from French Polynesia alone, more than six times the average for the Pacific as a whole.

A highly toxic dinoflagellate (*Pfiesteria piscicida*) has recently been discovered on the United States east coast. The cause for its appearance is not known and many questions remain about its natural history. Apparently, its two dozen life-history stages exhibit both animal and alga characteristics, allowing it to change habits, depending on circumstances. *Pfiesteria* is able to photosynthesize, but becomes a predator when fish are present. In its killing stage, *Pfiesteria* releases a neurotoxin that stuns the fish and facilitates consumption of the flesh. *Pfiesteria* is implicated in fish kills, organic pollution, and severe human ailments.

1.4.1.3 Anoxic bottom water

Low oxygen concentration (hypoxia) is characteristic of some deep-ocean waters. When hypoxic waters become anoxic (lacking oxygen) the result is a "dead zone," and all organisms that require dissolved oxygen perish. Hypoxic conditions are exacerbated by human activities, such as fertilizers from farms and lawns, and appear to be increasing in coastal and estuarine areas worldwide (Fig. 1.9).

A prominent example of a hypoxic zone occurs annually in the northern Gulf of Mexico. Hypoxic water was first observed off Louisiana–Texas shores during the mid-1970s, probably due to nutrient enrichment from the Mississippi and Atchafalaya rivers. By summer 1989, hypoxia extended over 9000 km^2 of the inner continental shelf. Recently, hypoxia has cut a lethal swath through about 18 000 km^2 of these coastal waters. Hypoxia–anoxia also occurs in Chesapeake Bay during summer months, when deep waters become nearly devoid of life. On New Jersey's continental shelf, localized oxygen depletion occurred in 1968, 1971, and 1974; in 1976, severe oxygen depletion began in July and persisted until October, creating an oxygen-depleted corridor 5–85 km offshore and extending 165 km southward to cover 12 000 km^2. The Black Sea is naturally anoxic below its halocline; this condition has been increasing, causing massive mortality of demersal fishes. The Baltic and Mediterranean seas also appear to be on long-term trends toward anoxia, should current trends of urbanization and pollution continue.

1.4.1.4 Mass mortalities, epidemics, and pandemics

Many diseases now afflict coastal-marine organisms in epidemic proportions (Table 1.11). Whether these

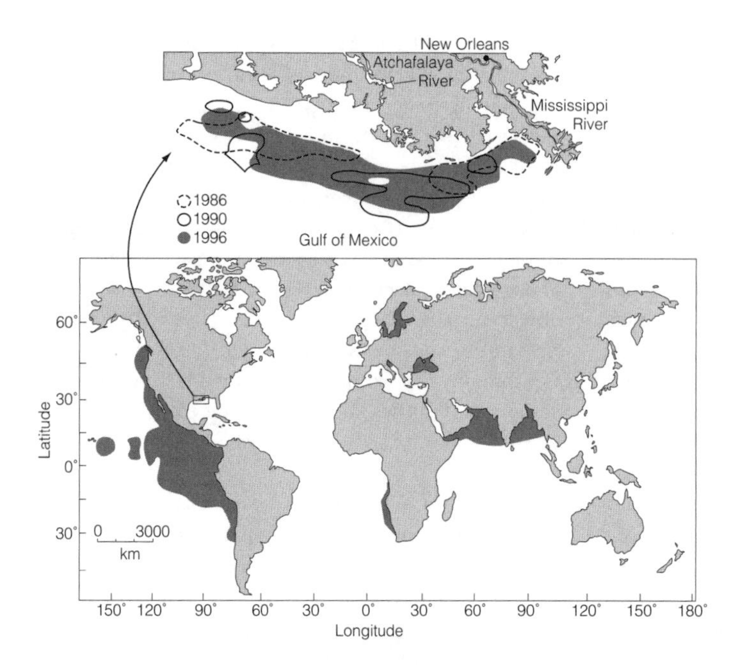

Fig. 1.9 Areas of the world experiencing marine hypoxia. *Lower figure*: portions of the Pacific and Indian oceans have lower dissolved oxygen concentrations and greater areas exposed to hypoxia than the Atlantic. These conditions may have existed for long periods of time. Hypoxia has been exacerbated in several other areas, e.g., the Black Sea, due to human activities. From Diaz & Rosenberg (1995), with permission from R.N. Gibson. *Upper figure*: northern Gulf of Mexico "dead zone" encompassing different areas of shelf water in recent summers. This seems to be a new condition, probably caused by increased nutrient inputs from the Mississippi and Atchafalaya rivers. From Ferber (2001), with permission from N. Rabalais.

are new or re-emergent is difficult to determine. In 1931–2, North Atlantic eelgrass (*Zostera marina*) was struck by a "wasting disease" caused by a slime mold (*Labyrinthula macrocystis*), which extirpated eelgrass in many areas. Eelgrass has returned to many areas, but complete recovery has not yet occurred. In the Caribbean Sea, a sea fan (*Gorgonia ventalina*) has been infected by a fungus (*Aspergillus sydowii*), evidently carried on airborne dust from Africa (page 239). On Florida reefs in June 1995, some 16 species of corals and a hydrocoral bore symptoms of an unusual coral disease – a readily transmissible, virulent bacterium (*Sphingomonas*) first described in 1990. In the 1980s, abundant Caribbean sea urchins (*Diadema antillarum*) that graze algae on reefs suddenly became rare due to disease. Such repeated, severe infestations have the potential to restructure reef communities.

In 1987, many freshwater Baikal seals (*Phoca siberica*) died from canine distemper. Forms of phocine distemper virus (PDV) killed approximately 25 000 North Sea harbor seals (*Phoca vitulina*) in 1988, after which the population recovered, only to experience large die-offs from PDV in 2002. Viruses have also caused mortalities among striped dolphins (*Stenella coeruleoalba*) and endangered Mediterranean monk seals (*Monachus monachus*). Fin whales (*Balaenoptera physalus*) have been reported with morbillivirus infection. In 1998, more than 1600 New Zealand sea lion pups (*Phocarctos hookeri*) died at Auckland Island during a bloom of toxic algae. Viral infections and pollutants also have been implicated in the deaths of more than 700 bottlenose dolphins (*Tursiops truncatus*) in 1987–8 off U.S. mid-Atlantic shores and in excess of 500 harbor seals in New England waters in 1979–80. Pollutants may increase susceptibility of marine mammals to infections by reducing immunity.

A host of diseases also occur in invertebrates, fishes, and marine reptiles, seabirds, and mammals in which populations are reduced and reproductive fitness is altered. Humans also face diseases caused by marine pathogens, for example cholera (*Vibrio cholerae*). This deadly disease is associated with consumption of contaminated shellfish and widespread problems of sewage treatment and water quality. Ships' ballast water can carry the pathogen and some species of plankton can act as hosts for dormant bacteria. In January 1991, residents of a town near Lima, Peru, suffered from a virulent Asiatic strain, perhaps introduced by an Asian freighter anchored in the harbor. The pathogen quickly spread to other South American and Central American countries. By the end of 1991, more than 3000 Peruvians and more than 1000 people from other countries had died. In 1992, a new form of the pathogen was identified in ten Asian nations. By February 1993, Brazil had become the focus of infection, with 32 313 cases and 389 deaths. By December 1993, Latin

Table 1.11 Selected examples of marine diseases associated with mass mortalities, defined as > 10% mortality within a population or group. Taxa that suffered mortality are listed under Biota; their location listed by Ocean/sea.

Group/Location	1930s	1940s	1950s	1960s	1970s	1980s	1990s
Biota							
Kelp							•
Red algae							•
Sea grass	•					•	
Sponge	•						
Coral						•	•
Clams						•	
Scallops						•	
Abalone						•	
Oysters		•		•	•	•	
Starfish					•		
Sea urchins						•	•
Fish			•				•
Seals			•			•	
Cetacea						•	•
Ocean/sea							
North Atlantic	•			•	•	•	
Northeast Atlantic					•		
Gulf of St. Lawrence			•				
North Europe	•					•	•
Mediterranean							•
Gulf of Mexico, USA		•					
North Caribbean	•						
Caribbean						•	
West Caribbean						•	
North Pacific							•
Northeast Pacific					•		
South Pacific							•
Lake Baikal							•
Australia							•
Antarctica			•				
Southern Ocean							•

Adapted from Harvell et al. (1999).

America and the Caribbean had experienced 700 000 cases and 6400 deaths.

1.4.2 Altered communities and ecosystems

Community and ecosystem change comprise the most far-reaching issue facing conservation, for they incorporate all other issues and are often associated with unpredictable and unexpected outcomes, including the emergent phenomena described above. Dramatic changes in biotic communities have occurred during the past several centuries, mostly because of human-initiated resource exploitation and physical alterations. Such changes go far beyond direct influences, affecting global chemical cycles and climate, and dramatically accelerated during the 20th century. Most scientists now concur that chemical cycles and climate continue to change on a global scale. Uncertainty remains about cause or direction, that is, to what extent human activities are a cause, where nutrients may be sequestered, or even whether global warming or cooling may result. One major problem is that only short-term changes are amenable to direct study. Longer time scales of change can be derived only from retrospective studies during hundreds to thousands of years. Nevertheless, as research continues, anthropogenic activities are ever more implicated.

One type of ecosystem change that may be occurring at the hands of humans concerns global biogeochemistry. Life on Earth is based on cycles of carbon, nitrogen, and phosphorus, among others. Carbon dioxide (CO_2)

Fig. 1.10 Baltimore Harbor is a major U.S. port and industrial location, representing the transformation of a subestuary of Chesapeake Bay into a conurbation. Photograph by the authors.

is especially critical as it is essential for photosynthesis, the basis for most life on Earth. The flux of CO_2 across the air–sea interface seems to be the greatest influence on the global carbon cycle. Fossil fuel combustion today adds billions of tons of CO_2 to the atmosphere annually, mostly from economically developed regions of temperate zones. This input has poorly understood effects on coastal-marine life. Alteration of the global nitrogen cycle also has consequences for ocean systems. Total net human nitrogen inputs to rivers and oceans amounts to many million tons per year. Such massive inputs have the potential to alter the nitrogen cycle.

Another major ecosystem alteration concerns watersheds, which act as Earth's arterial systems by delivering freshwater, materials, and nutrients that help sustain coastal-marine ecosystems. Freshwater is rapidly becoming Earth's most endangered resource. Only about 0.26% of global freshwater is available for human consumption. As demand increases, overdrafting is occurring in most parts of the world. This situation is made complex because water flows across jurisdictions, overland and underground, and uneven distributions are among the most acute and complex of social problems. The combined effects of construction of dams and reservoirs, natural water supply variations, water diversions, land use change, groundwater withdrawal, shoreline erosion, and saltwater intrusion all result in significant changes of coastal ecosystems, their productivity, and their biotic communities.

1.5 Conclusion

All the issues related above have roots in complex natural processes and are exacerbated by human activities to some extent. They are best viewed through the lens of human social trends, rates of change, and momentum – for example, population, consumption, technology, and efforts to increase human wellbeing. Tertiary issues are most difficult for human institutions to address, and require an ecosystem perspective. This is especially the case in the coastal realm, where most of the human population resides and where cities are spreading into conurbations (Fig. 1.10). World population reached six billion on October 12, 1999. About 60% of this population is concentrated within 100 km of coasts. In the future, higher rates of population increase are projected for coastal areas than inland. And, as coastal populations and cities expand, demands for resources expand exponentially. One major consequence is that the most productive areas for humans on Earth – estuaries, lagoons, deltas, and the coastal ocean – are being transformed into urban space or are becoming dominated by human activities, with consequences that can be difficult to predict.

Succeeding chapters will illustrate mechanisms for addressing current coastal-marine issues, the minimum knowledge base that must be attained, how the nature of change itself may be understood, and how the environmental debt may be confronted.

Chapter 2

Mechanisms

Sic utere tuo ut alienum non laedas. [Use what is yours so that others are not injured]
Legal axiom in common law

If the misery of our poor be caused not by the laws of nature, but by our institutions, great is our sin.
Charles Darwin

2.1 Introduction

Evidence of coastal-marine environmental degradation has lately awakened the need for conservation. A host of mechanisms has evolved at every level, from local to global, in attempts to balance social demands with environmental health and resource sustainability. A "mechanism" is a process or technique for achieving a result. For conservation, this involves policies, strategies, and complex arrays of tactics. The purpose of this chapter is to present an overview of concepts and agreements that have achieved widespread significance for conservation of the coastal realm. Agents of social and environmental change continue to influence and alter these mechanisms and their effectiveness.

2.2 Species and habitat conservation

Species and habitat conservation mechanisms include a broad spectrum of protection, restoration, and sustainable use, extending from local traditions, to regulations perpetrated by states and nations, to international agreements.

2.2.1 Species conservation

Species conservation is widely endorsed by society and by national and international agencies and organizations, and by thousands of traditions and laws. Numerous international agreements have entered into force in response to crises and the need for species' protection (Table 2.1). For example, early in the 20th century the Fur Seal Treaty forestalled extinction of fur seals and sea otters, and the Migratory Bird Treaty forbad market hunting and the feather trade that were decimating shorebirds, waterfowl, and others. An example of the need to regulate industry is the International Agreement on the Regulation of Whaling. Another approach is reflected by the U.S. Marine Mammal Protection Act of 1972. This Act uniquely mandated that marine mammals be maintained at "optimum sustainable population." This concept was, at first, controversial as it placed management in the still-unresolved context of ecosystem "health," challenging the management paradigm of "maximum sustainable yield." A host of agreements is directed toward commercial fisheries, in which species depletion can carry high economic costs. These agreements take two forms: first, for species or species groups, such as tunas and anadromous fishes, and second, for fishing methods or gear, such as longlining, drift nets, and trawling. All together, these approaches illustrate that highest priority is given to species that are depleted, endangered, or of economic worth, with a strong bias toward groups that arouse public sympathy, almost exclusively vertebrates. Only very recently has attention been devoted to non-commercial species or to threatened or endangered marine plants and invertebrates.

For depleted and endangered species, two primary conservation mechanisms are advocated: *in-situ* conservation, wherein species are protected within natural habitats, and *ex-situ* conservation, which occurs in gene banks, tissue culture, and captive breeding programs in zoos, aquaria, and wildlife centers. The latter institutions also stress education about ecological relationships of rare, endangered, and economically valuable species, which has resulted in greater public support for, and understanding of, coastal-marine life. Until the advent of large public aquaria and of recreational snorkeling and diving, marine life went unseen and unknown to most people. Even marine mammals that are so beloved by the public today were misinterpreted and ignored. For example, the killer whale (*Orcinus orca*) was presented in aquaria in the 1960s literally as a "killer," a perception that changed after the public viewed its

Table 2.1 Examples of national and international agreements for species protection.

Protection	Instrument (date in force)	Intended action
Fur seals	North Pacific Fur Seal Treaty (1911)	Ended fur seal harvest; protected sea otters
Birds	Migratory Bird Treaty (1918)	Regulates taking, selling, transporting, and importing of migratory birds
Whales	International Convention on the Regulation of Whaling (1946)	Regulates harvest and industry; moratorium on commercial whaling established, 1985–6
Marine mammals	U.S. Marine Mammal Protection Act (1972)	Established concept of "optimum sustainable population" and moratorium on taking most marine mammals
Antarctic seals	Convention on Conservation of Antarctic Seals (1972)	Adopted standards for conservation of Antarctic seals
Polar bears	Agreement on Conservation of Polar Bears, Oslo (1973)	Limits hunting to sustainable levels
Dolphins	Agreement to Reduce Dolphin Mortality in the Eastern Tropical Pacific Tuna Fishery (1992)	Regulates dolphin "by-catch" to the lowest possible level, with an objective of zero take
Endangered species	Convention on International Trade in Endangered Species of Wild Fauna and Flora (CITES, 1973)	Prohibits or controls all trade of endangered or threatened species; controls trade of wild species that have market value
Endangered species	U.S. Endangered Species Act (1973)	Forbids jeopardizing endangered or threatened species or adversely modifying critical habitats
Endangered migratory species	"Bonn" Convention on the Conservation of Migratory Species of Wild Animals (1979)	Forbids take of listed migratory species of wild animals, including sea turtles, birds, and marine mammals
Sea turtles	Inter-American Convention for the Protection and Conservation of Sea Turtles (2000)	Protects sea turtles and their nesting habitats in the Americas
Fisheries	Convention for the Establishment of an Inter-American Tropical Tuna Commission (IATTC, 1949)	Regulates catch of tunas in the tropical Pacific Ocean
Fisheries	International Convention for the Conservation of Atlantic Tuna (1966)	Regulates catch of tunas in the Atlantic Ocean
Fisheries	Convention for the Conservation of Salmon in the North Atlantic Ocean (1982)	For regulation of catch of salmon and for their conservation in the Atlantic Ocean
Fisheries	Convention for the Prohibition of Fishing with Long Driftnets in the South Pacific (1989)	Prohibits use of long driftnets in designated areas
Fisheries	Convention for the Conservation of Anadromous Stocks in the North Pacific Ocean (1992)	Principally directed toward regulation and conservation of salmons

Compiled from Crystal (1998); IUCN (1995); UNEP (1997).

intelligent behavior in aquaria. Even jellyfish and sharks have reached public sentiment *via* close-up presentation and education. The appeal of marine life now promotes a major ecotourism industry.

Thus, greater familiarity with coastal-marine life has transformed species conservation into a broader context, from single, charismatic, commercially valued species to marine life in general. However, present agreements too often leave the bulk of coastal-marine species unprotected. Also, considerable difficulties exist in implementation of agreements; few are sufficiently enforced and, for international agreements, only signatory nations participate, leaving the door open for non-signatories to continue the depletion. Responses to crises prevail, and insufficient mechanisms exist to prevent depletion or to protect habitat.

2.2.2 Habitat conservation

Habitat conservation is based on the assumption that protected habitats of sufficient size and design can allow depleted species to recover and environments to become sustainable. Protected areas serve several functions, most fundamentally to protect species together with their environments, and as control areas for research on natural history, community interactions, ecological function, and responses to environmental change.

Many international instruments have urged establishment of coastal-marine protected areas (CMPAs), and several thousand now exist worldwide (Table 2.2). During past centuries, CMPAs were established to perpetuate traditional uses, such as hunting or fishing, or for cultural reasons. During the early to mid-20th century, governments established CMPAs for species and habitat conservation and for scenic or cultural values; examples are Glacier Bay National Monument in Alaska (1925), Fort Jefferson National Monument in Florida (1935), and Green Island in Queensland, Australia (1938). In 1962, the International Union for the Conservation of Nature and Natural Resources (IUCN) hosted the First World Conference on National Parks and Reserves. A prescient recommendation was adopted, hinting at several concepts that have become paramount today, for example the land-sea concept of the coastal realm, no-take reserves, and the essential role of protected areas for research and habitat protection (Box 2.1).

CMPAs incorporate management objectives that range from strict habitat protection to zoning for multiple use, research, and recreation. In some cases, CMPAs have been justified by specific issues, for example to prevent oil and mineral extractions or ocean dumping. CMPAs have also been established for

Table 2.2 Examples of international agreements for environmental protection.

Region or type of habitat to be protected	Instrument (date in force)	Purpose
Oceans: continental shelf, high seas, territorial sea, contiguous zone	UN Convention on Law of the Sea (UNCLOS III, 1982)	Ocean jurisdiction and conservation
Antarctica	Annexes to Protocol on Environmental Protection to the Antarctic Treaty (1991): Annex II, Conservation of Antarctic Flora and Fauna; Annex V, Area Protection and Management	Specially protected land and ice habitats areas for scientific research and conservation; extends Antarctic Treaty to the Antarctic Convergence; adopts ecosystem management
Southern Ocean	Convention on the Conservation of Antarctic Living Marine Resources (CCALMR)	Part of the Antarctic Treaty System; sets forth guidelines and standards for conservation of marine living resources for the Southern Ocean
Arctic Ocean	Oslo Convention (1972, 1983); Paris Convention (1974, 1986); Arctic Environmental Protection Strategy (1991); Global Program of Action for Protection of Marine Environment from Land-based Activities (1995)	Protect Arctic ecosystems: protect, enhance, restore natural resources; recognize traditional–cultural needs of indigenous peoples; review state of Arctic environment; address pollution
Mediterranean	Protocol Concerning Mediterranean Specially Protected Areas (1982)	For establishment of a network of protected areas for habitats and wildlife
East Africa	Protocol Concerning Protected Areas and Wild Fauna and Flora in the Eastern African Region (1985)	For establishment of a network of protected areas for habitats and wildlife
Southeast Pacific	Protocol for Conservation and Management of Protected Marine and Coastal Areas of the South-East Pacific (1989)	For establishment of a network of protected areas for habitats and wildlife
Caribbean	Protocol Concerning Specially Protected Areas and Wildlife (1990)	For establishment of a network of protected areas for habitats and wildlife
Wetlands	"Ramsar" Convention on Wetlands of International Importance Especially as Waterfowl Habitat (1971)	Protect wetlands and waterfowl habitat
Heritage Sites	"Paris" Convention Concerning the Protection of the World Cultural and Natural Heritage (1972)	For protection of already-protected sites of global significance; nominated by governments as natural and cultural heritage

Compiled from IUCN (1995).

Box 2.1 First World Conference on National Parks: Recommendation Number 15

The First World Conference on National Parks, held in Seattle in 1962, was a gathering of more than 60 countries and represented the first international exchange of ideas on protected areas. The following Recommendation was adopted:

> WHEREAS it is recognized that the oceans and their teeming life are subject to the same dangers of human interference and destruction as the land, that the sea and land are ecologically interdependent and indivisible, that population pressures will cause man to turn increasingly to the sea, and especially to the underwater scene for recreation and spiritual refreshment, and that the preservation of unspoiled marine habitat is urgently needed for ethical and esthetic reasons, for the preservation of rare species, for the replenishment of stocks of valuable food species, and for the provision of undisturbed areas for scientific research.
>
> THE FIRST WORLD CONFERENCE ON NATIONAL PARKS invites the Governments of all those countries having marine frontiers, and other appropriate agencies, to examine as a matter of urgency the possibility of creating marine parks or reserves to defend underwater areas of special significance from all forms of human interference, and further recommends the extension of existing national parks and equivalent reserves with shorelines, into the water to the 10 fathom depth or the territorial limit or some other appropriate off-shore boundary.

Source: Adams (1962).

sea-bird rookeries and marine mammal breeding areas to protect individual species at vulnerable periods of their life cycles. These narrowly based restrictions are now being modulated by growing awareness of widespread declines of habitats and the need to protect entire biotic communities. The result has been that justifications for CMPAs have broadened to concern for biodiversity and maintenance of functional environments at regional scales.

These broader goals have raised issues of site selection. Initially, CMPA establishment was based on expediency, opportunism, and pragmatism. As comprehensive networks of CMPAs began to be developed, two sequential processes evolved: first, identification of sites based on information about species, natural history, and ecological function, and second, selection among identified sites in accord with social, political, and management considerations. The need for guidelines and criteria to guide rational, defensible choices soon became obvious. The first international guidelines and criteria arose out of the decade-long International Biological Programme (IBP, 1962–72), which was devoted to ecosystem-level research and which sought areas for research and monitoring on the effects of humans on ecosystems. In 1968, an international conference on Man and the Biosphere developed an action plan, which became the United Nations Educational, Scientific, and Cultural Organization (UNESCO)'s Man and the Biosphere Programme (MAB). A major portion of MAB was the development of an international network of "biosphere reserves." In 1974, a MAB Task Force proposed five general selection criteria: biogeographic representativeness, diversity, naturalness, uniqueness, and effectiveness as a research and conservation unit. These criteria were useful, but also vague, in practice. Further, MAB was overwhelmingly terrestrial in orientation. Thus, in 1975, an international conference on Marine Parks and Reserves expanded on concepts for selection. Finally, in 1981, comprehensive criteria and guidelines emerged from UNEP's Mediterranean Action Plan (Table 2.3). These remain applicable today, and have been widely adopted and modified although specific procedures differ, depending on the case in question.

Despite considerable progress and increasing public support for CMPAs, meeting goals of habitat protection has proved challenging. A global program of marine conservation, "The Seas Must Live," was initiated in 1977–8 by the International World Wildlife Fund, in cooperation with IUCN, to maintain earlier momentum for conservation. At that time, only a few species (some marine mammals, sea turtles, and fewer than a dozen fish) were recognized as endangered, and the seas still seemed "healthy" to most people. The two-year programme was short-lived and establishment of CMPAs was fraught with social and legal difficulties, and with lack of a strong scientific basis. A major impediment was the tradition that the oceans are open to all; "freedom of the seas" had driven marine law and commerce for centuries. Many CMPAs were created, but most were small and poorly enforced; in very few cases was fishing restricted. Even today, the global

Table 2.3 Criteria for selection of coastal-marine protected areas. Criteria for site identification are distinct from those for selection.

Identification criteria	Measure
Biological, ecological	**Degree to which:**
Dependency/criticalness	• important life-history stages of a species depend on area, or ecosystem depends on processes within region
Naturalness	• area is undisturbed by human activities
Representativeness	• area represents a habitat type, ecological process, biological community, physiographic feature, or other natural characteristic
Uniqueness	• endemics occur or habitat is "one-of-a- kind" in region
Diversity	• area is diverse in species, community types, or habitats (land-seascapes)
Autonomy	• area is an effective, self-sustaining ecological entity
Productivity	• ecosystem production contributes to species survival or human welfare
Landscape, cultural	**Degree to which a natural area:**
Land-seascape	• also contains features of outstanding natural beauty
Cultural/esthetic	• also contains important cultural, artistic, or historic features
Regional	**Degree to which an area:**
Representativeness/inclusiveness	• represents major regional characteristics, and/or includes all habitat types within the region
Research, monitoring, and education	• may contribute information relevant to conservation and management, is suitable for monitoring environmental change, or may contribute toward public or formal education
Conflict and compatibility	• may help resolve conflicts or enhance compatibility among natural resource values and human activities
Selection criteria	**Measure**
Urgency/threat	**Degree to which:**
Opportunism	• immediate action can be taken, lest species or area are lost
Ease of protection	• existing conditions or actions underway support goal
Defensibility	• area can be properly safeguarded
Accessibility	• area can be protected by regulation, enforcement, or other means
Restorability	• area is accessible to management
Redundancy	• area may be returned to a former valued, natural, or productive state
	• more than one area of a kind can be included
Research, training, and education	**Degree to which an area:**
Accessibility	• is accessible for research, education, or training and appropriate institutions
Benchmark	• serves as "control" or "baseline" for scientific monitoring and comparisons
Demonstration	• serves to exemplify techniques or scientific methods
Scientific interest	• represents ecological characteristics of interest for research and monitoring
Social-economic policy	**Degree to which an area:**
Policy	• carries out or contributes to policy development
Economic benefit	• provides protection that will benefit society in the long term
Social acceptability	• has assured local support of people
Public health	• enhances nutrition and/or diminishes pollution or disease agents
Recreation	• provides local community with use, enjoyment, and education
International importance	• ranks high as a site of international significance: e.g., Galapagos Islands, Great Barrier Reef, etc.
Tourism	• benefits tourism compatible with conservation

Compiled from IUCN (1981); Ray (1976).

extent of CMPAs is estimated to cover only about 0.5% of the coastal and open oceans, with a strong bias to species and areas that are most socially valued. Only a small portion of CMPAs is "no take"; that is, fully protected from extraction, unlike many terrestrial PAs. The Ramsar Convention for wetlands (Table 2.2) is the only international agreement specific to a particular habitat type, and this agreement came into being not to protect wetlands *per se*, but to protect birds that use wetlands as major habitats.

During the 1990s, crises of depletions and habitat deterioration became obvious. The scientific community accumulated evidence that depleted aquatic populations can be fully restored only if species and their habitats are conserved as a unit, and further, that the scope of conservation must be large, even regional. Scientists took the lead in urging that CMPAs cover at least 20% of the ocean, for both biodiversity protection and as no-take fishery reserves. Coral reefs have now attained international status *via* a global agreement, becoming the marine equivalent of tropical forests for biodiversity protection. An ambitious regional initiative, the European Union's Habitats Directive, is in the process of implementation, requiring the establishment of a network of reserves for biodiversity conservation, backed by political support and resources to do so (Box 2.2). Other examples illustrate increased momentum.

2.3 Governance

Governance concerns the establishment of rules of conduct that define practices, assign roles, and guide interactions designed to grapple with collective prob-

Box 2.2 The European Community Habitats Directive

John M. Baxter

In 1992, the European Community (EC) Council, the governing body of the European Union, issued Directive 92/43/EEC on the Conservation of Natural Habitats and of Wild Fauna and Flora ("Habitats Directive"). This was a major contribution by the European Community to the Convention on Biological Diversity. In conjunction with the earlier EC Council Directive 79/409/EEC on the Conservation of Wild Birds ("Birds Directive"), a network of designated sites, known as Natura 2000, will be established across the European Community to conserve natural habitats and species of wildlife that are considered rare, endangered, or vulnerable. As with all EC Directives, each member state must enact national legislation to bring into force and deliver the aims of the Directive. In the United Kingdom the Habitats Directive is implemented through the Conservation Regulations of 1994. These regulations place special responsibilities on all authorities with existing powers in the marine environment to use such powers to protect the features of interest of designated sites.

The implementation of the Habitats Directive by all member states is to follow a timetable culminating in confirmation of the designation of the network of Special Areas of Conservation by the year 2004. Thereafter these sites will be managed as appropriate and monitored to ensure that favorable conservation status is maintained. In the UK, implementation of the Habitats Directive has provided the first real opportunity for the designation and subsequent protection of a wide range of marine protected areas or marine Special Areas of Conservation (SACs). In 1992, when the Directive was ratified, there were only two small Marine Nature Reserves in the UK: Lundy Island off the southwest coast of England and Skomer Island off the south coast of Wales. By the year 2000, a network of more than 70 marine SACs had been identified for a wide range of habitat features and a smaller number of species. Marine habitats for which SACs can be designated include reefs, large shallow inlets and bays, sandbanks slightly covered by seawater at all times, sandflats and mudflats, lagoons, estuaries, and submerged and partly submerged caves. Marine species for which SACs can be designated include bottlenose dolphin (*Tursiops truncatus*), harbour porpoise (*Phocoena phocoena*), Atlantic grey seal (*Halichoerus grypus*), harbour (common) seal (*Phoca vitulina*), and the otter (*Lutra lutra*). Seabirds and waders are covered under the Birds Directive, but very few invertebrates are included for SAC designation.

The implementation of marine, nature-conservation management measures associated with SACs has presented many difficulties and has raised a number of issues, not least those relating to traditional and historical rights to exploit natural resources and concern over increased control and restrictions on already hard-pressed communities. The development of management schemes for various sites has been taken through an inclusive-partnership approach involving all those who are dependent on the marine environment for their livelihood and enjoyment. In developing such management schemes, the Habitats Directive requires that they take appropriate measures to avoid damage or disturbance to features of interest and that all new plans and projects are assessed prior to their implementation.

Until recently the Habitats Directive has been implemented only in coastal territorial waters of European Union (EU) member states within the 12 nautical-mile (nmi) boundary. As a result of a challenge to the UK government in the High Court by Greenpeace International there is now an expectation that all EU member states will implement both the Habitats and Birds directives in waters out to the limit of community interest – effectively the 200 nmi EEZ limits. This presents considerable additional challenges, not only for identifying sites, but also for establishing a legislative framework for actually protecting sites and features of interest.

lems. Governance encompasses a social group, nation, state, or collective interest of disparate entities that recognize common needs to meet agreed-upon goals. Members also recognize an interdependent relationship, where conflicts can defeat mutual needs. Thus, governance is about decision-making, legitimate authority, policies, and principles that involve a suite of social decisions within the context of the whole. Governance is a social contract that people have with their leaders, and includes social institutions, rules, and practices that provide a framework for decision-making and cooperation. Environmental governance focuses on such issues as pollution, climate change, fisheries, transboundary situations, coastal zone management, and environmental conservation.

Coastal-realm governance is extraordinarily complex because the issues are so varied and interrelated. A serious concern is that the coastal realm has typically lacked legal "distinctiveness," having passed unrecognized as a unit of management. The coastal realm hosts many difficult conflicts among fisheries, industrial expansion, offshore mining, mineral extraction, dredging, waste disposal, and property rights, where efforts to govern a heterogeneous body of people are seldom easy. Land and sea portions of the coastal realm are subject to separate customs and policies. Maritime law includes often-conflicting laws and traditions that separately govern submerged lands, the water column, pollution, and resource use. Marine resources fall mostly within the public domain, where incentives provided by private property rights are largely absent. These aspects complicate coastal-realm conservation and demand collaborative environmental approaches.

Attempts have been made to address conservation issues through mechanisms undertaken by states and nations, which combine traditional, command-and-control regulatory measures with collaborative agreements. The number and complexity of regulations and agreements have vastly increased in recent years. Highlights of major mechanisms range from those that protect and regulate species and habitats, to national and international regimes and ocean law. Today, as conditions change and science advances, governance is rapidly evolving, while being based upon – and often hampered by – long histories of allocation and tradition.

2.3.1 Governance hierarchy

Regulatory and voluntary mechanisms for governance exist at every level, from local to international. Governments have traditionally engaged in top-down, command-and-control regulatory measures to achieve conservation goals. Agreements and traditions, on the other hand, usually do not have such power, are bottom-up, largely voluntary, and depend on consent among parties.

2.3.1.1 National governance

National governments hold sovereign rights over lands, adjacent ocean space, and resources. What makes a nation sovereign is its claim on the environment and its legal right to determine norms of behavior and conditions of life, both human and non-human, within it. National governments retain the basis for power, without which international agreements would be powerless. In a functioning civil society, national government administers central power to implement policy and environmental laws, and to guarantee citizens certain rights, usually under national constitutions. Modern civil law originated almost three centuries ago; it largely follows Roman law, and today dominates most legal systems. In modern western society, national customs, traditions, common law, customary and religious laws, individual property rights, and communal rights exist side-by-side. For other societies, customary rules and communal non-authority rules may more strongly influence social life, and consensus among stakeholders determines decisions and their outcomes.

When nations claim territorial seas or otherwise apportion jurisdictions, legal authority is usually allocated hierarchically, from national government to provinces, native groups, and local authorities. Nations hold primary responsibility for regulation, research, and monitoring of resources, and national policies give legitimacy to power and define social practices that guide interactions among participants. Legal authority resides in government agencies that are authorized by legislation or decrees, and that hold regulatory powers and powers to control certain activities. Nations typically administer their responsibilities in a top-down, vertical flow of power, but when integration among powers is required, government operates horizontally.

Land and sea fall within distinct public and private jurisdictions. Use of coastal intertidal areas and shorelands immediately landward of the water's edge is usually a customary right of citizens. This public right varies considerably among the world's dominant legal systems. The English "public trust doctrine," for example, guarantees rights of traditional use to citizens for fishing, navigation, commerce, and recreation; the central idea is that the government holds uses of coastal waters, submerged soils, and their resources for public benefit. Because the doctrine is grounded in property ownership and "best use," it is intimately connected with the economy of society, family structure, and the political system. Also, under "nuisance law," legislative bodies can declare that certain activities constitute public nuisances, in many cases requiring zoning and planning. Scandinavian "rights of common access," on the other hand, allow public access to lands, beaches, and intertidal areas and run counter to property rights. Some societies do not base laws on forms of property ownership (e.g., Polynesian and Inuit cultures), and where these traditions co-exist with western law, confusion and conflict can result.

Thus, resource conservation among a variety of different interests often results in highly variable and evolving social and legal mechanisms. Command-and-control systems function effectively when addressing sector-based environmental problems (pollution control, waste management, and fisheries regulations), but may not function effectively when dealing with customs, traditions, and communal decision-making. This becomes especially problematic when, for example, commercial fisheries use harvest techniques that may be detrimental to traditional and recreational fishers, or *vice versa*, and when government allocates resources counter to traditional practice.

Differences among nations in perception and economic status contribute in fundamental ways to how resources are either exploited or conserved. A present example concerns "sustainable development," which originated with the World Commission on Environment and Development in 1984. Developed nations of the "north," many with already depleted resources, promote sustainability, whereas developing nations of the "south" seek economic development. Sustainable development becomes problematic when developing nations exploit their forests or fisheries for export to developed nations. This discrepancy has erected a dilemma that pits economic groups and nations against one another.

2.3.1.2 Regional governance

Regional agreements are attractive because they focus management on nations that share similar conditions, space, and resources. Regional initiatives offer comprehensive institutional frameworks for international cooperation by serving as links for governance and by having the capacity for evolving various agreements and technical assistance programs in response to environmental and resource problems.

One of the first regional agreements was the Antarctic Treaty (Tables 2.2 & 2.6), which grew out of a scientific program – the International Geophysical Year (1957). The Treaty ensures that Antarctica is used for peaceful purposes and international cooperation in science, and does not invite international discord. An important innovation was that national sovereignty was suspended. This Treaty soon became a model for international cooperation, especially in scientific research. The Treaty's Agreed Measures for the Conservation of Antarctic Fauna and Flora are an important consequence. However, the Treaty failed to include the Southern Ocean surrounding the Antarctic continent. Therefore, in 1980, the Convention for the Conservation of Antarctic Marine Living Resources (CCAMLR) endorsed ecosystem management and, in effect, extended the jurisdiction of the Antarctic Treaty to a natural oceanographic boundary, the Antarctic convergence that is the northern boundary of the Southern Ocean.

The United Nations, particularly, has facilitated regional programs for information exchange, response options, and national ocean-management strategies. In the 1970s, the United Nations Environment Programme's Regional Seas Programme began to play a catalytic role in developing and implementing regional seas programs (Table 2.4). Regional arrangements have the advantage of matching the geographic scale of many marine resource and environmental issues. Conventions, protocols, and action plans have been adopted by governments of some regions to address legal matters, transboundary resource problems, pollution, management, institution-building, protected areas, and finance. The concept of Large Marine Ecosystems (LMEs) has been similarly conceived to address ocean areas characterized by distinct hydrography, productivity, and trophic interactions important to major commercial fisheries. Fifty LMEs have been identified, which *inter alia* annually produce 95% of the world's fish catch (Fig. 2.1). LMEs form a powerful basis for discussion and are becoming recognized management units. They also provide a regional focus for addressing degradation of coastal environments from pollution and habitat loss.

Regional governance is of particular concern for global issues such as climate and highly migratory species. In the 1960s, the Food and Agriculture Organization of the United Nations (FAO) inaugurated a series of regional commissions and councils, including the Intergovernmental Oceanographic Commission

Table 2.4 United Nations Environment Programme (UNEP) regional seas conventions. Most have many aspects in common; protocols have been added to address priority concerns, of which samples are listed. Dates are for adoption of conventions; many are not yet in force (i.e. ratified). Several other regional seas programs exist, but many have not yet achieved conventions.

Region/sea	Instrument/administrator	Some major objectives
Baltic Sea	Convention on the Protection of the Marine Environment of the Baltic Sea Area (Helsinki Convention, 1974)	Pollution; protection of biodiversity; alien species; monitoring program; protected areas; integrated watershed management
Northeast Atlantic	Convention for the Protection of the Marine Environment (Oslo and Paris conventions, 1974; revised as OSPAR, 1992)	Formulated regional consensus for cooperative actions on resource management and biodiversity
Mediterranean Sea	Convention for the Protection of the Marine Environment and the Coastal Region of the Mediterranean (Barcelona Convention, 1976)	"Blue Plan" for long-term regional, coastal management, emphasizing pollution control and including a protected-area network
Arabian Gulf	Kuwait Regional Convention for Co-operation on the Protection of the Marine Environment from Pollution (Kuwait Convention, 1978)	"Kuwait Action Plan" for combating pollution and for transboundary movements and disposal of hazardous waste
Eastern Africa	Convention for the Protection, Management and Development of the Marine and Coastal Environment of the Eastern African Region (Nairobi Convention, 1985)	Framework strategy for comprehensive approach to coastal-area development; major concern for wild fauna and flora; pollution in cases of emergency
West and Central Africa	Convention for Co-operation in the Protection and Development of the Marine and Coastal Environment of the West and Central African Region (Abidjan Convention, 1981)	Comprehensive strategy for conservation and development; pollution in cases of emergency
Red Sea and Gulf of Aden	Regional Convention for the Conservation of the Red Sea and Gulf of Aden Environment (Jeddah Convention, 1982)	Comprehensive strategy for conservation; principal concern is for pollution by oil and other harmful substances
Wider Caribbean	Convention for the Protection and Development of the Marine Environment of the Wider Caribbean Region (Cartagena Convention, 1983)	Action Plan for Caribbean Environment Program; specially protected areas for wildlife; oil spills; reduction and control of land-based sources of pollution
South Pacific	Convention for the Protection of Natural Resources and Environment of the South Pacific Region (Noumea Convention, 1986)	Multilateral cooperation on protection of natural resources, dumping, and pollution

Compiled from UNEP Regional Seas Programme (2000, online); for additional information see Borgese et al. (1994); Dutton & Hotta (1995); Hinrichsen (1998).

(IOC), which has initiated studies on phenomena such as El Niño, which has major effects on climate, fisheries, marine diseases, and wildlife. Fisheries, especially, require establishment of international–regional commissions for management. An exemplary case is the Atlantic bluefin tuna (*Thunnus thynnus*), one of the world's most valuable fishes, managed by the International Council for Conservation of Atlantic Tuna (ICCAT). Atlantic bluefin tuna range from Labrador and Scandinavia to Brazil and Africa. Individuals can weigh as much as 700 kg, and are prized by international markets. Between 1970 and 2000, the western-Atlantic population declined to approximately 20% of its spawning stock biomass. ICCAT set quotas based on the assumption that eastern and western populations are distinct. Recent evidence indicates that these populations intermix. Nations must now acknowledge that Atlantic bluefin tunas might be a single population that may not recover until and unless fishing is restricted throughout the entire Atlantic range. But, who owns the fish and how can they be managed? Information, experience, and management perspectives differ among ICCAT members, and non-ICCAT members may have no conservation incentive because overexploitation can gain them considerable short-term benefit. If overfishing continues in the eastern or western Atlantic, fishers elsewhere may not have incentives to decrease fishing effort.

Another example of regional governance is the Caribbean. This sea is surrounded by two different sets

Fig. 2.1 Large Marine Ecosystems. From NOAA (2001) online, courtesy of K. Sherman.

of nations: small island nations speaking English, French, Spanish and Dutch, and continental nations that mostly speak Spanish. The nations have agreed to the Convention for Protection and Development of the Marine Environment of the Wider Caribbean Region (Cartagena Convention, 1983). This Convention contains two major elements: first, the Protocol Concerning Cooperation in Combating Oil Spills in the Wider Caribbean Region, and second, the Protocol on Specially Protected Areas and Wildlife of the Wider Caribbean Region. The latter calls for creation of a regional network of protected areas to conserve and restore ecosystems and to maintain essential ecological processes. However, implementation of this agreement is difficult due to lack of coordination, information, financial support, research, and monitoring capacity. The UNESCO-sponsored program on Caribbean Coastal Marine Productivity (CARICOMP), initiated in 1985, responds to the need to understand regional phenomena such as coastal eutrophication, diseases of marine life, and coral bleaching. CARICOMP is a regional network of Caribbean marine laboratories, parks, and reserves (> 25 sites in 18 countries, including Florida) open to any institution under a Memorandum of Understanding with its Steering Committee. In 1991, CARICOMP instituted a synoptic, standardized coastal ecosystem monitoring program, with centralized data management and communications dedicated to discrimination of human disturbance and natural variation. The CARICOMP program holds regular regional training workshops and scientific seminars and facilitates the funding of directed research programs.

2.3.1.3 International governance

International governance is global in scope and guided by cooperative agreements, similar to regional governance, but more encompassing. International governance mechanisms take several forms under international law (Box 2.3). Treaties and international customs are primary, the former being explicit and documented, the latter being derived from actual practice. The scope of international law embraces sovereignty, property rights, environmental standards, labor, safety, training, scientific documentation, and much more.

International regimes usually are directed to specific topics, such as fishing, navigation, trade, and scientific research. An example is the General Agreement on Tariffs and Trade (GATT), which led to the formation of the World Trade Organization. Most regimes rest on one or more (not necessarily legally binding) constitutive documents. Regimes are particularly necessary for transboundary environmental issues that arise when activities that occur wholly within the jurisdiction of one nation produce consequences that affect other nations. For example, fishing in one area can affect fishing in another area when a fish population migrates between both areas. This situation creates uncertainty about resource uses, as both the level of take and species distributions may not be known. Inevitably, national sovereignty becomes a major consideration. To deal with this, a general trend is to reallocate jurisdiction from national, to transnational, to supranational authorities.

International governance advanced greatly with the establishment of the United Nations (UN) in 1945, when its Charter gave it authority to harmonize actions of nations through diplomatic fora and international mechanisms for resolving common problems. The UN was initially constrained by national sovereignty. As needs arose to address international development, pollution, resource exploitation, and other transnational matters, the UN achieved preeminence by providing specialized agencies for problem resolution (Table 2.5). Each agency is composed of a policy-making body that represents the views of member states, and each has research and management expertise to which issues may be referred. Thus, the UN and its agencies play a critical role in consensus-building among nations. Nevertheless, the UN lacks regulatory power.

Many important environmental agreements and events have been coordinated by the UN (Table 2.6). One landmark was the UN Conference on the Human Environment (1972), held to guide a comprehensive approach to environmental problems. This conference brought global recognition to human–environment

Box 2.3 Some definitions in international law

Agreement: a compact entered into by two or more nations or heads of nations; in the wide sense, any act of coming into conformity; in the narrow sense, an accord between states, but less formal than a treaty; may or may not be obligatory; includes convention, treaty, protocol, accord, act, declaration, pact, provision, etc.

Convention: agreement concluded among states on matters of vital importance; often used *in lieu* of treaty, but usually restricted to agreements sponsored by an international organization; intended to be legally binding, but requires ratification.

Declaration: a document whose signatories express their agreement with a set of objectives and principles; may not be legally binding, but carries moral weight.

International law: the body of legal rules and norms that regulates activities carried out by agreement among nations; intended to be legally binding, but requires ratification.

Protocol: agreement that completes, supplements, amends, elucidates, or qualifies a treaty or convention; has the same legal force as the initial document.

Ratification: final confirmation of a treaty, convention, or other document by a nation's competent body (legislature or head of state), thereby becoming legally binding and securing that country's commitment to it; there is no prescribed length of time for ratification.

Regime: arrangements that contain agreed-upon strategies, principles, norms, rules, decision-making procedures, and programs that govern interactions of participants in specific areas, such as fishing, navigation, trade, and scientific research.

Resolution: text adopted by a deliberative body or an international organization; may or may not be binding.

Treaty (from Latin *tractere*, to "treat"): an agreement entered into by two or more nations or heads of nations; intended to be legally binding; requires ratification.

Sources: Modified from Fox (1992); Gamboa (1973); Gleick (2000); University of Virginia School of Law (website).

Table 2.5 United Nations (UN) environmental agencies with international marine programs.

UN agency (year formed)	Mandate	Examples of commissions or relevant programs
Food and Agriculture Organization, an autonomous agency within the UN system (FAO, 1945)	To improve nutrition, food production, and distribution; alleviate hunger/malnutrition; food standards; long-term strategy for conservation and management of natural resources	Intergovernmental Oceanographic Commission (IOC); Fisheries Department involved in Convention on Biological Diversity and UN Convention on Law of the Sea (UNCLOS); programs on environmental quality
UN Educational, Scientific, and Cultural Organization (UNESCO, 1946)	To advance universal respect for justice, rule of law, human rights, fundamental freedoms of all peoples; emphasizes interdisciplinary approach; promotes understanding; encourages scientific research and training	Man and the Biosphere Programme (MAB); World Heritage Sites; Coastal Regions and Small Islands Initiative; promotes international ocean science: works closely with International Council of Scientific Unions (ICSU) and Scientific Committee on Ocean Research (SCOR)
International Maritime Organization (IMO, 1958); preceded by Intergovernmental Maritime Consultative Organization (IMCO, 1947)	To develop policies for international shipping; to facilitate technical cooperation; concern for marine environment, maritime safety, efficiency of navigation; prevention and control of marine pollution from ships	Administers London Convention, and subsequent conventions, e.g., International Convention for the Prevention of Pollution from Ships (MARPOL) and protocols for pollution; develops guidelines for ballast-water control of exotic species introductions; measures to prevent accidents; maritime legislation and its implementation
UN Development Programme (UNDP, 1965)	To aid infrastructure development; to provide developing countries with knowledge-based policy advice on issues pertaining to poverty, building institutional capacity, and managing the challenges of globalization	Assists and financially supports developing countries and territories; *Capacity 21* launched at UN Conference on Environment and Development (UNCED, 1992) to help countries implement *Agenda 21*; by 2001, *Capacity 21* supported national and local *Agenda 21* efforts in 75 countries under a trust fund
UN Environment Programme (UNEP, 1972)	To coordinate environmental agreements and activities within UN system; to aid nations to develop and adopt environmental policies, strategies, and actions	Helps with formation of environmental treaties and agreements; funds and guides environmental strategies and action plans; coordinates regional seas programs and shared environmental problems of multicultural nations that border those seas; aids environmental negotiations, conventions (e.g., for biodiversity, climate change)

Compiled from Mangone (1988); UN agency websites.

interactions and declared: "Man has the fundamental right to freedom, equality and adequate conditions of life, in an environment of a quality that permits a life of dignity and well being." In essence, this asserts that when resources and environment are involved, a nation as a sovereign entity should not engage in activities that negatively affect the sovereignty – that is, rights – of other nations. The United Nations Environment Programme (UNEP) was established following this conference, its task being to help integrate international environmental negotiations and agreements. One of the most significant for conservation of life is the Convention on International Trade in Endangered Species of Wild Fauna and Flora (CITES), which prohibits all trade in species listed as endangered or threatened. CITES was ratified by 126 nations as late as 1994.

Another major turning point was the establishment in 1984 by the UN General Assembly of the World Commission on Environment and Development (WCED) – the "Brundtland Commission" – whose report, *Our Common Future* (1987), proposed "sustainable development" to protect the environment, while not constraining economic growth or development. This controversial concept demanded reconciliation of economic development with environmental protection. The United Nations Conference on Environment and Development – the "Earth Summit" (UNCED, 1992) – was held for this purpose. UNCED was the largest gathering of world leaders, concerned scientists, and conservation organizations ever held, being attended by representatives of 178 nations, 117 heads of state, and thousands of delegates and participants.

Table 2.6 Major international policy-setting, environmental agreements and events coordinated by United Nations (UN) agencies.

Policy-setting instrument	Environmental issue
Antarctic Treaty (1959)	Model for international cooperation; national claims of sovereignty set aside to ensure peaceful purposes and international cooperation in science
UN Conference on the Human Environment, Stockholm (1972)	First development of international environmental policy, including transboundary and pollution issues
Convention for the Conservation of Antarctic Marine Living Resources (CCAMLR, 1980)	Conservation of Antarctic marine living resources not covered by the Antarctic Treaty; adopted principles of ecosystem management
UN Convention on Law of the Sea (UNCLOS III, 1982, preceded by UNCLOS I, 1958, and UNCLOS II, 1960)	A "Constitution for the oceans"; extended jurisdictions of nations linked to duties to conserve living resources and protect the marine environment; set standards; raised international importance of islands and coastal nations; contained provisions for enforcement and settlement of disputes
World Commission on Environment and Development (WCED, 1983); *Our Common Future* (1987)	Placed concept of "sustainable development" into the global environmental lexicon; reexamined critical environment and development issues; proposed new forms of international cooperation; raised understanding and commitment to action
UN Conference on Environment and Development ("Rio Conference" or "Earth Summit"; UNCED, 1992)	Developed international agreements on global environmental issues (biodiversity, climate change, ozone depletion, desertification, deforestation); approved *Agenda 21* as an action plan
Framework Convention on Climate Change (1992)	Framework for addressing global warming, including concern for sea-level rise
Convention on Biological Diversity (1992)	For conservation of genetic, species, and ecosystem diversity; advanced policies for equitable sharing of benefits from use of genetic resources

Two principal environmental agreements emerged: the Convention on Climate Change and the Convention on Biological Diversity. The latter recognized biodiversity as the "common heritage" of humankind, following overwhelming scientific information that rates of human-caused species extinction were accelerating. Both conventions conceived new directions for environmental policy and conservation and endorsed the "precautionary principle," which urges that resource uses err on the side of caution. Principle 1 of the *Rio Declaration* placed humans at the "center of concerns for sustainable development." *Agenda 21* outlined strategies for environmental action and set forth rights and obligations of nations (Table 2.7).

Among all international conservation agreements, the Convention on Biological Diversity is the most critical and comprehensive. The Convention defines "biological diversity" (biodiversity) as "variability among living organisms from all sources including, *inter alia*, terrestrial, marine and other aquatic ecosystems and the ecological complexes of which they are part; this includes diversity within species, between species and of ecosystems." The Convention commits signatory nations to conserve all forms of biodiversity, to use biological resources sustainably, and to share equitably in benefits that arise from genetic resources. The Convention also obliges contracting parties to develop national biodiversity strategies and to integrate conservation of biodiversity into relevant plans, programs, and policies.

International agreements reflect the combined agendas and values of nations, and require a continual process of adjustment and clarification to meet changing social, political, and environmental conditions. As the international community has no counterpart of a national legislature, adjustments and clarifications become products of multilateral conferences and deliberations. For example, "Marine and Coastal Biodiversity" is among the major themes of the Convention on Biological Diversity, and was specifically considered by a Conference of the Parties to the Convention in 1995. This resulted in the Jakarta Mandate (1995), whose "Thematic issues" circumscribe many of today's major conservation challenges (Table 2.8). Additionally, the First Global Ministerial Environment Forum in Malmö, Sweden (2000) charted a vision for addressing major emergent environmental issues of the 21st century; its Declaration has special relevance for the coastal realm, by mention of urbanization, depletion of resources, and land-based sources of pollution (Table 2.9).

Another example of the development of international agreements concerns marine pollution (Table 2.10).

Table 2.7 The *Rio Declaration on Environment and Development* proclaimed 27 principles to guide the behavior of nations in fulfilling their environmental obligations; Principle 1 is quoted below. Goals and objectives are contained in *Agenda 21: Program of Action for Sustainable Development*. Program areas listed below are for section 2, chapter 17; other sections are summarized.

Rio Declaration on Environment and Development
Principle 1: "Human beings are at the center of concerns for sustainable development."

Agenda 21: Program of Action for Sustainable Development
Sec. 1 *Social & Economic Dimensions*: cooperation to accelerate sustainable development; combating poverty; changing consumption patterns; demographic dynamics and sustainability; protecting and promoting human health; promoting sustainable human settlement development; and integrating environment and development into decision-making.
Sec. 2 *Conservation & Management of Resources for Development*: atmosphere; land resources; deforestation; desertification and drought; mountains; agriculture and rural development; biological diversity; biotechnology; freshwater; toxic chemicals; hazardous wastes; solid wastes and sewage; and radioactive wastes.
Chapter 17: Protection of oceans, all kinds of seas, including enclosed and semi-enclosed seas, and coastal areas and the protection, rational use and development of their living resources
(A) Integrated management and sustainable development of coastal areas, including exclusive economic zones
(B) Marine environmental protection
(C) Sustainable use and conservation of marine living resources of the high seas
(D) Sustainable use and conservation of marine living resources under national jurisdiction
(E) Addressing critical uncertainties for the management of the marine environment and climate change
(F) Strengthening international, including regional, cooperation and coordination
(G) Sustainable development of small islands
Sec. 3 *Strengthening the Role of Major Groups*: women; children and youth; indigenous people; non-governmental organizations; local authorities; workers and trade unions; business and industry; scientific and technological community; and farmers.
Sec. 4 *Means of Implementation*: financial resources; transfer of environmentally sound technology; science for sustainable development; education, public awareness, and training; capacity-building in developing countries; institutional arrangements; legal instruments and mechanisms; and information for decision-making.

From UN (1992).

Table 2.8 The Conference of the Parties (COP) to the Convention on Biological Diversity addressed conservation and sustainable use of marine and coastal biological diversity. The Jakarta Mandate on Marine and Coastal Biological Diversity (1995) identified six key "Thematic issues".

Thematic issue	Operational objectives
1 Integrated marine and coastal area management (IMCAM)	• Review existing instruments • Promote development and implementation at the local, national, and regional levels • Develop guidelines and indicators for ecosystem evaluation and assessment
2 Marine and coastal living resources	• Promote ecosystem approaches to sustainable use of marine and coastal living resources • Make available to Parties information on marine and coastal genetic resources
3 Marine and coastal protected areas	• Facilitate research and monitoring activities on value and effects of marine and coastal protected areas, or similarly restricted areas, on sustainable use of marine and coastal living resources • Develop criteria for establishment and management of marine and coastal protected areas
4 Mariculture	• Assess consequences of mariculture for marine and coastal biological diversity and promote techniques to minimize adverse impacts
5 Alien species and genotypes	• Achieve better understanding of causes and impacts of introductions of alien species and genotypes • Identify gaps in existing or proposed legal instruments, guidelines, and procedures, and collect information on national and international actions • Establish an "incident list" of introductions
6 General	• Assemble database of initiatives on program elements, particularly integrated marine and coastal area management • Develop database of experts for development and implementation of national policies on marine and coastal biological diversity

From SBSTTA (1998).

Table 2.9 Malmö Ministerial Declaration. Edited from original text to highlight major points.

We, Ministers of Environment and heads of delegation meeting in Malmö, Sweden from 29 to 31 May 2000, on the occasion of the First Global Ministerial Environment Forum, held in pursuance of United Nations General Assembly resolution 53/242 of 28 July 1999 to enable the world's environment ministers to gather to review important and emerging environmental issues and to chart the course for the future . . .

Recalling the Stockholm Declaration of the United Nations Conference on the Human Environment, the Rio Declaration of the United Nations Conference on Environment and Development, the Barbados Declaration on the Sustainable Development of Small Island Developing States . . . Declare that:

Major environmental challenges of the twenty-first century
- There is an alarming discrepancy between commitments and action . . .
- The evolving framework of international environmental law and the development of national law provide a sound basis for addressing the major environmental threats . . .
- Environmental stewardship is lagging behind economic and social development, and a rapidly growing population is placing increased pressures on the environment.
- Environmental threats resulting from the accelerating trends of urbanization and the development of megacities, the tremendous risk of climate change, the freshwater crisis and its consequences for food security and the environment, the unsustainable exploitation and depletion of biological resources . . . are all issues that need to be addressed.
- It is necessary that the environmental perspective is taken into account in both the design and the assessment of macro-economic policy-making . . .

The private sector and the environment
- A greater commitment by the private sector should be pursued to engender a new culture of environmental accountability through the application of the polluter-pays principle, environmental performance indicators and reporting, and the establishment of a precautionary approach in investment and technology decisions . . .
- The potential of the new economy to contribute to sustainable development should be further pursued, particularly in the areas of information technology, biology and biotechnology . . .

Civil society and the environment
- Civil society plays a critically important role in addressing environmental issues . . . It provides a powerful agent for promoting shared environmental purpose and values . . .
- The role of civil society at all levels should be strengthened through freedom of access to environmental information to all, broad participation in environmental decision-making . . .
- Science provides the basis for environmental decision-making. There is a need for intensified research, fuller engagement of the scientific community and increased scientific cooperation on emerging environmental issues, as well as improved avenues for communication between the scientific community, decision makers and other stakeholders . . .
- We must pay special attention to threats to cultural diversity and traditional knowledge, in particular of indigenous and local communities, which may be posed by globalization . . .

The 2002 review of UNCED
- The 2002 review of the implementation of the outcome of the United Nations Conference on Environment and Development (UNCED) should be undertaken by an international conference at the summit level. The objective should not be to renegotiate Agenda 21, which remains valid, but to inject a new spirit of cooperation and urgency based on agreed actions in the common quest for sustainable development . . .

From UN (2000) online.

These agreements illustrate how nations enter into the bargaining process in pursuit of common interests. Each nation agrees to conditions as it sees fit, and each agreement becomes a form of social–political instrument that governs a nation's actions through collaboration with others. The agreements must be compatible with international law and are binding in the sense that participating or signatory nations act according to agreed-upon policies or goals.

2.3.1.4 Ocean law

The UN Convention on Law of the Sea (UNCLOS) exemplifies binding international lawmaking. Ocean law emerged in Roman times when nations began to expand ocean jurisdictions (Table 2.11). Prior to World War II, most nations claimed a 3–6 nautical-mile (nmi), approximately 6–12 km, territorial sea. As maritime activities increased, jurisdictions crept farther into oceans. Several conferences on law of the sea were

Table 2.10 Major international conventions on vessel pollution. Many have been followed by protocols on specific aspects. Dates indicate signing of agreements; ratification has occurred up to a decade later.

Date	Instrument	Intent
1963	Treaty Banning Nuclear Weapons Tests in the Atmosphere, in Outer Space and Under Water	To prevent nuclear pollution, globally
1969	International Convention Relating to Intervention on the High Seas in Cases of Oil Pollution Casualties	To prevent or mitigate oil pollution by accidents involving ships outside territorial waters. A protocol extending to other hazardous substances (chemicals) entered into force in 1983
1969	International Convention on Civil Liability for Oil Pollution Damage	To ensure adequate compensation from damage of oil pollution. Placed liability compensation on ship owners releasing or discharging oil
1971	International Convention for the Establishment of an International Fund for Compensation for Oil Pollution Damage	To provide further compensation to oil pollution victims. Placed the burden of compensation on ship owners, with time limits on amount payable. Funded by oil importer contributions
1973/1978	International Convention for the Prevention of Pollution from Ships (MARPOL) and its 1978 Protocol (MSARPOL) supersede the International Convention for the Prevention of Pollution of the Sea by Oil (1974)	Addressed pollution by oil, noxious liquid substances, harmful substances carried in packaged forms, sewage, and garbage. Widely regarded as the most important instrument of its type. Almost all other agreements since this time have depended on the principles therein
1974	Paris Convention for Prevention of Marine Pollution from Land-based Sources	Further restrictions on dumping of wastes at sea

Compiled from Mangone (1986); Borgese et al. (1994); IUCN (1995); Coe (1996); Crystal (1998).

held, culminating in the 1982 UNCLOS III, to which 159 nations became signatories. However, several industrialized countries, notably the United States, Great Britain, and Germany, objected to sea-bed mining and other provisions and did not sign. By 1965, a total of 32 nations claimed 12 nmi; by the late 1970s, 67 nations did so. The United States claimed a 200 nmi Exclusive Economic Zone in 1983 by Presidential Proclamation. UNCLOS III was ratified in 1994, and by the beginning of the 21st century, 27 countries had claimed territories greater than 12 nmi, with 14 extending claims to 200 nmi.

UNCLOS III represents by far the most ambitious, most historic, and most far-reaching of international agreements. It consists of 320 articles and eight annexes. At least 59 of the articles and three of the annexes have environmental significance, clustered around five sectors: (i) living resources; (ii) pollution; (iii) transportation; (iv) pollution-related activities; and (v) general provisions, such as for protection of the marine environment and codes of conduct for research. Each of these considers prescribed behavior, allocation of jurisdiction, criteria and standards, and institutional mechanisms. UNCLOS III was made even more complex with the adoption of *Agenda 21*, which outlined fundamental elements of environmental governance for the 21st century, including the concept of sustainable development. The full implications of *Agenda 21* for UNCLOS III are still emerging.

Three major features of UNCLOS III that make it different from its predecessors are:

- expansion of coastal states' rights to develop and manage marine resources are linked to duties to conserve living resources and protect the marine environment;
- advancing scientific knowledge and the pace of political, economic, and technological change require radical new approaches to the progressive development and implementation of ocean law;
- strengthening scientific, technical, and management capabilities is essential if the Convention is to become meaningfully implemented.

UNCLOS III has become a "constitution of the oceans" that gives nations opportunities to evolve their own management strategies, although not compelling them to do so. UNCLOS III's extension of national Exclusive Economic Zones (EEZs) seaward to 200 nmi, wherein nations enjoy sovereign rights over all resources, living and non-living, represents the greatest transfer of resources in recorded history, and divides most of

Table 2.11 Major events that influenced international maritime law and facilitated "creeping" offshore jurisdictions.

Date	Event	International agreement
AD 450	Codification of Roman law. Evolved from law of ancient Rome (753 BC–5th century AD)	Legal system forming western law. Established Law of Procedure and absolute ownership, unlike Germanic systems and English law
529–535	Freedom of the Seas	First legal document of the sea
1493	*Mar clausum*, by Papal bull *Inter Cetaera*	Gave Spain exclusive rights to land and sea west of the Azore Islands
1588	Defeat of Spanish Armada by England	Saved England from invasion, Dutch Republic from extinction. Delivered heavy blow to Spain
1625	Hugo Grotius *Mare Liberum*	Defense for Holland and Dutch East India Company
1702	Von Bynkershoek's cannon-shot rule: *De Dominio Maris*	Codified national claims within the range of shore-based artillery, approximately 3 nmi
1793	President Jefferson claims 3 nmi territorial sea for USA	Ripened into globally accepted standard over which a nation could assert ownership of the seas
1938	U.S. exploration of Gulf of Mexico outside territorial sea	Under international law, ocean beyond 3 nmi is common property of all nations. Drilling technology and discovery of oil in Gulf of Mexico spurred USA to extend its jurisdiction
1945	Truman Proclamations: USA establishes offshore control	Other nations assert their claims. Constricts freedom of navigation
1958	First UN Convention on Law of the Sea (UNCLOS I); not ratified	Produced four separate conventions: Territorial Seas; Fisheries; Continental Shelf; High Seas. Coastal nations have limited jurisdiction over foreign ships in contiguous zone, a zone of 12 nmi from baseline of territorial sea. Ratification failed over Rights of Innocent Passage
1960	UNCLOS II: not ratified	Nations claim 12 nmi territorial sea
1976	U.S. Fisheries Conservation and Management Act	Expanded American jurisdiction over fish, from 12 to 200 nmi, and signaled eroding political power of distant-water fishing fleets and growing voice of coastal fishing interests
1982	UNCLOS III: not ratified	Resources of high seas become *mare nostrum* (our seas), establishing 200 nmi Exclusive Economic Zone (EEZ)
1983	Reagan Presidential Proclamation 5030	Declared 200 nmi EEZ for USA, in line with central provisions of UNCLOS III
1994	UNCLOS III: ratified	All coastal nations claim 200 nmi EEZs
1999	Clinton Presidential Proclamation	Extends U.S. contiguous zone from 12 to 24 nmi offshore, for enforcement of environmental, customs, and immigration laws

Compiled from Archer et al. (1994); *Encyclopædia Britannica* (1999–2000) online; Wilder (1998).

ocean space into segments that force nations to agree *inter alia* on management regimes for shared resources. For island nations, EEZs represent jurisdictional extensions that are often many times the size of the nation's land area. UNCLOS III contains provisions for enforcing international pollution standards and for fisheries, including binding, dispute-settlement procedures. Conflict resolution is placed under the aegis of the signatories themselves, making UNCLOS III a unique instrument in international law, with far-reaching implications. For example, the dispute among nations of the South China Sea over the Spratly Archipelago – a conflict that has plagued the region for decades – reflects a unique set of history and geopolitics. The archipelago had no economic importance prior to 1982. Under UNCLOS III, islands that can sustain humans or an economic life are entitled to a 200 nmi limit. Title to the Spratlys could determine ownership of significant oil and gas resources. The outcome rests with the nations involved.

In 1995, UNCLOS III moved toward an ecosystem approach to fisheries, with adoption of the UN Agreement on Straddling and Highly Migratory Fish Stocks, which adopted sustainable use, conservation of marine biodiversity, and the precautionary principle. Accordingly, the FAO sponsored a voluntary Code of Conduct for Responsible Fisheries (Table 2.12), potentially signaling a major advance in fisheries management. However, obligations under the Code for developing coastal states may be more imagined than real, as these states often lack the capacity to undertake resource assessments, to develop management systems, and to monitor user activities effectively.

Table 2.12 FAO Code of Conduct for Responsible Fisheries, March 2001.

The Committee on Fisheries (COFI) at its Nineteenth Session in March 1991 called for the development of new concepts which would lead to responsible, sustained fisheries. Subsequently, the International Conference on Responsible Fishing, held in 1992 in Cancûn (Mexico) further requested FAO to prepare an international Code of Conduct to address these concerns. The outcome of this Conference, particularly the Declaration of Cancûn, was an important contribution to the 1992 United Nations Conference on Environment and Development (UNCED), in particular its Agenda 21. Subsequently, the United Nations Conference on Straddling Fish Stocks and Highly Migratory Fish Stocks was convened, to which FAO provided important technical back-up. In November 1993, the Agreement to Promote Compliance with International Conservation and Management Measures by Fishing Vessels on the High Seas was adopted at the Twenty-seventh Session of the FAO Conference (Annex 1). Text adopted by the Twenty-eighth Session of the FAO Conference on 31 October 1995.

This Code sets out principles and international standards of behaviour for responsible practices with a view to ensuring the effective conservation, management and development of living aquatic resources, with due respect for the ecosystem and biodiversity. The Code recognizes the nutritional, economic, social, environmental and cultural importance of fisheries and the interests of all those concerned with the fishery sector. The Code takes into account the biological characteristics of the resources and their environment and the interests of consumers and other users. States and all those involved in fisheries are encouraged to apply the Code and give effect to it. The Compliance Agreement is an integral component of the Code.

- FAO Ministerial Meeting on Fisheries, FAO Headquarters, Rome, 10–11 March 1999, at the invitation of FAO's Director-General. Ministers responsible for fisheries met in Rome as a sign of their attachment to the implementation of the Code of Conduct for Responsible Fisheries. Inter alia, the Ministers were especially appreciative of FAO's role in promoting the application of the Code of Conduct and the increasingly wide adoption of the Code by States and concerned organizations. The Rome Declaration on the Implementation of the Code of Conduct for Responsible Fisheries, unanimously adopted by the FAO Ministerial Meeting, was attended by 126 Members of the Organization.
- FAO has widely disseminated the Code and is actively promoting its implementation.
- FAO's Fisheries Department has elaborated a draft strategy for the promotion and implementation of the Code.

From FAO (2001) online.

2.4 New directions

As will become evident in Section II, knowledge about ecosystems and the roles species play in structuring environments clearly illustrates the need to address species and environments together as systems. Coastal zone management and adaptive ecosystem management are two new approaches that are widely being promoted.

2.4.1 Coastal zone management

The "coastal zone" (coastal realm in this book) incorporates the coastal plain and the coastal ocean. Coastal zone management (CZM) is intended to treat this area as a whole, with the principal objective of resolving conflicts among users. CZM is a management-planning process, distinguished by efforts in consensus-building, and is generally conceived as having three basic components: (i) a set of public goals or policies; (ii) a framework of procedures for carrying out those policies; and (iii) a set of organizations or agencies to implement procedures. However, most CZM programs are narrowly conceived, being mostly concerned with zoning and accommodating users *inter alia*.

A majority of nations identify coastal issues as paramount concerns, where goals for economic growth frequently conflict with resource conservation – for example, when shipping, harbor development, oil and gas exploitation, and housing conflict with fisheries, recreation, and conservation. These conflicts are inherently difficult to resolve, especially when agencies are involved that employ separate regimes to manage land or sea, fish or water quality, and public health or environmental health. In such cases, CZM becomes a collection of fragmentary procedures involving institutions, each with command-and-control legislative and regulatory processes, assembled under a new banner, and driven by responses to crisis events, such as hurricanes, industrial developments, or rights to resources.

Despite these difficulties, CZM has made considerable progress in establishing priorities, zoning, resolving user conflicts, and gaining partnerships among multiple levels of government and the public. The ambitious U.S. program attempts to balance competing land and water uses while also protecting sensitive resources. This program depends on cooperation and incentives among federal and state levels of government, being administered through federal grants to states, and is largely voluntary. There is only indirect federal control over coastal uses, except for situations of national significance. The program encourages states to develop plans and to take appropriate actions under mutually acceptable agreements, such as establishment of research-oriented estuarine protected areas under the National Estuarine Research Reserve System (NERRS). As such, approaches differ among states and regions and fragment management along continuous coasts. Atlantic maritime states emphasize wetland preservation; Pacific states focus on improving public access to beaches and controlling land development; Alaska Native land claims overshadow coastal management issues; and Gulf of Mexico states are oriented toward improving navigation and waterfowl habitat and resolving disputes over resource ownership. Also, ports are generally excluded from CZM programs. Additionally, the U.S. program is vulnerable to political swings and can be substantially weakened when federal spending is reduced, as occurred during efforts to increase America's energy independence by developing domestic offshore oil and gas resources.

Integrated coastal zone management (ICZM) is a conceptual extension of CZM, but attempts to be less top-down and more participatory. ICZM has been strongly influenced by UNCED, and attempts to overcome single-sector management, fragmented jurisdictions, and hierarchical command-and-control procedures by shifting toward a predominantly bottom-up approach among stakeholders, and by increased emphasis on long-term intergenerational sustainability. While ICZM aims to achieve sustainable resource use through harmony and balance among various economic and environmental sectors, it is challenged by insufficient authority to deal with integration of economic, social, and political forces. However, ICZM remains highly attractive as a unifying concept, providing a framework for collaboration, communication, and coordination, as it brings together people and information from multiple disciplines and various sectors to achieve coherence and cooperation.

2.4.2 Adaptive ecosystem management

Adaptive ecosystem management is derived from interdisciplinary, ecological principles and has evolved as a consequence of failures to manage related issues interdependently and to accommodate to change. Ecosystem management attempts to understand how species relate to ecosystem dynamics, how whole natural systems work, how they change over time, and how humans have affected them. As such, ecosystem management is directed toward issues that are large-scale, long-term, and interconnected, rather than short-term and fragmentary. The premise is that ecosystem processes play fundamental roles in species' distributions and behavior, that species also influence ecosystems through various feedbacks, and that management must accommodate to social and environmental change.

Coastal-marine ecosystem management depends heavily on integration of numerous disciplines, strong research, and monitoring programs that can integrate conservation of living resources with oceanography, climate, coastal processes, and society. Ecosystem management recognizes that knowledge of any ecological system is always incomplete, and addresses moving targets that require an adaptive approach, that is, management that can be adjusted as new information becomes available and as new social or environmental conditions arise. This form of management also recognizes that decisions are made under conditions of great scientific uncertainty, therefore urging precaution until adequate knowledge is available and monitoring procedures are in place. In this sense, ecosystem management tends to shift the burden of proof to the user.

Ecosystem management is becoming a universal goal for conservation. In some instances it is an explicit or implicit mandate, as in several laws and agreements (e.g., the Marine Mammal Protection Act, CCAMLR). Ecosystem management is being attempted in many arenas to resolve conflicts, as between forestry and fisheries (Box 9.4, page 277). It has also been proposed for ocean management in general, on the premise that all forms of resource use are interrelated. In all cases, ecosystem management has the potential to provide processes for developing environmental strategies at a hierarchy of scales and for identifying cost-effective and efficient means for meeting environmental goals. The assumptions that uncertainty requires precaution and that incomplete information requires adaptive management strongly suggest that ecosystem management is experimental and that achieving positive results will involve strong collaboration among biological, environmental, and social sciences, as well as enhanced agency interaction and public understanding.

2.5 Agents of change

Individuals, governments, organizations, public-interest groups, and scientists all have helped set new courses for coastal-marine conservation. Events play an important role, calling immediate attention (oil spill) or bringing gradual concern (biodiversity loss, fishery declines) to issues. Agents of change influence policy, connecting local needs to international circumstances, watersheds to coastal oceans, and living resources to environments.

2.5.1 Social history and coastal-marine conservation

Coastal-marine conservation stems from tradition. Tradition is a form of social glue that reflects a heritage when nature and humans existed in parity. Tradition honors the long-established maxim, "first in time, first in right." Many traditions continue a close relationship with nature, following specialized knowledge and practices that evolved over time – observing species' habits, harvesting at the proper time and place, and the best ways of doing so. Some traditions may appear archaic in a modern context; modern commercial fishing is a technological extension of hunting and gathering and depends almost entirely on nature's abundance. Co-ownership of resources is another tradition among many cultures, wherein individual, private-property ownership often loses relevance. Agreements among nations over shared resources reflect traditional co-ownership practice. Protected areas, too, may stem from tradition: Manuae Island in the Cook Islands has been a protected cultural site since the 5th century, and a lagoon in Western Samoa was protected to honor the defeat of the Tongans in battle centuries ago (Fig. 2.2).

Fig. 2.2 Site of a protected area at Manono, Western Samoa, where the Tongans were defeated in battle. As the Tongans left, their chief cried out "Malietoa melietau" ("Brave heroes, well fought!"). Photograph by the authors.

Tradition is preserved in today's conservation concepts. Modern conservation emerged with industrial society during periods of Enlightenment and Romanticism in the 18th and 19th centuries, when environmental change and species depletions became increasingly obvious. Scientists and explorers documented losses posed by unrestrained human activities. Zealous, anti-capitalist French reformers opposed destruction of the hardwood forests of Mauritius in the late 1700s. Nature took on new meaning, captured by the Romantics in paint and print. Yet the sea remained shrouded in mystery until the mid-1800s when ships could probe ocean waters and collect unknown creatures that caught public attention and inspired more scientific study. The American naval officer Matthew Fountain Maury published *Physical Geography of the Sea* in 1855, which became an immediate best seller. The British Admiralty and Royal Society of England in 1872 sponsored the voyage of HMS *Challenger*, a 3.5 year study that tracked 68 890 miles of ocean; leading scientists published volumes that documented thousands of new marine species and 715 new genera, demonstrating that the oceans teemed with life. Oceanography was born, resource conservation was emerging, but still, the ocean's abundance was thought to be limitless.

Tradition has imposed itself on uses of living resources with two concurrent themes. One was that humans could improve on nature. Thomas Jefferson in a letter to George Washington (1787) expressed a widely held belief: "The greatest service which can be rendered any country is to add a useful plant to its culture." This expressed a motivation to meet human needs by growing food, improving nature's yields, increasing profits, and encouraging industry. To help repopulate depleted Atlantic salmon, trout, shad, and others in New England waters, Commissioner Spencer Fullerton Baird of the new U.S. Commission of Fish and Fisheries (1871) advocated fish culture. Baird's assistant, George Brown Goode, promoted fish culture to make fish so cheap and abundant that fewer fishery restrictions would be needed. Fishes (notably salmon) and shellfish (notably oysters) were transferred around the world to increase local supplies and improve on nature.

The second theme justified resource exploitation. Commercial fisheries expanded greatly during the 1800s, driven by the prevailing concept that once fishing became unprofitable in one area, fishing could move to other areas, allowing recovery of depleted fish in the first area, where fishing could then resume. The ocean was conceived as resilient and sea resources inexhaustible. This concept was expressed by the formidable Thomas Henry Huxley, defender of Darwinism, at the International Fisheries Exhibition (London, 1883): "Any tendency to over-fishing will meet with its natural check in the diminution of the supply . . . this check will always come into operation long before anything like permanent exhaustion has occurred." Huxley's view was widely held, and drove many fisheries toward further depletion, such as those for seals, whales, sea turtles, cod, herring, shad, salmons, halibuts, tunas, swordfishes, lobsters, and oysters. Whales suffered an especially prolonged attack by humans due to their valuable oil, meat, baleen, and other products.

The traditional concepts that nature could be improved upon, resources were limitless, and oceanic resiliency were in full swing when the concept of "wise use" for sustainable forestry was proposed by President Theodore Roosevelt and colleagues early in the 20th century. Scientists and naturalists increasingly issued warnings as they observed decimated environments and species loss, for example those of the American bison (*Bison bison*), passenger pigeon (*Ectopistes migratorius*), waterfowl, and shorebirds. These losses made evident the need to place certain species and areas "off-limits." In 1917, the Ecological Society of America formed a Special Committee for the Preservation of Natural Conditions, marking the beginning of the natural-areas movement in the United States. Wise-use and natural-areas concepts resulted in protected areas on land and protection for fur seals and sea otters in the ocean. Nevertheless, fisheries exploitation accelerated and marine conservation was neglected.

Not until the post-World War II period did issues of marine pollution, migratory species, fisheries, trade, and rights of ownership become major concerns. The oceans played increasingly important roles in military operations, global trade, and resources, dominated by developed nations who were best able to pursue ocean resources. The oceans, traditionally perceived as a global "commons" and open to all nations under the principle of Freedom of the Seas, became a major diplomatic arena. The coastal realm became a most-valued place to live, work and industrialize, and to gather resources. Increasing prosperity, leisure time, and disposable income boosted seashore development, expanding marinas, homes, recreation, and international tourism. Island and coastal nations became foci for development. Harbors became industrial complexes that knitted nations into a global economic web, with energy and petroleum gaining political preeminence.

During this period, scientists gathered evidence that awakened the public sector to the effects of overexploitation, urbanization, and industrial growth on natural areas and resources. The creeping domestication of coasts and seas by commerce, fishing, pollution, and oil and gas was forcing a squeeze on traditional practices. Coastal development crossed national jurisdictions, bringing rapid change that required new rules. As scientific information about human activities and their consequences mounted, the complexity and magnitude of issues also became recognized. But as new rules for conservation were advocated, tradition and past practices hung on; species' introductions and overexploitation continued, justified by the needs of a growing human population. Thus, a complex patchwork of laws and agreements evolved at international, national, provincial, and local levels to address human and environmental needs, while also preserving traditional practices of resource exploitation, waters as receptacles for wastes, and legal–jurisdictional allocation to resolve conflicts.

2.5.2 Bottlenecks

Present-day conservation is marked by a baffling array of agencies, government sectors, laws, agreements, and regimes. The proliferation of international agreements relevant to the oceans has created "treaty congestion," in which mandates for living resource exploitation often conflict with those for environmental health and conservation. For nations, the situation is similar. As a result, bottlenecks are created, and agreements and regulations are often underutilized and poorly enforced.

One bottleneck is reflected in trade and environmental policy. Trade is driven by criteria of efficiency that are internal to economic productivity, economies of scale, and cost reduction. Environmental policies aim toward results that are external to the economic system, such as environmental quality, biodiversity, and human wellbeing. Conservationists, among others, raise persistent fears that trade liberalization will promote measures that increase economic benefits at the expense of resources and environments. For example, the western tropical Pacific tuna fishery formerly involved a huge porpoise by-catch. Vigorous debates among nations, agencies, scientists, and the public resulted in national legislation (the U.S. Marine Mammal Protection Act) to reduce porpoise by-catch and to forbid tuna imports that killed porpoises. Results have been positive: from 1972 to 2000, porpoise by-catch dropped from 368 600 to almost nothing in the eastern tropical Pacific Ocean, with the voluntary commitment of tuna-fishing nations. However, legislation to protect porpoises has been challenged before the World Trade Organization by nations that interpret the legislation as interference with free trade in tuna.

Similar bottlenecks confound regional programs. International agreements may or may not take environmental needs into account, making it difficult for governments and the general public to obtain a clear picture of environmental objectives and commitments. Mechanisms to coordinate regional goals are widely sought, but for adjacent communities with different social histories and different legal–political systems, progress is often difficult. In such cases, confusion may center around social wellbeing, maintenance of environmental quality, tourism revenue, living resource sustainability, public health, or any combination of these.

Within nations, goals and procedures separately set for coastal activities, military installations, navigable waters, regulation of commerce, habitat quality, and resources also result in conflicts. Legislation may reflect different philosophies, as is apparent among the U.S. Marine Mammal Protection Act, Endangered Species Act, and Fishery Conservation and Management Act (Box 2.4). Conflicts also occur when national policy differs from provincial levels of government. For example, in many nations national and local agencies responsible for waste disposal, energy policy, and resource extraction often differ strongly on approaches, leading to legal debates about "consistency." Also, traditional practices with long histories may be forced to compete with commercial uses and technical advances outside local control. Traditional practices can be viewed as anachronistic when scientific evidence reveals a mismatch in perceptions of resource abundance. In these cases, traditional uses almost inevitably suffer.

Finally, bottlenecks reflect jurisdictional conflicts. In developed societies, jurisdictional allocations rarely

Box 2.4 U.S. legislation relating to marine mammals

Robert V. Miller

Marine Mammal Protection Act of 1972 (MMPA). During the late 1960s and early 1970s, three contentious issues spurred passage of the MMPA: (i) incidental take of large numbers of porpoises by the eastern tropical Pacific purse-seine tuna fishery; (ii) depletion of large whales by commercial whaling; and (iii) take of harp seal pups for their fur, by clubbing them on the pack ice.

To resolve these issues, the MMPA incorporated several radical measures and a few innovative features. It placed jurisdiction over marine mammals with the federal government and established a moratorium on take with few exceptions: for Eskimos, Aleuts, and certain coastal Indian tribes to take marine mammals for subsistence needs and to allow continuation of tuna fisheries while requiring reduction of the take of porpoises to the lowest practicable level. The MMPA also allowed for return of management authority to states under certain stringent criteria; established a system of permits for scientific research and for display of marine mammals in oceanaria; established the Marine Mammal Commission to provide advice to the regulatory agencies, to oversee research activities, to interact with the scientific community, and to report annually to Congress on the state of marine mammal issues and activities. The MMPA established two innovative concepts:

- Optimum sustainable population: the level of a population that will result in maximum productivity of that population or species, keeping in mind the optimum carrying capacity of the habitat and the health of the ecosystem of which it is a part. This was intended to determine the population level that must be maintained, whenever a take of a species is being considered.
- Added protection for depleted species, such that take is prohibited for any species deemed depleted, even where the moratorium is waived.

On balance, the Act has been beneficial for marine mammals, nationally and also internationally, as it represents national policy *vis-à-vis* marine mammals worldwide. Nevertheless, much ambiguity remains for certain provisions, such as optimum sustainable population, which has led to protracted regulatory reviews seeking an operational definition. Furthermore, the interpretations of the Act have at times placed many obstacles in the paths of returning management to the states; none have regained management authority – a situation unlike any other wildlife management concept, except those for endangered species and migratory species. Obstacles have also been placed before conduct of science, and even of conservation, such as abrogation of the Fur Seal Treaty at the urging of some conservation organizations.

Endangered Species Act of 1973 (ESA). This Act was passed because of increasing public concern over extinction of plant and animal species, and to correct the shortcomings of the earlier Endangered Species Conservation Act (1969). The ESA also provided for protection of habitat critical for a species' survival. The need to preserve biodiversity was pioneered in the Act's recognition that species are of esthetic, ecological, educational, historical, recreational, and scientific value to the nation and its people. The Act further declared the purpose of providing a means whereby the ecosystems upon which species depend may be conserved.

The ESA established several enforceable features for species conservation. First, it established categories of Endangered and Threatened species, and five broad criteria for listing such species: (i) present or threatened destruction, modification, or curtailment of habitat or range; (ii) overutilization for commercial, sporting, scientific, or educational purposes; (iii) population reduction due to disease or predation; (iv) inadequacy of existing regulatory mechanisms; and/or (v) other natural or man-made factors affecting the species' continued existence. Second, it enacted the legal concept of Critical Habitat, intended to provide protection for a species' environment and allowing regulatory agencies to control development or other activities that may threaten species or their habitats. Third, it adopted recovery planning, forcing responsible federal agencies to convene a recovery team composed of federal, state, and other representatives to develop a plan to reverse declining trends in populations or habitats, ideally leading to recovery so that a species can be removed from the listing. Finally, it required consideration of certain key features of the National Environmental Protection Act, such as biological assessment and an environmental impact statement if proposed actions by other federal agencies may affect a listed species.

The ESA has contributed significantly to the conservation of marine mammals. A prime example is the extensive consultative process over all aspects of outer-continental-shelf energy development. Because of the existence in coastal waters of Alaska of several listed species of cetaceans, all phases of development from initial leasing to production are scrutinized by agencies and the public. The Threatened category provides some measure of protection for species or their populations before they are in danger of extinction throughout all or a significant portion of their range. This has provided for gradual reduction of protection when a species has recovered to the extent that it is no longer considered threatened. An outstanding success story has been the California gray whale, which has benefited from the ESA to the extent that its population is now estimated to be at or larger than its pre-exploitation level. Consequently, it has been delisted. However, identification of critical marine habitat remains difficult and has led to some contentious proceedings, for example the reduction of fisheries to provide for Steller sea lion recovery in the absence of certainty about the actual effect of fishing.

Fishery Conservation and Management Act of 1976 (FCMA), as amended in 1996 as the Magnuson–Stevens FCMA (MSFCMA). This Act was the product of concern over increasing foreign fishing outside U.S. territorial waters (3 nmi) and exclusive fisheries jurisdiction (12 nmi), which resulted in depletion of fishes utilized by U.S. fishermen. Also, slow progress of the Law of the Sea (LOS) negotiations gave little cause for hope that an agreement would be reached on several contentious issues being discussed. This landmark legislation came into being just as LOS negotiations were reconvening.

The FCMA established jurisdiction over all fisheries within the U.S. 200 nmi Exclusive Economic Zone (EEZ). The Act also established Regional Fishery Management Councils that have authority seaward of states' 3 nmi jurisdictions. Councils are composed of representatives from states, the federal government, industry, and scientists. They prepare fishery management plans for individual species and deal with seasonal restrictions, allowable take, and effects of fishing on other species. Recently, councils have faced contentious issues of preventing or reducing overcapitalization by various means, such as limited entry.

The Act has been amended, or has caused amendments to be made, as follows: (i) to MMPA, that foreign fishermen within the Fishery Conservation Zone have to obtain permits for incidental take of marine mammals if there is any likelihood that this might occur; (ii) to MSFCMA, that the National Marine Fisheries Service convene an Ecosystem Principles Advisory Panel to recommend how best to develop ecosystem-based management and research, and also to require identification of essential fish habitat.

The 200 nmi EEZ has extended jurisdiction dramatically, but has also caused new boundary problems. For example, the Bering Sea is now under two-nation jurisdiction except for the "donut hole," which falls outside of either U.S. or Russian jurisdiction and has been subject to overexploitation and serious enforcement problems. By agreement, the United States and Russia jointly manage the fisheries there, but allow other nations to fish only when the population of a target species exceeds the sustainable level and an allowable take can be calculated. The MSFCMA and MMPA may conflict. The MSFCMA is driven by yields to the fishery, and defines optimum yield so broadly that it can apparently justify almost any quantity of catch. The MMPA, on the other hand, is protection-oriented, and requires marine mammals to be maintained at optimum sustainable populations, including from the indirect effects of competition for food resources by the fishing industry.

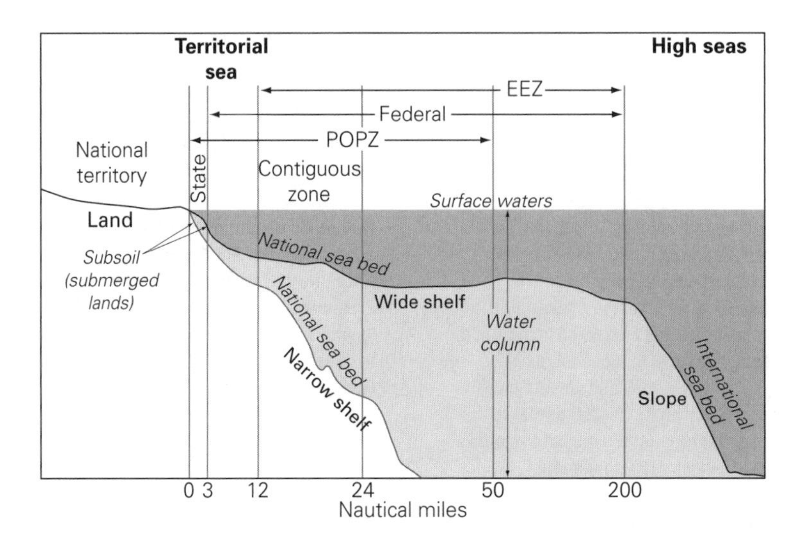

Fig. 2.3 Ocean jurisdictions recognized under international law. The ocean is subdivided vertically into subdivisions to which differing regulations apply: surface water, water column, sea bed, and subsoils (considered to be submerged coastal lands, usually under state or province control). Horizontally, ocean space is allocated to: national sovereignty out to 12 nmi, shared between federal authority and states/provinces; a contiguous zone where unimpeded access is allowed, but where nations can exercise control over customs, immigration, and fiscal or sanitary (pollution) matters; an Exclusive Economic Zone (EEZ) where most nations exert control over living and non-living resources; and high seas (open ocean) which remain outside any jurisdiction, but where certain restrictions may apply. Some nations have established a Prohibited Oil Pollution Zone (POPZ) under the International Convention for the Prevention of Pollution from Ships (MARPOL). Compare with ecological subdivisions, Fig. 3.8 (page 67).

incorporate biological or ecological relationships or recognize natural boundaries (Fig. 2.3). Underneath land or water, one agency may manage mineral strata, another surface shipping, a third air rights, a fourth land use, a fifth water quality, a sixth living resources within the water – and so forth. In many nations, different agencies manage wildlife, commercial fish, and habitats (e.g., forests, wetlands). Such divisions are especially problematic for coastal-realm species with home ranges that pass through several international, national, state and local jurisdictions during their lifetimes. The consequence is that various life-history stages of many species become segregated making conservation especially difficult.

The overall result of bottlenecks is that environments and species' natural histories are fragmented in multifaceted and conflicting ways. When conflicts are approached sector by sector, agencies compete for limited financial resources and coordinate poorly; incentives to participate and mechanisms for implementation may be lacking or inadequate. Numerous decisions are made each day, but their mutual effects may not be anticipated. The pace of change adds another dimension to these bottlenecks. International agreements have developed so rapidly during the past half-century that developing nations are placed under disadvantages. As developing nations emerged following World War II, they were forced to evolve legal mechanisms within a decentralized power base, and had to manage multi-faceted problems before governmental structures and processes were in place. The same situation can occur within nations.

2.5.3 Central place of strategy

The major challenge for coastal-marine conservation concerns turning from crisis tactics to a logical progression – from policy to strategy to tactics. Policy responds to a future vision, sets goals for a desired outcome, and evolves by consensus or bargaining among constituencies. Good policy involves forecasting and is based on explicit and testable theories, verifiable information, and exchange of information. Bad policy is reactive and depends on implicit theories, assumptions, unanalyzed data, and closed-door negotiations.

Making environmental policy is, above all, about politics and government (Fiorino, 1995). Optimally, environmental policy is based on information and logic, in which decision-makers are helped, through efforts of natural and social scientists and the public, to solve complex problems. However, making environmental policy is not straightforward, as it also involves economics, ownership, and difficult choices regarding ethics, consumption, life-styles, and commitment to resource sustainability. Policies for sustainable development, biodiversity protection, social equity, and resource conservation are in a constant battle with those for economic growth, pursuit of wealth, energy acquisition, and consumption. In most democracies, constitutional systems of checks and balances intend to prevent rash action and maintain social equity. However, these systems pre-date environmental awareness. When policies collide, as among public trust, property rights, and depletion of shared resources, the interaction between policy and politics can result in a series of compromises. Power (or priority) is no sooner granted to one person, institution, or resource when another steps in to limit it. Hence, environmental policies may owe little to environmental paradigms and much to political expediency, taking the form of piece-meal efforts without a focus on ultimate outcomes, and tending to happen as much by default as by directed action.

Since the 1970s, several hundred bilateral and multilateral international agreements have become legally binding, and call on nations to develop policies and action programs; these include climate change, pollution, biodiversity, wetlands, endangered species, and trade, among others. Although these agreements are clearly interrelated and demand anticipatory treatment, they are almost always treated independently. Policies may also be developed to defend long-term resource sustainability, only to be reversed by economic priorities. In other situations, a policy may be developed for long-term sustainability, but stakeholders perceive differences in mutual benefit. The need for environmental policies is now widely perceived among the public. However, implementation remains a challenge.

Strategy provides a framework for action, falling logically between policy and tactics to form a guide for setting action priorities (Fig. 2.4). Strategy is a military term defined as ". . . the art of defeating the enemy in the most economical and expeditious manner" (Morison, 1958). Successful strategy provides incentives, establishes institutional capacity with clear accountability, exposes errors and inefficiencies, and identifies true costs. It incorporates: (i) a prioritized set of objectives that addresses critical local and regional

Fig. 2.4 The central place of strategy between policy and tactics.

issues; (ii) an analysis of how obstacles may be identified and overcome; (iii) gaps in knowledge and means for acquiring an adequate information base; (iv) guidelines and criteria for management; (v) proposals for support; and (vi) a means for adaptive adjustment. A successful environmental strategy includes preventive action, precautionary approaches, public participation, research, and monitoring so as to incorporate feedbacks and improve managers' ability to lead and agencies' to adjust.

Tactics concern the art of deploying and maneuvering forces to carry out strategy. Tactics comprise the on-the-ground and in-the-water activities, and currently dominate conservation. Tactics are also unequivocally the most costly portion of conservation. They involve legal and non-legal measures (including zoning, resource quotas, facility provision, development controls, pricing mechanisms, sanctions, permit requirements, equipment controls, codes of behavior, environmental assessments, research and monitoring, establishment of protected areas, etc.). Government agencies – national to local – carry out the greater portion of conservation tactics, often with the collaboration of international agencies and conservation groups. Government tactics are carried out with: (i) authority by legislation or decree; (ii) enforcement responsibility; (iii) public accountability; and (iv) taxpayer support. Conservation groups (non-governmental organizations, NGOs) and scientific organizations also carry out tactical actions through a variety of mechanisms, such as habitat protection through land ownership and influence.

A logical progression from policy to strategy to tactics in environmental conservation is difficult. Tactics often lead to conservation "success stories," and are a means whereby agencies and conservation organizations enhance their reputations, build influence, and raise funds. Strategy is least glamorous and most difficult, often being perceived as a delay, even though strategy is necessary to increase efficiency, reduce costs, address uncertainties, set alternative scenarios, and turn crises to advantage by learning to avoid them. Nonetheless, strategy is often the "missing link" in conservation. When strategy is lacking, policies are immediately followed by hosts of tactics that can compete with one another, especially when agencies and organizations target particular issues with different intended outcomes. Without strategy, tactics may be "quick-fix" solutions that invite errors that can reverberate in unintended directions. Tactical approaches are critical when unexpected events require unforeseen, immediate actions to stop the "hemorrhaging" of a situation. When tactical approaches work, policy changes may occasionally occur. However, occasional successes do not lessen the risks incurred by tactics undertaken in the absence of a well-defined strategy.

The need to fill the strategy gap between policy and tactics for global conservation was foreseen in the 1970s. The goals of the *World Conservation Strategy* (IUCN, 1980) were: (i) to maintain essential ecological processes and life-support systems; (ii) to preserve genetic diversity; and (iii) to ensure the sustainable utilization of species and ecosystems. The emphasis on "processes" and "systems" represented a significant strategic shift from the past. A revision followed a decade later – *Caring for the Earth: A Strategy for Sustainable Living* (IUCN, UNEP, WWF, 1991) – containing nine "Principles" and a chapter on Oceans and Coastal Areas that recommended 12 actions (Box 2.5). Finally, the *Global Biodiversity Strategy* (WRI, IUCN, UNEP 1992) outlined the goals and actions required for this aspect of conservation alone (Box 2.6). While incorporating the comprehensive and complex nature of conservation, these documents lack many elements of true strategies, being largely goal-setting instruments. A few nations have developed elements of coastal-marine conservation strategies, yet no global strategy exists that integrates the many biological, ecological, and social issues facing coastal-marine conservation.

Box 2.5 Principles and Priority Actions for Oceans and Coastal Areas

Principles for Sustainable Living

- Building a sustainable society
- Respecting and caring for the community of life
- Improving the quality of human life
- Conserving the Earth's vitality and diversity
- Keeping within the Earth's carrying capacity
- Changing personal attitudes and practices
- Enabling communities to care for their own environments
- Providing a national framework for integrating development and conservation
- Creating a global alliance

Priority Actions for Oceans and Coastal Areas

- Develop a national policy on the coastal zone and ocean
- Establish a mechanism to coordinate the planning and allocation of uses of the coastal zone
- Allocate marine resource user rights more equitably among small-scale, large-scale and sport fisheries, and give more weight to the interests of local communities and organizations
- Conduct information campaigns to raise the profile of coastal and marine issues; and include a strong marine component in environmental education in all countries
- Promote marine protected areas
- Conserve key and threatened marine species and gene pools
- Place high priority on preventing marine pollution from land-based sources
- Adopt procedures for effective prevention of pollution from ships and offshore installations, and for rapid response to emergencies such as oil spills
- Ratify or accede to the United Nations Convention on Law of the Sea (UNCLOS) and other international legal instruments and develop an effective regime for sustainable use of open-ocean resources
- Expand and strengthen international cooperation both regionally and among funding agencies and intergovernmental organizations
- Promote interdisciplinary research and exchange of information on marine ecosystems

Source: From *Caring for the Earth: A Strategy for Sustainable Living* (IUCN, UNEP, WWF 1991).

Box 2.6 Selected elements of the *Global Biodiversity Strategy*

Catalyzing action through international cooperation and national planning

- Adopt the Convention on Biological Diversity
- Adopt a resolution designating an International Biodiversity Decade
- Establish a mechanism to provide guidance on priorities for the protection, understanding, and sustainable and equitable use of biodiversity
- Establish an Early Warning Network to monitor potential threats to biodiversity and mobilize action against them
- Integrate biodiversity conservation into national planning processes

Establishing a national policy framework for biodiversity conservation

- Reform existing public policies that invite the waste or misuse of biodiversity
- Adopt new public policies and accounting methods that promote conservation and the equitable use of biodiversity
- Reduce demand for biological resources

Creating an international policy environment that supports national biodiversity conservation

- Integrate biodiversity conservation into international economic policy
- Strengthen the international legal framework for conservation to complement the Convention on Biological Diversity
- Make the development assistance process a force for biodiversity conservation
- Increase funding for biodiversity conservation, and develop innovative, decentralized, and accountable ways to raise funds and spend them effectively

Creating conditions and incentives for local biodiversity conservation

- Correct imbalances in the control of land and resources that cause biodiversity loss, and develop new resource management partnerships between government and local communities
- Expand and encourage the sustainable use of products and services from the wild for local benefit
- Ensure that those who possess local knowledge of genetic resources benefit appropriately when it is used

Managing biodiversity throughout the human environment

- Create the institutional conditions for bioregional conservation and development
- Support biodiversity conservation initiatives in the private sector
- Incorporate biodiversity conservation into the management of biological resources

Strengthening protected areas

- Identify national and international priorities for strengthening protected areas and enhancing their role in biodiversity conservation
- Ensure the sustainability of protected areas and their contribution to biodiversity conservation

Conserving species, populations, and genetic diversity

- Strengthen the capacity to conserve species, populations, and genetic diversity in natural habitats
- Strengthen the capacity of off-site conservation facilities to conserve biodiversity, educate the public, and contribute to sustainable development

Expanding human capacity to conserve biodiversity

- Increase appreciation and awareness of biodiversity's values and importance
- Help institutions disseminate the information needed to conserve biodiversity and mobilize its benefits
- Promote basic and applied research on biodiversity conservation
- Develop human capacity for biodiversity conservation

Source: Modified from WRI, IUCN, UNEP (1992).

2.5.4 Development and assistance organizations

Conservation actions, especially those requiring international and regional cooperation, require considerable financial support. Funds come from development banks, national agencies, private foundations, and other sources. The International Monetary Fund's Directory of Economic, Commodity, and Development Organizations lists dozens of sources of support. Most development and assistance organizations work closely with UN agencies (Table 2.13). Their main objectives are to assist developing nations in policy development, national strategies, infrastructure, and specific conservation projects.

Development banks have often supported projects with deleterious environmental consequences. Under pressures from governments, NGOs, and the public, they have gradually become more concerned with conservation. Most multinational banks have adopted

Table 2.13 Selected multilateral development banks and funds. The World Bank and the United Nations (UN) have almost the same membership. International funding is competitively available through international, intergovernmental organizations such as the Global Environment Facility. Coastal-marine resources receive only a small portion of total funds available.

Name	Formation goal	Functions
World Bank (WB, 1944)	Integrates nations into wider world economy; promotes long-term economic growth to reduce poverty in developing countries	Largest single source of development lending; exerts policy leadership; trustee for Global Environment Facility (GEF) Trust Fund, an independent international financial entity created (1991) by UN Environment Programme, UN Development Programme, and World Bank to help developing countries deal with environmental concerns
International Monetary Fund (IMF, 1944)	Monitors world currencies; helps maintain orderly system of payments between countries; lends money to members with serious imbalance of payments	Major influence on development policies of developing countries; monitors transactions in international trade and investment
UN Development Programme (UNDP, 1965)	Provides developing nations with policy advice on a range of issues pertaining to poverty, institutional capacity, and globalization	Assists nations and territories; *Capacity 21* was launched at UN Conference on Environment and Development (1992) to assist nations implement *Agenda 21*; as of 2001, *Capacity 21* supported 21 efforts in 75 nations
Development banks	Provide financial support and professional advice for economic and social development in developing countries	*Five regional development banks*: African; Asian; European Bank for Reconstruction and Development; Inter-American Development Bank Group; World Bank Group *Multilateral financial institutions*: European Commission (EC) and European Investment Bank (EIB); International Fund for Agricultural Development (IFAD); Islamic Development Bank (IDB); Nordic Development Fund (NDF) and Nordic Investment Bank (NIB); OPEC Fund for International Development (OPEC Fund)

From IMF (2001) online.

policies of sustainable development, with specific goals for biodiversity, fisheries, ocean law, shipping, pollution, global climate change, regional seas, freshwater, and related issues. The Global Environment Facility (GEF) has been influential in supporting conservation of biodiversity in developing nations through cooperative efforts of UNEP, the United Nations Development Programme (UNDP), and the World Bank.

Assistance organizations are promoting economic instruments that increasingly are being accepted as means to change human behavior. Economic incentives are being applied to protect forests and fisheries and to establish and manage protected areas. Many economic incentives seek to internalize the "externality" costs of resource depletion and pollution so that producers, transporters, and consumers face full social and environmental costs of pollution and resource extraction. For example, the "polluter pays principle" reflects a shift in the burden of proof. Assistance organizations do not generally support research, but may support assessment and monitoring programs that influence management.

2.5.5 Non-governmental organizations

People power has achieved national and international importance through non-governmental organizations (NGOs). International conservation NGOs arose during the post-World War II period, with the founding of the International Union for the Conservation of Nature and Natural Resources (IUCN, the World Conservation Union) in 1948. IUCN has grown into a large organization with worldwide influence and strong connections with the UN and national governments. Other national and international NGOs emerged especially from the 1960s onwards, and mostly in the developed world. NGOs are exceedingly diverse in their interests and methods, and together form effective communication channels among policy, politics, science, and the public. Tens of thousands of NGOs now exist worldwide, only about half of those in developing nations being older than 15 years.

Separately, or occasionally *en masse*, NGOs lobby government, publicize information of strategic importance, and influence international conferences. They can be major actors in negotiation, as reflected at the Rio Conference in 1992, where more than 1400 NGOs were accredited to participate in discussions leading to the Convention on Biological Diversity. Coalitions of NGOs are now present at many international meetings, such as those of the International Whaling Commission, London Dumping Convention, and Convention on International Trade in Endangered Species. NGOs have helped negotiate environmental issues and have pressured international institutions to enlarge their environmental activities. Significantly, the UN has devised a system of formal relationships with several NGOs. Thus, NGOs have become a powerful force, influencing the direction of environment and development policies around the world through advocacy and "on-the-ground" action.

NGO interests are overwhelmingly directed at crisis situations, such as endangered species, depleted fisheries habitat protection, and environmental pollution. The survival of NGOs often depends on the courage and persistence of dedicated individuals. Only the largest and most powerful have scientific expertise. To confront this latter problem, some have developed ties with universities and government research organizations. Programs often lack comprehensive strategies. Notable exceptions concern strategies for protected areas and biodiversity protection *via* land acquisition and "hot spots"of species richness.

NGOs are constrained to raise most of their funds by marketing issues that are attractive to the public, such as protection of charismatic, endangered species, and establishment of protected areas. Coastal-marine programs have remained relatively neglected until recently. Programs for coral reefs, wetlands, sea turtles, marine mammals, and most recently "no-take" fisheries reserves have gradually intensified during the 1990s.

2.5.6 Science

Scientists have been instrumental in conservation at least from the 1700s, and have assumed leadership roles in many aspects of environmental policy, strategy, and tactics. Scientists have long been the main discoverers of resource depletions and environmental problems, and have provided information and strategies needed to solve them. This is especially true for marine environments, which are least accessible to the public among all of Earth's ecoregions.

International coastal-marine science arose during late-19th-century explorations, and soon became concerned with resource conservation, pollution, and similar issues. Scientists have long recognized that pollution and fisheries declines are interrelated. At the turn of the 20th century, for example, the International Council for Exploration of the Seas (ICES) was formed as a scientific advisory group to northeast Atlantic governments on the issues of fisheries and pollution. For a century, ICES has helped immeasurably in fisheries management and pollution reduction, and became a model for a sister group recently formed for the North Pacific (PICES).

Another noteworthy example of scientific involvement is the Antarctic Treaty (1959). The Scientific Committee on Antarctic Research (SCAR) is the advisory group to the Treaty. SCAR reports to the International Council of Scientific Unions (ICSU), which represents the scientific academies of many nations, worldwide. Many other international scientific programs report to ICSU. The Scientific Committee on Oceanic Research (SCOR), formed in 1957, is an interdisciplinary, non-governmental scientific organization charged with promotion of international oceanographic activities. The International Geosphere–Biosphere Program (IGBP) is coordinated through ICSU. This study of global change includes a program on Land–Ocean Interactions in the Coastal Zone (LOICZ). The international program *Diversitas*, under both ICSU and UNESCO, is concerned with research on and conservation of the ecosystem function of biodiversity. Additionally, the Intergovernmental Oceanographic Commission (IOC) of UNESCO provides the Secretariat for the Joint Group of Experts on Scientific Aspects of Marine Environment Protection (GESAMP). Many other examples exist at all levels from international to local.

Coastal-marine scientists are presently involved in conservation mainly through research, education, publications, and advice to governments, managers, and conservation groups. Scientists play a catalytic role by advancing knowledge, addressing gaps, and bringing issues to the forefront, mostly *via* peer-reviewed publications and open discussions. During past decades, the scientific community was mostly concerned with basic information gathering. More recently, this community has shifted toward more involvement in applied science to address conservation needs. Science is constrained by the nature of funding, most of which supports research not immediately relevant to conservation, although essential in the long term. This situation is ripe for change, as most conservation agreements, such as those for fisheries, pollution, biodiversity, and protected areas, call for a scientific basis.

2.6 Conclusion

Three features of mechanisms are essential for comprehending major problems, as well as advances, in coastal-marine conservation. First, mechanisms and issues (Chapter 1) are unevenly matched in time, space, and degree. Second, the need for integration of issues and mechanisms *inter alia* strongly indicates a systems approach. And third, increased use and support of science is essential for solutions. Succeeding chapters illustrate these features.

Section II

Science

Porcupinefish (*Diodon hystrix*), on a Bahamas reef.
Photograph by the authors.

The purpose of this Section is to describe the coastal realm holistically, to describe its basic components and properties, and to outline attributes and interactions that qualify the coastal realm as an ecosystem of global significance. Within this realm, air masses deliver precipitation and aerosols, watersheds deliver nutrients and sediment to deltas and estuaries, and the energies of ocean waves, currents, and tides are modulated by continental geometry and coastal-ocean bathymetry. This realm is also characterized by exceptionally high biological diversity and productivity where diverse assemblages of invertebrates, fishes, reptiles, birds, and mammals congregate to carry out their life histories and, in doing so, significantly affect coastal-realm structure and function. The result is a system of extraordinary ecological complexity, signified by strong internal interconnectances among organisms and environments. The question is: how can the relationships among living organisms and their physical environments become better understood, and lead to understanding of the relationships between the smaller-scale instabilities and large-scale order of the coastal-realm ecosystem.

Top: Sandy shore and waves between headlands. Hat Head National Park, Australia. Photograph by the authors.
Bottom: South polar skua (*Catharacta maccormicki*), flying over emperor penguin (*Aptenodytes forsteri*) rookery, on shore-fast ice. Cape Crozier, Antarctica. Photograph by the authors.

Chapter 3

The coastal-realm ecosystem

The observer imposes a perceptual bias, a filter through which the system is viewed.
S.A. Levin (1992)

3.1 Introduction

The coastal realm is where land, ocean, and atmosphere interact with great intensity. This realm is also extraordinarily productive and biologically rich. Its land–seascapes are more varied and variable than for any other major realm of the Earth. Special properties relate to exchanges of heat and materials, bioenergetic transformations, and evolution in which antecedent conditions determine paths of adjustments and rates of change. These properties make apparent that the coastal realm functions as a system.

The coastal realm's global-marine extent may be visualized as an irregular ribbon 500 000–1 000 000 km long bordering continents and islands, varying in width from hundreds of meters to a thousand kilometers, and corresponding in extent to continental plains and shelves (Fig. 3.1); where continental shelves are narrow, the marine boundary may also be defined by the 200 nautical-mile Exclusive Economic Zone. Its inland boundary is defined by the extent of continental plains, and subject to the landward penetration of marine aerosols. These boundaries may be modified by the extent to which terrestrial processes, watersheds, coastal upwellings, and human activities have measurable effects on biogeochemistry and marine ecology.

3.2 Global dynamics

Two long-term, global processes are responsible for the physical conformation of the coastal realm: tectonic

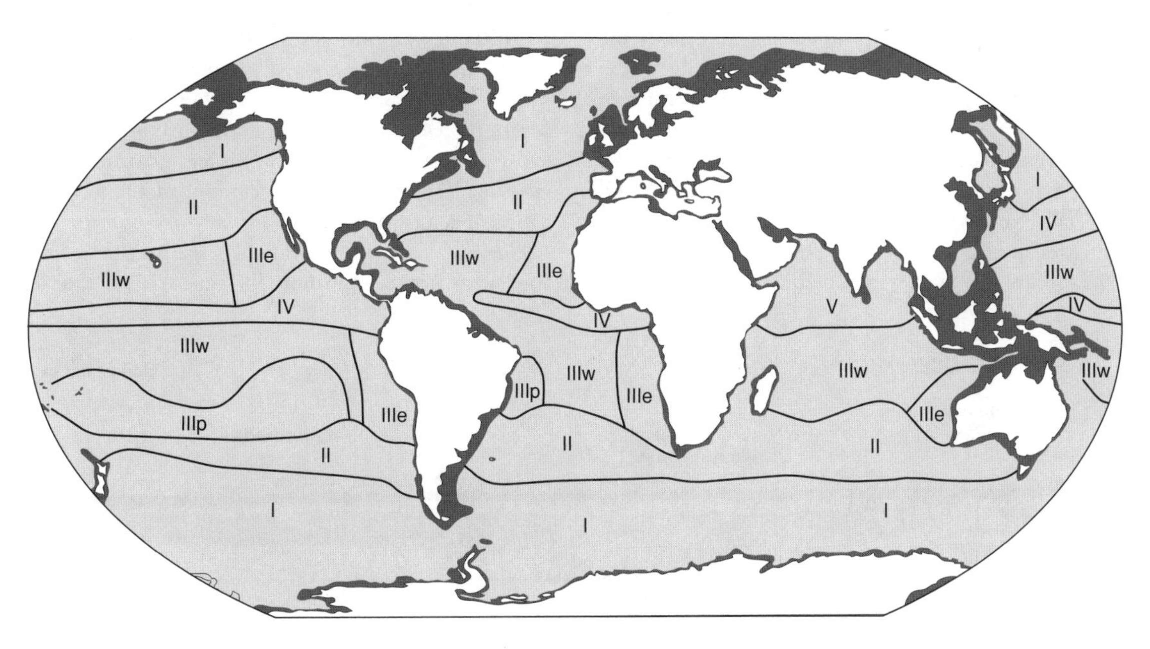

Fig. 3.1 Continental shelves (dark shading) help define the marine extent of the coastal realm. Ocean provinces are distinguished by physico-oceanographic features. Surface currents may be classified into: (I) variable eastward; (II) weak and variable; (III) trade-wind; (IV) strong westward and equatorward; and (V) monsoons with seasonal reversals. Category III is further subdivided into: (IIIe) strong equatorward; (IIIw) westward; and (IIIp) strong poleward. Adapted from Hayden et al. (1984); Holligan & Reiners (1992).

movements of the Earth's crust and climate change. These have sculpted the coastal realm into complex erosional and depositional features and resistant geological facies. Smaller-scale regional processes, such as loads of sediment or ice sheets that increase rates of land or continental-shelf subsidence or that cause coastal mountains to rise also have substantial effects. As these processes are continuous, today's coastal realm is young and ephemeral, and no part is without continuous and relatively rapid change.

3.2.1 Tectonics

About 500 million years ago, plates that compose the Earth's surface were arranged in a pattern very different from today. By about 200 million years ago, the plates had drifted together and the Earth was divided into one supercontinent, Pangaea, and one ocean, Thalassa. The coast was almost unbroken and shallow, shelf water was probably warm throughout. The plates began to separate about 135 million years ago and the Sea of Tethys intercepted two land masses, Laurasia to the north and Gondwanaland to the south. As the plates continued to drift apart, the continents took up their present positions.

Coastal-realm features relate to their tectonic-plate locations. Collision coasts occur where plates come together, usually involving subduction of one of the plates (Fig. 3.2). These coasts characteristically have a relatively straight, mountainous terrestrial aspect that fronts a steep, narrow continental shelf, and often contain active volcanic and earthquake zones. Trailing-edge coasts occur where plates spread apart, and generally have wide shelves, barrier islands, and large estuaries and lagoons. Both coastal types are modified by exposure to waves, winds, ocean currents, and tidal range. Neo-trailing-edge coasts have recently been formed and exhibit rifting (Red Sea, Sea of Cortez). Afro-trailing-edge coasts occur where both coasts of a continent are trailing (Atlantic and Indian Ocean coasts of southern Africa). Amero-trailing-edge coasts occur on the trailing edge of continents (east coasts of North and South America). Marginal-sea coasts occur on the back-arc basins of marginal seas (seas included within island arcs; e.g., the Aleutians and Kurile Islands); because these coasts are more protected from the open ocean than other types, they usually are the most diverse, with curved coastlines that are frequently modified by large rivers and deltas.

3.2.2 Sea-level change

Sea-level fluctuations have sculpted the coastal realm on time and space scales different than those of tectonics. Cycles of climate cooling and warming, most recently during the Pleistocene period of the past few million years, have alternately led to formation of massive ice sheets and glaciers on land and land almost bare of ice. Ice-sheet formation appropriates a substantial portion of the world's freshwater, causing sea levels to fall; when ice sheets melt, sea level rises. Presently, the upper limit of the coastal plain stands at approximately 50–100 m elevation, representing maximum sea level during interglacial periods. The outer limit lies at the outer continental shelf, between about 100 and 200 m depth, representing sea levels during maximum glacial and ice-sheet extents (Fig. 3.3). Ocean warming also raises sea level, by thermal expansion, adding to inundation of coastal land.

However, eustatic (worldwide) rise and fall of sea level may not be apparent in some places, due to

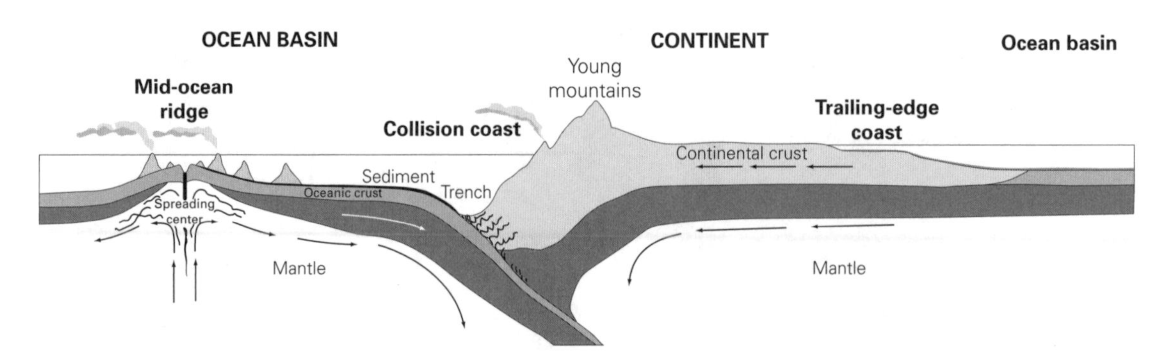

Fig. 3.2 A cross-section from a spreading center at the eastern Pacific Rise through the Peru–Chile trench and across South America illustrates formation of collision and trailing-edge coasts. Adapted from Inman & Nordstrom (1971), with permission.

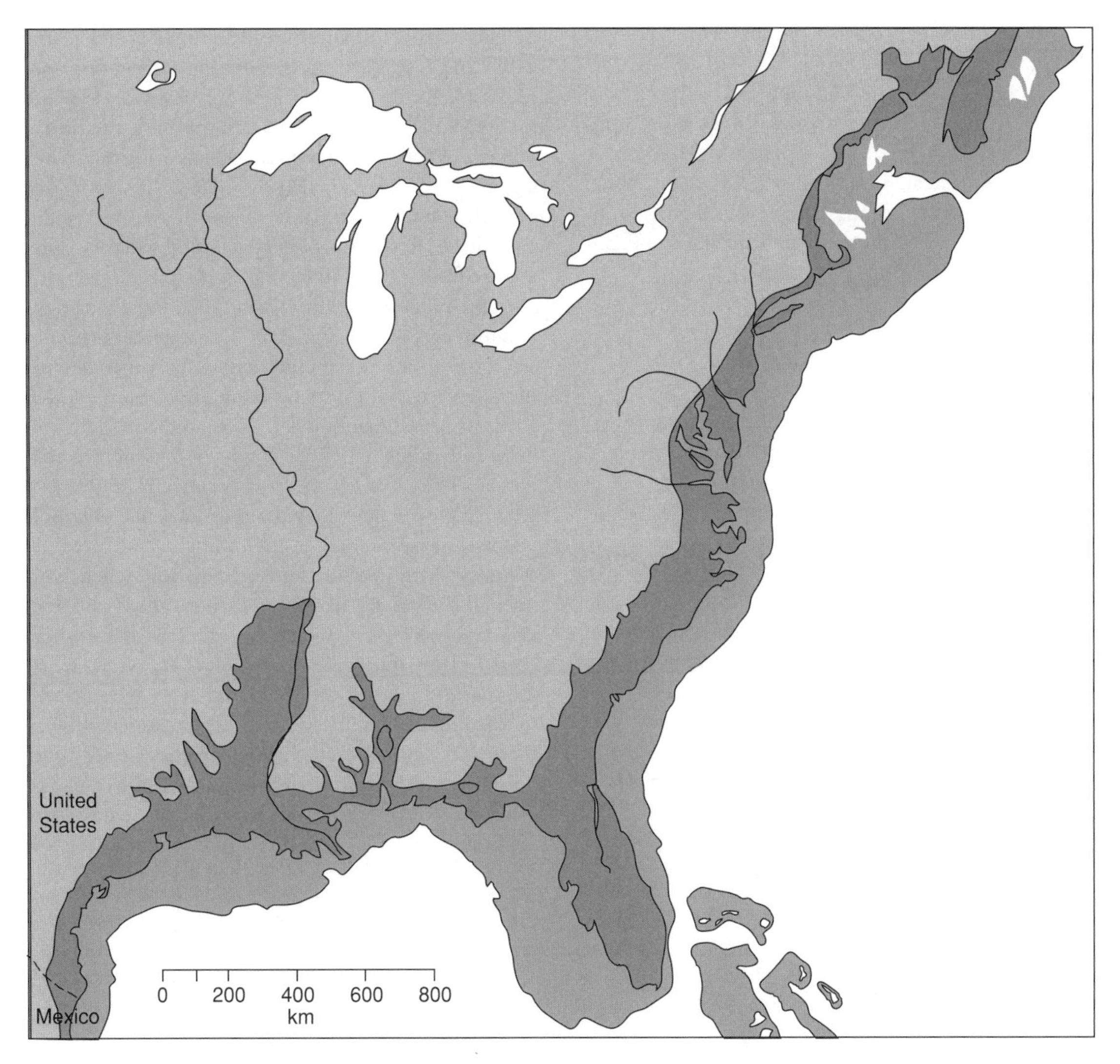

Fig. 3.3 Sea-level change helps define the coastal realm. Light gray represents uplands beyond the reach of historic maximum sea-level rise; dark gray defines present-day coastal plains; medium gray represents the extent of land during minimum sea level during the latest Pleistocene ice age. Adapted from Emery (1969), with additional information courtesy of R. Dolan, University of Virginia.

isostasy (equilibrium changes of the Earth's crust). For example, sea-level rise is accompanied by changes in movements of water, wind, and sediment. Storm surges increase in height and intensity and storm waves move water farther inland to increase erosion. Thus, massive amounts of sediment are channeled onto continental shelves; the added weight causes shelves to subside. Shelf subsidence is compensated by uplift of coastal land, resulting in rugged coastlines with narrow shelves. Ice sheets cause similar isostatic changes. In Antarctica, the weight of ice is so great that it has depressed the entire continent hundreds of meters. If the ice sheet were to melt, the continent would rebound, the shelf would be shallower, and global sea levels would rise. Today, upland river flows and ice continue to influence the continental-shelf edge.

Therefore, eustasy and isostasy combine to displace coasts vertically and horizontally and to facilitate new coastal geometries and topographies. If land surface subsides at the same time that ocean volume increases, the rate of submergence will be greater than that due to changes in ocean volume alone. Lagoons and marshes also transgress across the landscape to keep pace with change. As sea level has alternately exposed and submerged sedimentary coastal plains, shorelines have advanced and retreated many times across mid- to inner continental shelves. Thus, continental shelves (submerged continental plains) bear embedded scars of continental riverbeds and glaciated lands.

3.3 Defining attributes

Attributes of the coastal realm define conditions for a unique, diverse, and abundant biota that has evolved mechanisms to cope with variable conditions.

3.3.1 Major coastal-realm subdivisions

Five major coastal-realm units result from large-scale physical processes (Fig. 3.4). The terrestrial portion consists of uplands, coastal plain, and tidelands, and the marine portion consists of shoreface and offshore entrainment volumes. Uplands are not strictly part of the coastal realm, but are included where freshwater input is significant to coastal-realm processes. The coastal plain is formed of recent sediments and is presently above sea level, with gentle topography and highly variable width from a few meters to hundreds of kilometers. Tidelands are delimited by the inland extent of saline, spring-tide waters; tidal freshwaters are included. Marine subdivisions are receiving seasheds for terrestrial watersheds. The shoreface entrainment volume, where sediment may oscillate with wave conditions, is narrow, generally less than 20 m depth. The offshore entrainment volume, where surface waters influence the sea floor only indirectly, is subdivided by water masses, fronts, gyres, and complex bottom topography.

Many permutations involving these five fundamental units are possible (Fig. 3.5). The way that processes are coupled among the units defines fundamental types of watershed and receiving-basin associations. Water, sediment, and organisms are key ingredients. The volume and origin of either fresh- or marine water is critical. The influence of uplands concerns freshwater, which delivers energy, particles, nutrients, and pollutants to coastal-marine environments and provides migratory pathways for many species. And, where continental shelves are narrow and rivers short, different associations of biota would be expected than where coastal plains and continental shelves are wide, gently sloping, and more turbid. Many fishes are well adapted for feeding on benthic fauna in turbid waters; others, those that depend on sight for detecting food, avoid turbidity. Reef-building organisms, such as oysters and corals, are variably tolerant of sediment. Sedimentary island systems and the continent of Antarctica have few or no rivers, and watersheds are not significant.

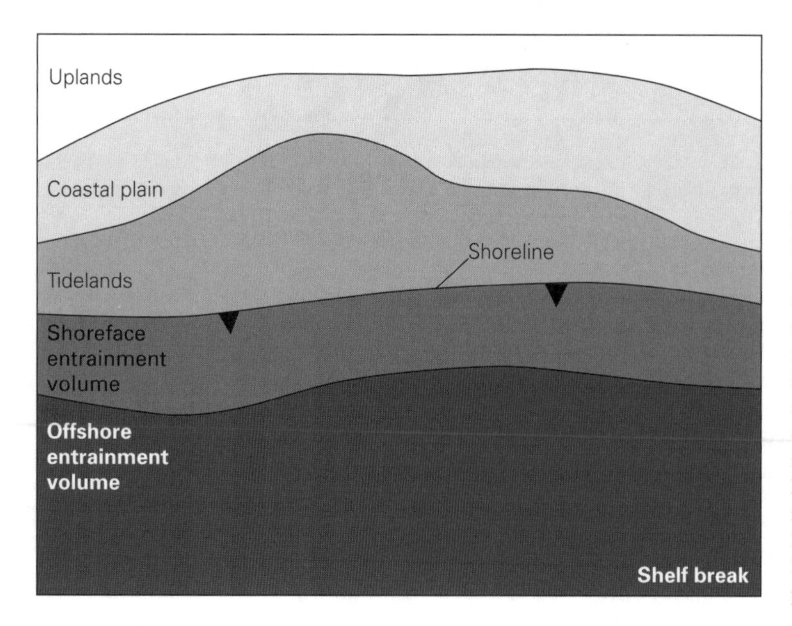

Fig. 3.4 A five-part scheme defines boundary conditions of the coastal realm. Longshore boundaries (dark triangles) can be defined by: headlands, capes, and shoals; changes in shore orientation; circulation discontinuities; and submarine topography. Controls on the system include: watershed and receiving-basin morphology; terrestrial and marine climate; wind, waves, currents, and tides; fluvial discharges, bedload, suspended load, and dissolved load; terrestrial and marine biota, especially biogenesis; and human land or sea uses. From Ray & Hayden (1992), fig. 21.2, with permission.

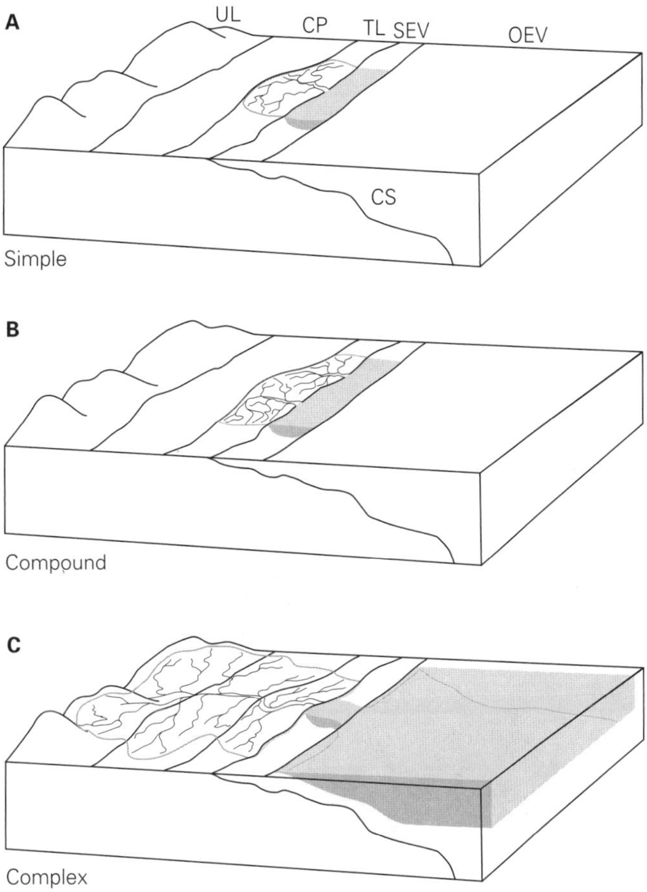

Fig. 3.5 Functional types of coastal watersheds–seasheds may be defined using the five-part scheme of Fig. 3.4 (UL, uplands; CP, coastal plain; TL, tidelands; SEV, shoreface entrainment volume; OEV, offshore entrainment volume; CS, continental shelf). Simple systems (A) have terrestrial–marine exchanges through only one watershed unit; in this case, tidelands drain into the shoreface entrainment volume. Compound systems (B) include more than one watershed, here where multiple streams or estuaries drain into a common shoreface volume within a longshore reach of coast. Complex systems (C) involve two or more units; here, a large drainage that includes all three terrestrial subdivisions has sufficient flow to bypass the shoreface volume to drain directly into the offshore entrainment volume. Many other permutations of this five-part scheme are possible. From Ray & Hayden (1992), fig. 21.3, with permission.

3.3.2 Solar radiation

The driving force for physical and bioenergetic processes is solar radiation. The sun emits a radiation spectrum that includes ultraviolet (300–400 nm), visible "light" (400–700 nm), and infrared (700–3000 nm). Light is fundamental for photosynthesis, animal vision, and photoperiodic responses in both plants and animals. Photosynthesis in plants and a few animals captures radiant energy to package elements (oxygen, carbon, nitrogen, phosphorus, sulfur, etc.) into complex living matter. Radiation also creates conditions of heating, cooling, and energy exchanges that drive ocean climate, atmospheric winds, currents, and the water cycle. Seasonal differences in solar radiation occur as a consequence of the Earth's tilt on its axis. Areas between the Tropics of Cancer and Capricorn are the only places that receive perpendicular radiation from the sun at some time of year. Seasonal surpluses of heat at lower latitudes and deficits at higher latitudes affect the land and ocean differently, causing an uneven distribution of energy and thermodynamic differences in the global land–ocean–atmosphere system. Differential heat absorption and re-radiation, combined with landmass configuration, the Earth's rotation, and gravitational forces, help explain the ocean's physical behavior.

3.3.3 Atmosphere–ocean interactions

Interactions among resistant features of the land and fluid motions of water and air are critical for distribution of coastal life. Fluid motion introduces spatial and temporal complexity that is difficult to predict and to interpret.

3.3.3.1 Coastal climate

Of particular importance for understanding coastal-marine life is the atmospheric boundary layer, which is

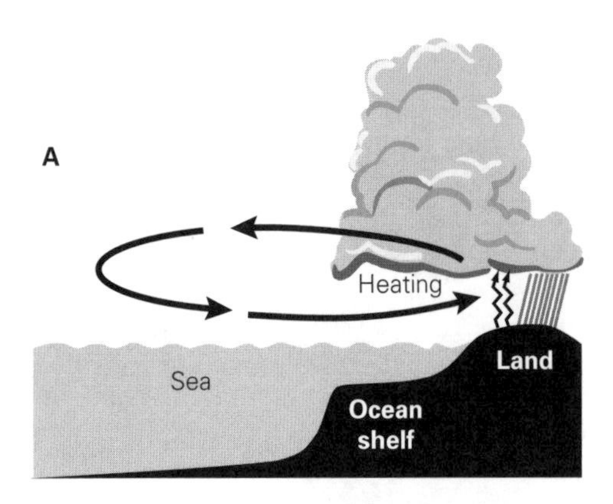

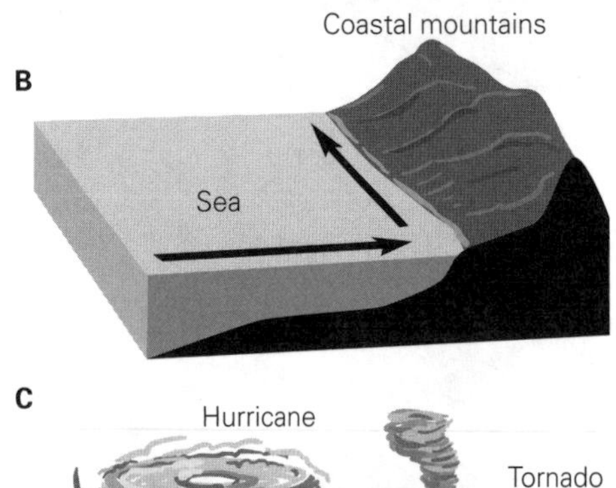

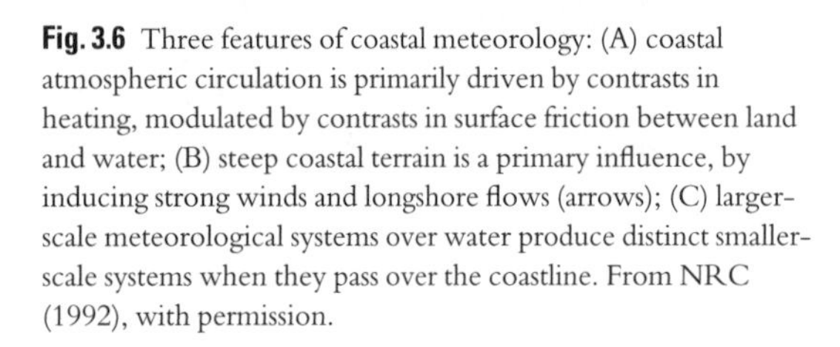

Fig. 3.6 Three features of coastal meteorology: (A) coastal atmospheric circulation is primarily driven by contrasts in heating, modulated by contrasts in surface friction between land and water; (B) steep coastal terrain is a primary influence, by inducing strong winds and longshore flows (arrows); (C) larger-scale meteorological systems over water produce distinct smaller-scale systems when they pass over the coastline. From NRC (1992), with permission.

a layer of air of variable thickness that lies immediately over the land and sea. The boundary layer is the medium through which transfers of heat, energy, and materials occur among land, ocean, and atmosphere. This layer has mostly been studied over homogeneous surfaces of land and ocean. However, the coastal realm is strongly inhomogeneous.

Coastal meteorology illustrates unique properties (Fig. 3.6). The most basic concerns are exchanges of heat, momentum, and water vapor between the atmosphere and underlying surfaces. When underlying surfaces are heterogeneous, thermal contrasts contribute to formation of unique phenomena, such as land and sea breezes, thunderstorms, atmospheric and oceanic fronts, winds associated with elevated and uneven landscapes, sea-surface waves, atmospherically induced coastal-ocean currents, upwelling, and fog, haze, and stratus clouds. Formation of stratiform clouds modifies the radiation balance, affecting water temperature. Convergence of marine air over coastlines can result in strong convection, heavy precipitation, and runoff, in turn affecting erosion and dispersal of sediment. Changes in heat flux can also produce instabilities in boundary layers, cloud cover, and winds.

Understanding coastal climate involves the formidable challenge of measuring interrelated processes and events on multiple scales. At large scales, coastal climate has dramatic effects on CO_2 exchange, photosynthesis, and biological productivity. At smaller scales, rapid atmospheric events, such as passage of storms, wind events, and interactions caused by pressure gradients and topography, are also significant.

3.3.3.2 Geostrophic and gravitational forces

All fluids that move over the Earth's surface (i.e., air and water) are shifted by the Coriolis effect. Because the Earth is a sphere and rotates eastward once a day, a point at 30° latitude moves east at a velocity of about 1500 km h^{-1}, but a point at 60° latitude moves at only 800 km h^{-1}. Thus, if a particle not attached to the Earth moves north from 60°, it initially moves eastward at about 1500 km h^{-1}; as it moves north, the Earth beneath it moves more slowly. Hence, the particle deflects to the right because of its high initial velocity; deflection is to the left in the southern hemisphere.

The Coriolis effect also affects the water column. Surface water to depths of about 100 m, when set in motion by wind, is deflected right in the northern hemisphere and left in the southern hemisphere, theoretically at 45°. Deeper layers of water are progressively deflected, producing a phenomenon known as the Ekman spiral. When movements of all layers are combined, the net deflection of surface waters is approximately 90°. At large scales, this results in oceanic Ekman circulation, expressed as gyres over large oceanic basins. This circulation elevates lighter, less dense water in the gyre's center, resulting in outward flow due to gravity, a phenomenon known as geostrophic flow. Thus, oceanic currents occur in balance between the Earth's rotational and gravitational forces. A dramatic example is the clockwise-flowing North Atlantic gyre, which is modified by continental boundaries and strongly influenced by the northwest-Atlantic Gulf Stream. This powerful current attains speeds of up to 1.5 m s^{-1}, with a volume of 100×10^6 $m^3 s^{-1}$ – 1000 times the Amazon, the world's largest river. The Gulf Stream's initial course is set by the North American continental slope. Warm- and cold-core rings spin off it and help form a huge gyre in the mid-Atlantic. Gulf Stream circulation strongly affects coastal climate from North America to Europe. It is a boundary zone of the coastal realm and is a transport medium and habitat for many forms of life.

The Earth's rotation and Earth, moon, and sun gravities send massive volumes of ocean water and kinetic energy to the coastal realm as tides and tidal currents. Because the moon passes over any point on Earth every 24 hours and 50 minutes, two low and two high tides generally occur. In some areas, basin size results in only one tide cycle per day. In other areas, a mixed pattern occurs. Thus, spatial arrangements of land and sea, the Coriolis effect, coastal geomorphology, and the depth and size of tidal basins all affect the hydrodynamic regime. Because the moon's orbit relative to the Earth's equator shifts from 28.5°N to 28.5°S, the alignments of sun, moon, and Earth vary considerably. When the sun, Earth, and moon are in line, their combined gravitational forces produce strong spring tides with maximal range. When the sun and moon are at right angles relative to the Earth, their gravitational forces partially cancel each other, and minimal neap tides occur. The rise and fall of tides create intertidal gradients of exposure to air and seawater and give rise to tidelands, tide pools, and tidal currents that affect sediment, larval transport, hydrology, and oxygen levels. In intertidal zones, pulses of submergence and exposure affect local biodiversity and productivity. Where funnel-shaped basins occur, tidal ranges are amplified, and strong tidal bores can preclude the occurrence of some organisms. The strongest tidal bores and greatest tidal ranges occur in the Bay of Fundy, Canada, and Cook Inlet, Alaska, where tides can exceed 10 m in height.

3.3.4 Biotic attributes

Physical attributes of the coastal realm set conditions that have resulted in a distinctive biota with exceptionally high productivity.

3.3.4.1 Biota

The biota of the coastal realm are distinct from those of the open ocean or freshwater. For example, fishes constitute about half of all vertebrates worldwide. An estimated 26 300 species have been described, and perhaps an additional 4000 have yet to be discovered or described. About 10 000 known species, or 54%, live in freshwaters. About 44% (~ 11 300 or 80% of marine species) live within coastal-realm waters. Only about 12% (around 2900) are oceanic – that is, live in the open ocean beyond the continental shelf.

Marine mammals are another example. They have adapted to all aquatic environments; 123 modern species have been described, of which three have become extinct as a result of human exploitation: two (beaked whales, Family Ziphiidae) are known only from dead specimens, and six are freshwater. The natural histories of the remaining 112 species are well enough known that a reasonable assessment of habitats is possible (Table 3.1). Of these species, 98 (~ 80%) spend significant portions of their lives in the coastal realm, 45 (40%) are entirely coastal; 67 (56.8%) occur in the open ocean, and only nine (7.6%) are fully oceanic. This finding is particularly significant, as marine mammals are endothermic ("warm-blooded") and require much more food in proportion to their weight than do ectothermic ("cold-blooded") fishes or invertebrates. Therefore, the presence of abundant marine mammals in the coastal ocean confirms this realm's high productivity. Seabirds (also endothermic) are also extraordinarily abundant in the coastal realm. These examples confirm the global uniqueness of the coastal realm. Other groups of species would be expected to follow similar patterns. However, many, if not most, of the Earth's coastal-realm species remain to be discovered or described.

3.3.4.2 Production

On land and in water, plants use the sun's light to convert carbon and other essential elements into food for animal consumers. Primary production is generally expressed as grams of carbon fixed per meter squared per year (g m^{-2} yr^{-1}). Gross production is the total amount of organic matter produced in an area over a given time. Net production expresses the amount of organic matter produced in excess of species' physiological needs (growth, maintenance, reproduction) and therefore available to the ecosystem. Alternatively, production may be expressed in terms of biomass, the total weight of living material in a specified area.

Net primary production in the coastal realm is generally greater than for either the open ocean or the land (Fig. 3.7). Production in the open ocean is similar to that in terrestrial deserts and tundra; its overwhelming

Table 3.1 Numbers of species of marine mammals that occur in coastal-marine habitats: TE, terrestrial; FW, freshwater; CO, coastal; OC, oceanic. Some occur in more than one habitat. Subspecies and vagrants are not included. Coastal species include: 17 (TE/CO) + 8 (FW/CO) + 20 (CO) = 98. Oceanic or coastal-oceanic species include 5 (TE/OC) + 53 (CO/OC) + 9 (OC) = 67.

Species	TE/CO	TE/OC	FW/CO	CO/OC	FW	CO	OC
Carnivora							
Pinnipedia:	17	5	2	6	1	3	
Seals, sea lions, and walruses							
Cetacea							
Mysticeti:				9		2	1
Baleen whales							
Odontoceti:			5	38	3	14	8
Toothed whales, dolphins, and porpoises							
Sirenia							
Sea cows and dugongs			1		2	1	
Total							
Species numbers	17	5	8	53	6	20	9
Per cent of total (rounded)	14.4%	4.2%	6.8%	44.9%	5.1%	16.9%	7.6%

Data from Rice (1998).

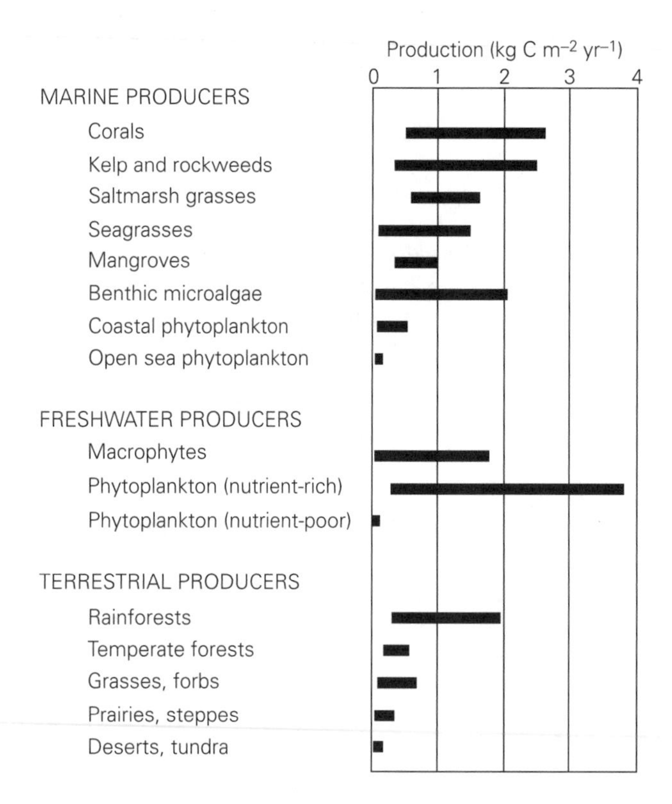

Fig. 3.7 Annual net production in freshwater, marine, and terrestrial environments. With few exceptions, coastal-realm environments are most productive. From Valiela (1984), fig. 1-14, with permission.

area and volume, however, yield more than 60% of total marine production. Reasons for high primary coastal production relate to high nutrient input from watersheds and coastal upwelling. Turnover time also enhances overall productivity. Turnover rates of aquatic organisms can be measured in hours, as in the case of bacteria, and up to a day for some phytoplankton, compared with 1–2 times a year or even decades for most terrestrial plants. Most large marine plants such as kelps grow new leaves annually at exceptional rates, in contrast to terrestrial plants that devote much of their biomass to hard tissue for support against gravity, meaning that production is low relative to biomass. Measurement of production is more difficult for animals than for plants, mostly because most animals are mobile, feed at various trophic levels, and energy demands for growth and reproduction vary greatly.

Temporal and spatial variability create patterns of high and low production that make measurement difficult. Coastal-marine production varies with oceanic and atmospheric forcing, topographical discontinuities, atmospheric nutrient loading, upwelling, density gradients, seasonal heat exchanges, differences in light penetration, nutrient distributions, river- and stormwater inputs, interactions among biota, and benthic processes. High- and low-production areas are interspersed among coastal environments, creating patchiness of organic matter. High-energy systems, such as wave-dominated open coasts, differ significantly from low-energy deltas, estuaries, and lagoons as sources of nutrients and materials (e.g., detritus) for nearshore and continental-shelf environments. Furthermore, short-term fluctuations and events (e.g., storms, precipitation) often drive productivity in the coastal realm, and multiple interacting variables make long-term averages poor indicators of nutrient conditions.

3.4 Physical components

Physical components form the setting of the coastal realm. They have arisen from processes that have modified the global scale structures imposed by tectonics and sea-level change. These components give the coastal realm its diverse and distinctive character and set the stage for species' natural history. Each component contains characteristic species assemblages and ecological processes. Gradients in exchanges of nutrients, energy, materials, and biomass result from interactions among adjacent components.

3.4.1 Coastal ocean

The coastal ocean is the source of more than 90% of the world's commercial fisheries. It is a highly dynamic and complex mass of shallow water that influences, and is influenced by, the adjacent land and open ocean, while retaining its own character.

Coastal oceans are characterized by distinct water masses (Fig. 3.8). Complex current regimes and tidal rhythms result from variations in water depth, solar heating, and salinity. Fronts create boundary conditions and result from interactions among freshwater, seawater, the atmosphere, and bathymetry. Escarpments, deltas, estuaries, capes, bays, and topography complicate nearshore water regimes. Breaking waves, violent storms, and beats of tides subject the coastal ocean to chaotic stirring. In addition, movements of water over rough, shallow benthic environments create turbulence that moves sediment and disturbs benthic communities. Although surface temperatures, solar radiation, and freshwater inflow change seasonally, internal water-column boundaries and other characteristics exhibit a degree of order to which life has adapted.

Coastal oceans vary regionally, according to differences in salinity, temperature, density, circulation patterns, fronts, depths, material exchanges, degrees of turbulence, and dynamic exchanges. Shelf size, water depth,

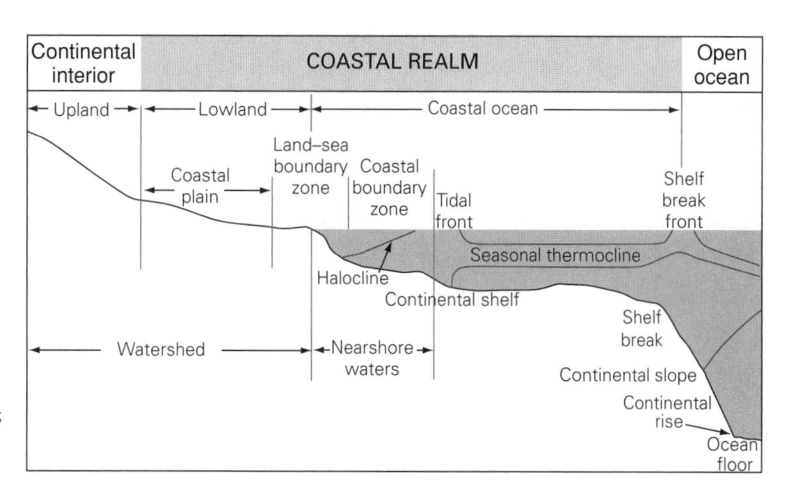

Fig. 3.8 A diagrammatic representation of coastal-realm boundaries that strongly influence distribution of the biota. Adapted from Holligan & Reiners (1992); Pernetta & Milliman (1995). Compare with Fig. 2.3, page 50 of this volume.

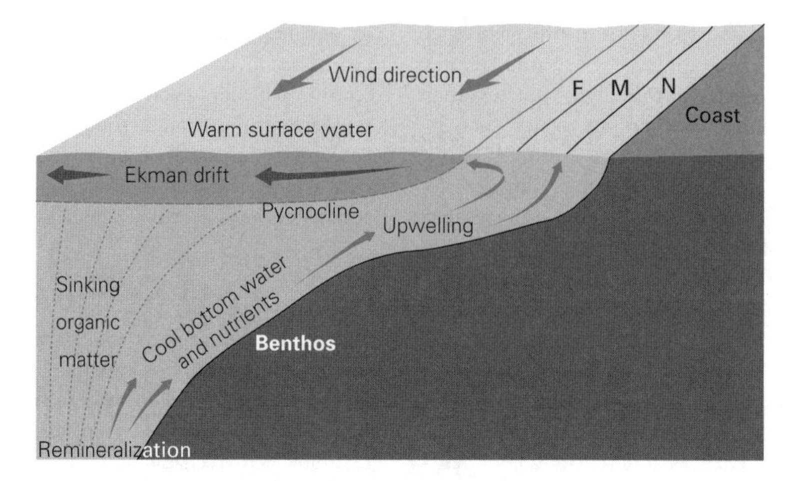

Fig. 3.9 Coastal upwelling brings nutrient-rich water to surface waters, thus enhancing coastal production. Coastal upwelling is usually generated by longshore winds, which push waters offshore, causing warmer, nutrient-poor, oxygen-rich waters to be replaced by cold, high-nutrient waters low in oxygen content. N: Nearshore waters of high primary productivity, containing phytoplankton of small cell size, and with few zooplankton organisms and fishes; mid-zone waters. M: Waters with abundant, large phytoplankton, large zooplankton, very abundant small, filter-feeding planktivorous fishes, and numerous sea birds. F: Frontal waters containing few plankton and planktivorous fishes, but with plentiful carnivores (fishes, sea birds, marine mammals). Adapted from Bakun (1996); Gross & Gross (1996); Mann & Lazier (1991).

topography, and ocean gyres play distinct roles. Proportions of materials received from land, air, and open ocean also differ among coastal oceans. Some coastal oceans are exposed to nutrient-rich, ocean-upwelled water generated offshore by winds and currents. Upwelling, a notable feature that usually occurs on western sides of continents with narrow continental shelves, supports high plankton and fish production (Fig. 3.9). Nutrient-poor downwelling occurs when the wind shifts direction. Downwelling can also occur when surface-water density increases or cools in winter, or when high evaporation increases salinity and density in warm, dry climates. Heating, evaporation, and salinity vary among regions because solar radiation varies, being greater at the equator than toward the poles. Coastal oceans are also subject in varying degrees to seasonal pulses of freshwater. Proximity to land can make coastal waters turbid, which affects light penetration and restricts phytoplankton productivity to shallow, well-lighted waters. In clearer water, phytoplankton production may occur throughout the water column.

All coastal oceans are continually subjected to internal and external forces that expose organisms to changing conditions. Circulation patterns, eddies, and currents can change direction, delivering unseasonably cold water from higher latitudes or warm water from equatorial regions. Disturbances from storms and flooding can alter "normal" conditions, causing organisms to move, die, or be replaced by others. Understanding how life has been able to adapt to this highly variable water body is key to conservation efforts.

3.4.2 Coastal benthos

Coastal lands extend beneath the coastal ocean as submerged continental shelves, which comprise a complex mix of igneous and metamorphic rocks and sediment. The benthos receives a rain of organic and inorganic materials, such as dust, detritus, debris, carcasses, hard parts, and fecal pellets from rivers, the atmosphere, and fallout from the water column. The benthos is also a source of biological and physically or chemically altered materials to the water column and to the atmosphere as gases and aerosols. Thus, the coastal benthos and coastal ocean are ecologically more strongly connected than is the case for the deep, open ocean.

Benthic sediment varies with latitude. Carbonate sediment is more common in the tropics than in polar regions. Mud is common in regions with slow water movement, as at mouths of large rivers. Organic mud deposits are most common in areas with high temperature and rainfall, such as the humid tropics; fine sandy

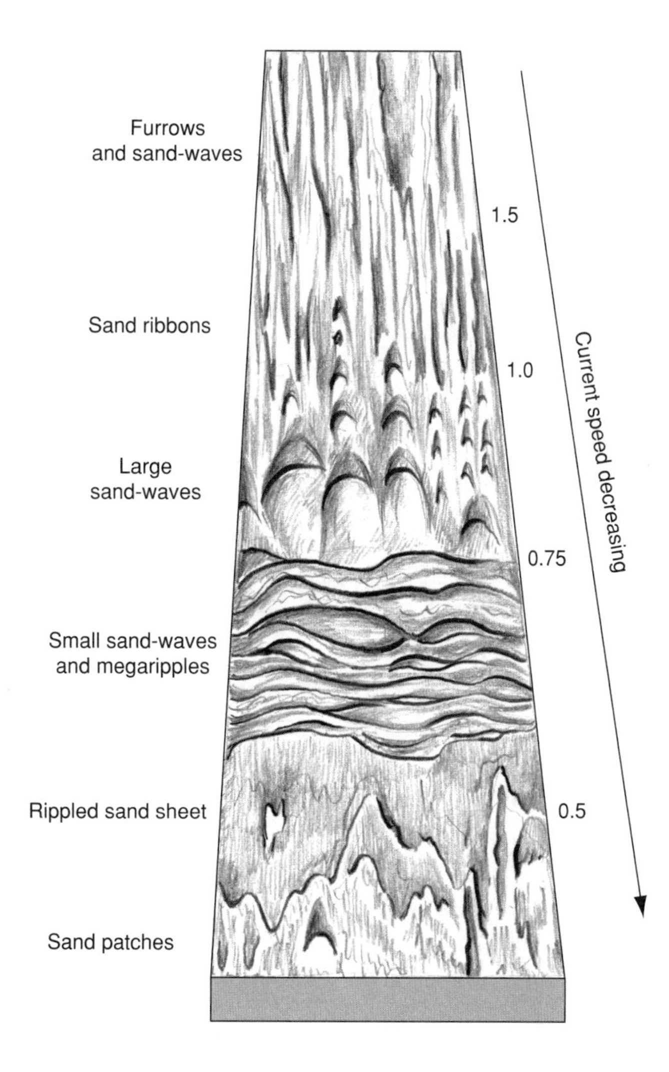

Fig. 3.10 Sedimentary bed-forms indicate speed and direction of sediment transport. They develop as a result of decreasing current speed offshore, as well as tidal reversals and irregular non-tidal currents and waves. Turbulence and drag also act to mix, disperse, and resuspend bottom particles. The distance covered in the diagram can span hundreds of kilometers. From Belderson et al. (1982), fig. 8.6, with permission.

bottoms are more common in regions with moderate temperatures and rainfall. Coarse sandy and gravel deposits are common at high latitudes due to glaciation and ice rafting, and rocky bottoms occur extensively where sediment supply is reduced and rock is bare.

The sediment–water interface is distinguished by hydrological processes that alter benthic structure and influence exchanges of materials, nutrients, and gases. Unidirectional currents and waves in shallow water form ripples on the bottom that influence the distributions and types of benthic organisms; farther offshore, complex currents create different patterns (Fig. 3.10). The turbulent layer of water flowing over benthic surfaces varies in thickness, and depends on benthic roughness, frictional shearing, current velocities, and water depth. Generally, the rougher the substrate, the lower the velocity needed to maintain a turbulent boundary layer. Where waves influence the benthos, fine-grained sediments are winnowed out by turbulence and particles are inhibited from settling. Boulders, rocks, gravel, consolidated bottom, and benthic organisms such as algae and coral resist frictional drag imposed by water movement, except when subject to the strongest storms. Turbulence within the benthic boundary layer is important for sediment gas exchange and availability of oxygen, depending on flow rate, particle size, and interstitial sediment spaces. Greater turbulence deepens the oxygenated zone, below which an anoxic zone might occur in the sediment. The effect of gas exchange on benthic life can be considerable, extending from food webs among benthic and demersal consumers to biogeochemical cycling.

Biological activity plays a prominent role in benthic processes. Primary producers may attach directly to the bottom, depending on light penetration. Rooted, shallow-water plants, such as seagrasses, absorb and recycle nutrients, intercept water flows, and help maintain substrate stability. Bacteria convert organic nitrogen into nitrogen gas. Planktonic diatoms and foraminiferans contribute tiny shells to sediment, and animal plankton contributes fecal pellets, shell, and body parts. Colonial organisms create biogenic structures, such as reefs, which influence currents and create habitat for other species. Other organisms burrow, thereby aerating and redistributing sediment. Burrowing and feeding by many animals – from worms to walruses and whales – also can have important consequences for nutrient cycling, biogeochemistry, and biotic distributions.

Benthic consumers are exceptionally abundant and varied, represented by protozoa, small interstitial animals, suspension-feeders that consume suspended particles, deposit-feeders that consume detritus, benthic invertebrate predators, and demersal fishes that utilize both the water-column and benthic environments at different stages of their life cycles. Benthic biological richness creates extraordinarily complex production processes, many of which are poorly understood.

3.4.3 Watersheds and rivers

Watersheds consist of networks of channels defined by geomorphic divisions of the landscape, and rivers are the outlets that deliver billions of tons of dissolved materials and suspended sediment to the coastal ocean. Discharges vary greatly with season, year, and latitude. Rivers in wet, tropical regions contribute slightly more than half the world's river water to the oceans annually. The Amazon (South America) alone accounts for almost 15% of the total annual discharge of the world's major rivers. The Amazon has been estimated to discharge 6300 km^3 of water and 1100 million mt of sediment every year; the Yellow River and the Ganges/Brahmaputra discharge roughly equal amounts of sediment (about 1000 million mt yr^{-1}), but discharge only 49 km^3 and 970 km^3 of water, respectively. These and other rivers with sediment loads of more than 15 million mt are estimated to contribute a total of about 7 billion mt of sediment to the coastal ocean annually. Smaller mountainous rivers also account for a large proportion of sediment discharge, and high-latitude rivers contribute large loads of glacial sediment in summer. Extrapolation of sediment discharge by all the world's rivers, large and small, yields 13.5 billion mt a year; bedload and flood discharges may add another 1–2 billion mt. Because of the impact of geology, morphology, precipitation patterns, and human activities, there is no direct relationship between water volume and sediment discharge.

A few terrestrial regions entirely lack watershed flow to oceans, for example the Dead Sea, Death Valley, Great Salt Lake, and the Caspian and Aral seas. Furthermore, not all water reaches the coastal ocean through rivers; freshwater also reaches the ocean in sheet flows or through flooding. Groundwater is another important, but poorly known, contributor of freshwater and materials. A classification of rivers that is relevant to delivery of water, sediments, and other materials, and their influences on coastal processes, may be calculated as the ratio between hydrological discharge in $km^3\ yr^{-1}$ and watershed area in km^2. One classification is: arid rivers deliver less than 100 $km^3\ yr^{-1}$ of water, subarid rivers 100–250, humid rivers 250–750, and wet rivers more than 750 (Milliman & Farnsworth, In preparation).

River discharges continuously change and are difficult to measure. The Yellow River in China was near the top of the list in sediment discharge to the coastal ocean only two decades ago; by 2000, it had dropped below the first 25 due to entrapment of sediment by numerous dams and diversions of water for agriculture. Yangtze River sediment has been so reduced by more than 45 000 dams, large and small, that its delta is apparently affected. The Colorado River once transported 135 million mt of sediment a year through the Grand Canyon; now it contributes less than 0.1 mt to the Gulf of California. Similar tales may be told for the Nile, Mississippi, and many other rivers. Northern hemisphere rivers are most affected; one estimate is that 77% of total water flow of the 139 largest northern rivers has been affected strongly or moderately by dams and other forms of water regulation.

Large watersheds may extend plumes far out to sea. The Amazon and Orinoco rivers of South America together create a sediment and nutrient plume that extends into the southeastern Caribbean Sea. Freshwater and sediment have a two-fold effect on marine systems, by introducing dissolved and particulate materials and by modifying structure and dynamics of the environment. The extent and nature of these effects are strongly influenced by river-mouth geometry and tidal dynamics, where fluvial input, wave energy, and tidal energy all interact. River plumes and the areas they

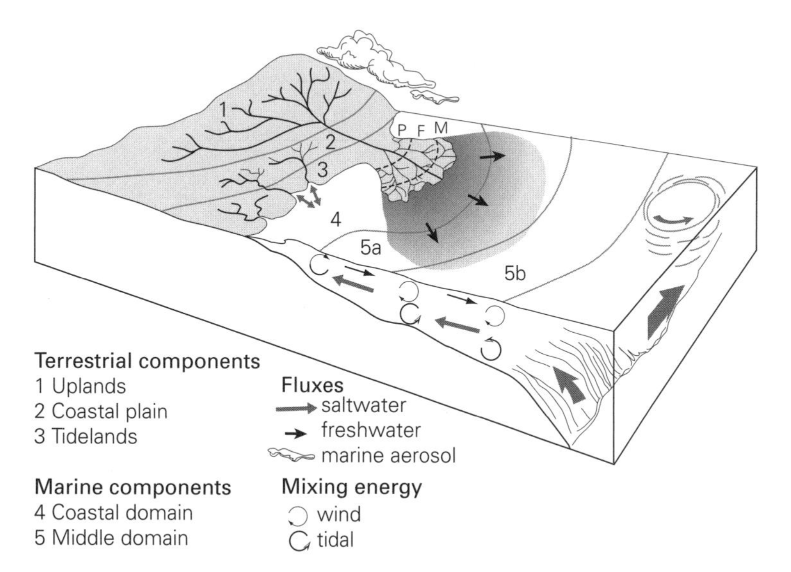

Fig. 3.11 Deltas affect, and are affected by, many properties of the coastal realm. P, delta plain; F, delta front; M, marine prodelta. The delta plume is indicated by the shaded area. The numbers (1–5) reflect the subdivisions of Fig. 3.4; 5a and 5b are subdivisions of the offshore entrainment volume that are separated by fronts and develop as a result of water-column circulation patterns (circles). Modified from Ray & McCormick-Ray (1989).

influence usually support higher production and carry higher biomass than other coastal waters. However, biological activity generally increases only after a sufficient amount of the fluvial sediment has settled, such that light transmission can allow phytoplankton to utilize dissolved nutrients. Annual or short-term changes brought on by storm events or periods of drought exert significant effects on rivers and their delivery of materials to coastal waters. Periods of drought affect the distribution and reproductive success of many species, resulting in strong or weak year classes and significant demographic alterations for long-lived species, and possible extirpations of short-lived species.

The recent history of significant changes to watersheds dates primarily from the glacial maximum of the late Pleistocene, when six periods of glacial melting and retreat exposed much of the coastal landscape. During glacial retreat, landscapes rapidly revegetated, which greatly reduced erosion and sediment discharge by rivers. Human occupation began during this period, at first with little effect. But with the development and expansion of agriculture and pasture for domestic livestock, sedimentation increased by two orders of magnitude. During the industrial age and especially during the past 50 years, dams began to restrict water flow, trapped sediment, and reduced sediment discharge. Coastal–ocean interactions as such, have changed continually, both in response to natural climate change and to human activity. Any future changes undoubtedly will reflect both natural and greenhouse-related climate change, as well as changing land use and river management.

3.4.4 Protruding coasts

Deltas and headlands protrude into the coastal ocean. Deltas result when accumulations of sediment exceed processes that carry it away or extensively rework it. Deltas are important depositional systems that strongly influence coastal circulation and hydrodynamics (Fig. 3.11), and often support extensive agriculture, fisheries, and habitats for juvenile fish, shellfish, and birds. Rich deposits of coal, oil, and natural gas are often derived from subsurface deposits. Many fishing fleets and ports are located on tributaries of major deltas.

Deltas grade into three main environments from land to sea: (i) a delta plain dominated by fluvial processes; (ii) a delta front reflecting river–ocean interaction; and (iii) a fully marine prodelta. Thus, deltas reflect a balance among interactions of the fluvial system, tides, and waves, which serves to classify them (Fig. 3.12). The Mississippi River delta is the best studied. Its watershed drains more than 3.2 million km^2 of the United States, and delivers large amounts of sediment, nutrients, and debris into the Gulf of Mexico every year. Nearly two-thirds of that material settles on the shallow shelf south and west of the delta. Because Mississippi plumes are turbid, primary productivity in inshore areas is severely light-limited. Offshore, a zone of intermediate salinity occurs, where light penetration is sufficient for high phytoplankton production and biomass, which can contribute to anoxia.

Headlands are rocky and/or sandy protrusions of land. They vary in scale from small projections to large

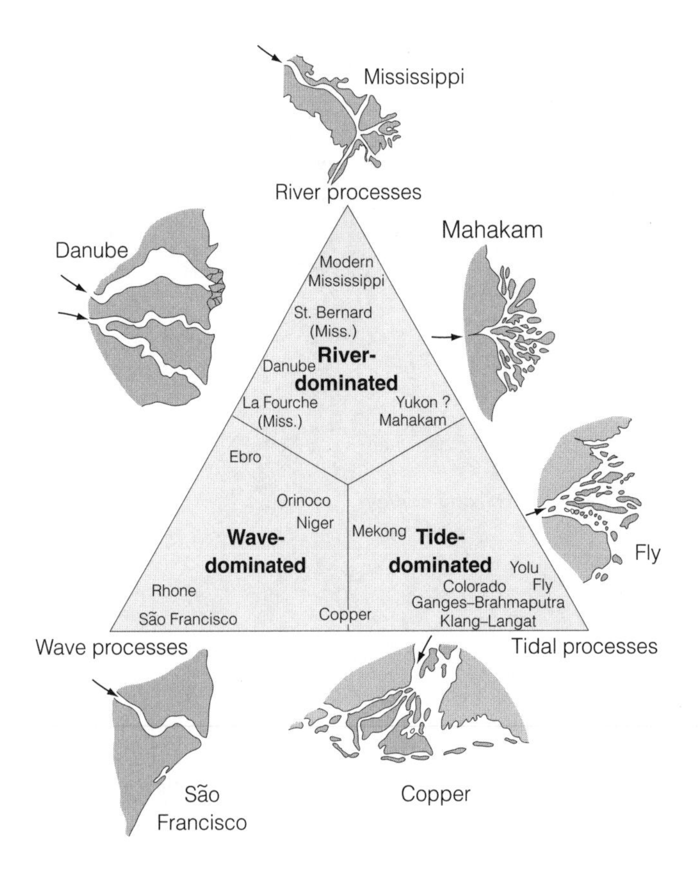

Fig. 3.12 Classification of deltas based on relative intensities of river, wave, and tidal processes that influence sedimentation and transport. Some examples are named. From Galloway (1983), fig. 7.12, with permission.

peninsulas, and can result from erosion, isostasy, tectonics, or glacial processes. Nearshore wave convergence and downstream gyres and turbulence distinguish all headlands, with important influences on tides, turbulence patterns, sediment dispersal and deposition, shoreline evolution, transport of food and nutrients, and larval retention and dispersal. Hydrologically, a headland forms a convex lens that tends to focus wave energy, in contrast to a bay that forms a concave lens and spreads wave energy. When waves converge, the energy per unit length of wave is increased, and this energy impinges on the headland region; thus, headlands tend to be high-energy environments. On rapidly eroding coasts, sediment may be trapped in a relatively sluggish environment downstream from a headland, thus building shoals that affect flow patterns.

3.4.5 Embayed coasts

Embayed coasts have free connections to the coastal ocean and receive varying amounts of freshwater from land that dilutes seawater. Embayed coasts include estuaries, bays, and lagoons, all of which are intense venues of land–sea interaction and are extraordinarily productive and ecologically complex. Many of the world's largest cities are built on or near drained marshes and filled land of embayed coasts.

Estuaries are semi-enclosed bodies of water that have free connections with the open ocean within which seawater is measurably diluted by freshwater derived from land drainage. Estuaries are among the most varied and complex, and often most productive, of the Earth's environments. Present-day estuaries have been built by the latest major episodes of sea-level rise and fall; thus, they are geologically ephemeral, estimated to be only about 1% of the age of the shelf, which itself is a relatively young portion of the Earth. It follows that estuarine biotic communities are equally young. Estuaries can be extensive on trailing-edge coasts. On collision coasts, where coastal plains are steep and narrow, estuaries are relatively small.

Tides, river flows, floods, storms, erosion, deposited sediment, and biogenic structures are major controls of estuarine geometry. Estuaries tend to evolve toward

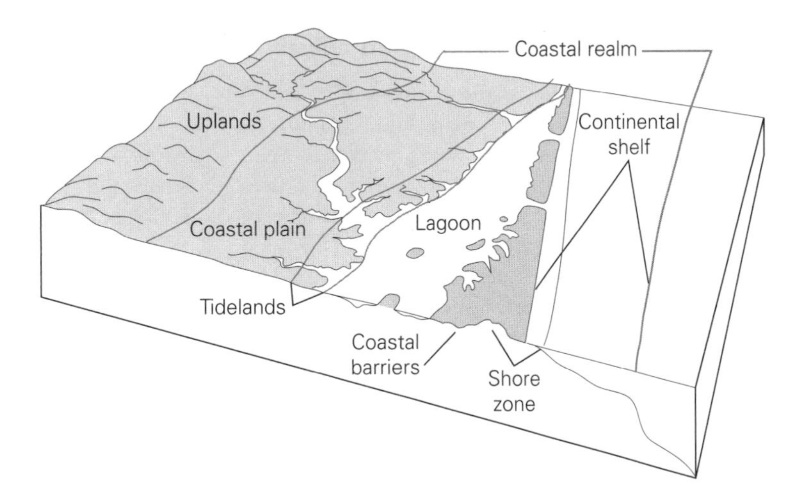

Fig. 3.13 Lagoons lie at the interface between land and sea. Most shallow-water lagoons are enclosed by a coastal barrier. From Ray & Gregg (1991), with permission.

trumpet-shaped forms where tidal energy is forced into a narrowing channel, which increases frictional drag and dissipates tidal energy. The interaction between ocean tides and freshwater inputs causes estuaries to grade from a marine zone to a middle estuary subject to strong salt and freshwater mixing, to an upper, fluvial, freshwater or mildly brackish portion that may or may not be subject to tidal action. Salinity is useful for subdividing estuaries into zones: limnetic (= freshwater), 0.5 parts per thousand (ppt); oligohaline, 0.5–5 ppt; mesohaline, 5–18 ppt; polyhaline, 18–30 ppt; and euhaline, more than 30 ppt.

Estuaries can also be classified by basin geomorphology:

- *Coastal plain estuary* (drowned river valley): usually confined to areas with a wide coastal plain where seawater has invaded existing rivers due to sea-level rise.
- *Fjord*: generally U-shaped in cross-section, with steep sides resulting from glaciation. These estuaries usually have a deep basin, and a shallow sill may be present near the mouth.
- *Bar-built estuary*: occurs in flat, low-lying areas, where sand tends to be deposited in bars lying parallel to the coast. These estuaries are usually shallow and wind-mixed and can occur when offshore sand barriers are built between headlands into a chain to enclose the body of water.
- *Tectonic estuaries*: a miscellaneous collection of types formed from faults or folding of the Earth's crust. These estuaries often have excess freshwater flow.

Many estuaries are combinations of these structures, which results in highly varied hydrology, distribution of sediment, production, aquatic vegetation, and biota.

Lagoons, bays, gulfs, and sounds are loosely applied terms for other embayed coasts. Freshwater input can be considerable, but is not usually as much a controlling factor as in estuaries. Coastal lagoons are most often oriented parallel to the coast, with water depths averaging no more than a few meters. Lagoons are separated from the ocean by barriers, such as reefs and shoals, thus providing relatively benign environments. Bays and gulfs can be large enough to be considered regional seas: for example, bays of Bengal (Indian Ocean) and Biscay (eastern Atlantic) are similar in scale to gulfs of California (eastern Pacific) and Guinea (eastern Atlantic). Sounds combine bay and estuary features: an example is Long Island Sound (western Atlantic).

Lagoons are important and widely distributed features along coasts and are often associated with estuaries (Fig. 3.13). However, lagoons react differently than estuaries to hydrological and meteorological driving forces, mostly because of their lesser freshwater inputs and smaller tides. Lagoons fall into three general types, defined by their hydrology: (i) where inflow of seawater equals outflow during a tidal period; (ii) where inflow of seawater exceeds outflow, due to evaporation within the lagoon; and (iii) where outflow of seawater exceeds inflow, due to the addition of freshwater. The significance of these exchanges is that different salinity regimes result, which help determine the types of biological communities that can exist within the lagoons. A hypersaline lagoon is a special type that occurs along arid coastal regions, for example in the Middle East and Australia and seasonally in Texas and Mexico; species diversity is very low in these lagoons.

3.4.6 Wave-dominated open coasts

Exposed coasts are where the coastal ocean, atmosphere, and land exchange energy most intensively. Most energy is delivered by waves, which form in response to a generating force, most commonly tides and wind, but also from earthquakes, fallen objects, mud slides, and other disturbances. Waves in open water cause water to oscillate, but the water itself does not move. Restoring forces such as water's surface tension and Earth's gravity cause waves to subside. However, when waves approach shallow water and "feel" the bottom, the oscillating pattern becomes asymmetrical and the wave breaks, creating a surf zone of high turbulence and longshore currents. At the fluid–sediment interface, waves can exert tremendous forces. Waves 12 m high may reach velocities of approximately 16 m s^{-1}, and accelerations can reach 1000 m s^{-2}, or about 100 times the acceleration of gravity. A wave 3 m high can transmit 100 kW of energy per meter of its crest line. Sediments of up to about 20 m depth can be set in motion by this magnitude of wave (Inman & Brush 1973).

Wave behavior is modified by coastal structure. Surface waves may travel along the coast or oscillate as a shelf seiche. As shallow water is encountered, waves are refracted at an angle, bending until their crests are parallel to the shore. If waves encounter an object, such as an island or headland, they diffract and wrap around the object. Waves may also be reflected from relatively linear shorelines or obstacles, and if the reflected wave travels in exactly the opposite direction from the original, a standing wave can develop. Variable characteristics of wave motion and the complex nature of open coasts make wave behavior complex and often unpredictable.

3.4.6.1 High-energy beaches

Open, unconsolidated, sedimentary coasts are among the most dynamic of coastal environments. Beaches usually form crescent shapes, often interposed between rocky headlands. Their structure represents a net balance between wave energy and sediment supply. Extreme energetic forces, such as storms, hurricanes, tsunamis, and earthquakes, can profoundly alter them. Nearshore bars form at places where bottom-water velocities are relatively low or where sediment concentrations are sufficiently high. Reflective beaches resist wave-generated forces and are characterized by steep, linear faces and well-developed beach cusps and berms. Dissipative beaches have low angles, concave beach faces, wide surf zones, and one or more offshore bars.

Sand dunes form behind beaches and result from marine sand being delivered by waves. Dunes buffer waves and wind and are stabilized by vegetation. Sediment exchange between beaches and dunes is an important process for maintaining morphological stability. Dunes also provide shelter for many species and help regulate the water table. Beaches and dunes are home to unique biological communities that are adapted to cope with poor soils, drought, exposure to heat, and desiccation. They are also favorite recreational areas for humans, which places beach-inhabiting species at risk; one such species is the endangered piping plover (*Charadrius melodus*) of North America that nests only on exposed beaches.

3.4.6.2 Low-energy unconsolidated shores

Mud- and sandflats form under low-energy, tidal conditions. Mudflats are common features of shallow, estuarine-channel banks, and represent a low end-member of a spectrum of wind and wave energy. Some tidal flats serve as local reservoirs for heat energy and as sources of dissolved salt and nutrients for offshore waters.

Sand- and mudflats are biologically rich, contrary to their appearance. A wide variety of epifauna and infauna – crabs, snails, shrimps, burrowing worms, mollusks, fishes, and others – inhabit sand- and mudflats. Horizontal and vertical zonation patterns typically result from sediment sorting, tidal movements, and benthic chemistry. Mixtures of algae and vascular plants grow in, or are anchored to, sand and mud bottoms in shallow waters, and many organisms, including unicellular or filamentous algae and small animals, may be very abundant on their fronds and leaves. Microscopic green, red, and brown algae, flagellates, and blue-green and photosynthetic bacteria can also be very abundant on muddy–sandy surfaces.

3.4.6.3 High-energy rocky coasts

Rocky coasts are complex, high-energy environments, usually exposed to the full forces of currents, waves, and wind. Wave projectiles consisting of water and spray are greatest at low-tide fringes of rocky intertidal zones. The highest physical disturbances are breaking waves in intertidal and subtidal zones; physical disturbances decrease with depth.

Five major factors appear to be important for the development of rocky-shore, biological communities:

(i) selection by metamorphosing larvae for sites of attachment; (ii) seasonal fluctuation in larval abundance; (iii) biological interactions within and between species; (iv) size of the substrate; and (v) physical disturbance. Rocky environments support a high diversity of macroalgae and many sessile invertebrates. The sea palm (an alga, *Postelsia palmaeformis*) attaches to rocky environments of the temperate eastern Pacific, and thrives under strong wave action. Biotic zonation patterns are often distinctive: in temperate latitudes, these are often characterized by an upper zone of littorinid snails and lichens, a middle region of barnacles and mussels, and a lower zone of algae. Zonation and competition–predation relationships of rocky coasts have long been intensive subjects of research, and have been a foundation for the science of community ecology.

3.4.7 Islands

Islands are microcosms of continents as islands exhibit all features described above. Two continents are themselves "islands": Australia and Antarctica. Islands are classified as continental, marginal, or oceanic. The first of these border continents: e.g., Canadian Archipelago and archipelagos off the west coasts of Chile and Alaska. Marginal islands circumscribe regional seas and are formed by tectonic movements, e.g., the Aleutians and the Caribbean Antilles. Oceanic islands are surrounded by deep oceanic water and may be formed by vulcanism or sedimentary processes, e.g., Fiji and the Azores. Some islands are formed in a linear sequence where tectonic plates pass over hot spots of the Earth's mantle in mid-ocean, e.g., the Hawaiian chain. Where tectonic plates are spreading in mid-ocean, the Earth's mantle material continuously rises and solidifies into islands, as occurs on the mid-Atlantic Ridge from Iceland in the north to Bouvet Island near the Southern Ocean. However, most large island systems are formed by a complex of processes. Indonesia and the Philippines are the results of combinations of tectonics, vulcanism, sedimentation, coral-reef formation, and other processes. The Bahamas in the western Atlantic and the Maldives in the Indian Ocean have been formed by combinations of chemical and physical processes and biotic reef development, which have resulted in deposition of carbonate sediments. Atolls are a special case, in which ring-like reefs surround submerged oceanic platforms.

The biota of islands are as different as are the islands themselves. Sedimentation and freshwater inputs from island watersheds create conditions similar to continents. Most carbonate karst formations, on the other hand, contain sparse freshwater and few, if any, streams, which largely eliminates estuarine conditions and anadromous and catadromous species. Island size is presumed to be important for biodiversity; some data suggest that as island area increases, so do numbers of species. This "island biogeography" hypothesis has been developed for terrestrial species, but is not well understood for coastal-marine environments. Rather, habitat variety may have greater influence on species numbers surrounding islands, independent of island size.

3.4.8 Ice-influenced coasts

Ice-influenced coasts result from actions of glaciers, ice sheets, and sea ice. Glaciers and ice sheets form when precipitation in the form of snow is so thick that it compacts into ice, whereas sea ice is frozen seawater. In both cases, climate has a great influence, increasing ice in cold times, and decreasing it in warmer periods.

Glaciers have left a legacy of ice-scoured landscapes and fjords. Glaciers and ice sheets have also deposited extensive mounds of poorly sorted gravel and large boulders on the landscape. Drumlins form headlands in numerous sites in southeastern Canada and the northeastern United States; moraines form the spine of Long Island, New York. Boulders are also prominent constituents of glacial deposits, sometimes occurring as isolated erratics many tens of kilometers from their sources, and often on headlands. Also, erosion of coastal, boulder-rich tills has resulted in boulder-lag platforms. Many formerly glaciated, mid- to high-latitude areas with large supplies of boulder-size debris are exposed to sea ice, resulting in the development of ice-rafted and ice-pushed accumulations that constitute a distinctive class of deposits.

Both polar oceans (the Arctic Ocean and the Southern Ocean that surrounds Antarctica) and their adjacent seas are dominated by sea ice, much of which is seasonally ephemeral. Arctic sea ice area changes by about 35% between summer and winter; that of the Southern Ocean changes by about 85%. The lower seasonal variability of Arctic ice is due to the fact that the Arctic Ocean is surrounded by land and is colder than the Southern Ocean. Sea ice forms an important habitat in both regions; on its underside, diatoms accumulate and fish, crustaceans, and others find refuge. Sea ice also forms substrate for marine birds and mammals (penguins, seals, and walruses) to rest and to breed.

Although sea ice hosts much biota, it can be inimical to benthic and shore life. When sea ice is moved toward shore by currents and wind, it scours the shoreline, thus inhibiting growth of algae or attachment of sessile organisms. This scouring often extends deep into subtidal areas, due to the thickness of compacted ice. Conversely, the abutment of sea ice on shores can be of advantage to some forms of life. Polar bears (*Ursus maritimus*) hunt on sea ice for seals and may swim and walk hundreds of miles on ice in pursuit of prey. Different sea-ice conditions also result in different prey availability. Arctic foxes (*Alopex lagopus*) do not swim, but frequently forage far onto sea ice when it abuts shores and is relatively continuous.

3.5 Biotic land–seascapes

Biotic land- and seascapes are among the coastal-realm's most productive components. They occur variably within the larger-scale physical components. Biotic land- and seascapes result from self-organizing biogenic and biophysical processes. Biogenesis is the formation of environments by biological activity; examples are seagrass beds, macroalgal beds, oyster and coral reefs, and carbonate platforms. The species that form these structures generally depend on currents to deliver food and nutrients and to disperse wastes. Many biogenic structures are stationed on mobile substrates, such as silts and sands, and create complex hydrological environments.

Taken together, biotic seascapes occupy only a small fraction of the total area of marine and estuarine systems. However, they are important out of scale to their size, as they add dimensional complexity to coastal systems and are locations for intensive biotic activity and high production (Fig. 3.7; page 2). Reef-forming species (corals, mollusks, algae, sapporellid and vermetid worms, and others) are physical engineers that directly or indirectly control the availability of nutrients and sediment, and physically modify, maintain, or create habitats for other species.

3.5.1 Wetlands

Wetlands result from interactions of water, substrate, and biota. Coastal wetlands usually occur at the interface of aquatic and terrestrial systems and share characteristics of both. Wetlands depend on constant or recurring shallow inundation of or saturation by water, but surface water may not always be present. Wetlands frequently have acid soils and anoxic substrates, and their plants are distinctive.

Coastal wetlands are among the most productive of all ecosystems. Net production can reach about 1000 to more than 3000 g C m^2 yr^{-1} (dry matter) or even more. Pulsed tidal action is one cause for especially high production of coastal wetlands. Nutrients and pollutants carried into wetlands by watersheds are sequestered, altered, recycled, or exported as organic carbon at rates substantially higher than for adjacent terrestrial ecosystems. Wetlands are sources, sinks, and transformers of nutrients and are important for maintenance of a host of organisms and food webs far beyond their boundaries, as much of their production is exported. Wetlands also play an important role in bacterial denitrification.

Differences among wetlands relate to geomorphic setting, water sources, and hydrodynamics. Wetlands vary with latitude and include wet tundra of boreal, austral, and polar environments, temperate marshes, tropical and subtropical mangroves, and a host of other associations. Coastal marshes commonly occur along estuaries and lagoons and play important roles in maintaining and improving water quality, trapping sediment from upland runoff, and reducing silt that would otherwise settle on shellfish and submerged aquatic vegetation beds or fill channels. The interactions among wetland plants and soils are complex, especially where tidal and interstitial water motion and rainfall affect plant growth by maintaining water levels and flushing. Less frequent wetting may result in buildup of toxic sulfides, drying, and subsequent increases in interstitial salinity.

Coastal marshes are dominated by grasses and low vegetation. High marshes are irregularly flooded and low marshes and tidal flats are regularly flooded. Varying degrees of salinity modify these marshes, and estuarine plants are good indicators of different salinity regimes. Coastal swamps, on the other hand, are dominated by trees and shrubs and are more structurally complex than marsh communities; they mostly replace salt marshes along protected shorelines in warmer climates. Mangroves are represented by a few dozen tree species of several families that share saltwater tolerance. Mangroves form distinctive zones. In North America, red mangrove (*Rhizophora mangle*) is more closely associated with open water than black mangrove (*Avicennia germinans*) and other mangrove species. The Old World has many more mangrove species than the New World.

3.5.2 Algal forests

Algal beds are formed by non-vascular, non-rooted plants ("seaweeds"). These plants usually grow attached to solid, intertidal and subtidal substrates; some species float unattached. Temperate kelp forests support a great diversity and abundance of life, including many fishes of commercial and sporting importance. Tropical sargassum weed (*Sargassum*) attaches to hard substrates, but when detached by storms, drifts into the mid-North Atlantic "Sargasso Sea" where it continues to grow and becomes a habitat for several species that are camouflaged to imitate the seaweed's fronds.

Seaweeds differ in pigmentation and photosynthesize in response to different spectra of light. Light intensity is reduced logarithmically with depth, and spectral components from red to blue are sequentially absorbed, which influences the depth at which various algae can occur. Green algae (Chlorophyta) usually occur in shallow water and utilize red and blue light most efficiently. Brown algae (Phaeophyta) also use red and blue, but have a wider extension into the blue-green region. Red algae (Rhodophyta) grow deepest and make maximum use of green light, less of red, and least of blue.

Algae vary greatly in size. The largest are the kelps (*Macrocystis, Laminaria, Elktonia* and others); *macrocystis* can reach more than 60 m in length. Kelps form extensive forests in cool-temperate waters. Mid-size brown algae (*Ascophyllum*, *Fucus*, etc.) occur on lower portions of the intertidal zone in all temperate seas. Smaller macroalgae have recently begun to dominate coral reefs that have lost grazing fishes due to overfishing; their presence can impede the settlement of corals and the development of reefs.

3.5.3 Seagrasses

Seagrasses are vascular, rooted, flowering plants, not true grasses. Seagrasses belong to only two families, containing in total about 12 genera and 50 species, in contrast to the 50 families and 500–700 species of freshwater vascular plants. Several seagrass species form extensive beds that thrive under a range of conditions, from sheltered to fully exposed shallow waters and sand or mud bottoms, but not in areas where intense wave action constantly resuspends sediments. Beds of eelgrass (*Zostera*) can cover extensive intertidal and subtidal areas in cool-temperate marine and estuarine waters. A variety of others also occur in temperate latitudes, such as surfgrass (*Phyllospadix*), widgeon grass (*Ruppia*), and others that are important waterfowl foods. Among the dominant tropical species are turtle grass (*Thalassia*) and manatee grass (*Syringodium*).

Most seagrasses are highly productive and are specialized to take nutrients from benthic substrates, in much the same way as terrestrial plants. Their abundance varies greatly with species, light availability, sediment depth, nutrients, and circulation. Seagrasses are important foods for sea turtles, sea cows, and a variety of fishes and invertebrates. Seagrass beds form important habitats for a variety of biota, and are especially important nurseries for juveniles of many species.

3.5.4 Calcium carbonate reefs

Carbonate reefs are diverse in species composition and structure. Their major feature is calcium carbonate deposition by plants and animals. Carbonate reefs may occur as discrete rocky structures or as broken rubble of dead shells, hard body parts, feces, detritus, and sediment that accumulate into a hard bottom. Despite great variety of form, reefs have the common properties of high production and many interstices for attachment and shelter for diverse arrays of species. They are "hot spots" of biological activity that add dimensional complexity to the seascape. Oysters and mussels build extensive temperate reefs, and a diversity of corals, algae, and worms build reefs in the tropics. Oyster reefs have been exploited for centuries and are the most diminished of reefs.

Coral reefs are circumtropical, and occur widely in shallow waters. Presently, they draw great conservation concern. In environments where conditions permit, coral reefs can form massive structures; Australia's Great Barrier Reef has been called "Earth's largest living organism." Nevertheless, global reef extent is not large: total coverage is only 284 000 km^2 worldwide, less than 0.1% of the total ocean surface. This small area belies coral-reef biodiversity, which may be as great as for any other environment on Earth. Coral reefs are formed by a wide variety of species that secrete calcium carbonate; species include hermatypic (reef-building) corals, coralline algae, annelid worms, molluscs, and a few sponges (making the term "coral reef" something of a misnomer). Coral-reef growth is limited to water temperatures above 20°C; most reefs are sensitive to temperatures greater than 30°C. Corals and some other coral-reef organisms, such as giant clams (*Tridachna* species), host tiny, single-celled, photosynthetic algae, called zooxanthellae, in their tissues, which give corals their distinctive colors and supply nutrients to their hosts. Therefore, availability of light is also important, and coral reefs are sensitive to turbidity, eutrophy, and sedimentation.

Tropical coral reefs take three basic forms. Fringing reefs form directly on rocky shores, barrier reefs are separated from shores by lagoons, and atolls consist of a circular reef surrounding a central lagoon. Charles Darwin correctly hypothesized that atolls form when corals grow upwards as the seamounts to which corals attach sink, thus creating a ring-like structure. Most of the world's 400–500 atolls occur in the Indian and Pacific oceans; only a few occur in the Caribbean, notably in Belize and the Bahamas.

In most reefs of any kind, the entire reef community contributes to energetic efficiency. High species diversity and a wide range of morphologies and life histories result in a system wherein primary production is linked to the calcification process, and complex food webs allow for efficient recycling. For tropical coral reefs, the result is that gross production can be among the highest of all the Earth's environments. However, diversity and complexity carry high energetic costs, creating high energy demands for maintenance and respiration, such that little energy remains for export. Thus, coral reefs, counter-intuitively, have low net production (Fig. 3.14), making them especially vulnerable to overfishing and other forms of exploitation.

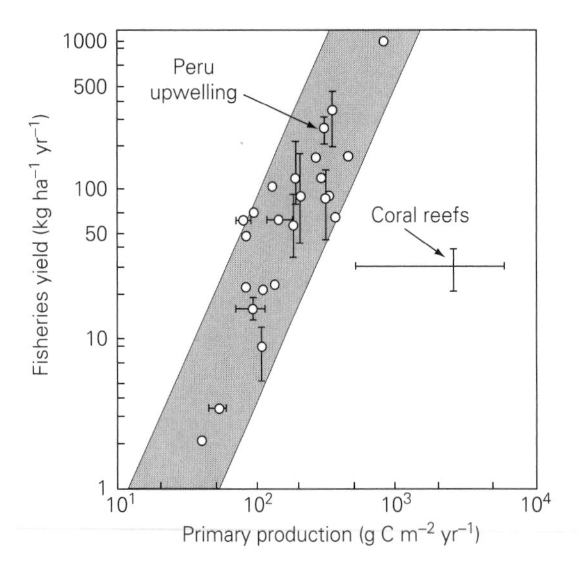

Fig. 3.14 The relationship between fisheries yield and gross primary production for coral reefs and other marine ecosystems. Coral reefs are especially vulnerable to fishing due to their low net productivity. From Birkeland (1997), with permission.

3.6 Coastal-realm properties

Individual components of the coastal realm complement one another and provide a setting for ecological processes that transform, recycle, and store energy and chemicals, as well as for evolutionary processes that alter land- and seascapes. Properties give relevance to component parts and observed phenomena, such as wetlands production, or reefs that occur in response to a given set of environmental conditions. When the coastal realm is studied as a whole, distinctive properties emerge that are different than is suggested by individual components alone.

3.6.1 Energetic transformations

The coastal realm is characterized by high biomass of large and small organisms, which results from the efficient conversion and transfer of energy. The transformation of one kind of energy into another relates to laws of thermodynamics. The first law, conservation of energy, states that energy may be transformed, but is never created or destroyed. The second law, entropy, states that energy transformation always involves some energy degradation. Heat and work are two forms of energy, and when work is performed, a significant portion of energy is lost as heat.

Energy acquisition, storage, and transfer are central to all species' life histories and survival, because work is involved in biological metabolism, movement, growth, and reproduction. Food webs represent transfers of energy. At their base is photosynthesis, which reaches a thermodynamic limit of approximately 3–5% from conversion of solar energy to energy required for maintenance of plant physiological processes. Transfers from plant producers to consumers to predators have efficiencies ranging from about 10 to 20%, but with considerable variability among herbivores and carnivores, and between cold-blooded (ectothermic) invertebrates and fishes and warm-blooded (endothermic) birds and mammals.

Marine food chains tend to be longer than terrestrial ones and food webs are more complex. For example, for a wolf, three trophic levels (vegetation to herbivore to wolf) are involved, but for a tuna, four or five trophic levels (phytoplankton to zooplankton to small fish and/or to larger fish to tuna) are required. Thus, less total annual production is apportioned to top marine predators than to terrestrial ones, theoretically at least. This is partially compensated by the fact that turnover rates of marine primary producers (phytoplankton) are much higher than for land plants; that is, phytoplankton may have low biomass at any one time, but high production when compounded over seasonal or annual periods. Furthermore, energy demands of most large terrestrial predators are greater than for marine ones, as the former are mostly endothermic birds and mammals, whereas most top predators in the sea are ectothermic invertebrates and fishes. The exceptions are marine birds and mammals, many of which have compensated for high energetic demands by a number of morphological and behavioral adaptations, as well as by adopting short food chains; sea cows (Sirenia) are herbivores and baleen whales eat zooplankton. Killer whales (*Orcinus orca*) are an exception; their position at the very top of marine food webs forces them to be few in number and relatively low in biomass.

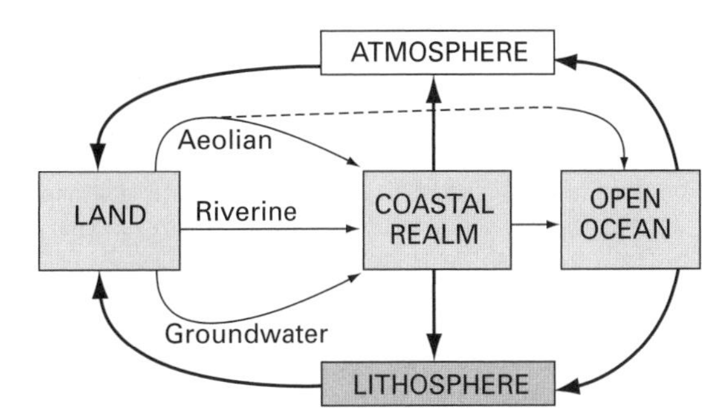

Fig. 3.15 The central importance of the coastal realm is shown by the main global pathways for the transport and transformation of matter. From Holligan & Reiners (1992), with permission.

3.6.2 Biogeochemical cycling and retention

The coastal realm plays an essential role in global biogeochemistry by mediating the transformation and transport of materials that are interchanged between land and sea (Fig. 3.15). Also, land–sea–atmosphere interactions are important for redistribution of bioactive chemicals. At intercontinental scales, important trace elements are carried in dust particles by winds across oceans (Box 3.1). However, availability and cycling of these materials involves complex processes of capture, storage, and export. Organisms play central roles, which are difficult to assess without knowledge of the natural histories of species with regard to chemical transport, sequestration, and burial. At small scales, individual species import, rework, and export materials (Fig. 3.16) that, in the aggregate, have important consequences at landscape and regional scales.

Nutrients are chemicals needed for life. Their pathways describe roles that organisms play in their communities. In terrestrial environments, nutrients required by land plants are usually generated nearby, from the decaying remains of previous generations. In water, decaying matter can be sequestered some distance from sources, as by currents or by sinking from the sunlit euphotic layer, to become unavailable for

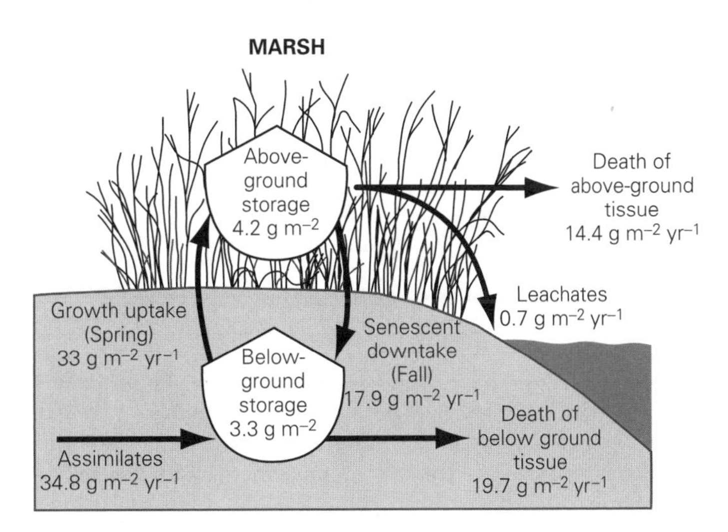

Fig. 3.16 Marsh-grass systems are among the most productive ecosystems on Earth, and are mostly nitrogen limited. Saltgrass (*Spartina alterniflora*) enhances bacterial, nitrogen-producing activity by reoxygenating anerobic marsh soils. In winter, storage is in roots; in the summer growing season, most storage is above ground. When young and growing vigorously, saltgrass is a net carbon importer. As it matures, it is a major exporter, thus providing an energy subsidy for surrounding systems. Here, nitrogen storage is almost half an order of magnitude less than throughput. From Carter (1988) after Hopkinson & Schubaver (1984), with permission.

Box 3.1 Dust-to-dust: wind-blown material

Michael Garstang and Amber J. Soja

Upwards of a billion tons of terrogenous material is lifted into the atmosphere and distributed around the globe every year (Fig. 1). Most of this material originates over the global deserts. Almost half of the above amount leaves the west coast of North Africa, originating over the Sahara and Sahel. Vast quantities of dust pour off the Gobi desert into the Pacific Ocean. Saharan dust has been found over the Near East, in the Alps, in Finland, and in Scotland and pervasively in the Caribbean, the Amazon Basin, the coastal southern United States, and Mexico.

Sources of the airborne particles or aerosols are numerous. A large fraction is soil dust, but organic matter from dry vegetation, products from biomass burning, and input from human activity including cooking fires, agriculture, and industry all contribute to the total airborne load. Particle sizes range from 1–200 μm for smoke from intense fires to submicron sizes, which include pollen and spores. Silicon is often the dominant element, with abundant iron originating from the iron-rich, red lateritic (oxisol) soils of the subtropical deserts. As many as 22 elements have been found in African dust captured in the middle of the Amazon rainforest. Phosphates, potassium, and nitrogen are present in amounts which when deposited reach kilograms per hectare per year.

The large particles and the greater part of the mass of aerosols are deposited in the coastal waters of the source region, and constitute a significant input of nutrients to these waters. Primary productivity in the coastal waters is dependent upon a suite of abundant (e.g., iron, carbon, nitrogen, and phosphorus) and trace nutrients (e.g., silicon, copper, and zinc), all present in the airborne load. Iron limits productivity in some oceanic regions and is abundant in the airborne soil dust. Evidence from the Pleistocene period suggests that increases of export production, which may have contributed to lower CO_2 concentration in the glacial atmosphere, were accompanied by a greater supply of iron from wind-blown aerosols. While upwelling is conventionally believed to provide the nutrients required for phytoplankton production, aeolian events may play a far greater role than has been previously recognized.

Patterns of phytoplankton blooms detected from the Coastal Zone Color Scanner carried on the Nimbus 7 satellite show expansion and contraction of blooms off the west coast of Africa which are spatially and temporally synchronous with dust outbreaks. As dust is transported by the atmosphere away from source regions, the total airborne load and deposition rates decrease. Both, however, remain significant. Annual transport through a hypothetical wall erected from the ocean surface to 4 km altitude and extending from 10 to 25°N latitude along 60°W longitude is estimated at 25–37 mt per year. This dust load enters the Caribbean Sea and continues onward across the Florida Peninsula and into the Gulf of Mexico. Similar transports extend northwestwards over the Sargasso Sea and southwestwards into the Amazon Basin. Estimates based upon calculations made over land in the nearly closed system of the Okavango Delta in northern Botswana show that dust from the atmosphere can on an annual basis contribute between 6 and 60% of the nutrient load. Previous studies had assumed all of the nutrients in this delta to be waterborne. Similar conditions may exist over coastal waters remote from the major sources of airborne material. Waters such as the Sargasso Sea, known to be largely oligotrophic, may receive significant nutrient supplies from atmospheric deposition. The same transport and deposition processes that deliver airborne nutrients to coastal waters can import trace elements from industry and agriculture (e.g., pesticides, fungicides, other organics). Such deposition has been suggested as a possible cause of coral die-off in the Caribbean.

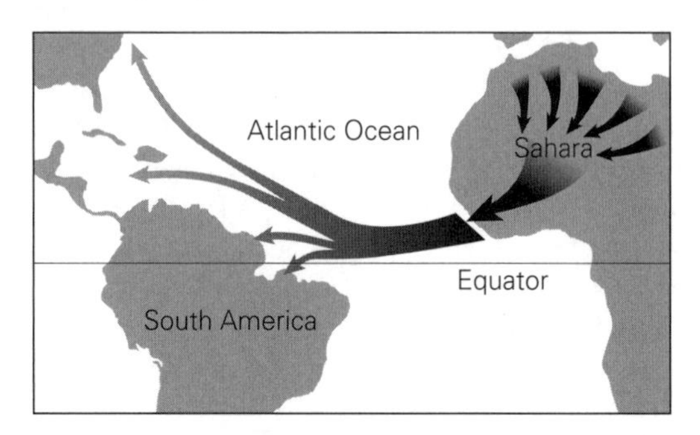

Fig. 1 Aerial transport of dust from continent to continent and ocean. Courtesy of M. Garstang.

Sources: Garstang et al. (1998); Prospero (1999); Swap et al. (1992).

plant growth. Therefore, nutrients can become limiting unless captured and recycled by organisms or returned by upwelling, bioturbation, river flow, or other mechanisms to areas where they can be used.

Although organisms package only a small fraction of the Earth's total materials at any one time, much ocean chemistry is mediated biologically. Ninety-nine percent of all living matter is composed of carbon, hydrogen,

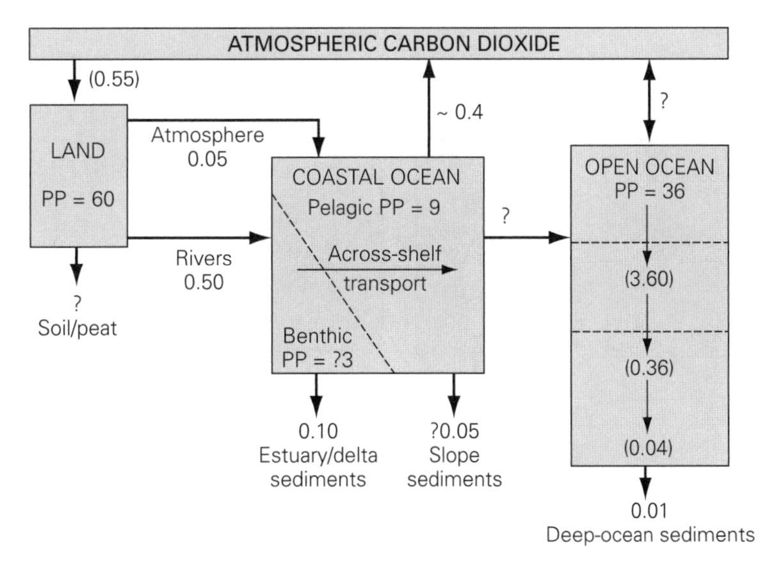

Fig. 3.17 Coastal oceans play an important role in the global carbon economy. Although mechanisms are not well understood, carbon is mostly mediated through living organisms. In order to maintain the global cycle of carbon, the ocean must be a net source of CO_2 and the main source to the atmosphere is almost certainly from inshore, coastal waters.
Numbers are gigatons per year
PP = primary production.
From Holligan & Reiners (1992), with permission.

oxygen, and nitrogen. Cycling of carbon is of particular interest, as the capacity of the world's oceans to absorb and retain this important element is largely mediated by organisms (Fig. 3.17). Phytoplankton take up carbon during photosynthesis, and release CO_2 during remineralization. Heterotrophic bacteria can metabolize carbon and other organic compounds dissolved in ocean water, and ubiquitous unicellular cyanobacteria are also producers of organic carbon. Organic photosynthesis and respiration influence atmospheric CO_2 composition, and the air–sea exchanges of CO_2 are driven by physiological and chemical interactions. For every molecule of calcium carbonate created during calcification and growth, a molecule of carbon dioxide is released into surface waters. Therefore, carbon in surface water should be in balance with the atmosphere.

However, the carbon cycle is complicated by remineralization of carbon and by availability of nitrogen, phosphorus, and oxygen. Where concentrations of these elements are high, net carbon production is high as well. Also, organisms of terrestrial, coastal-ocean, and open-ocean environments process materials at different rates and from different sources, with different sinks, and in different forms (e.g., dissolved or particulate). Therefore, the eventual transport of carbon is a consequence of these processes and the accumulation of carbon compounds in surface water. Much of this transport occurs as a series of processes referred to as the "biotic pump," in which large amounts of calcium carbonate exoskeletons of marine plants and animals (such as microscopic protozoan foraminiferans and coccolithophores) are exported out of surface waters to deeper waters, to be sequestered there or transported out. This sinking of biogenic particles has long been understood to drive respiration in the ocean interior and help maintain a strong vertical ocean gradient of inorganic carbon. A reverse biological pump occurs from the ocean floor to surface waters when deep-sea fishes spawn on the bottom and their eggs rise to the thermocline to hatch.

A further factor concerns differences in behavior among bioactive elements that are critical to their availability for organisms. Nitrogen and oxygen can escape to the atmosphere, but carbon and phosphorus remain essentially in the water. Under aerobic conditions, phosphate (Fig. 3.18A) is readily absorbed onto calcium carbonate and clay mineral particles, forming insoluble compounds with certain metals. Phosphate seldom enters deep into sediments because it encounters anaerobic conditions where bacteria act to dissolve it. In contrast, the nitrogen cycle is dominated by a gaseous phase – nitrous oxide (NO_2). The nitrogen cycle includes a pathway by which nitrogen gas is fixed into organic compounds by bacteria and bluegreen algae (Fig. 3.18B). For both phosphorus and nitrogen, the biota alter uptake and remineralization processes. Furthermore, nitrogen and phosphorus limitations on productivity differ among estuaries, coastal ocean, and open ocean, and shifts in phytoplankton community composition can shift the amount of these essential nutrients.

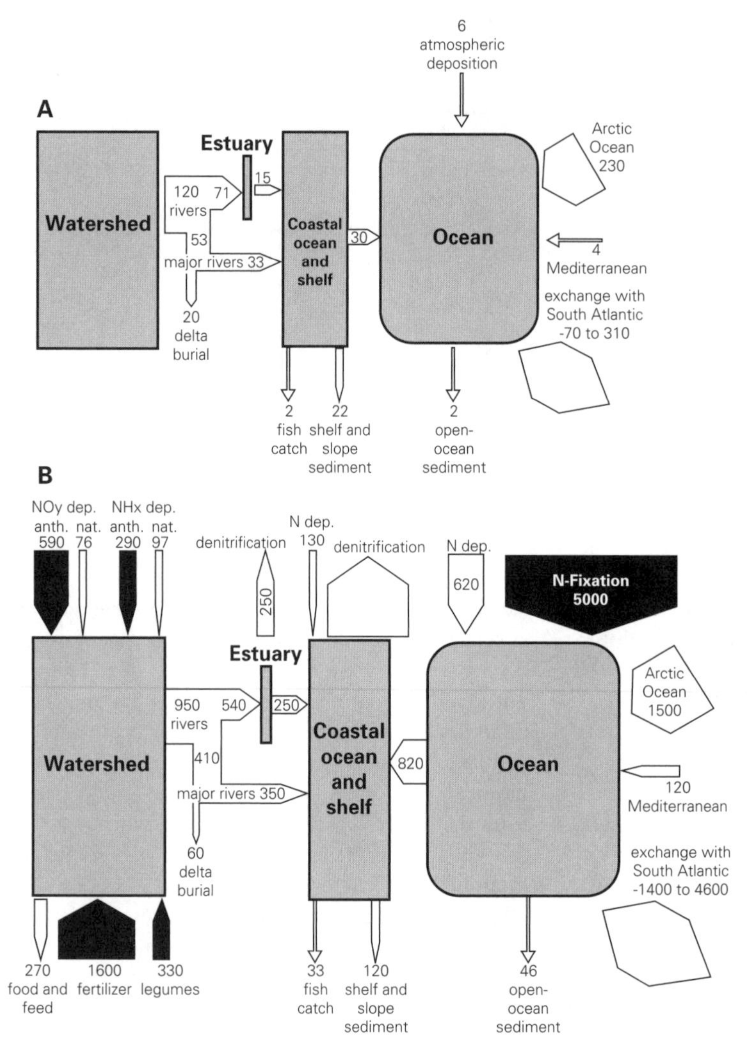

Fig. 3.18 Nitrogen (N) and phosphorus (P) budgets illustrate the central importance of the coastal ocean, including the continental shelf. Distributions of N and P through the environment are quite different. N has a significant gas phase; P does not. N is dispersed by atmospheric and hydrological processes; P transport is on particles, which limits its dispersion. In these diagrams, arrow width is sized according to median flux magnitude (in gigamoles per year). (A) P budget for the North Atlantic Ocean and its watershed; most imports are through rivers, there is no gaseous export, and most storage is in biomass or sediment. (B) N budget for the North Atlantic Ocean and its watershed. Black arrows indicate introduction of new sources of reactive N: 5000 gigamoles from bacterial N-fixation; deposited oxidized forms of N (NOy, anthropogenic); deposited reduced inorganic N (NHx, anthropogenic); anthropogenic fertilizer and legumes. Anthropogenic extractions occur as food and fertilizer. From Galloway et al. (1996), figs 2, & 3, with permission from Kluwer Academic Publishers.

Large, coastal-marine invertebrates, birds, mammals, and migratory fishes also play important roles in orchestrating biogeochemical cycles. Significant amounts of organic matter are transported during animal migrations. Migratory fish, including salmon (*Salmo* and *Oncorhynchus* species), menhaden (*Brevoortia* species), and American shad, hickory shad, blueback herring, and alewife (*Alosa* species), transport substantial amounts of nutrients between coastal-ocean and estuarine environments when they return annually from the sea to spawn in freshwaters. In doing so, they contribute significantly to the productivity of their spawning sites by depositing nutrients, thus possibly also enhancing productivity important to the development of their own young. Transfers of organic matter from sea to land by birds and marine mammals are especially significant in high-latitude systems.

Biotic transfers of nutrients and materials are rarely accounted for in studies of biological oceanography. While biotic transfers may pale in comparison to quantities transported in water, timing and location may be as important as relative quantities.

3.6.3 Evolution

Continual change is a condition of the coastal realm, an evolutionary sequence that results from antecedent conditions and large- to small-scale interactions.

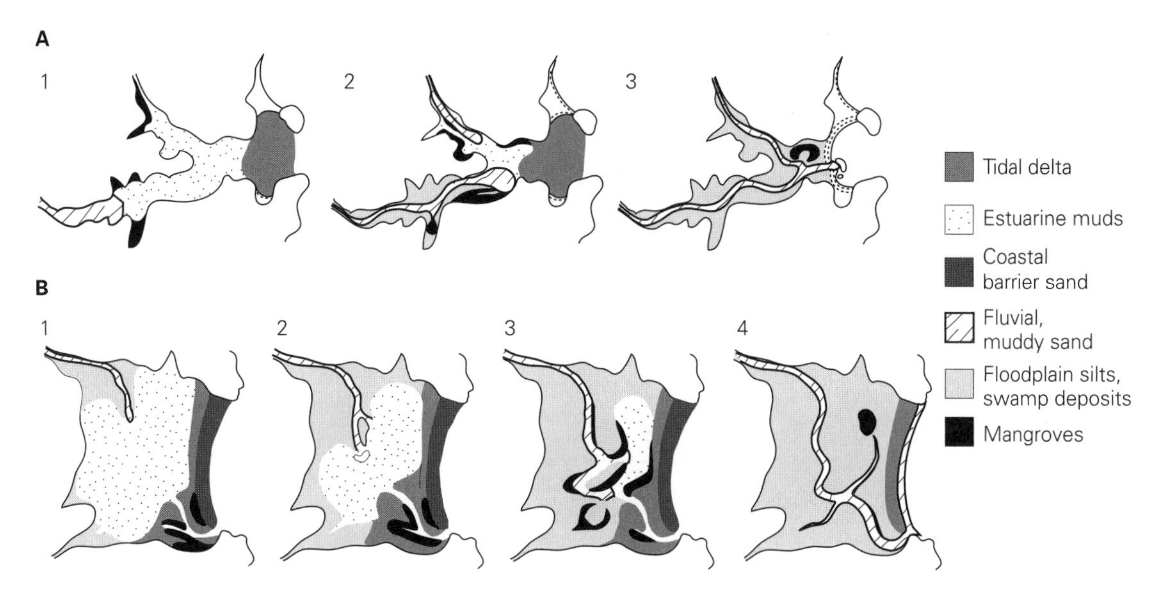

Fig. 3.19 Stages in estuarine evolution of: (A) drowned river valley; (B) barrier estuary. In both, the deeper-water portion, underlain by estuarine muds, becomes smaller, resulting in channel formation and redistribution of mangroves. This process can be altered or reversed by storm events. Water covers estuarine muds and fluvial muddy sand. From Roy (1984), with permission.

Coastal sediment change, for example, is the product of morphodynamic, hydrological, and biotic processes that occur in response to changes in external conditions and to feedbacks between topography and fluid dynamics. Landforms (deltas, beaches, barrier islands, and rocky shores) are created from small particles of silt and sand (2–125 μm diameter) to boulders (> 256 mm). These landforms are subject to macroscale forces of sea-level rise, glaciation, tectonics, storms, winds, and climatic events. Physical boundaries move and plants and animals strive to keep pace; if they do not, or cannot, they perish. Estuaries also typically change according to interactions among physical and biotic processes (Fig. 3.19). Sea-level rise could produce significant changes in lagoons and marshes, forecast by different scenarios of sea-level rise that would also require biota to change (Fig. 3.20). Thus, the coastal realm is a dynamic place, where boundaries move, forces are altered, and prediction is difficult.

How organisms influence land–seascape and community evolutionary processes is revealed by studies of biotic changes. For example, changes in tidal conditions require biotic adjustments that lead to reorganization and formation of new biotic patterns. The biota also change in response to deposition of new sediment or to sea-level rises. Opportunistic plant species, characterized by high levels of energy utilization, high fecundity, low survival rates, and widespread seed dispersal, migrate into new sediment. Over time, biotic associations are replaced by new and presumably more stable communities.

The evolution of coastal-realm systems is driven at all scales by a variety of adjustments and interactions. In simplest terms, this is illustrated by net balances and feedbacks among physical and biotic processes and their responses to changing factors. In real-world biotic–abiotic interrelationships, however, such intervening factors as biogenesis, bioturbation, human activities, and stochastic physical events complicate coastal-realm evolution. Processes of change are inherently nonlinear and time-dependent, modified by stochastic events that impose new conditions on antecedent morphology. Thus, long-term knowledge is important for understanding evolution of coastal-realm components and their land–seascapes.

3.7 The coastal realm: a complex ecosystem

Geomorphologists have long treated the land–sea coastal realm as a system. More recently, ecosystem ecologists have become concerned with transformation of land–seascapes, to focus on large-scale physical

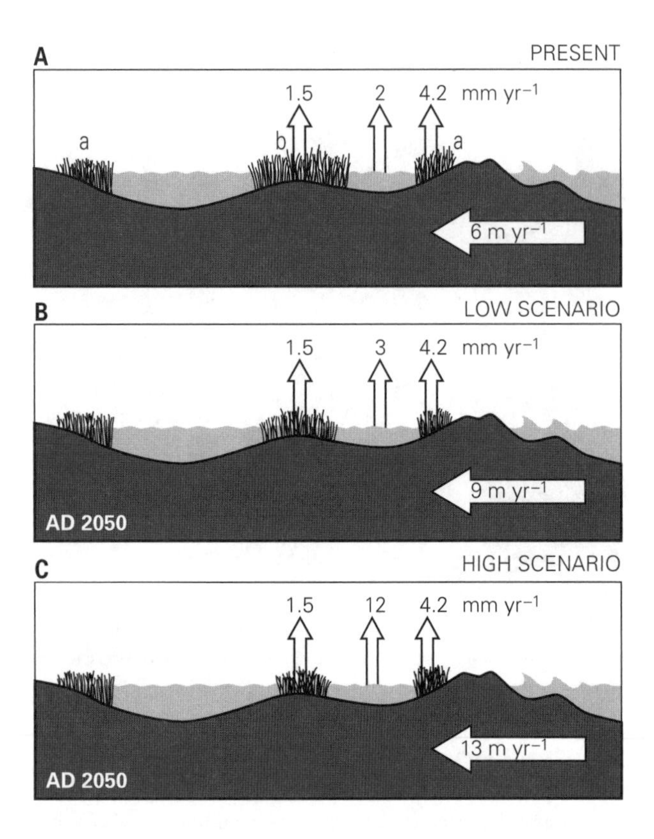

Fig. 3.20 Different sea-level rise scenarios in rates of change of marsh growth at Virginia Coast Reserve, Long-Term Ecological Research Site. Horizontal arrows indicate landward erosion rate of the barrier island. Upward arrows indicate constant rates of marsh growth, and sea-level change. (A) Under present sea-level rise, mid-lagoon marshes will disappear; barrier fringing marshes will persist because sea-level rise of 2 mm yr^{-1} is more than 1.5 for mid-lagoon marshes (b), but less than 4.2 for barrier fringing marshes (a). (B) Under a low sea-level rise scenario, mid-lagoon marshes will more quickly disappear; fringing barrier marshes will still persist. (C) Under a high sea-level rise scenario, all marshes disappear and the barrier will quickly erode. From Hayden et al. (1991), with permission.

processes, accumulation of matter, and flows of energy. A focus on interactions among these processes suggests considering the coastal realm as a complex ecosystem.

3.7.1 The ecosystem concept

Ecosystems are products of their past and are defined by their inputs, outputs, internal organization, and hierarchical structure. Their components are their "parts" or "things." Interactions among components give rise to emergent properties. Just as bones and flesh give rise to fish or whales, interactions among components give rise to ecosystem resiliency, persistence, and vigor. Thus, an ecosystem is a complex of interacting subsystems that persists through time due to interaction among its components. An ecosystem exhibits definable organization, spatial and temporal continuity, and functional properties, which can be viewed as distinctive to the system rather than of its individual components. Ecosystem properties of biogeochemical transformations, energetics, and evolution are key in maintaining ecosystem function. Under favorable conditions, ecosystems may be said to evolve toward greater efficiency, complexity, and resiliency. But even under presumably stable conditions, ecosystems are dynamic and may undergo a variety of trajectories in both time and space (Fig. 3.21).

The ecosystem has long been recognized as a fundamental unit of nature – a special kind of system that is open, in a quasi-equilibrium state, recognized by its structure, and defined by its organization, production, and biodiversity. Ecosystem boundaries are established by the observer; thus, by definition, ecosystems are subject to selection of the scale of observation and are seen to be nested within one another. Thus, land–seascapes are nested within coastal physical components, and coastal components within the coastal realm. At the largest scale, the coastal realm's origin, characteristic biota, integrated sets of components and processes, and properties mark it as a global-realm ecosystem of equal rank to terrestrial and open-ocean realms.

3.7.2 Inputs and outputs

Ecosystems are open, input–output systems, not simply flow-through systems that utilize and process

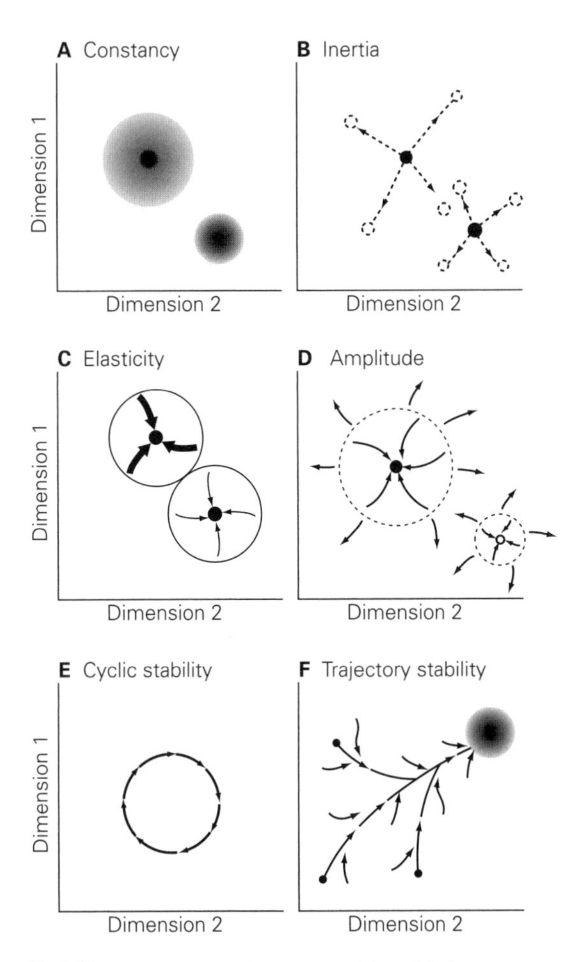

Fig. 3.21 Some concepts of ecosystem stability. (A) Constancy: lack of change in some system parameter (e.g., species number); shading intensity indicates frequency of system-state occurrence. (B) Inertia: ability to resist external perturbations; dotted circles represent system states following perturbations. (C) Elasticity: rate at which the system returns to its former state following a perturbation; speed of return is proportional to thickness of the arrows. (D) Amplitude: area over which the system is stable; dotted lines enclose regions within which systems return to initial states following disturbance. (E) Cyclic stability: system property cycles about some central point or zone; stable limit cycle is represented by a doughnut indicating the probability of finding the system in a particular state. (F) Trajectory stability: system property moves toward some final endpoint or zone despite differences in starting points; system converges to a particular state from a variety of starting positions. From Orians (1975), with permission from Kluwer Academic Publishers.

continuous fluxes of energy and materials across their boundaries (Fig. 3.22). The coastal realm exhibits ample evidence of an input–output system at many levels, wherein materials and energy are transformed,

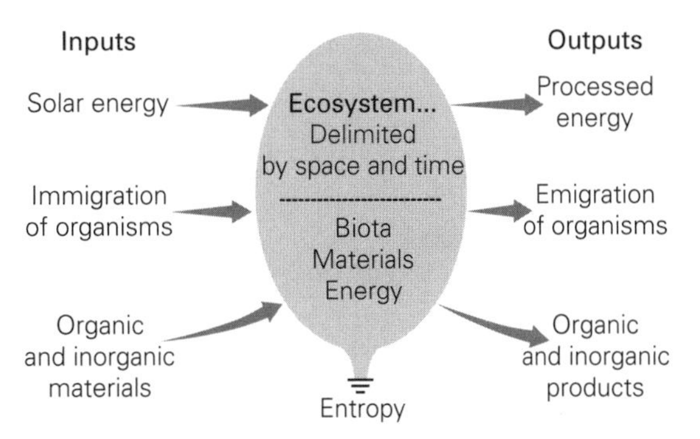

Fig. 3.22 Ecosystems are open, thermodynamically non-equilibrium, and spatially explicit, for which boundaries may be natural (watershed contours, coastal-ocean fronts, interface of water–benthos) or arbitrary (whatever feature is convenient or of special interest, e.g., species range).

regulated, and cycled. Inputs and outputs include pulsed dispersals, migrations, and energy exchanges carried out by species that enter the system, feed, deposit materials, and leave. Species' dispersals and migrations may be viewed as transfers of energy and materials from one ecosystem to another. In some cases, feedbacks from species' natural histories enhance conditions for their own survival (e.g., self-stereotaxis in oysters and anadromous fishes that transport nutrients to natal streams).

Temporal variability influences inputs and outputs. For example, tidal pulsing and water-flow changes are key to delivery of nutrients and materials and to exporting waste in estuaries. Intertidal plants and animals that are adapted to these pulses gain advantages, expressed by enhanced productivity and performance. Predictable phases of the moon stimulate many species to breed and predators to feed. Biological communities also display timed oscillations, such as cycles of abundance between predator and prey. If biological and physical systems are in harmony, performance can be amplified; if not, it can degrade. That is, if a species or ecosystem receives pulses consistent with its own internal pulse, it can benefit. But if such pulses grow too large or too random, chaos can result. In some cases, systems have evolved mechanisms to defend against random events; for example, coastal marshes survive infrequent storm events by maintaining loose structures whereby wave energies are dissipated, sediment that may smother the marsh is exported, and nutrients are accrued that stimulate growth and reproduction. The

Table 3.2 Patterns of disturbance on Caribbean coral reefs at different spatial and temporal scales.

Process	Spatial extent	Duration	Frequency
Predation	1–10 cm	Minutes–days	Weeks–months
Damselfish gardening	1 m	Days–weeks	Months–years
Coral collapse (bioerosion)	1 m	Days–weeks	Months–years
Bleaching or disease of individual corals	1 m	Days–weeks	Months–years
Storms	1–100 km	Days	Weeks–years
Hurricanes	10–1000 km	Days	Months–decades
Mass bleaching	10–1000 km	Weeks–months	Years–decades
Epidemic disease	10–1000 km	Years	Decades–centuries
Sea-level or temperature change	Global	10 000–100 000 years	10 000–100 000 years

From Jackson (1991), with permission.

duration and timing of these processes are critical; growth and recovery take time, and lags in timing can influence reorganization.

3.7.3 Hierarchical scales of organization

Ecosystems operate over wide spectra of time and space. Hierarchical organization implies that each ecosystem is composed of interacting, lower-level entities that are components of higher levels. For example, organisms are parts of populations, populations compose communities, and communities together with their physical environments constitute ecosystems. The single most important consequence of hierarchical structuring is the concept of constraint, which emphasizes that behavior of an ecological system is limited by (i) the potential behaviors of its components, and (ii) the environmental constraints imposed. Constraints may be imposed either by biota or by environment. Some constraints are imposed by biotic capacity from lower hierarchical levels. For example, a school of fish cannot swim faster than the slowest individual and still remain a school of fish. A coral reef cannot perform its functions or endure as a reef if the requisite organisms are not present. Other constraints are imposed from higher levels; fishes and invertebrates breed only when temperature, phases of the moon, or other conditions permit. Available food and its nutritional value may constrain population growth, and nutrient remineralization processes and quality of light may constrain plant growth.

That hierarchical levels influence one another becomes apparent by grouping behaviors into similar temporal and spatial classes. Many patterns of disturbance co-occur in time and space on coral reefs (Table 3.2), for example, in which dynamics at any one level are affected by what happens at other levels and by antecedent conditions. That is, each level in a hierarchy may act to stabilize or disrupt other levels, and prior events may set initial conditions for future events. Another way to view ecological hierarchy is by means of species assemblages that occur among associated habitats. For example, seascape patterns among coral reefs, seagrasses, and mangroves define scaled habitat relationships, which are useful for identifying hierarchies of biotic assemblages (Fig. 3.23). Assemblages within adjacent habitats are useful for elucidating community interactions.

Interpreting the scale properties of ecosystems, and their consequences, is among the greatest challenges faced by ecologists. The hierarchical order of ecosystems is not necessarily apparent by observation of individual parts, nor are hierarchies among different classes of components necessarily consistent (Fig. 3.24). Coastal-realm structure and function are direct consequences of the manner in which components and processes at different scales are organized and interact. The extent and frequency of ecological patterns and the processes that created them are most apparent at regional scales, but at small scales the variation may be so great that chaos results. That is, no pattern may be apparent when many small areas within the same habitat are examined, due to small-scale differences in life histories and sensitivities to local disturbances. Thus, the structure of most natural ecosystems is determined by a scaled array of historical events, life histories, and physical and biological processes that act over a variety

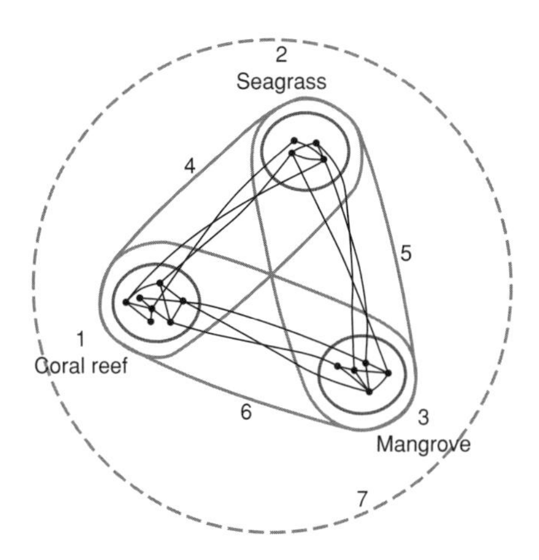

Fig. 3.23 Food-web associations among coral reefs, sea grasses, and mangroves result in a three-tiered hierarchy of relationships. First, some species inhabit any one of the three systems (1, 2, 3), e.g., certain coral-reef gobies and blennies. Second, other species require two of the three systems during their life histories (4, 5, 6), e.g., grunts (Family Haemulidae) that rest in daytime on coral reefs and feed during the night in seagrasses. Third, some species are ubiquitous (7), e.g., barracudas. From Ray & McCormick-Ray (1992).

of temporal and spatial scales, and are most recognizable and more predictable at larger scales.

3.7.4 Equilibrium, persistence, and resiliency

Henry Louis Le Châtelier's (1850–1936) basic principle about a system in equilibrium is that when a stress is brought to bear, the system tends to react in such a way as to achieve a new equilibrium state. In biological systems, this is called "homeostasis." In ecosystems, homeostasis depends upon feedbacks among components and properties that serve to achieve a dynamic equilibrium between order and disorder, growth and dissipation. That is, the ecosystem is presumed to exist in a reasonably stable – yet dynamic – state or condition.

Living systems are subject to change, day to day, year to year, decade to decade, with responses to disturbances that are usually non-linear, and with uncertain outcomes. In biotic communities, numbers of most populations do not fluctuate so greatly that the ups and downs drift preponderantly in one direction or the other, until the species either dies out or entirely overruns its environment. Rather, populations tend to

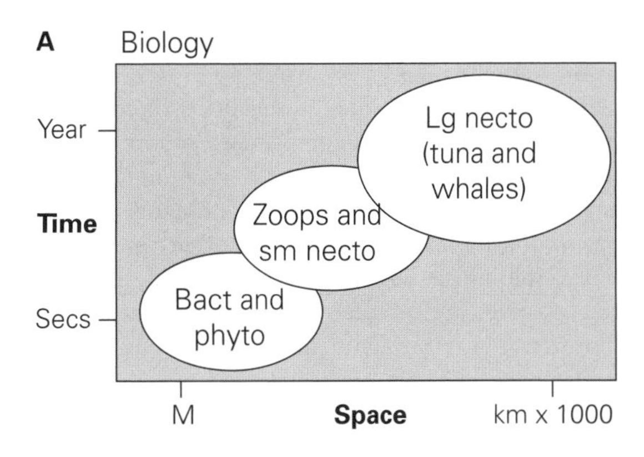

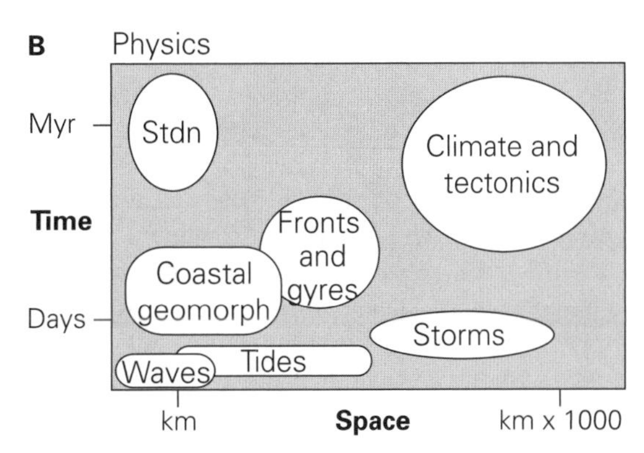

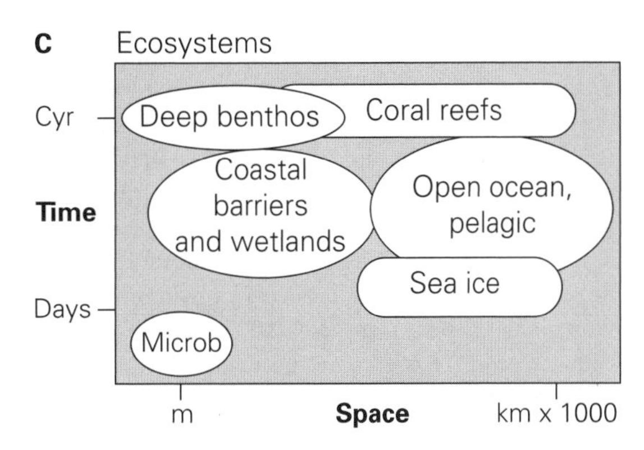

Fig. 3.24 Biota, physical properties, and ecosystems appear incongruous when spatial and temporal scales are compared. Sizes of biota are relatively linear, from small primary producers to higher consumers. Physical relationships are highly variable, as are ecosystems. These inconsistencies make cross-disciplinary studies difficult. From Ray & Hayden (1992), fig. 21.7, with permission.

Bact = bacteria
Zoops = zooplankton
Sm necto = small nekton
Lg necto = large nekton
Stdn = sedimentation
Geomorph = geomorphology
Microb = microbes

fluctuate or oscillate around quasi-equilibrium, so long as the habitat retains its essential features, and competition or predation are not fundamentally altered. This assumes that the interactions among organisms and their environments lead to some kind of mutual identity, and that rates of immigration and extinction are monotonic functions of the number of species present. The result is a predictable number and kind of species for any given area – at least theoretically. But in fact, persistence or resiliency of species, communities, and ecosystems are rarely predictable and rarely, if ever, stable in the literal sense. However, when living components are observed as individual units, each may be seen to respond to perturbations in a unique way, affecting or being affected by others.

Homeostatic feedback mechanisms serve to maintain biological systems at all levels of organization. Should ecological organization become perturbed, individual species may vary in numbers, but the system tends to return to its characteristic range of conditions. Persistence is a measure of how long an ecosystem may endure in a quasi-equilibrium state, and resiliency is a measure of how fast an ecosystem that has been displaced from quasi-equilibrium returns to it. If a deviation occurs in one direction, negative feedbacks force the system in the opposite direction and the system maintains its identity. In the case of positive feedbacks, responses to an initial deviation act to reinforce the change in the direction of the deviation. Homeostatic mechanisms no longer operate. Amplification can then occur and the system can be driven into a new regime or state. For example, when sediment turns water turbid, light is reduced, photosynthesis is impeded, and plants die. When rooted plants that stabilize sediment die, more sediment occurs in the water column, further reducing light, more plants die, and so on. This positive feedback process has become chronic in some estuaries, especially where dredging disturbs the benthos.

3.7.5 Emergence and complexity

A view of the world as a hierarchy of interacting, input–output systems with homeostatic feedback mechanisms means that ecosystem equilibrium is appropriate only in a dynamic context. The dynamic equilibrium concept of ecosystems implies a steady state, with reference to a point, or identity, about which fluctuations occur. Disturbance, discontinuities, and disorder threaten this identity, as instabilities open new opportunities, in which case ecosystems may bifurcate in new directions. Thus, non-equilibrium thermodynamics theoretically explains the appearance of high-level order from low-level chaos (Prigoqine 1980).

Emergent properties arise from feedbacks among a complex array of components and processes. Consider water as an analogy: if one knew all about the properties of hydrogen and oxygen, one could not predict water's properties. Similarly, emergent ecosystem properties cannot be predicted from studies only of individual components. Emergent properties have allowed the coastal realm to be extraordinarily productive and resilient – within limits! Anticipating the outcomes of emergent properties remains a major challenge for both science and conservation.

New levels of organization are the fingerprints of complexity. Ecosystems constitute a special type of system, designated by Weinberg as "organized complexity" (Box 3.2). Complexity increases as the number of components in a system increases, and correspondingly, as the number of variables and parameters in a mathematical model describing the system increases. The coastal realm exhibits a high degree of complexity because of its dizzying array of biodiversity, structural–functional components, and processes that occur at different spatial and temporal scales and frequencies. In this situation, physical laws, boundary conditions, and changes in biodiversity may help describe the outcomes of long sequences of probabilistic events, each of which could have turned out differently. It is precisely because of the high number of coastal-realm components and interactions and their variability that a concept of the coastal realm as an adaptive complex ecosystem is required in any attempt at management or conservation.

Coastal-realm ecosystem complexity reflects basic ecosystem concepts. First, ecosystems are far from equilibrium because they are open with respect to energy, matter, and information. Second, coastal-realm ecosystems are bounded, in the sense that their varying states cannot exceed limits imposed by thermodynamics and the characteristics of their biotic communities. Third, components of the coastal realm behave nonlinearly. Fourth, coherence and persistent features of the coastal realm suggest that complexity and self-organization are its major hallmarks. Adjustments to altered conditions involve process-linked, chain reactions of responses that appear to operate on the domino

Box 3.2 What is an ecosystem?

A "system" can be thought of as a construct of the human mind, and its delineation can be somewhat arbitrary. Weinberg (1975, 2001) conceived three types of systems (Fig.1). He characterized *small-number systems* as "organized simplicity." Examples include mechanical systems (automobiles) whose parts follow basic physical laws, e.g., Newton's laws. These laws explain interactions and predict outcomes by means of mathematical equations. At the other extreme are *large-number systems*, which display "unorganized complexity." Identical components are large in number, random in behavior, and yield to overall statistical averaging. An example is the random motion of molecules of a perfect gas, on the order of 10^{23} molecules. The individual motions of the molecules are virtually unknowable, as they collide and rebound unpredictably. Yet, their overall motions at a given temperature follow the predictive power of gas laws.

Ecosystems do not fit either of these categories. Rather, they are *medium-number systems*, characterized by structural and functional interactions among an intermediate number of components. Weinberg's term for these is "organized complexity." The components express a wide spectrum of process rates and spatial characteristics. Ecosystems are not easily amenable to either mechanical or statistical solutions. Some degree of abstraction is required to study ecosystems, but they are not merely abstractions. Rather, they are best understood as organized units of nature. A key to understanding them is to make their space- and rate-dependent organization and linkages explicit. That is, ecosystems cannot be understood on any single spatio-temporal scale, and no single type of observation can be extrapolated to define the nature of the underlying processes.

Ecosystems are in constant change, whereby they may be said to lose identity, exhibit trajectories, or exist in a state of dynamic equilibrium. The systems that exist at any one time have been selected from all systems of the past; they are the best "survivors." System survival thus refers to the length of time a system exists and its evolutionary history. Thus, an ecosystem's identity becomes synonymous with its resiliency and viability.

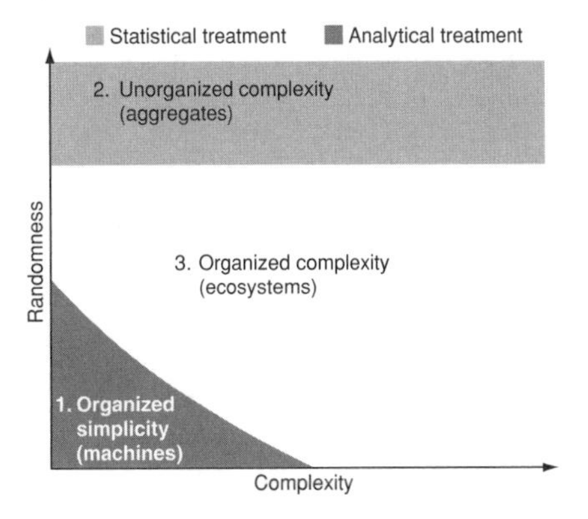

Fig. 1 Three types of systems differ in degree of complexity and organization. (1) Organized simplicity: small-number systems contain so few components that analytical treatment applies, e.g., machines. (2) Unorganized complexity: large-number systems contain so many randomly interacting identical components that their statistical properties appear to be deterministic and statistical treatment can be applied, e.g., molecules in a gas container. (3) Organized complexity: medium-number systems, which contain an intermediate number of components interacting non-randomly, cannot be analyzed by traditional methodology and often appear to be stochastic. From Weinberg (1975; 2001), fig. 1.9, with permission from Dorset House Publishing Co., Inc. (all rights reserved).

Sources: O'Neill et al. (1986); Weinberg (1975, 2001).

principle, meaning that the changes that occur in one component during an adjustment period often initiate responses among other components. A myriad of effects may cascade with removal of key species or structural features – for example, removal of oyster reefs from estuaries or urchins from tropical reefs, or massive removals by fisheries, all have resulted in major structural and functional ecosystem changes. Linkages can also act in reverse, such as when non-native species are introduced and break long-adapted ecological relationships.

3.8 Conclusion

The coastal realm is a major subdivision of the Earth. Viewed hierarchically, physical and biotic attributes acting over short and long time spans have resulted in a global ecosystem exhibiting organized complexity. Knowledge of coastal-realm interrelationships, natural histories of critical species, and mechanisms of physical transport and storage is essential for interpreting the outcomes of natural or human-caused change, and for coping with cascading effects.

The perception of the coastal realm as a global ecosystem is fundamental for coastal-realm conservation and management. Attributes, components, and properties of ecosystems are interdependent, and changes in any one have the potential to alter the entire ecosystem. History shows that coastal-realm ecosystems are particularly resilient to natural change, but the extent to which they can recover from massive unnatural (human-caused) change is not readily predictable. Alterations of the Earth's arterial system, its watersheds and rivers, have dramatically affected delivery of freshwater, sediment, nutrients, and biotic materials – as well as pollutants – to the coastal ocean and presently exert major influences on biotic and abiotic processes. Extraction of living resources, pollution, and other human activities have been equally effective in producing change.

The complexity of observing the time–space attributes and functions of the whole coastal realm has forced compartmentalized approaches. Crucial questions are: Can the compartmental approach be overcome? What functions do various species contribute to the whole? And, are presently altered states of coastal-realm ecosystems reversible?

Chapter 4

Natural history of coastal-marine organisms

Not only are biological systems staggeringly complex, they have an evolutionary history shaped by chance and natural selection, which is both stochastic and blind to future consequences . . . Indifference to a phenomenon's natural context can result in a paralyzing mismatch between the problem and the questions put to it.

G.A. Bartholomew (1986)

4.1 Introduction

Biodiversity, at the simplest level, is concerned with the variety of species and their evolutionary relationships. A second level concerns natural history – feeding, reproduction, symbioses, adaptations to habitat, and the like. The third level concerns how species assemble to form functional communities, how these communities are distributed in time and space, and how they contribute to ecosystem function.

The Convention on Biological Diversity (see Section 2.3.1.3) is targeted toward all of these levels, but approaches depend on the user. For ecologists, ways that species function help explain ecosystem pattern and process. For conservation biologists, interest is on community persistence, resistance, resilience, and vulnerability to disturbance. For managers, maintenance of particular species is a mandate, mostly for those of commercial or social value. For social scientists, biodiversity is seen as a social "good." For businessmen, species and habitats may represent economic value, biotechnology, trade, or tourism – or, conversely, an impediment to development. For politicians, biodiversity can represent conflicts between nature preservation and social or economic welfare. For all, the key to understanding how life manages to survive is natural history.

4.2 Diversity of coastal-realm life

Conservation typically focuses on individual species. Therefore, the first task is the accounting of species and their status. Most coastal-realm species are small and poorly known; most remain to be discovered and many known species have yet to be named. But, more fundamental are the questions: Why have so many species evolved? And, in what ways have species accommodated to the environmental heterogeneity and dynamics of the coastal realm?

4.2.1 Taxonomic diversity

Taxonomy examines species' genetic and morphological characters to produce a classification that reflects evolutionary history. The most widely accepted method uses Charles Darwin's criteria of inheritance (geneology) and similarity (morphology) to describe phylogeny (relationships). "Systematics" is often used interchangeably with "taxonomy," but is a broader term that integrates all attributes, including physiology and behavior, that contribute to understanding of relationships. The resulting classification is hierarchical (Table 4.1).

Hierarchical taxonomic grouping suggests evolution from common ancestors. The taxon of phylum is most

Table 4.1 The taxonomic hierarchy of the "true" herrings. Subgroups (subclass, subfamily, etc.) are not included.

Taxon	Example	Definition
Kingdom	Animalia	The most general level; includes all animals
Phylum	Chordata	Major animal subdivision
Subphylum	Vertebrata	Animals with backbones
Class	Osteichthyes	Bony fishes
Order	Clupeiformes	Herrings and anchovies
Family	Clupeidae	Herrings
Genus	*Clupea*	Herring
Species	*harengus*	Atlantic herring
	pallasi	Pacific herring

Adapted from Nelson (1984); Robins et al. (1991).

basic. At the other end of the spectrum, the taxon of species includes individuals that form one or more interbreeding populations and produce fertile offspring. Species' populations are the essential units of evolution. New species result when sufficient genetic change occurs within one population of a species that it is no longer able to breed with other populations; this usually occurs when species' populations become separated by physical boundaries.

Genetic techniques are increasingly being used to differentiate species. DNA (deoxyribonucleic acid) is a complex organic molecule that occurs in all cells. DNA encodes genetic information in genes, the units of inheritance. Genetic variation and recombination during sexual reproduction provide a basis for variation among individuals. The problem is where to draw the line; that is, at what point does a population become sufficiently different from its ancestral one that it deserves a different taxonomic identity? Another problem is that the species concept is derived from higher animals and plants, and may be invalid for microbes, which have complex sexual and asexual exchanges.

That a consistent definition of "species" is probably not possible for all forms of life is not surprising. The living world is immensely richer, more diverse, and more dynamic than classification suggests. In 1866, Ernst Haeckel suggested that the tree of life had only two main branches, plants and animals, and that some Protozoa-like species, such as flagellates, exhibit both plant and animal characteristics. Most scientists now recognize five kingdoms of life (Table 4.2). Two are "microbes:" the Bacteria are the most ancient; the other is the extremely diverse Protoctista, which probably arose by symbiogenesis, in which genetic material of two or more bacteria combined to form new organisms. Many protoctists are so different that many distinct kingdoms have been proposed.

4.2.2 Why so many species?

One of the first questions biologists ask is: How many species exist? About 2–3 million species have been given scientific names. However, scientists estimate that fewer than a tenth of existing species have been described. The proportion of scientifically named marine species is much less than for land, reflecting logistic difficulties. Assessing marine life is a formidable challenge; the ocean depths average about 3800 m, so that its living volume exceeds the thin skin of terrestrial life by 100 : 1. Recent finds hint that many more species remain to be discovered (Box 4.1).

Environmental controls that regulate the numbers of coastal-marine species are poorly understood. One factor is the physical size of organisms. Large size limits the number of individuals, and therefore the number of species, that may be present in a habitat (there is only so much space). Large size also influences energetics (there is only so much energy to share). Food-web studies indicate that each species interacts with a limited number of others, theoretically placing an upper limit on species numbers. Relative abundances may be a further factor; usually a few "dominant" species account

Table 4.2 The five kingdoms. The total numbers of species are not known for any kingdom. Estimates of total species numbers vary from about 5 to 60 million. The Animalia appear to be most phyletically diverse and speciose.

Kingdom	Description
Bacteria (14 phyla)	Prokaryotes (lack membrane-bounded nuclei); all major modes of metabolism represented (autotrophs, heterotrophs, chemosynthesizers); reproduce non-sexually by binary fission; conjugation to pass genetic material is not associated with reproduction; e.g., bluegreen algae, chemobacteria, etc.
Protoctista (30 phyla)	Microorganisms with membrane-bounded nuclei; evolved from symbiosis of at least two different bacteria in which different characteristics emerged; remarkable variation in cell organization, cell division, reproduction, nutrition, and life cycles; e.g., microalgae, flagellates, etc.
Animalia (37 phyla)	In form, the most diverse of organisms; diploid (two sets of chromosomes) as adults; embryos develop by fusion of haploid (one set of chromosomes) egg and sperm; most ingest nutrients through mouth; some acquire nutrients by absorption; a few acquire symbiotic photosynthesizing plants.
Fungi (3 phyla)	Form spores with chitinous walls; most have a sexual reproductive stage involving conjugation; lack embryonic stage; non-motile at all life-history stages.
Plantae (12 phyla)	Adults are haploid complementary sexes produced by spores; spores produced by diploid embryo retained in female, usually in flowering body; most adapted for life on land or in shallow water; vast majority are photosynthetic.

Compiled from Margulis & Schwartz (1998).

Box 4.1 A sampling of marine discovery

- The bacterium *Prochlorococcus* was described only a little more than a decade ago. It thrives at the bottom of the photic zone all over the world's oceans, and may be the most abundant organism on Earth. It was formerly known as an "unidentified green coccoid."
- Brown tides caused by a previously unknown protoctist led to the demise of the bay scallop industry of southern New England in the 1980s.
- Major estuarine fish kills have been associated with a previously undescribed "phantom" dinoflagellate (*Pfiesteria piscicida*) whose existence and identity were announced in 1992.
- Deep-sea, hydrothermal vent communities have been known for only about a quarter-century.
- The deep-water, 5 m-long, plankton-feeding megamouth shark (*Megachasma pelagios*) was discovered in the Pacific Ocean in 1976.
- In 1982, a tiny, benthic-dwelling invertebrate (*Nanaloricus mysticus*) was described, adding Phylum Loricifera to the catalog of marine diversity.
- A group of bacteria, discovered only two decades ago, was recognized in 2001 to be critical to the cycling of both organic and inorganic carbon in the ocean, necessitating redrawing planktonic food chains and reconsidering the extent to which the oceans act as sinks for atmospheric carbon.

for most individuals. Space has also been proposed to affect biodiversity. The concept of island biogeography assumes an equilibrium between rates of colonization and extinction, and that smaller islands (or isolated habitats) are able to support fewer species than larger ones. This concept has not been sufficiently tested for marine systems. In shallow-water environments, disturbances occur frequently enough to prevent equilibrium from being attained. Environmental heterogeneity may play a larger role than space. For example, even small coral reefs provide a great variety of microhabitats for life, that is, numerous nooks and crannies for invertebrates and fishes and substrate for sessile organisms.

Marine species' diversity appears to be highest in or near the benthos. A host of species lives within the sediment (infauna), on the sediment or benthic structure (epifauna), or near the bottom (demersal). Additionally, many benthic species support epibionts (species that grow attached to other species), which adds to benthic diversity. Most benthic fauna are small, and are classified by size. Microfauna are 0.001–0.1 mm (bacteria and protoctists are even smaller); meiofauna range from 0.1 to 1.0 mm; and macrofauna are greater than 1 mm. Megafauna are largest, and include oysters, crabs, sponges, anemones, and demersal squids and fishes. The megafauna are reasonably well known. However, high rates of discovery of small benthic fauna challenge the assumption that tropical forests and coral reefs are the most species-diverse systems on Earth. Unfortunately, insufficient data exist to confirm this.

Another measure of diversity concerns phyla. In this case, marine systems are by far the most diverse systems on Earth, as they host all animal phyla (Fig. 4.1). Furthermore, because the coastal realm includes both terrestrial and aquatic components, all plant phyla are also represented. Air-breathing terrestrial animals that evolved for life on land and have re-entered the sea have amplified marine diversity, possibly to take advantage of the high productivity there. Some re-entrants retain a land connection, breeding on land and feeding at sea – sea turtles, some sea snakes, sea birds, sea otters, polar bears, and pinnipeds. Others are entirely marine, surfacing only to breathe – most sea snakes, cetaceans, and sea cows. Very few insects have adapted to marine environments, while disproportionately contributing to terrestrial biodiversity.

4.2.3 Form and function

A species' form and anatomical features are the best indicators of its function. Evolutionary convergence and parallelism confirm the relationship between form and function. Convergence occurs when unrelated groups of organisms evolve similar morphologies to common advantage; for example, gills of mollusks (clams, oysters) and fine, filter-feeding mechanisms of fish and whales function similarly. Parallelism occurs when related species follow similar evolutionary courses; examples are claws among different orders of crustacea, or similar streamlined body shapes within and among many groups of fishes.

However, clarification about what constitutes an "organism" is necessary for clarifying taxonomic and functional relationships. Generally, organisms have (i) genetic identity, (ii) discrete body form, and (iii) physiological integrity and autonomy. Colonial animals (e.g., protozoans, corals, sponges, tunicates, etc.) are composed of many genetically identical (clonal) individuals that act together as a single organism. In some colonial sponges and coelenterates, different individual

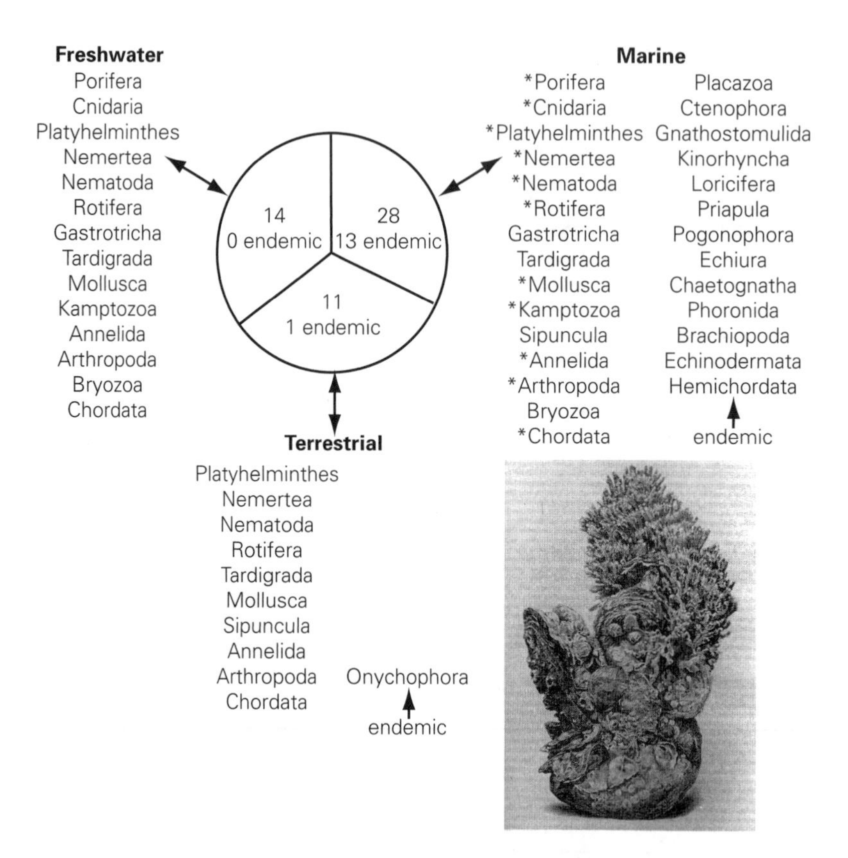

Fig. 4.1 Phyletic diversity is richest in marine environments and poorest terrestrially. Eleven phyla (*) have symbiotic species, and an additional four phyla (not indicated: Orthonectida, Dicyemida, Nematomorpha, and Acanthocephala) are exclusively symbiotic and often parasitic. *Note*: Some phyla are listed elsewhere under different names, and different classifications exist; as many as 37 animal phyla have been proposed (Table 4.2). Adapted from May (1988). *Inset*: Oyster patch. As many as 15 phyla may occur within a clump of eastern oysters (*Crassostrea virginica*) – more than for all terrestrial environments together. Drawing from Winslow (1882).

colony members perform different functions that, taken together, constitute an individual organism. Other colonial species are plastic in the form they express; that is, they take different forms under different environmental conditions. In other cases, individuals of a species may be morphologically identical, but are genetically distinct; that is, they are not one species, but several "sibling" species, as is the case for the common edible mussel (*Mytilus edulis*), which is at least three "sibling" species. This raises questions about the degree to which sibling species are functionally the same. Many cases of sibling species are being discovered, causing some scientists to estimate that if all such cases were to become known, species diversity would dramatically increase.

These problems can be partially solved by lumping species into functional categories, for example trophic levels that express feeding and energetic relationships. Food chains are trophically categorized into primary producers (autotrophs), consumers (heterotrophs), and decomposers that remineralize dead and decaying organic matter. Even this simplification can be misleading. Coastal-realm animals have a much greater size range than terrestrial animals, extending approximately 21 orders of magnitude in weight from the smallest bacteria at 10^{-13} g to blue whales at 10^{8} g. In general, larger animals consume larger prey, whereas smaller animals and plants are better able to assimilate nutrients through external tissues, due to higher surface-to-volume ratios. In aquatic systems, primary producers (autotrophs) may be small (phytoplankton) or large (for example, near shore where seagrasses and giant kelps form underwater forests). Marine consumers (heterotrophs) may be small or very large, and possess a great variety of food-finding features; some of the largest species (whales and whale sharks) eat zooplankton, and some protozoans are voracious predators. Mixotrophs, which can be both autotrophic and heterotrophic, are also common.

Form can also lead to non-intuitive conclusions about function. Walrus (*Odobenus rosmarus*) tusks are not used to dig for their favorite food, clams; huge whale sharks (*Rhincodon typus*) are docile and eat small plankton. Even for protozoa, form may not predict function. Some dinoflagellates are autotrophic, yet can switch to being fish predators (e.g., *Pfiesteria piscicida*). Others are symbionts with animals, for example coral zooxanthellae. Some multiply rapidly in the water column to form toxic red and brown tides (*Gyrodinium*, *Pyrodinium*), and still others can be deceptively toxic; for example, *Gambierdiscus toxicus* causes ciguatera poisoning, which can be fatal to humans.

4.3 Life in water

Coastal-realm organisms have evolved survival mechanisms very different from those required for terrestrial existence. Terrestrial life is confined to interfaces between atmosphere and land, marine life is exposed to the vastly different physical properties of water, and intertidal life is exposed to both conditions. Terrestrial organisms need strong skeletal support to resist gravity, while marine life is buoyed toward weightlessness. Terrestrial life faces desiccation and the combined effects of heat, wind, and low relative humidity, while marine life faces relatively low oxygen concentrations. In terrestrial environments, oxygen is usually abundant and temperature is less of a factor for oxygen availability. Marine life must also contend with dissolved chemicals, buoyancy, fluid motion, gas diffusion, and osmotic losses of body fluids. Intertidal life is subjected to submergence and air exposure, extremes of heat and cold, desiccation and drowning, salt- and freshwater, movement and collisions, turbulence, and wave and current forces.

4.3.1 Water as a control on life

Properties of water act in varying combinations to produce boundaries that can restrict marine life to physiological domains. Marine organisms are affected by water's temperature, salinity, density, and varying oxygen concentrations. Water conducts heat much faster than air; its high specific heat allows temperature to vary relatively little and much more slowly compared with air. Oceans absorb most of the Earth's heat; temperatures vary over a range of approximately 40°C the world over (exceptions are hydrothermal vents, which reach temperatures near the boiling point). Conversely, air temperature varies over a range of more than 80°C. In air, oxygen is usually abundant, with approximately 9 moles O m^{-3} at 10°C. In air, temperature is not a significant factor for oxygen concentration. In seawater, maximum oxygen concentration at 10°C is only about 0.35 moles m^{-3}, but at 30°C, oxygen concentration falls to around 0.2. Because cold water holds more oxygen than warm water, polar water is relatively oxygen-rich. High water temperatures occur near land in shallow tropical and intertidal areas, and these have low oxygen concentrations. Therefore, many tropical organisms live at or near their upper thermal and lower oxygen limits.

Water density varies with temperature, salinity, and depth. Density decreases with higher temperatures and increases with higher salinity. Undiluted seawater freezes at about −2°C; the freezing point for pure freshwater is 0°C. With depth, water temperature generally decreases, from the warm, solar-heated surface to deeper, colder, darker, denser water. When surface water is colder, fresher, and therefore less dense, the reverse can be true; that is, deeper water can be warmer. The zone of rapid vertical temperature change is known as the thermocline and an area of steep density change is the pycnocline. The water column is strongly stratified when the thermocline and/or pycnocline signal abrupt temperature or density differences. Both the thermocline and the pycnocline break down when the water column is turbulent and well mixed. Temperature, salinity, and density interact to create water masses that strongly affect the distribution of organisms.

Another property of water that creates boundaries is light, which is critical for photosynthesis and for vision. Light intensity just below the surface is at least halved, and decreases rapidly with depth. The clearest water transmits only about 1% of ambient surface light to depths of 200 m. Light penetration is strongly affected by turbidity, water density, turbulence, and the sun's angle relative to the surface, and is also affected by back-scattering from particles and planktonic organisms. In the euphotic (well lighted) surface zone, light is sufficient for photosynthesis and oxygen production, but below this zone few photosynthetic plants exist, and those that can exist have special photosynthetic

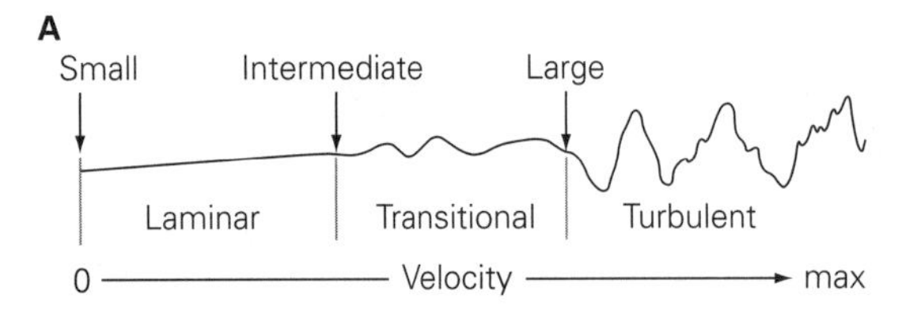

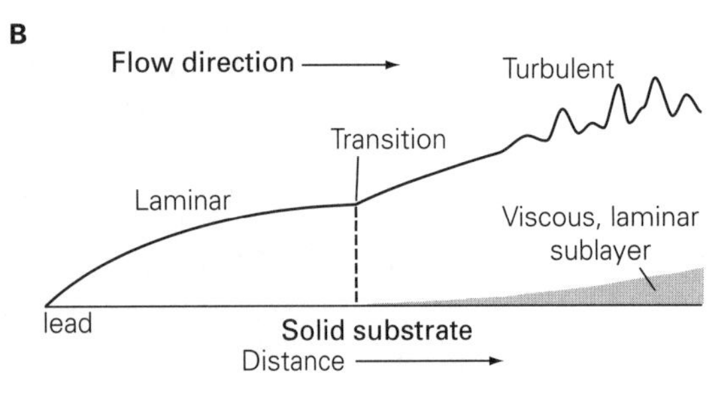

Fig. 4.2 (A) Streamlines in an unbounded fluid follow smooth paths at low velocity. As velocity increases, flow becomes transitional then highly irregular (turbulent). (B) Within the boundary layer, starting flow over a substrate is laminar. Over an increasing distance, the substrate retards velocity and a viscous, laminar sublayer is formed that increases turbulence in the boundary layer. Modified from Denny & Wethey (2001); Munson et al. (1994); Vogel (1994).

pigments. Where light does not penetrate and plant photosynthesis is not possible, microbial chemosynthesis may dominate production. And if respiration dominates the unlighted aphotic zone, especially during warm summers when oxygen concentration is low, hypoxia or anoxia can prevail, restricting or eliminating many forms of life.

Water density affects the coherence of water molecules and their "stickiness," that is, viscosity. Viscosity is a property of all fluids, and increases with decreased temperature. Seawater is approximately 50 times more viscous than air, and is about 25% more viscous at 0°C than at 20°C. Viscosity allows fluids to resist deformation, thus slowing water movement over an object or the sea bed (Fig. 4.2). Water is slowed the most nearest the object, forming a boundary layer with a turbulence gradient away from the object. Viscosity is also a property every organism feels when moving through water. Viscosity affects feeding, locomotion, and survival of organisms the size of phytoplankton, whose motility and inertial forces are very small. It is a problem particularly for small organisms, which are surrounded by relatively "thick" water. At sizes greater than about 10 μm, organisms are not strongly affected by viscous drag and generally can overcome its constraints. The viscous boundary layer around predators and prey complicates prey capture and affects feeding. If the inertia of a small organism is negligible relative to the viscous forces in the fluid carrying it away, it may escape capture.

Water is also a polar solvent, and dissolves more substances than any other liquid. When salts dissolve in water, they dissociate into charged ions called electrolytes. Due to dissolved carbon dioxide, carbonate, calcium carbonate, and other salts, the ocean is mildly basic (pH ~ 8.2). Electrolytic salts play a major role in sea-life distribution, creating physiological barriers that require adjustment of extra- and intracellular fluids through osmotic and ionic regulation. Salts of metallic elements – sodium, calcium, and trace metals such as iron, zinc, and nickel – are necessary for life, although exposures to high concentrations of trace metals can be toxic or cause physiological stress. Solubility of many metal complexes is dramatically affected by changes in pH, with acidification leading to dissolution of some metals (zinc, iron, aluminum, manganese) from sediments and particulates. Cellular metabolism (photosynthesis, respiration), organic degradation, and anoxia also affect pH. Because of aquatic life's requirement for oxygen, carbon, nitrogen, sulfur, phosphorus, and other elements, biological activity and elemental cycles are intimately linked to water's properties.

Water's most notable characteristic is that of a fluid in motion. Fluid behavior is complex, mostly nonintuitive, and involves dynamics, force, momentum, and energy, expressed in terms and concepts that relate to its motion: density, viscosity, streamlines, velocity gradients, drag, continuity, and a theoretical equation for water flow (Bernoulli's principle). These contrast with descriptions for solids: mass, elasticity, interfaces, shearing planes, friction, conservation of mass, and conservation of energy. Newtonian principles of force and momentum pertain to water in motion; force is the product of mass and acceleration, and momentum is the product of mass and velocity. Force causes a body to move, and inertia is a body's natural resistance to change, at rest or in motion. Once in motion, a net force is required to make a body move faster or slower, that is, to change momentum. In a collision between bodies, momentum is conserved. So long as the appropriate level of driving energy persists, turbulence is maintained and energy is effectively extracted by forming large eddies. In a general circulation pattern, energy is progressively dissipated through successively smaller scales, wherein acquired energy is discharged as heat. At molecular levels, the smallest eddies are overwhelmed by viscosity. Fluid motion in aquatic media thus covers a wide spectrum of scales, from general circulation in whole basins, to turbulence, and to small, viscous ranges of molecular motion. Behavior of many organisms, and almost all small ones, falls within the turbulent-to-viscous range of motion.

Water movement and forces of drag, acceleration, and lift strongly affect organisms. Flowing water tends to form streamlines that follow parallel paths. Under natural conditions, flows are almost always turbulent to some degree and streamlines break down, mixing the water column. Over solid surfaces or when encountering a non-streamlined object, streamlines become unstable and break down into turbulence, forming into random eddies that move in the net direction of flow, and velocity gradients form over the surface. The presence of an object, or organism, imposes a drag, measured as removal of momentum by that object from the moving fluid (Fig. 4.3). When water passes over a curved object, compressed streamlines accelerate velocity and produce lift. The size and shapes of organisms and their flexibility aid them in overcoming these complex, interacting forces (Fig. 4.4). The behavior of moving water is made even more complex by winds and tides, freshwater flows, basin geometry–topography, bottom depth and roughness, and more. As fluid momentum increases, drag, acceleration, and lift become too great, moving and toppling objects from their locations. Benthic organisms and sediment are moved, banks are eroded, and objects in the water column can be thrown into chaotic motion.

Finally, consider aspects of flow. Water is an almost incompressible fluid. If a volume of water is forced through a narrowing passage, its velocity must increase

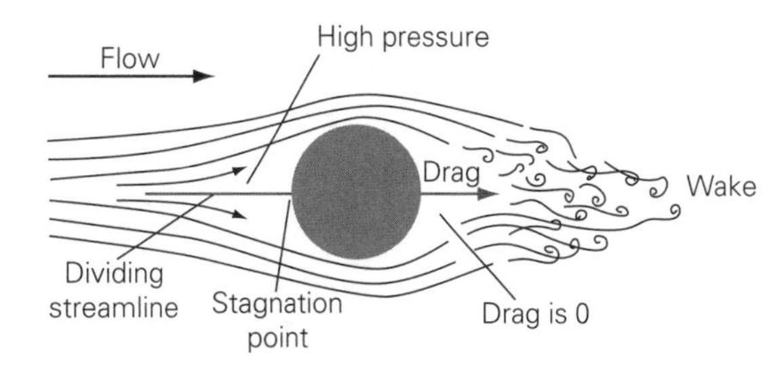

Fig. 4.3 Frictional drag results when an object interrupts flow. Streamlines separate upon encountering an object, compress over the object, and finally break down into a turbulent wake. Drag results from relatively high pressure on the object's upstream face and relatively low pressure in a turbulent downstream wake. This pressure difference tends to push the object downstream in the direction of flow. Modified from Munson et al. (1994); Vogel (1994).

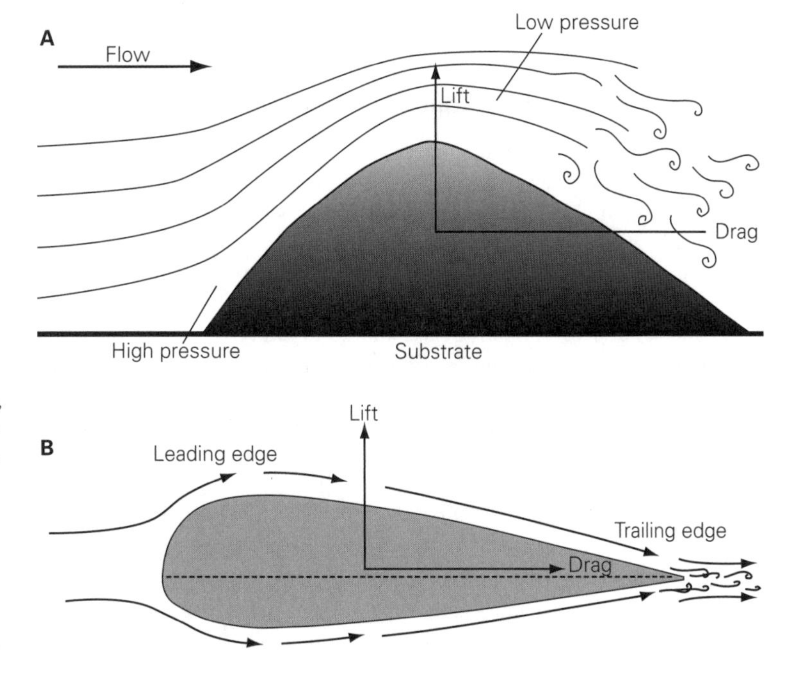

Fig. 4.4 (A) A limpet attached to a solid substrate experiences both lift and drag. According to the Bernoulli principle, when the pressure on one face of an object is greater than on the opposite side, a force called lift acts at right angles to the direction of flow. To avoid being moved, the limpet must cling tightly to rocky shores, where flows and turbulence can be strong. (B) A streamlined body with different curvature above and below experiences lift due to increased fluid velocity on the upper surface. Modified from Munson et al. (1994); Vogel (1983).

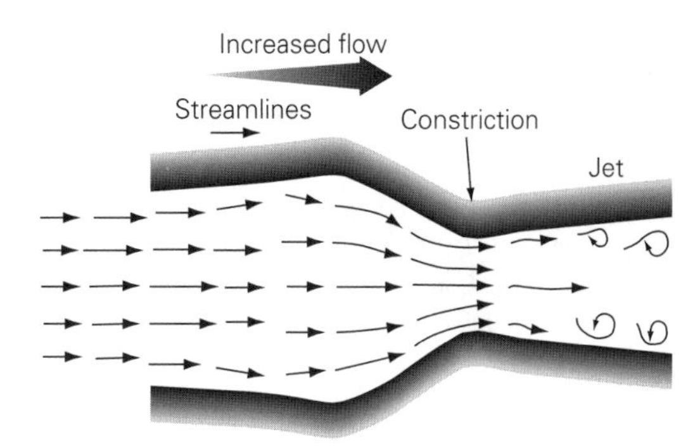

Fig. 4.5 The principle of continuity is illustrated by an analogy of water flow through a tube constricted at one end. Streamlines of fluid converge and flow volume does not change. As water enters the restriction, flow rate increases, and the emerging jet is turbulent. Modified from Denny (1988).

to allow that same volume to pass – the principle of continuity (Fig. 4.5). If a given volume of fluid enters at one end of a pipe, that same volume must exit the other end. The old London Bridge once rested on a set of boat-shaped piers so wide that almost half the Thames River was blocked, nearly doubling current speeds downstream. To protect the piers from scour, engineers constructed "cut-waters," which made the situation worse. Small boats entered at considerable risk in attempts to "shoot the bridge" (Vogel, 1983). When fishes approach channels, dams, and other blockages, they also face similar problems of flow.

4.3.2 Life adapts

Aquatic organisms live throughout the water column and contend successfully with water's properties and its constant motion. Life's responses (adaptation, behavior) can be immediate, happen during days or weeks (acclimation or acclimatization), or involve many generations (genetic). Behaviors vary from automatic taxes and kineses to complex social and community interactions. Sensory perception and communication make use of visual, auditory, and chemical signals, which may be stereotyped and genetically programmed or learned. Feeding cycles and timing of reproduction and migration often mimic major tidal cycles, solar days, and lunar months.

Avoidance behaviors are basic. Fish gather in large schools to confuse predators. Many bivalve mollusks avoid predators or noxious stimuli by immediately shutting their shells. Predator avoidance often results from sensing of pressure waves; for example, tubeworms withdraw quickly into their tubes when they sense alterations of pressure created by a predator's swimming motion. Some species turn pressure changes to advantage. Sharks possess sensitive organs on their heads to detect swimming prey. Some seals have recently been shown to follow hydrodynamic wakes by means of their sensitive whiskers, presumably enabling them to follow prey fishes.

Acclimation and acclimatization allow organisms to adapt to new or varying environmental conditions; acclimation refers to adaptation during a shorter time, acclimatization to longer times. Bivalve mollusks may respond to alien salinities by immediate closure of their shells. However, many marine organisms are physiological conformers, that is, they remain isotonic with seawater by changing the chemical composition of body fluids. Teleost fishes are regulators, maintaining their body fluids at one-third to one-fourth the salinity of normal seawater by osmoregulation; in full seawater, their salts are eliminated and water conserved, but in freshwater, salts are conserved and freshwater eliminated. A few reptiles have successfully adapted to marine environments; most have a salt gland with secretory capacity consistent with their seawater environment. Sea turtles can maintain constant body-fluid salinity in full seawater, but estuarine diamondback terrapins (*Malaclemys terrapin*) have salt glands that are intermediate in secretory capacity between those of terrestrial and fully marine reptiles. Many seabirds and marine mammals also possess mechanisms to eliminate excess salt.

Most organisms require oxygen and give off carbon dioxide as they respire. The high oxygen consumption of "warm-blooded" species (homeotherms: birds and mammals) precludes them from living without access to air. Marine "cold-blooded" species (ectotherms: invertebrates, fishes, and reptiles) have variable body temperatures, determined by their surroundings; thus, they demand less oxygen and derive their oxygen from water (with the exception of marine reptiles). Some fishes (tunas, marlins, and swordfishes) are fast swimmers, which necessitates high oxygen demands; some have developed special tissues that allow them to function as quasi-homeotherms. Many forms of sulfur-reducing and other bacteria, on the other hand, live without oxygen.

Notable adaptations of coastal-marine life concern survival, reproduction, and growth under conditions imposed by water's motion. Flatfishes experience lift under conditions of Bernoulli's principle (Fig. 4.6). As

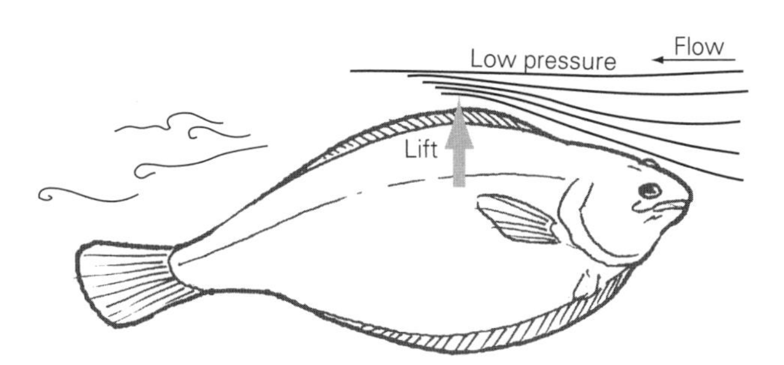

Fig. 4.6 A bottom-living flatfish lying flat on its side and exposed to flow experiences lift from pressures caused by fluid motion (Bernoulli principle). Modified from Vogel (1983).

reef communities grow, they interact with currents, turbulence, and the forces of lift, drag, and pressure (Fig. 4.7). On wave-swept shores, sea life must accommodate to moving water forces that commonly attain velocities of 14–16 m s^{-1} (> 50 km h^{-1}); when associated with large storms, wave speeds may reach 20 m s^{-1}. Although not nearly as fast as wind speeds, water's force is much greater due to its density, momentum, and viscosity. Organisms have evolved mechanisms to withstand such power. Algae and barnacles, for example, attach to rocky shores by holdfasts and cementation. Macroalgae anchor to the subtidal bottom, their flexible blades buoyed by air bladders that are exposed by water motion to light and nutrients.

Water's motion, turbulence, and other properties also strongly affect feeding, reproduction, and locomotion. Sessile organisms are totally dependent on water movements for delivery of food particles. Currents also aid dispersal of their larvae, remove wastes, and oxygenate benthic water. The continuity of water subsidizes energy costs, exemplified by sessile sponges that feed by using flagellae to draw water in through small incurrent apertures into an elaborate internal canal system that exits through one or a few large excurrent openings (Fig. 4.8). Some sponges pump water at rates many times that of a human heart. Many sessile organisms release eggs and sperm into the water column, where turbulence, eddies, and vortices facilitate mixing and enhance fertilization. Filter-feeding black corals and gorgonians living on seamounts, for example, benefit from accelerated flow and turbulence over rough surfaces which concentrate food particles.

Mobile organisms face different problems. Swimming is energetically costly. An organism can move only if it possesses sufficient energy to overcome resistant forces of drag, viscosity, and turbulence. Fishes detect water-pressure change by means of a lateral line system on their heads and/or sides, and move in a behavioral response (rheotaxis) to take refuge in eddies behind rocks and reefs, where a streamlined body helps reduce water forces (Fig. 4.9). Physiological, morphological, and behaviorial adaptations have allowed the pelagic yellowfin tuna (*Thunnus albacares*) and wahoo

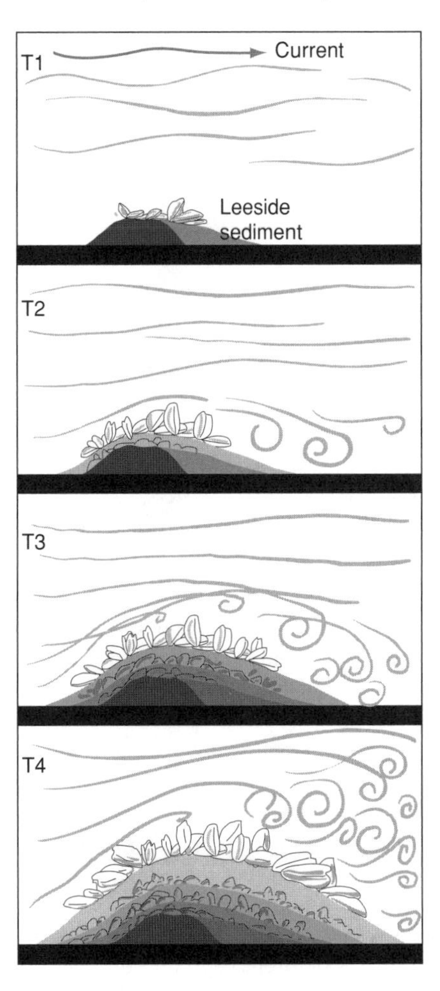

Fig. 4.7 Currents and benthos interact during growth of an oyster bed. As the bed grows over time (T1–T4), turbulence increases in the water column. Adapted from Kidwell & Jablonski (1983).

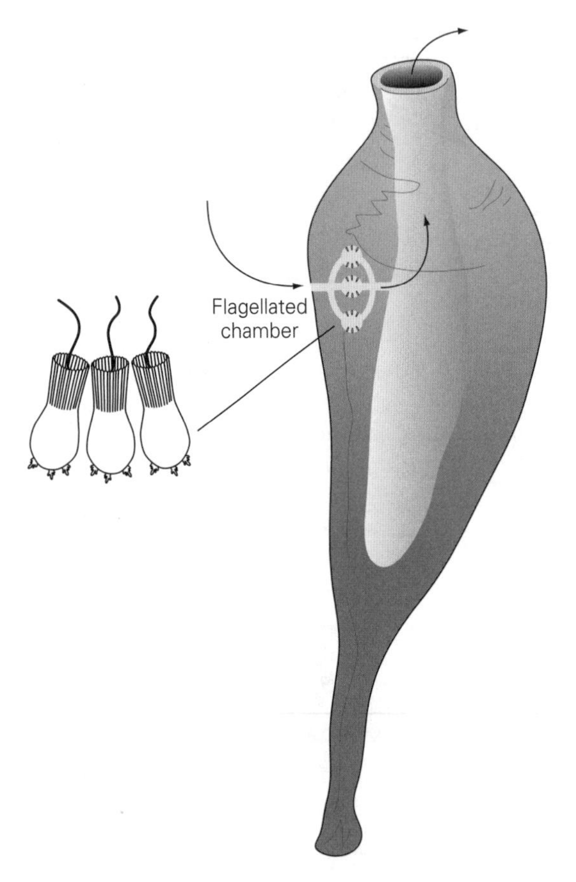

Fig. 4.8 A typical benthic sponge, with many small and one or a few large openings on its surface and an interior canal system, benefits from the principle of continuity. Small flagellated calls (exaggerated in illustration) every few seconds pump a volume of water through the larger chamber (simplified) equal to the volume of the entire sponge. Modified from Vogel (1983).

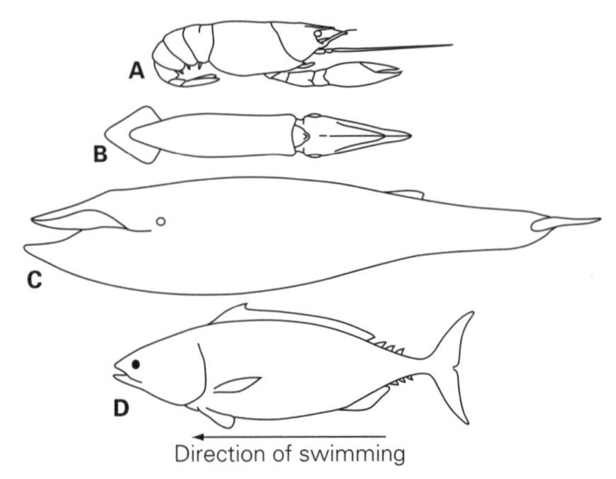

Fig. 4.9 Many swimming animals have evolved body shapes that reduce drag. Most swimmers, such as baleen whales (C) and pelagic fish (D), move forward; crayfish (A) and squid (B) swim backward. The usual method of locomotion of the crayfish is by crawling. Modified from Vogel (1983).

(*Acanthocybium solandri*) to attain speeds reported to approach about 20 m s^{-1} (~ 70 km h^{-1}), in 10–20 s bursts. Killer whales (*Orcinus orca*), reputedly the fastest marine mammal, may attain 15 m s^{-1}, in short-term bursts of speed. Some studies consider these speeds excessive; most recent estimates for cruising speeds of porpoises are in the order of only 8 m s^{-1}.

Water's density has resulted in many adaptations. Most organisms are denser (have higher specific gravity) than water, and must make adjustments to achieve neutral buoyancy. External appendages and internal devices such as oil droplets prevent or reduce sinking. Many phytoplankton species have oil droplets in their bodies or long skeletal spines to inhibit sinking. Algae are equipped with floats to hold their photosynthetic blades near the surface. Most fishes have evolved a membrane-enclosed swim bladder within the body cavity, which evolved first as a lung and secondarily became adapted as a hydrostatic organ. Also, because water is denser than air, sound travels about 10 times faster, therefore much farther, in water than in air. Many animals have taken advantage of water density by evolving mechanisms for acoustic communication. This allows many fishes and marine mammals to communicate for considerable distances in water, thus compensating for impeded vision caused by water's relative opacity. Some seals, walruses, and a few whales use stereotyped sounds (songs) for courtship purposes, similarly to birds, thus allowing communication over long distances. (Fig. 6.19, page 193). A few species of dolphins use sound production and reception for echolocation, as do bats; dolphins, however, use it over far greater distances due to water's density.

4.4 Life-history diversity

Understanding the nexus of natural history and ecosystem function is critical for biodiversity and conservation. Diversity is diminished when all life-history stages of a species are lumped under one taxonomic name. Species with direct development, such as marine mammals, have comparatively simple life histories, in contrast to most aquatic invertebrates and fishes which navigate through several life-history stages during their

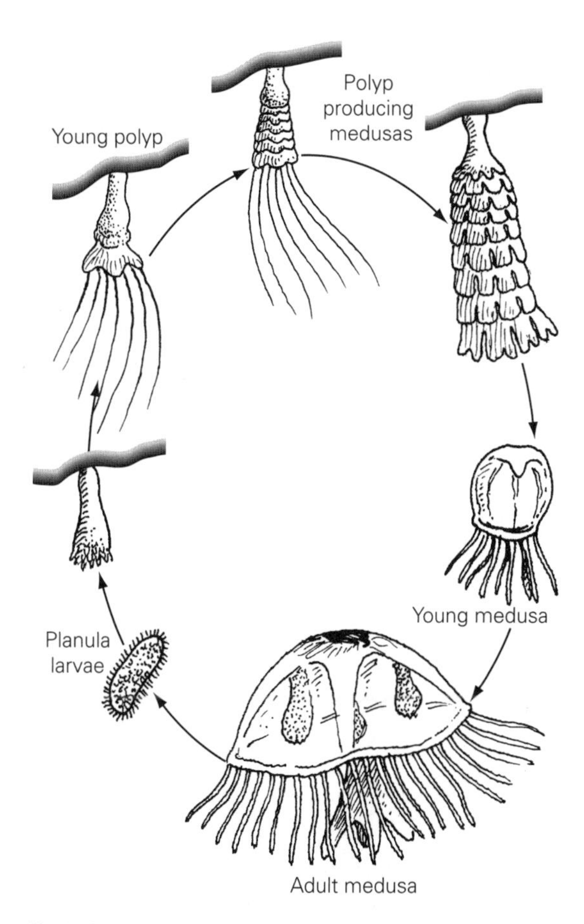

Fig. 4.10 Many coelenterates (jellyfish, corals, etc.) have complex life histories. The moon jelly (*Aurelia*) is taxonomically a single species, but exhibits several different stages. Its adult, sexual medusa and planula larva stages are pelagic; its polyp stage is sessile and asexually reproduces medusas. Modified from Buchsbaum et al. (1987).

lives. Each stage functions as though it were a different species, requiring different habitats for growing, feeding, reproducing, and larval settlement. These habitats may be physically far removed and temporaly distinct. Additionally, juvenile invertebrates and fishes sometimes resemble adults, but consume different-sized prey; their position in food webs is functionally different with age and different habitats are exploited. Consequently, life-history stages can be morphologically, behaviorally, and ecologically distinct (Fig. 4.10). In contrast, some large coastal-marine species have simple life histories, notably sea birds and marine mammals, whose young and adults lead very similar lives and often group together.

Different life histories involve constraints and biological trade-offs that optimize fitness, that is, increase each individual's chance of passing genes on to the next generation. Optimization of fitness involves a wide variety of mechanisms, more complex than is apparent from taxonomy, form, and natural history alone. G.E. Hutchinson (1965) visualized a species' "niche" as a hypothetical ecological hypervolume, not synonymous with habitat, which is the physical place where a species lives. A full portrait of an organism's niche includes requirements for space, food, physical conditions, and appropriate conditions for mating and other behaviors.

Coastal-realm species fall roughly into five categories:

- *structural*: sessile "engineers" that build biogenic structures, such as reefs, algal forests, and seagrass beds;
- *interstitial*: motile species that inhabit living structures and interstices or surfaces of abiotic rocky and sedimentary environments; that is, epifauna and infauna;
- *demersal*: free-swimming species that live near the bottom;
- *pelagic*: species that swim freely in the water column, sometimes for considerable distances;
- *symbiotic*: species that live in close association with other species; parasites, commensals, and mutualists.

Many species cannot strictly be assigned to these categories, as many life histories are a mix of these types; for example, larvae of structural species, such as oysters, corals, and crabs, are pelagic; pelagic species, such as jellies, have sessile stages, and some larvae are parasitic.

4.4.1 Feeding and food webs

Requirements for food connect organisms to their total environment. Due to the diversity of coastal-marine life-forms, feeding methods are more varied and food webs more complex than for other environments. Numerous environmental factors modify feeding behavior and food preferences. To acquire food, each organism must first find it, then employ mechanisms that override prey defenses, then possess appropriate processing mechanisms to extract materials and energy from it – all involving a suite of behaviors, morphology, and physiological mechanisms.

An early principle in ecology was that of linear food chains. Most plants are autotrophic, gaining energy from the sun and turning it into a usable form through photosynthesis. Some producers are chemotrophic and require fuel molecules as energy sources (H_2S, methane, ammonia, nitrite, sulfur, hydrogen gas, or ferrous iron). Most animals (heterotrophs) gain energy

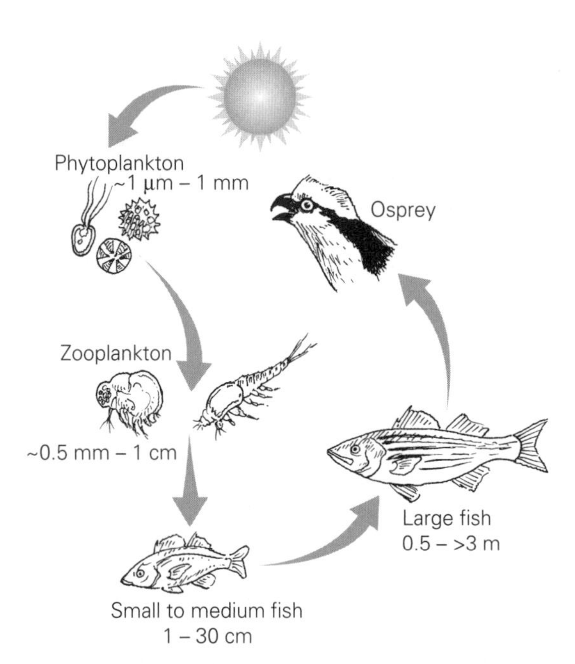

Fig. 4.11 A five-level food chain. Energy is initially passed from solar radiation to phytoplankton, with a transfer efficiency of about 3%. Up the chain, transfer efficiencies range from approximately 10 to 20% (generally more for ectotherms than for endotherms). Thus, most energy from the sun to the predatory osprey is lost up the food chain. However, intricate food webs and high overturn rates of primary producers and zooplankton increase efficiency, making predators more abundant than food chains predict.

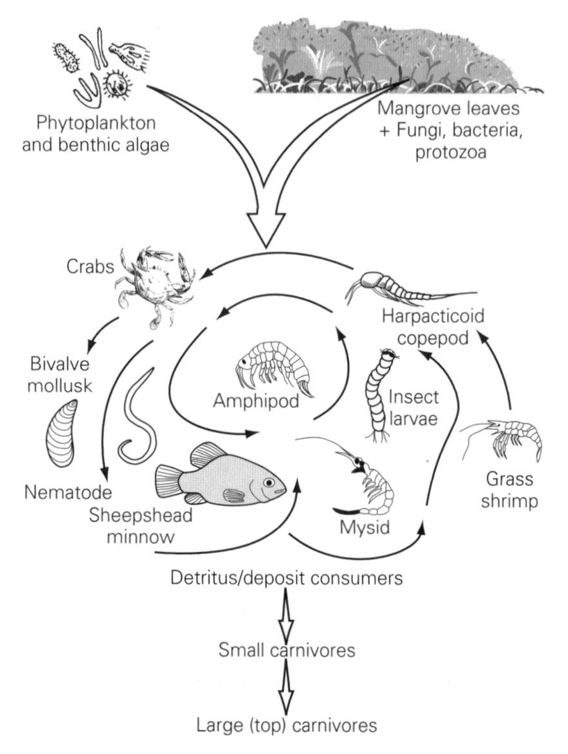

Fig. 4.12 Detritus particles derived from plants are cycled through inshore food webs when detritivores and deposit-feeders consume particles and expel them as feces. Particles and detritus are exported to small and large predators outside this system. Detritivores and predators and/or their larvae or juveniles also export organic nutrients to the system. From Odum & Heald (1975).

through food consumption. Consumers may pass energy through between two and five trophic levels to top predators (Fig. 4.11). Finally, decomposers (microbes, fungi) remineralize organic matter derived from dead organisms, which then becomes available again to producers.

However, this linear (or circular) trophic structure does not accurately reflect energy exchange. Rather, food webs exhibit a reticulate structure of feedbacks and connectivities involving complex pathways among plants and animals. For example, the detrital food web involves inputs from phytoplankton, algae, and coastal vegetation and contributions of materials and nutrients to offshore systems by means of a reticulum of interactions (Fig. 4.12). Furthermore, different life-history stages of the same species often consume foods that vary widely by size and taxon; pelagic larval fishes consume plankton, juveniles may eat small invertebrates, and adults may be top predators. Other species are omnivores, consuming both plants and other animals; the green turtle (*Chelonia mydas*) eats zooplankton during its pelagic juvenile stage, then switches as an adult to plants. Arctic ringed seals (*Phoca hispida*) may consume zooplankton or fishes, depending on sea-ice conditions, season, and water depth. The killer whale (*Orcinus orca*), a "top predator," consumes a variety of organisms, from fishes to seals, and even other whales, depending on season, location, and availability of prey. Food webs, therefore, illustrate that species can only rarely be assigned any one trophic level, as very few species exist as gustatory specialists.

Marine organisms possess a large variety of mechanisms to obtain nourishment. Microorganisms, many invertebrate parasites, and several free-living invertebrates engulf (phagositize) whole particles, sometimes larger than themselves, or absorb dissolved organic matter directly through the cell surface. The excretory plumes released from zooplankton can be an important source of nutrients for phytoplankton. Filter-feeding is an especially efficient mechanism used by many aquatic organisms, from sponges to whales. Several small invertebrates (e.g., copepods) produce feeding currents that aid in pulling particles into the mouth, where food morsels are identified, sorted, and eaten; the copepod's slow swimming speed and its use of feeding currents suggest that it has evolved a feeding procedure based on turbulent diffusion. Corals obtain dissolved food by a mutualistic association with zooxanthellae, but are also capable of predation. However, most consumers are herbivores that graze on plants, or predators that capture prey, or deposit-feeders that feed on detritus accumulated on the bottom. Predation is especially varied. Most predators engulf prey by using a variety of food-gathering mechanisms – claws and jaws, cilia, and predatory ambush – but some have evolved specialized mechanisms, such as mucus nets that trap and entangle prey.

Jellies, squids, fishes, seabirds, and marine mammals are dominant aquatic consumers due to their variety and biomass. They eat a wide variety of foods and transfer energy along food-web pathways and across ecosystem boundaries through migration, spawning, and death. Their feeding modes have evolved to optimize energy extraction. Three factors determine a successful search for prey: (i) relative mobility of predator and prey; (ii) size of the perceptual field of the predator relative to the density and size of the prey; and (iii) the proportion of attacks that result in a successful capture of prey. A variety of complex behaviors is used for successful search and capture. For example, humpback whales concentrate zooplankton by exhaling air to form bubble nets. Tunas and other pelagic predators have complex behaviors that allow them to capture fast-swimming, schooling fishes. The general rule is: "big fish eat little fish." However, sharks, some teleost fishes (blue fish, *Pomatomus saltatrix*), and killer whales are unique in their ability to tear flesh from prey larger than themselves. Similarly, herbivores consume large plants piece by piece, as when sea urchins graze on macroalgae or periwinkle snails consume saltmarsh grasses. A few snails and fishes seem to "cultivate" algal patches by means of selective feeding and exclusion of other predators.

Community pattern is affected directly and indirectly by food-web organization and variable connectivities among organisms. A predator can have an important "top-down" effect on community structure. Conversely, enhancement of primary production can have a "bottom-up" effect that amplifies consumer abundance. Observations of such effects have led to concepts of trophic cascades, in which strong connectivities among species influence properties of communities and ecosystems. Two sorts of cascades have been proposed. Species-level cascades occur within a subset of a community's species, for example when changes in predators affect prey, and changes in prey affect others. Species-level cascades are apparent in relatively simple systems, where a few species have strong effects on one another; an example is western Atlantic salt marshes, where crabs and terrapins limit herbivorous snails that otherwise overgraze saltgrasses (Fig. 4.13). Community-level cascades consist of changes among species or processes that substantially alter biomass distributions throughout an entire system. In both cases, strengths of species interactions are critical.

Both species and community cascades are abundantly represented in coastal-marine systems, but the effects of interactions among species on population dynamics, energy and material flows, and other ecological processes within whole ecosystems are poorly understood. When large numbers of species have multiple trophic relationships involving variable turnover rates of production, general predictions about how cascades may operate are usually precluded, a particularly difficult problem for fisheries management. Clearly, food-web dynamics are key to understanding coastal-marine community structure and ecological dynamics. However, simplistic concepts of linear trophic structures (producer to consumer to decomposer) are inadequate to explain how ecosystems operate.

4.4.2 Reproduction, survival, and recruitment

Reproductive rates, patterns, and life spans of organisms vary enormously among coastal-marine organisms. The life spans of many single-celled organisms and small plants and invertebrates are brief, with many generations a year. Some invertebrates, fishes, sea turtles, and mammals may take decades to double their numbers. Sharks and rays produce only a few, large offspring, and large whales and walruses usually produce only one offspring every two years.

Species face trade-offs between reproductive output and parental investment. Asexual reproduction by cell division or vegetative growth allows organisms to respond quickly to opportunities. Under optimal

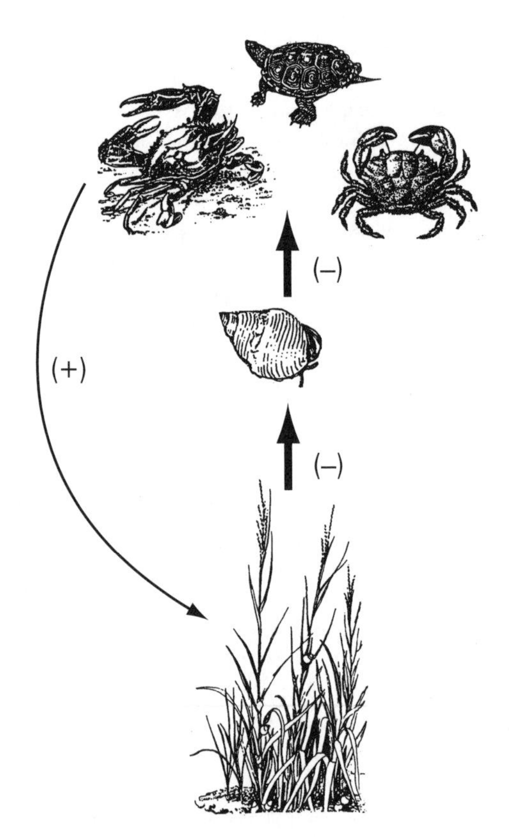

Fig. 4.13 Marsh-inhabiting predators – blue crabs (*Callinectes sapidus*), diamondback terrapins (*Malaclemys terrapin*), and mud crabs (*Penopeus* spp.) – exert top-down controls on western Atlantic salt marshes by controlling densities of herbivorous snails (*Littoraria irrorata*), facilitating luxuriant saltgrass growth. In the absence of predation, snail densities increase, resulting in runaway consumption of the marsh canopy and ultimately transforming the marsh into a barren mudflat. From Silliman & Zieman (2001). Diagram courtesy of B.R. Silliman.

conditions, bacteria can double population numbers in minutes. Some species produce large numbers of eggs and sperm, partaking in external fertilization. To enhance fertilization rates, many coastal-marine species rely on synchronized timing among individuals. Some species are cued by moonlight to aggregate and spawn during specific times of year. Most invertebrates and bony fishes produce small larvae that transform directly into adults, with little or no parental care. Others build nests, mouth-brood, or invest in parental care to ensure higher offspring survival. For sexually reproducing, long-lived species, age of maturity and/or first reproduction strongly influences rates of population increase. Different survival rates reflect the different ways that species meet environmental variables such as seasonal change, temperature change, food availability, rainfall, currents, winds, and lunar or day–night cycles. Where survival rates are low, each individual may have only a chance in a thousand, or a million, of surviving, necessitating high energy demands for egg production. Conversely, low egg or offspring production and parental care usually results in high survival rates, but involves high energy investment.

Various life-history strategies have evolved under complex and varied environmental conditions to maximize recruitment back into breeding populations (Box 4.2). Coastal fishes have evolved especially diverse reproductive strategies. Many are transient species that occasionally occur in estuaries, and others are estuary-dependent at some life-history stage (Table 4.3). Optimum habitats are critical at each stage of development, but factors that control recruitment operate on many time–space scales and often are so variable and unpredictable that they seem random. In cases where environments are ephemeral, such as shallow marshes, creeks, and nearshore ponds, some species produce eggs that are stored in a resting state in sediments. This egg bank functions as a time-dispersal system that allows rapid recolonization upon the appearance of an appropriate set of conditions. Similar cases occur among plankton that form spores then suddenly appear in the water column. Storage of reproductive products decreases the probability of population extirpation and increases recolonization rate when environmental conditions permit. This can be risky when generations do not overlap; a single year of failed reproduction can eliminate the population from the habitat. For the vast majority of species, even those of commercial importance, the precise factors controlling recruitment are speculative or unknown.

Reproductive strategies of re-entrant vertebrates (marine reptiles, birds, and mammals) display especially interesting cases of evolutionary adaptation. Such species face trade-offs between life on land or sea ice, where they breed, and water, where they feed. Most have evolved flippers for swimming, which restricts terrestrial locomotion, and which makes them vulnerable to predation when laying eggs (penguins, sea turtles) or bearing young on land or sea ice (pinnipeds). Eared seals (sea lions, fur seals) have retained terrestrial mammalian reproductive behaviors despite their need to dive for food and their ability not to breathe for periods of up to an hour or more. Thus, trade-offs

Table 4.3 Reproductive patterns of U.S. east coast fishes that enter estuaries. Only about a quarter of these species are estuary-dependent, that is, are obligate to estuaries at some life-history stage.

Group I. Facultative estuarine users: typically spawn in summer; nurseries either in estuaries or inner shelf; juveniles emigrate to deeper oceanic waters in fall.
Group II. Seasonal residents: adults migrate into estuaries to spawn in spring or summer; juveniles emigrate to ocean in fall.
Group III. Anadromous species: adults migrate through estuaries to spawn in freshwaters; young-of-year of most species emigrate before winter.
Groups IV. Early users: most species spawn over shelf in spring; at least some young-of-year stay a short time in estuaries, but spend the remainder of the year in shelf waters.
Group V. Delayed users: most species spawn over shelf in summer to fall; larvae spend first winter on shelf; enter estuaries in spring and emigrate in fall to shelf.
Group VI. Distant spawners: spawn in outer continental-shelf waters; young ultimately use estuaries as nurseries.
Group VII. Expatriates: larvae result from distant spawning; larvae arrive in estuaries in summer, but do not survive winter temperatures.
Group VIII. Summer spawners: resident species and the largest group; shallow-water species that spawn in summer; larvae develop in the immediate vicinity of spawning sites.
Group IX. Winter spawners: few resident species; spawn in the winter or spring.
Group X. Migrating spawners: resident species that undergo spawning migrations within estuaries.
Group XI. Unclassified: species for which some populations appear to be estuarine and other populations are not.

Adapted from Able & Fahay (1998).

between aquatic feeding and terrestrial breeding involve a remarkable suite of adaptations (Fig. 4.14). Some sea snakes, all sea cows (Sirenia), and all porpoises and whales (Cetacea) avoid these trade-offs by living their entire lives at sea.

4.4.3 Migration and dispersal

Migration and dispersal are means by which species seek appropriate locations for their life-history needs. Migration is a purposeful and repetitive movement of organisms for feeding and reproduction, by swimming, flying, and moving within or among water masses. Migratory distances vary, from transoceanic movements of whales, fish, sea turtles, and sea birds, to lobsters that take underwater "marches," and land crabs that march from mangroves to sea to lay eggs. Dispersal is a broader term. Mobile species may actively disperse to seek appropriate habitat at any life-history stage. For sessile species, dispersal is restricted to distribution of eggs and larvae by means of water currents. Both migration and dispersal influence population persistence, coexistence of species or genotypes, and genetic differentiation among local populations.

Dispersal plays key roles in demographics, survival, and colonization. If the site to be colonized is small or if dispersal leads individuals to unsuitable environments, recruitment may be unsuccessful. If suitable habitat is found, predation and competition by local residents may be too great to be overcome. Dispersal is important in small populations for maintaining genetic variability and preventing genetic drift. Individuals with greater genetic variation usually have greater fitness and better survival rates than those with less.

Environmental cues mediate migration and dispersal. Day length and temperature changes are well known examples. The Earth's magnetic field and water-mass attributes have also been implicated as cues to migration, and many species respond to lunar cycles. For example, California grunions (*Leuresthes tenuis*) migrate shoreward to spawn in beach sands only during highest tides; a lunar month passes for eggs to hatch and for larvae to develop before they are carried away by the next highest tide. The trade-off is that a month on the beach involves risks of predation. Endogenous timing (e.g., circadian rhythm) is also important, although how environmental cues and endogenous timing interact is often unclear.

Biological processes alone cannot account for migration and dispersal. All sea life is affected to some degree by hydrodynamics, for example residual currents, turbulence, gyres, winds, or river plumes. Ocean physics and the apparent continuous nature of the ocean has intuitively led to the notion that larval dispersal by major ocean currents is a form of passive dispersal over long distances. This perception is perpetuated by the term "plankton," derived from "planktos" (Greek for "drifting"), which implies that larvae are controlled by

Box 4.2 Life-history strategies of fishes

K.O. Winemiller

A remarkable diversity of reproductive strategies is observed among the fishes. For example, certain live-bearing sharks and the coelacanth (*Latimeria chulmnae*) produce one offspring at a time, whereas the ocean sunfish (*Mola mola*) releases over 600 000 000 pelagic eggs in a single spawning bout. Life-history strategies result from trade-offs among attributes that have either direct or indirect effects on reproduction and fitness. Comparative studies have yielded a robust pattern of fish life-history syndromes, with three primary life-history strategies defining the endpoints of a triangular continuum (Fig. 1). One endpoint, the *periodic strategy*, defines species that have delayed maturation at intermediate or large sizes, produce large numbers of small eggs, and tend to have short reproductive seasons and rapid larval and first-year growth rates. Another endpoint, the *opportunistic strategy*, characterizes species that mature rapidly at small sizes, produce relatively small numbers of eggs, and have long reproductive periods with multiple spawning bouts. A third endpoint, the *equilibrium strategy*, defines species that produce relatively small cohorts of large eggs or neonates, often in association with a long reproductive season, and have well-developed parental care.

Periodic strategists enjoy two benefits from delayed maturation and large adult body size: capacity to produce large numbers of eggs, and enhanced adult survival during periods of suboptimal environmental conditions. Periodic fishes often have synchronous spawning that coincides either with migration into favorable habitats or with favorable periods within the temporal cycle of the environment. These fishes cope with large-scale spatial heterogeneity by producing great numbers of tiny offspring, at least some of which thrive once favorable locations are encountered. Normally, early larval survival is very low in the marine environment. For the few fortunate larvae that encounter areas of high resource density, growth is rapid. At higher latitudes, environmental variation is cyclic. Periodic fishes exploit seasonal variation by releasing large numbers of progeny during periods favorable for their growth and survival. In tropical pelagic habitats, large-scale variation in space may represent a periodic signal as strong as the seasonal variation at temperate latitudes. As a result of upwellings, gyres, convergence zones, and other oceanographic features, physical parameters (salinity, temperature), primary production, and zooplankton densities are unevenly distributed in the open ocean. Massive cohorts of small pelagic eggs enhance dispersal capabilities of marine fishes during early life stages. Mortality due to settlement in hostile habitats (advection) is balanced over the long term by survival benefits derived from the recruitment of a certain fraction of larval cohorts into suitable regions or habitats.

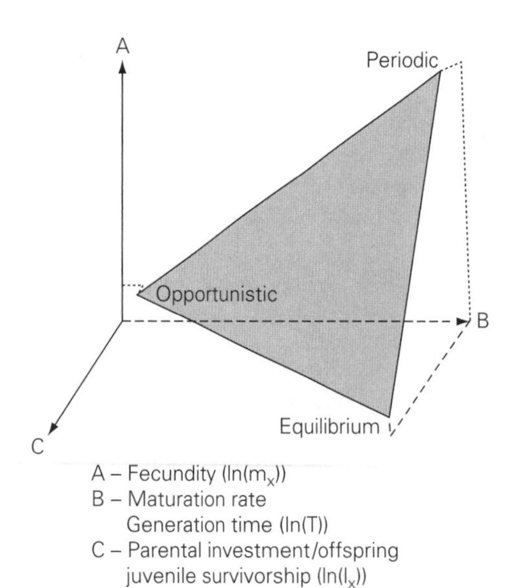

Fig. 1 Fish species life-history strategies reflect a trade-off between reproduction and fitness.

The opportunistic strategy yields a high intrinsic rate of population increase (r) and is associated with high population turnover. Although the size of egg cohorts tends to be small in these small fishes, reproductive effort is actually high because they reproduce early and often. In extreme cases, serial spawning results in an annual biomass of spawned eggs that greatly exceeds female body mass. Small fishes with early maturation, high reproductive effort, and high intrinsic rates of increase are efficient colonizers. These populations can quickly compensate for high adult mortality. The opportunistic strategy is observed in anchovies (Engraulidae), silversides (Atherinidae), and killifishes (Cyprinodontidae) – species often found in dynamic habitats or faced with high predation risk.

Equilibrium fishes have large eggs and parental care that results in larger, more advanced juveniles at the onset of independent life. Marine ariid catfishes (oral brooding of a few large eggs) and sharks, rays, and other live-bearing fishes with long gestation periods and large neonates provide extreme examples of the equilibrium strategy. Parental care seems to be more common in tropical nearshore and coral-reef fishes (e.g., pipefishes, seahorses, eelpouts, some gobies) compared with tropical pelagic and temperate marine fishes.

Of course, intermediate life-history strategies occur within the triangular gradient of life histories. For example, live-bearing is usually associated with few young, but rockfishes (Scorpaenidae) and other cool-water fish often have small numbers of large eggs. Also, divergent life-history strategies frequently co-exist in the same habitats. A species' ecological niche determines the individual's perception and the population's response to variation.

This triangular life-history continuum has a quantitative foundation. Fitness can be estimated by *r*, the intrinsic rate of natural increase of a population or genotype. The intrinsic rate of population increase can be approximated as

$$r \sim \ln(R_0)/T$$

where R_0 is the net replacement rate, T is the mean generation time, and

$$R_0 = \Sigma\, l_x m_x$$

In this equation l_x is age-specific survivorship and m_x is age-specific fecundity, resulting in

$$r \sim \ln(\Sigma\, l_x m_x)/T$$

Therefore, population growth rate depends directly upon fecundity, survivorship, and timing of reproduction. Averaged over many generations, the three parameters must balance, or the population eventually will decline to extinction or grow exponentially.

The three endpoint strategies result from trade-offs among age of maturation (positively correlated with mean generation time), fecundity, and survivorship. The periodic strategy corresponds to high values on the fecundity and age at maturity axes (the latter a correlate of population turnover rate) and low values on the juvenile survivorship axis. The opportunistic strategy of high *r* (*via* rapid maturation) corresponds to low values on all three axes. The equilibrium strategy corresponds to low values on the fecundity axis and high values on the age at maturity and juvenile survivorship axes.

Large body size in periodic strategists enhances adult survivorship during suboptimal conditions and permits storage of energy and biomass for future reproduction. The possibility of perennial reproduction represents a bet-hedging tactic whereby, sooner or later, reproduction coincides with favorable conditions that facilitate strong recruitment. Spawning tends to be periodic and synchronous, so that generations are often recognized as discrete annual cohorts that may dominate the population for many years. Correlations among parental stock densities and densities of recruits have been shown to be negligible in many of these fishes over wide ranges of parental abundances. Recruitment frequently depends on climatic conditions that influence water movement, egg/larval retention zones, and productivity, and on other environmental factors that determine early growth and survival. For periodic fishes, the variance in larval survivorship that serves as input for population projections lies well beyond our current measurement precision and accuracy. Even under pristine conditions, the fate of most larvae is an early death. Therefore, it follows that some minimum level of spawning must occur during each spawning period if strong cohorts are to develop during the unpredictable exceptional year. Management of periodic strategists requires maintenance of some minimum adult stock density so that periodic favorable conditions can be exploited, as well as protection of spawning and nursery habitats. Because recruitment is determined largely by unpredictable interannual environmental variation, this minimum density will be impossible to determine with any degree of precision.

Theoretical studies have shown that reducing mean generation time is the most effective strategy for maximizing the intrinsic rate of increase in a density-independent setting. Many opportunistic fishes are found in shallow marginal habitats, the kinds of environments that experience the largest and most unpredictable changes on small spatial and temporal scales. Tidal dynamics change water depth in shallow habitats such as tidal pools and salt marshes. In the absence of intense predation and resource limitation, opportunistic-type populations quickly rebound from localized disturbances, and these populations ought to show large variation in abundance, with infrequent strong density-dependence. Because they tend to be small and often occur in marginal habitats, opportunistic fishes usually are not exploited. Some important commercial fishes, like menhaden (*Brevoortia patronus*), are intermediate between opportunistic and periodic strategists.

The equilibrium strategy should be favored in density-dependent settings, and this may be why it is more common among coral-reef fishes than among estuarine and pelagic fishes. Compared with opportunistic and periodic strategists, equilibrium strategists tend to show moderate fluctuations in population density, and should conform better to stock–recruit models. Because equilibrium strategists produce relatively few offspring, early survivorship must be relatively high. Relatively few equilibrium fishes are commercially exploited on a large scale. Management of equilibrium fishes should stress habitat integrity and healthy adult stocks to promote surplus yields that can be harvested and replaced *via* natural compensatory mechanisms.

currents as if they were inorganic particles. However, evidence has accumulated that larvae possess behaviors that can, at least in part, determine their own dispersal. Blue crab (*Callinectes sapidus*) larvae depend on frontal systems and currents to enhance their dispersal from deep to nearshore estuarine water. Estuarine oysters may depend on residual currents to retain larvae in the estuary. These larvae undertake short vertical migrations or become incorporated into gyres to help direct their own dispersal. Turbulence may also act to

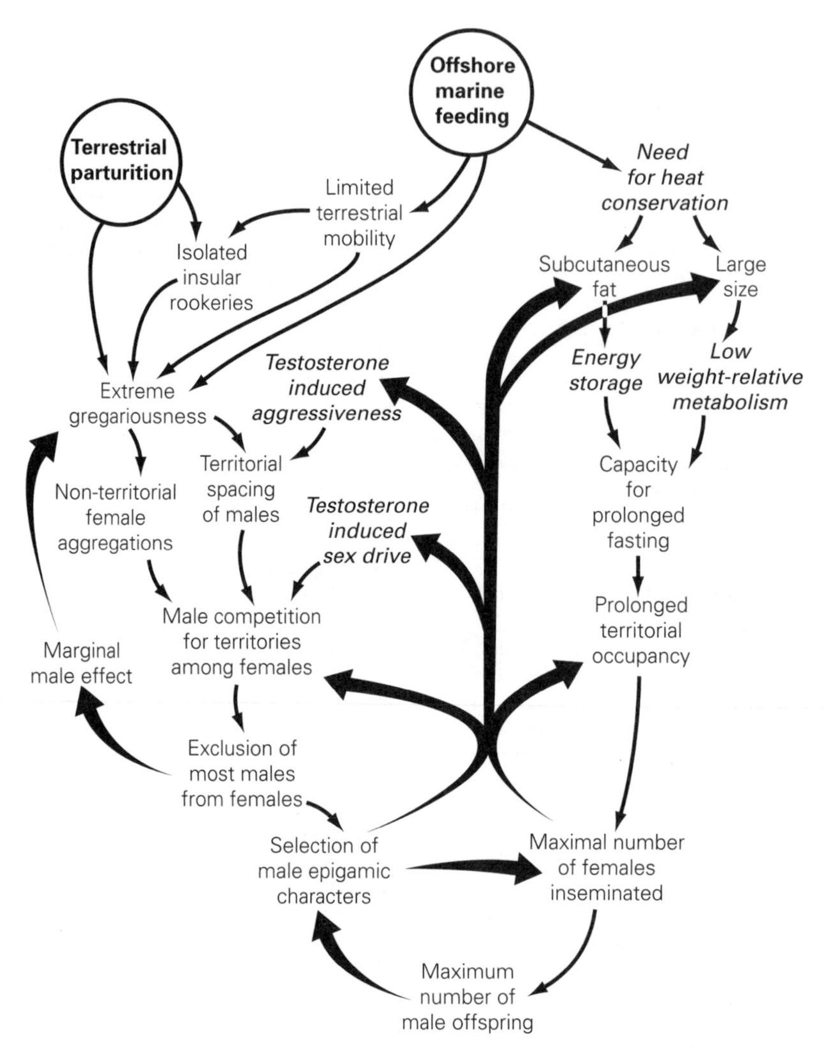

Fig. 4.14 Sea-lion life history combines terrestrial reproduction and marine feeding. Boldface italics indicates attributes or functions common to most mammals. Non-boldface indicates attributes and functions typical of polygynous sea lions (more than one female per male). Thin arrows indicate inputs and outputs. Broad arrows show positive feedbacks. This diagram combines demands of physiology, behavior, and environment, which have successfully been met by these animals. From Bartholomew (1970), with permission.

retain larvae in the same area as the population that produced them (self-recruitment). Passive transport could carry considerable costs to survival. Diffusion of passive particles in ocean currents and limited developmental time for most species of pelagic larva (days to a few weeks) suggest that many larvae may not survive long journeys to favorable environments. Rather, larval retention, which demands some degree of larval behavior, may be of greater importance for recruitment.

In short, one of the least understood aspects of coastal-marine life is planktonic larval dispersal. Whether organisms are passively dispersed in ocean currents or direct their own movement (self-recruitment) is at the state-of-the-art of natural-history research. In either case, survival depends on reaching an appropriate habitat. Both passive transport in currents and self-recruitment may be equally important, depending on the species.

4.4.4 Symbiosis

Competition and predation are dominant factors that

shape community structure. However, organisms also live together in symbiotic associations, and collaborate. Symbiosis takes three forms: (i) commensalism, when one partner benefits; (ii) mutualism, when there is a beneficial association between two or more kinds of organisms; and (iii) parasitism, which benefits one at the expense of the other. These associations are generally long-term rather than transitory (parasites that cause acute disease are an exception), and one or both partners usually have structural and/or behavioral adaptations that foster the association.

Symbioses, especially mutualism, are especially common among marine species. This relationship can be either facultative or obligate. In extreme cases, reproductive systems of the species involved are linked, so that continuation of the association is almost automatic. Many symbioses have trophic underpinnings. Several species of fish and shrimp gain food by relieving larger fishes of parasites at "cleaning stations." Mutualism also occurs between polyp-inhabiting, photosynthetic dinoflagellates (zooxanthellae) and many hermatypic (reef-building) corals. The zooxanthellae, which lend color to corals, gain protection from predation, while the corals benefit from nutrients derived from the zooxanthellae. Dinoflagellates also occur as symbionts in protozoans, turbellarians, and many other animals. Many sponges form associations with blue-green algae, and some anemones are colored green by chlorophyte symbionts. In spite of the great variety of potential animal hosts for algal symbionts, very few mollusks (the giant clam, *Tridacna*, is an exception) and no annelids, arthropods, or echinoderms have algal symbionts.

Whole biotic communities can form mutualistic associations. These communities consist entirely of living biota that use other plants and animals as substrates. For example, epiphytic algae grow on seagrasses, many mollusks and worms live on and within corals and other mollusks, and sponges and sea lilies (crinoids) host a variety of organisms. Sea lilies themselves are commensals on other organisms, such as sponges, corals, and octocorals. Some sea lilies even occur on other sea lilies. An especially striking example occurs on prop roots of red mangroves. These communities appear not to be random associations. Rather more species occur on the leeward sides of roots than on windward sides, and fronts of roots have more species than backs. Further, algae are most prevalent in well-lighted areas and on windward sites, while sponges and ascidians predominate in leeward areas. These observations indicate that larval supply may shape community composition, and physical factors may influence species distributions. To what extent the associations are obligate is unknown.

A form of mutualism occurs when a "death assemblage" affects a "life assemblage" by means of post-mortem feedback. Animal parts, such as living or dead shells, provide substrates that can have disproportionate impacts on species richness in benthic communities. Dead bodies of whales and shells serve as substrates for a wide variety of organisms. The former occur in the deep sea and may facilitate distributions of deep-sea animals. The latter form substrate for boring organisms and habitat for many others. Shells also provide refugia for opportunistic species and otherwise poor competitors. The initial accumulation of shells can result from either biological or physical processes (from mass mortality, gregarious behavior, delivery of allochthonous hard-parts, or seafloor reworking). Dead-shell substrate is important for adult survival and successful recruitment of some species. Many gastropods require hard substrate for attachment of egg capsules. Bodies of dead animals and shells also alter physical properties of sediment and distribution of epifaunal and infaunal organisms.

4.4.5 K–*r* selection

Some species are slow-growing, slow to colonize, and spend much time caring for their offspring. Others are prolific and rapidly colonize new areas. These two life-styles are referred to as K and *r*, respectively, or as equilibrium and opportunistic species. In the case of K-selected species, an essential feature is that populations respond to environmental influences under what is known as density-dependence. This means that as a species becomes more abundant, environmental resources (food, appropriate habitat) will increasingly be in short supply; a population feedback occurs in which reproductive rate declines. Conversely, if the species becomes less abundant, reproductive rate rises and population numbers increase. Populations of K-selected species appear to fluctuate around environmental carrying capacity, unlike *r*-selected species, which are not density-dependent. Thus, K and *r* adaptive complexes (Table 4.4) involve differing sets of interactions and feedbacks (Fig. 4.15).

The K–*r* adaptive complex cannot be separated into individual parts. For example, K- and *r*-species have often been differentiated on the basis of absolute body size, on the assumption that large body size buffers a

Table 4.4 The *r*–K adaptive complexes.

Correlates & attributes	*r* selection	K selection
Climate	Variable and/or unpredictable, uncertain	Fairly constant and/or predictable, more certain
Mortality	Often catastrophic, non-directed, density-independent	More directed, density-dependent
Survivorship	Juvenile survivorship low compared with adults	Juvenile survivorship high compared with adults
Population size	Non-equilibrium, variable in time; usually below carrying capacity, fills ecological vacuums, recolonization required	Equilibrium, fairly constant in time; at or near carrying capacity, recolonization not usually necessary
Intra- and interspecific competition	Variable, often lax	Usually keen
Selection favors	Rapid development, high *r*, early reproduction, smaller body size, semelparity (individuals reproduce only once)	Slower development, lower resource threshold, delayed reproduction, larger body size, iteroparity (individuals reproduce repeatedly)
Life span	Usually short	Longer

Modified from Pianka (1970).

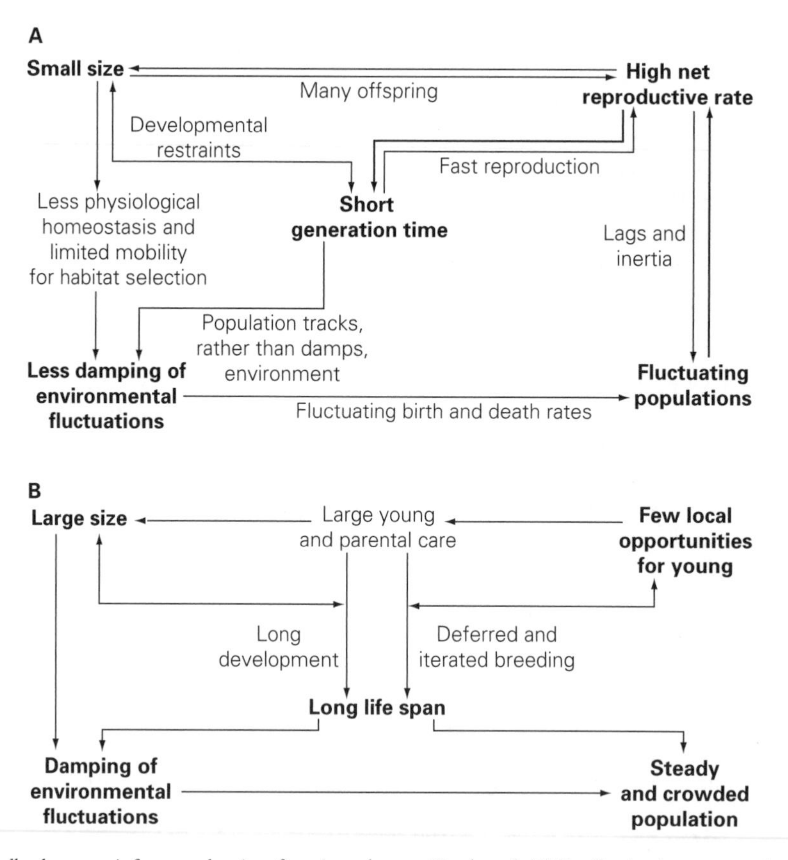

Fig. 4.15 Positive feedbacks may reinforce tendencies of species to be *r*- or K-selected. (A) Small animals are assumed to be more sensitive to environmental fluctuations and to suffer massive die-offs, leading to a premium on rapid reproduction and reinforcement of smaller body sizes and shorter generation times, thereby increasing vulnerability to environmental fluctuations. (B) If an animal is large, it may be more independent of environmental fluctuations, meaning that it will be less vulnerable to massive population reduction. Hence, these species have long life spans and reproductive rate may be deferred to parental care. From DeAngelis et al. (1986) after Horn & Rubenstein (1984), with permission.

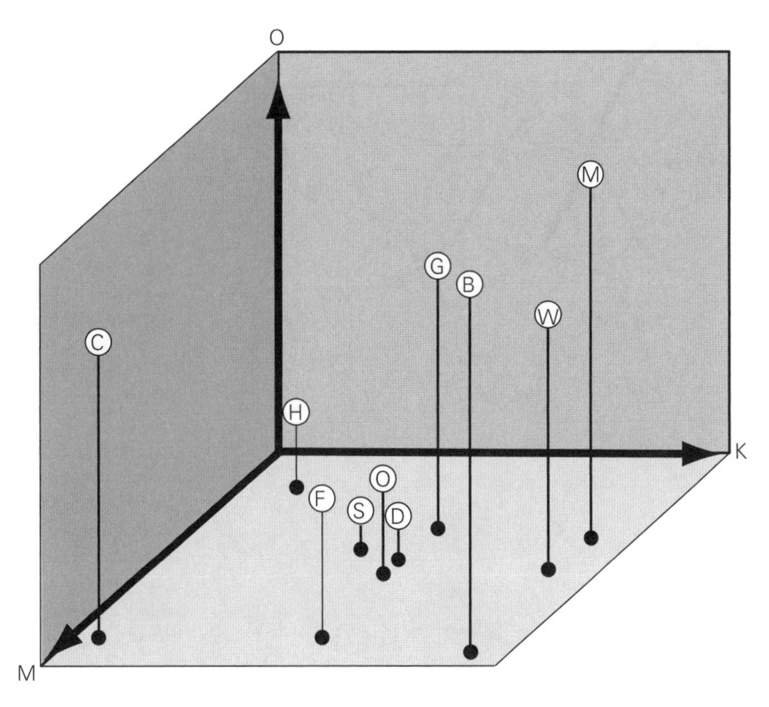

Fig. 4.16 Relative *r*–K positions of some marine mammal species. Three non-parametric axes illustrate: Food obligateness [O]; Maturation, reproductive rate, and/or length of life [K]; Environmental stability and predictability [S]. Within the box: [H] Harbor seals (*Phoca vitulina*) are small, live in variable environments, have facultative food habits, mature at a relatively early age, and bear young once a year; they are relatively *r*-selected. [B] Blue whales (*Balaenoptera musculus*) are very large, live in relatively predictable environments, are food restricted, mature late, are long-lived, and may bear young only every 4–6 years; they are relatively K-selected. Others are: [C] Crabeater seal (*Lobodon carcinophagus*) of the Southern Ocean; [W] walrus (*Odobenus rosmarus*) of Arctic shelves; [S] California sea lion (*Zalophus californianus*) of the temperate North Pacific; [F] northern fur seal (*Callorhinus ursinus*) of the temperate North Pacific; [O] sea otter (*Enhydra lutris*) of the temperate North Pacific; [D] belukha (white) whale (*Delphinapterus leucas*) of the Arctic–subarctic; [G] gray whale (*Eschrichtius robustus*) of the temperate to subarctic North Pacific; [M] bowhead whale (*Balaena mysticetus*) of the coastal Arctic. From Ray (1981).

species from harsh environmental conditions. However, large size is only one part of the adaptive complex, and can be misleading if comparisons are made among taxa with widely differing habitats and life histories. Also, low reproductive rate suggests that a population is density-dependent, therefore self-regulating. Insects are often assumed to be *r*-selected, as they are small, generally have short life spans, and reproduce rapidly. Conversely, marine mammals have been widely perceived as K-selected, due to their large size and low reproductive rates. However, marine mammals exhibit 4–5-fold differences in population doubling times. Rather than being assumed as K-selected in comparison to *r*-selected insects, marine mammals exhibit a range of K–*r* adaptations (Fig. 4.16). Similarly, species of insects are relatively K- or *r*-selected.

Whether a species is density-dependent or not strongly influences how it may be managed and/or restored. This also has ecological implications. Communities dominated by K-selected species appear more stable than communities dominated by opportunists. K–*r* concepts can also help determine community relationships. K-selected species, being strong competitors, have evolved for optimizing fitness; *r*-selected species can more readily overexploit resources. Unfortunately, little quantitative information exists on K–*r* relationships among coastal-marine species.

4.4.6 Phenology

Phenology refers to the duration, timing, and coordination of critical periods for a species' survival with reference to environmental conditions. Predators often compress into a limited area to feed, for example, when upwardly swimming plankton or small schooling fishes converge in windrows created by Langmuir circulation.

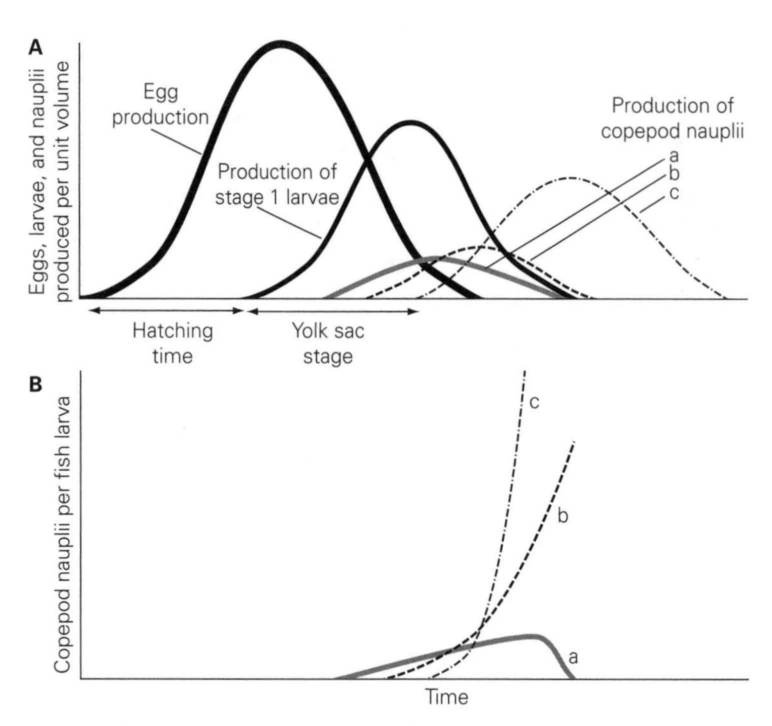

Fig. 4.17 An illustration of phenology. Larval fish production may be matched or mismatched in time with larval food (copepod nauplii) production. (A) Numbers of eggs varies in time. Numbers of feeding larvae vary with predation and as yolk sacs are absorbed. Survival of feeding larvae varies with predation and food availability under various conditions (a–c). (B) The ratio of nauplii to fish larvae represents the degree of feeding success. In A and B: a = an early production cycle in which food is available early and increases at a slow rate, and fish survival may be low because larvae do not grow fast and may be subject to predation; b = the average condition that provides larvae with food throughout the period, although perhaps not at a maximum quantity; c = the late cycle when food is available later and increases rapidly, although this scenario may be dangerous because larvae might not find food when the yolk sac is exhausted. From Cushing (1973).

Whales migrate long distances to take advantage of periodically high production at high latitudes. Environmental cues involve conditions of tides, moon phases, and seasonal temperatures that are synchronized with internal biological clocks and mechanisms important to successful feeding and reproduction. It would be of little use for a fish to produce young when larval food is missing. Rather, reproduction must be timed so young will have access to adequate nutrition (Fig. 4.17).

An exquisite example of phenology concerns ice-breeding seals (e.g., Arctic hooded seals, *Cystophora cristata*; Southern Ocean Weddell seals, *Lepotonychotes weddelli*, and others) that bear pups on sea ice. Newborn pups are covered with a thick coat of fine hair that insulates them when dry, but is not an insulator in cold, polar waters. Therefore, pups need quickly to acquire a thick insulating coat of subcutaneous fat before the ice breaks up. This is a race against time that depends on feeding rate and weight gain before the pups are forced to enter the water to feed. Climatic factors determine break-up, high phytoplankton production, and production of invertebrates and fishes upon which young seals feed. Therefore, ice seals have evolved mechanisms, including milk that can be more than 50% fat, for weight gain of more than 300% in less than a week to about a month, depending on the species. After weaning, the mother departs and the young seal enters the water well insulated by fat, but must learn to feed for itself. If the time and timing of climatic, biological, and ecological events are disassociated, the young seal may either be weaned at low body weight and poor insulation or food availability may be ill-timed, and the pup may not survive.

4.4.7 Keystone and foundation species

Some species disproportionately influence their communities or ecosystems. "Keystone" is a term originally intended to describe species that prevent resource monopolization; by consuming prey species, such species facilitate the coexistence of prey species by inhibiting competition, thus preventing competitive exclusion. A broader definition of "keystone" relates to a trophic cascade. For example, sea otter (*Enhydra lutris*) predation on sea urchins in the North Pacific inhibits overgrazing of macroalgae; in the otter's absence, urchins proliferate and overgraze kelp to cause the virtual disappearance of the kelp and the associated kelp community. Thus, the keystone concept extends to functions other than predation.

Foundation species are those that significantly modify environmental conditions, species interactions, and resource availability through their presence, but not necessarily their actions. "Foundation" is applied to species that are physically integral to the environment and may be viewed as "key" ecosystem architects or engineers that alter habitats and transform ecosystems by their physical impact. They alter their environment by creating structures, as is illustrated by mollusks, coralline algae, and corals that build reefs that form a habitat for other species.

Species differ markedly in their capacity to act as keystones and/or ecosystem engineers. Seven factors call attention to their role: (i) life-history activities (what each stage does); (ii) their numbers per unit area (density); (iii) spatial distribution of individuals; (iv) how long the population has been present; (v) durability of the species' constructs and impacts; (vi) the number and types of resource flows modulated by constructs and impacts; and (vii) the number of other species dependent upon these flows. Oysters, corals, saltgrasses, and mangroves qualify as species that can structurally dominate their communities and alter hydrology, nutrient cycles, and sediment stability and, therefore, qualify as foundation species. Feeding bioturbation by walruses (Chapter 6) and predation by sea otters qualifies these species as keystones.

Keystone and foundation species, because of their structural or functional importance to their ecosystems, may signal a condition or "health" of the environment. So-called "indicator" species are often sought for conservation and management, such a species becoming a surrogate for detecting environmental change. In the sea otter–kelp–sea urchin example, the sea otter may be viewed as the keystone, the kelp as the foundation, and the abundance of either of those as an indicator of ecosystem condition. Also, the sea urchin, which is affected by changes in either sea otters or kelp, may be an appropriate indicator of the state of this system. This illustrates that almost any species may indicate some environmental condition at some scale. The first step is establishing what is to be indicated. Second, an indicator species should be one for which sufficient knowledge of natural history is available to establish that the species is relatively obligate with respect to the feature being indicated. When it is established that a species clearly affects others in its community, or by its activities initiates a cascade of effects, its identification as an "indicator" can be extremely useful for ecosystem analysis and conservation monitoring.

4.5 Biological assembly

How species assemble in their communities or ecosystems is essential to their survival and is a critical issue for conservation. The first level of assembly is that of individuals in a population. The second level is the guild, which is a multispecies assemblage whose species share similar traits. The third level is the community, which is a unit in which predation, competition, and cooperation among species determine community structure and function.

4.5.1 Populations

A population is a group of individuals of a single species that mate and produce viable offspring. Much variation exists in population structure. Some species are territorial and live solitary lives except during reproduction. Others form groups of the same size and age that remain together throughout life (schools). A common population trait is the capacity to increase in numbers. However, most populations tend to preserve a characteristic level of abundance, being described colloquially as "abundant," "common," or "rare."

Density-dependence is the basis for management of many, if not most, populations of commercial fishes, but is pertinent only for equilibrium, K-selected species. This concept is not applicable to density-independent populations, nor for fishes that aggregate to breed (Chapter 7). For the latter, a minimal number of individuals may be necessary to trigger reproductive behavior and to ensure adequate fertilization. At low densities depensation may result; that is, the population can decline to extinction if its density has become lower than a critical threshold, even though a few individuals capable of reproducing remain. Non-equilibrium conditions can also occur when habitats are disturbed or when exotic or invasive species are introduced. Hence, assumptions of density-dependence depend on detailed knowledge of species, its environmental constraints, and its behavior.

Most species have more than one interbreeding population. Population richness describes circumstances where a number of sustainable populations occur within the species' range. In some cases, these populations gather to breed in a common location; for example, American eels (*Anguilla rostrata*) live as juveniles in rivers and estuaries and migrate to the mid-Atlantic to reproduce. Conversely, adults of other species may group at sea, then separate into several

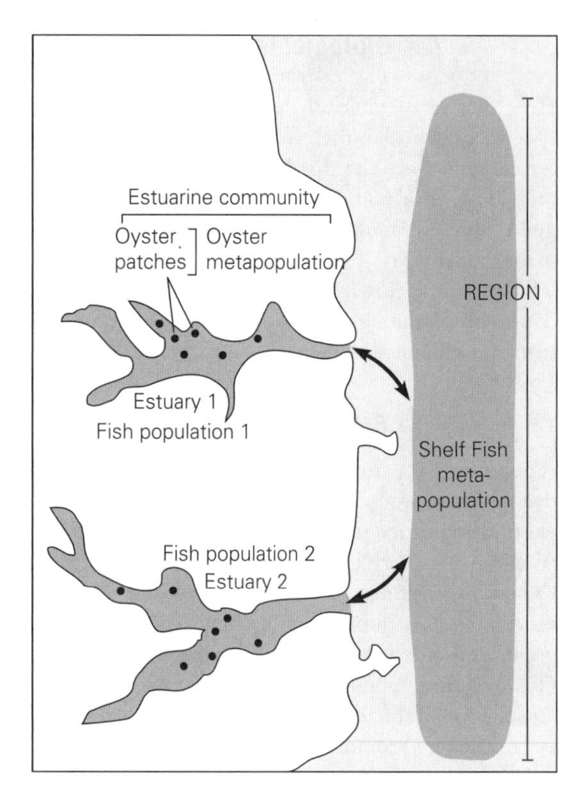

Fig. 4.18 Estuaries may include metapopulations of oysters and fishes. Oyster beds exemplify intra-estuarine metapopulations. Fishes exemplify extra-estuarine metapopulations, requiring estuaries at some portion of their life histories – especially those fishes that reproduce in estuaries and assemble as subadults and adults over the shelf. Fisheries management, therefore, involves multiple estuarine populations, whereas oysters may be managed in single estuaries. Modified from NRC (1995).

breeding populations in estuaries and freshwater (river herrings and salmons). For many reef fishes, juveniles move among populations. In all these cases, interacting populations are connected by movement to form a metapopulation – a population of populations. Metapopulations exist, to some extent, in almost all species.

Metapopulation dynamics has important implications for species' survival, as it describes relationships among species' populations and environments. Metapopulations can take several forms. One model assumes a large, relatively invulnerable source population from which individuals move to smaller habitats containing more transient populations. Another model assumes a set of equally sized habitat patches, with local populations frequently going extinct, and vacated patches being recolonized from currently occupied patches – the so-called "rescue effect." In either case, careful examination is required of the relationships between species' natural histories and environmental variability. Metapopulation theory and models often assume populations at quasi-equilibrium, which is not common for species in highly dynamic environments such as the coastal realm. Another factor concerns population recovery following catastrophe or major environmental disturbance. A high probability of risk exists for a metapopulation of an estuary-dependent fish if only a few estuaries maintain the metapopulation (Fig. 4.18). If a disturbance negatively affects those few healthy estuaries, the entire metapopulation may collapse. This circumstance can be unpredictable unless the dynamics of all subpopulations are considered together.

4.5.2 Guilds

Guilds are multispecies groups that share common resources and have similar behavior. Many guilds relate species according to feeding. For example, at least four guilds associate seabirds and marine mammals (Skov et al. 1995). Guild 1 is composed of many seabirds concentrating near their breeding colonies to feed; guild 2 contains seabirds and porpoises that feed together in shelf waters; guild 3 consists of several small- to medium-sized cetaceans and seabirds concentrated mainly along shelf edges and remote banks and ridges; guild 4 is comprised of large cetaceans and seabirds that feed on small, oceanic crustaceans. These groupings provide insight into consumer–resource dynamics that would be difficult to reveal through studies of single species.

Guilds often occur when several species gather in upwelling areas to feed. Explanations for why such guilds occur may have to do with resource benefits or adaptation to a particular habitat. This explanation may not be consistent among the guild's species for all life-history functions; that is, a guild may be a feeding association, but for reproduction the species usually disassociate into different guilds. Guilds of seals in polar regions can be grouped by their adoption of sea ice as habitat for reproduction, molting, and rest. However, this guild breaks down when examining feeding: some species feed mostly on zooplankton, others on fish, squid, or even penguins and other seals. Thus, any species may be assigned to several guilds according to various aspects of life history.

Guilds are especially useful for analyzing species-rich environments, such as coral reefs, where it would be almost impossible to examine each of dozens of planktivore species, or even each of less numerous herbivores and predators. One notable feature of coral-reef fishes is the formation of multispecies feeding guilds that likely form due to food preference. Thus, in complex communities, guilds can be used to simplify the diversity of ecological roles.

4.5.3 Communities

Communities are composed of interacting species, their populations, and guilds. Communities are not haphazard assemblages of individuals, but exhibit hierarchies of interactions in which individuals of a species compete and cooperate among themselves and among other community members for food, space, or optimum conditions. Thus, community structure and dynamics represent the summed fates of all organisms involved.

Communities are spatially bounded and referable to habitat type, and are commonly named for a dominant species or biotic structure (saltmarsh, seagrass, coral reef, mangrove) or physical feature (mudflat, sea ice, shelf, pelagic). Communities and their species are interconnected by exchanges of energy and materials, immigration and emigration, successional change, biogeochemical transformations, and physical change. A community is not equivalent to an ecosystem, as the physical environment is not specifically included in the former. Understanding any community requires study of its structure and function in time and space, and depends heavily on knowledge of species' natural histories, including predation, competition, reproduction, succession, disturbance, and habitat modification. Measurement of production, exploitation, and assimilation efficiencies allows reconstruction of energy flows that bind populations together. Interspecies relationships are especially important. However, influences of the natural environment (physical, chemical or climatic) are often omitted.

Natural selection acts on all species within a community, but necessarily affects and depends on the community as a whole. Co-evolution among species is often the glue that bonds the community together. Community relationships develop through competition, predation, and cooperation among species, during which some species are incorporated and some species are eliminated from the evolving structure. In this context, community complexity results from the product of the diversity of interacting organisms, and increases with space, temporal events, and environmental heterogeneity. Three interrelated concepts are important for understanding community behavior. *Persistence* is a measure of how long community structure and function lasts or how fast the community gains or loses species. *Resistance* is a measure of the ability of the community to resist change induced by intrinsic or extrinsic forces. *Resilience* is a measure of the rate that the community can recover from changes caused by perturbations. When environmental controls are included, these terms may also describe ecosystem dynamics and feedbacks. An example is the effect of sea-ice scour on maintenance of alternate states of rocky-shore communities (Fig. 4.19).

4.6 Patterns

Species and communities are not randomly distributed across the land and seascape. Biotic patterns of present-day Earth have resulted from more than four billion years of history, from the first small, anaerobic bacterial organisms, to a world rich in oxygen and diverse in life-forms. The outcome has been a biosphere with roots in the past, but one in which process and pattern are intimately interconnected.

4.6.1 Temporal patterns

At any one time, a community consists of permanent residents, temporary residents, and transitory individuals. However, species richness can change at any temporal scale, by hour or season, through environmental succession, or in response to climate change. Environmental stasis does not occur for long under natural conditions. Community composition changes when species leave or enter; some species are active at night, some in daylight, others in the crepuscular hours of morning and evening, and others seasonally. But the community remains functional. In polar regions, springtime brings an abundance of species; in tropical regions, lunar and weather cycles change species richness. On decadal time scales, El Niño–La Niña cycles cause major shifts in distributions of sea life and change biogeographic patterns. During Pleistocene periods of glaciation throughout all latitudes, environmental shifts changed species diversity and abundance.

Succession describes a temporal–functional change within a community. That is, a community's form and function progress over time in an apparent orderly

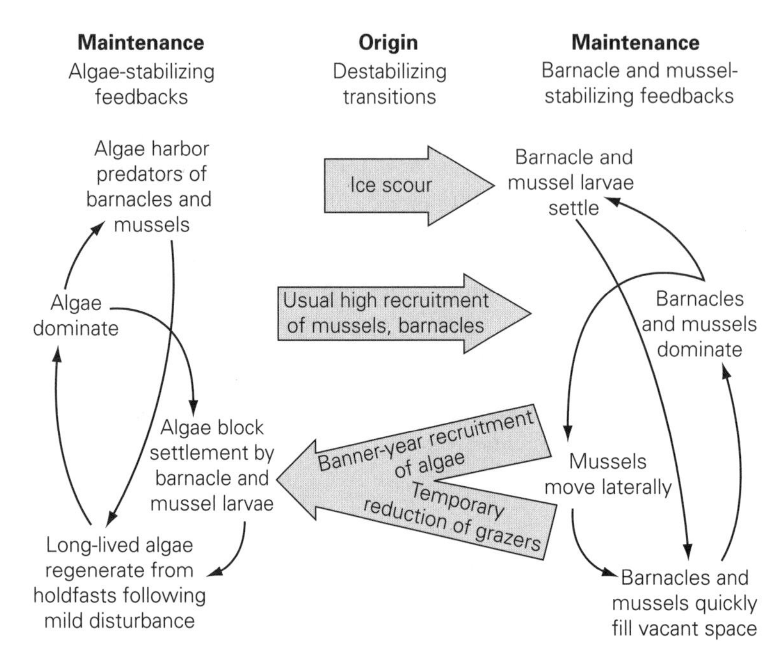

Fig. 4.19 Concept of positive-feedback-driven, alternate stable states on a rocky shore forced by change. From an algae-dominated community, ice scour initiates a switch to a community dominated by mussels and barnacles, or *vice versa*. Algae prevent mussels and barnacles from colonizing by creating settlement barriers and refuge for predators. Following ice scour, barnacles and mussels settle quickly into vacant spaces, facilitated by their gregarious settlement that prevents algal establishment. Another event, such as extraordinary algal recruitment and temporary reduction of grazers, can cause a switch from barnacle–mussel community back to algal bed. From Petraitis & Lantham (1999), with permission.

replacement of one species, or species guild, by another. In simple terms, succession rests on observations that certain species colonize new areas first, and set conditions for others that follow. Theoretically, successional communities move toward a "mature," more "stable" community of complex associations that functions differently from its predecessors. Exactly how succession operates is a matter of considerable research and debate. Nevertheless, succession is well established, for example, in observations of marine "fouling" communities on objects placed in the sea (e.g., piers, ships). Microbes colonize first. Within hours or days, bacteria and detritus particles are enmeshed in a matrix of chemical interactions. Within days to weeks, a thick bacterial film becomes mixed with diatoms and protozoa graze on microflora, prompting settlement of many species' larvae. Succession also occurs in seasonal changes of phytoplankton. These small, short-lived organisms respond quickly to environmental change. In mid- to high latitudes during the production season, small diatoms appear first and are soon replaced by larger diatoms as species richness increases. As nutrients decline, the decreasing diatom standing stock produces resting stages that sink. Subsequently, a community of mostly small, dinoflagellate species replaces the diatom community. Another example concerns successional stages that occur under conditions of overenrichment. If successional species are exposed to nutrient-rich water after a low-nutrient period, earlier successional species may reappear. But if oxygen concentration continues to decrease, a complex community of diverse heterotrophic organisms changes to a simple community, mostly of bacteria that withstand anoxia.

Diversity in early successional stages of community development is generally low and the species to appear first are not generally predictable. If a variety of redundant species (those with similar characteristics) is available, similar succession can proceed under similar environmental conditions, but with varying species compositions. At any stage, species may consume, overgrow, or disturb earlier species and replace them. Thus, succession can have several possible endpoints insofar as species composition is concerned, and is predictable only in a broad sense.

4.6.2 Spatial patterns

Spatial patterns are obvious at all scales. Species clumping occurs where clonal communities exist side-by-side or when species aggregate for purposes of breeding, feeding, or rest. Some predators aggregate to gain advantages for search and capture of prey; contrarily, prey may clump to avoid predation, as do fish in schools. Hard substrates (reef, floating algae) attract species and enhance biological production, resulting in spatial oases of productivity or shelter. At larger scales, biotic patterns form heterogeneous land- and seascapes, with productive areas interspersed among relatively unproductive regions.

Coastal-realm boundaries can be physical, as among water masses or watersheds, or biological and ecological, as for landscapes. Boundaries can also be perceptual, as between biotic communities or biogeographic provinces defined by the interests of the investigator. In all cases, boundary delineation depends on scale, and the nature of the boundary itself.

4.6.2.1 *Patterns in species richness and endemism*

Recognition of spatial boundaries influences the measurement of two related conservation interests – species diversity and endemism. Species are usually confined to one of a few biogeographic provinces or smaller areas, within which most are naturally uncommon or rare – that is, in most communities a few species dominate. When boundaries isolate populations from another, strong population separation results in shifts in genetic frequency, even for large species with ability to move thousands of kilometers along continental margins. Boundaries may not be obvious for species that may be almost cosmopolitan; many protozoans, algae, and fungi have extremely broad ranges because the organisms themselves or their tiny disseminules are long-lived and can be broadcast widely by water or wind. However, genetically unique species can be used to define the areas.

Biodiversity within a bounded space is measured in at least three ways: (i) *richness*, the numbers of species within a sampling unit; (ii) *abundance*, the numbers of individuals of a species within that unit; and (iii) *evenness*, a measure of the relative abundances of all species present. These three measures may be applied across three geographical scales. *Alpha* diversity expresses the number of species in a particular community – a measure of within-area diversity. *Beta* diversity represents the variability in species composition along an environmental gradient – a measure of between-area diversity. *Gamma* diversity refers to the relative numbers of species within large geographical regions. For each of these, and their various combinations, indices are calculated mathematically. Diversity indices rarely include all or even most taxa in a community, as a result of lack of information on the total species that may be present and their ranges; in fact, species inventories can never be complete due to natural variation, mobility, and logistics. Therefore, these indices are always statistical generalizations.

Endemism expresses the minimum range for a species that occurs nowhere else, and is defined by scale. The boundary decision is judgmental, as every species is endemic at some scale. Local endemics occur solely in small, isolated areas. The "living fossil" coelacanth fish (*Latimeria chalumnae*) apparently occurs only in isolated, relatively deep, ledgy areas of the Indian Ocean (population ranges are uncertain). In contrast, sea turtles and giant whales are endemic to whole ocean basins. Local endemics are often of highest conservation priority under the assumption that these species are most vulnerable to local disturbances. However, wide-ranging species can be equally as vulnerable when any one life-history function is restricted to a small area; for example, breeding areas and nurseries of wide-ranging fishes, sea turtles, albatrosses, and seals are among the most vulnerable habitats of coastal-realm species.

Assessment of endemism at regional scales is affected by Rapoport's Rule, which expresses a relationship between species ranges at high and low latitudes; that is, the mean size of species' ranges is assumed to decrease toward lower latitudes, perhaps because long-term environmental change has been less than for higher latitudes and/or because species that evolved to favor lower temperatures since the latest glacial period are relatively few. The result is that the fewer species of high latitudes have large home ranges but are relatively abundant within their ranges. Thus, if conservation favors local endemics, as is often the case, bias may be directed toward the tropics, even though many relatively abundant, high-latitude fishes, seabirds, and marine mammals are equally depleted or ecologically or economically endangered (cods, some salmons, several sea birds, some seals, and large whales).

4.6.2.2 *Gradients and zonation*

Transition zones between spatial patterns are known as ecotones. These are gradients between systems, characterized by unique time and space scales and by strengths of interactions between adjacent systems. Striking transitions occur between dry, hard land and fluid water, which create stresses for some forms of life and advantages for others. Wave-swept intertidal environments exert costs and benefits that affect the distribution and abundance of species (Fig. 4.20). Also, organisms living nearshore in benthic environments colonize appropriately sized benthic particles from intertidal shores to deeper water (Fig. 4.21). In high-energy, transitional environments, where sediment grain size is relatively large and organic and microbial contents are low,

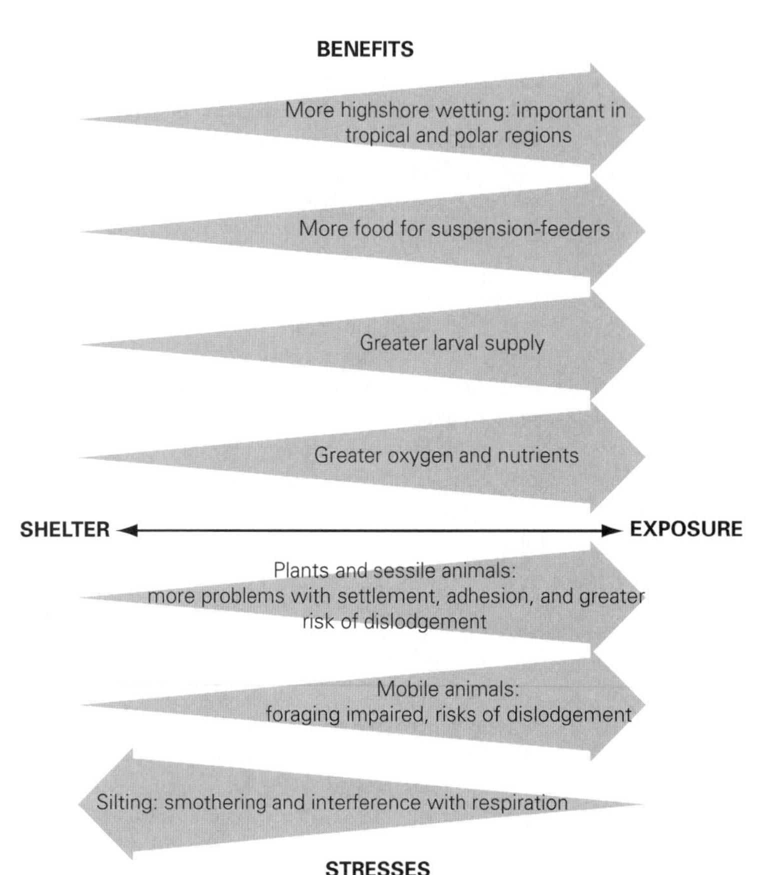

Fig. 4.20 Wave action produces a horizontal gradient that can be beneficial or stressful. From Raffaelli & Hawkins (1996), with permission.

suspension-feeders tend to be most abundant. In rocky inter- and subtidal areas where water is relatively clear, kelp beds form. In muddy, low-energy depositional areas with weak flows and greater vertical fluxes of food, fine sediments, chemicals, and larvae, deposit-feeders are most abundant. Additionally, biological structures (e.g., seagrass beds, reefs, patches of polychaete tubes, bivalve shells) exhibit environmental feedbacks that create gradients in water flow, food, larvae, and oxygen content which also affect the distributions of other species. Because of the location of ecotones between adjacent areas, species may benefit from both; thus, in ecotones, productivity and abundance may be enhanced. But also because of their placement, ecotones have strong propensities for change.

At a global scale, gradients in species richness may reflect latitudinal differences. Early explorers observed that the tropics teemed with life, temperate zones had fewer kinds of animals and plants, and the arctic and antarctic seemed stark and barren by comparison. This assumption holds only if all species are counted equally. That is, gradients in species richness depend upon which taxon or which type of ecosystem is examined. Although total fish diversity decreases toward the poles, some taxa increase and some are restricted to high or low latitudes. For example, butterfly- and angelfishes (Chaetodontidae and Pomacanthidae) are more diverse in the tropics, but rarely occur in colder waters, while true cods (Gadidae) are more diverse in cold-temperate waters, and icefishes (Nototheniidae) are restricted almost entirely to the Southern Ocean. In marine mammals, sea cows are restricted to the tropics, but most seals and large whales are most varied and abundant poleward; polar bears occur only at the poles. In the deep, open ocean, there is no gradient in higher taxa; that is, all major phyla are usually represented. Only at lower taxonomic levels do gradients appear, but their nature depends on the taxon under examination.

Another common assumption is that environmental heterogeneity is greatest in the tropics, partially

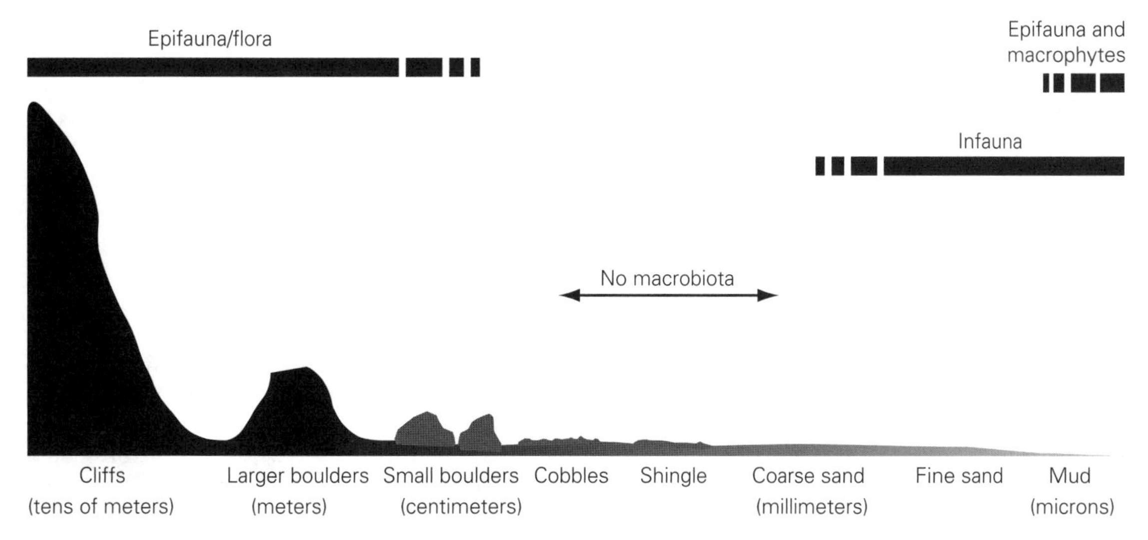

Fig. 4.21 Particle sizes generally decrease offshore or off reefs. Surface-dwelling species are present at both ends of the particle-size gradient. Infauna are mostly restricted to smaller particle sizes. Macrobiota are rare in the middle section. From Raffaelli & Hawkins (1996), with permission.

explaining greater species diversity there. However, at the land–seascape scale, habitat heterogeneity appears to be as high at high latitudes as elsewhere (Table 4.5). In fact, coastal-marine environmental heterogeneity at high latitudes may be increased temporally by seasonal variability and spatially by sea-ice formation. This does not imply, however, that within-habitat heterogeneity (smaller-scale patchiness) is greater or lesser among these environments. Unfortunately, comprehensive studies on relationships among species ranges, habitat heterogeneity, and latitude are rare and inconclusive for coastal-marine environments. At the very least, latitudinal gradients in species richness and land–seascapes are meaningful only in the context of closely related groups of species with similar natural histories.

Gradients in species richness are also longitudinal and radial. Among tropical Indo-Pacific reefs, fish and coral diversity decrease toward higher latitudes, but also decrease eastward from Indonesia toward Hawaii and westward toward Africa. Waters surrounding islands also exhibit radial gradients, in which diversity decreases from a center in all directions. For true islands, that is, a land area surrounded by the sea, a series of abrupt gradients usually occurs from the island's interior into shallow water and out to sea. For habitat islands, such as reefs in sedimentary or seagrass environments, the gradient may be abrupt or gradual, depending on the species association involved. For pelagic species, gradients among water masses can be obscure and are highly subject to rapid changes induced by climatic and oceanic variability. Also, gradients between water masses occur vertically with depth. In higher-latitude seas, thermoclines and pycnoclines are generally nearer the surface than in lower latitudes and deeper waters are physically and biologically distinct from surface waters.

Causes for gradients in species occurrence or richness are complex. Simple cause–effect relationships are rare. At least three classes of environmental factors are involved: (i) consumable resource gradients such as nutrients (for plants, light is a resource); (ii) regulator gradients (temperature, salinity), which affect rates of physiological processes; and (iii) gradients that may have no direct influence on organisms, but are correlated with both resources and regulators (latitude, water depth), which can also affect production and diversity. Species are most often assumed to be adapted to optimum physiological conditions along a regulator gradient – for temperature, neither too hot nor too cold, or for salinity, neither too fresh nor too salty. However, species often do best at high levels of a resource gradient – a nutrient, for example. If two or more species are similarly adapted, competition may mask the effects of these factors. This is apparently true for unrelated trees lumped together as "mangroves." As a group, these salt-tolerant plants are apparently not good competitors

Table 4.5 Presence and relative importance of nearshore and benthic coastal habitat types of North America (north of Central America): TR, tropical; SB, subtropical; TM, temperate; BR, boreal; AR, arctic; E, east; W, west. Rankings: 0, absent; 1, rare; 2, common; 3, extensive.

Habitat	Region				
	TR	SB	TM E–W	BR E–W	AR
Maritime forest woodland	3	3	3–3	3–3	0
Coastal shrublands	3	3	3–3	3–2	3
Coastal grasslands	3	3	3–2	3–2	3
Coastal tundra	0	0	0	2–3	3
Coastal cliffs	0	0	0–2	3–3	3
Coastal marshes	2	3	3–2	3–2	3
Coastal swamp forest	2	2	2–0	1–0	0
Mangroves	3	0	0	0	0
Intertidal beaches	3	3	3–3	3–3	3
Intertidal mud- and sandflats	2	3	3–2	2–2	2
Intertidal algal ecosystems	1	1	2–3	3–3	1
Subtidal hard bottoms	2	0	2–3	3–3	3
Subtidal soft bottoms	3	3	3–3	3–2	3
Subtidal seagrass beds	3	3	3–1	3–3	0
Subtidal kelp beds	3	0	0–3	3–2	0
Living reefs	3	1	2–0	1–0	0
Sea ice	0	0	1–0	2–1	3

Relative importance scores for habitat types among regions:
TR: 3s = 9; 2s = 4; 1s = 1; 0s = 3.
SB: 3s = 8; 2s = 1; 1s = 2; 0s = 6.
TM(E): 3s = 8; 2s = 4; 1s = 1; 0s = 4.
TM(W): 3s = 7; 2s = 4; 1s = 1; 0s = 5.
BR(E): 3s = 11; 2s = 3; 1s = 2; 0s = 1.
BR(W): 3s = 7; 2s = 6; 1s = 1; 0s = 3.
AR: 3s = 9; 2s = 1; 1s = 1; 0s = 6.
Adapted from Ray (1991).

with other tree species, so mangroves may be forced into saline conditions, where their differing salt tolerances result in zoned distributions; Caribbean red mangroves (*Rhizophora* sp.) occur in deeper nearshore water than black mangroves (*Avicennia* sp.). Thus, many environmental conditions act together to produce distinct patterns in species' occurrence and abundance. Any species, or group of species, may be variously affected by diminishing light with depth, increasing estuarine salinity seaward, increasing solar radiation with lower latitude, variable tidal strength, exposure to desiccation and wave action, and gradients in temperature, depth, biotic production, sediment, bed-flow processes, and other factors.

4.6.3 Disturbance

Biotic communities normally flourish under optimum conditions, oscillating between highs and lows in patterns of species dominance and production. When disturbance occurs, communities must reorganize. Disturbance is a normal feature of communities and environments, and an important determinant in colonization, species diversity, and ecological function. Disturbances from storms, dredging, pollution, or arrival of new (exotic or invasive) species can disrupt spatial structures and interfere with temporal oscillations, species' behavior, and community interactions, all of which require some community adjustment. In many communities, there appears to be an optimal disturbance frequency at which biodiversity is maximized, depending on the physical environment, type of community involved, and productivity of the surrounding environment. The rate, frequency, and magnitude of disturbance are important controls on biotic and land–seascape development. Intermediate levels of resources and disturbance appear to maximize biodiversity, at least theoretically (Fig. 4.22). Absence of disturbance may help explain why most deep-water

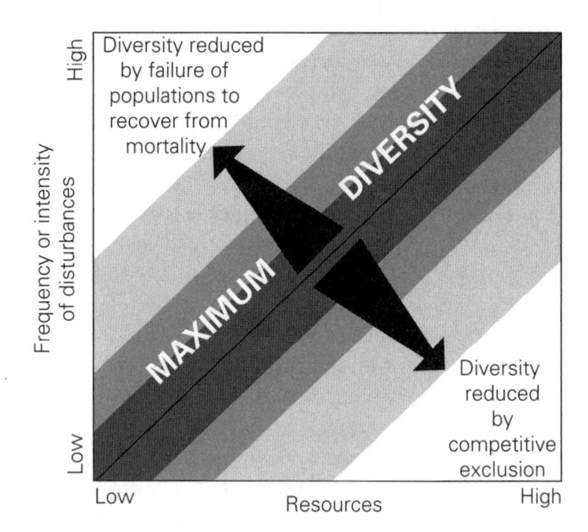

Fig. 4.22 Diversity is maximum at intermediate levels of competition and disturbance, and most likely to be influenced by landscape-scale and regional processes. Competitive exclusion under low disturbance frequencies and intensities reduces diversity. At high levels of disturbance, few species can exist. When resources (e.g., nutrients, food, water) are low, few species can exist; when resources are at high levels, populations may exceed carrying capacity, or self-pollute their environments, or hypoxia can occur. From Huston (1994), with permission.

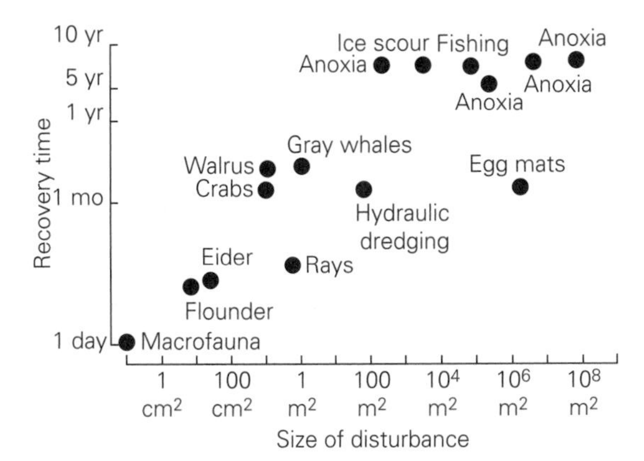

Fig. 4.23 Recovery time varies with size of disturbance of benthic sediments. Small-scale disturbances have shorter recovery times than extensive disturbances. When large animals feed together on the bottom, recovery time is dwarfed by that of anoxia; fishing effects can be massive and recovery time slow. From Hall et al. (1994), with permission.

environments appear to be more uniform, relatively stable, and less biodiverse than elsewhere.

Disturbance can be harmonic and regular (waves, tides) or disruptive and irregular (tidal waves, storms). In nearshore environments, disturbance of the substrate is a function of wave force, being directly proportional to substrate particle size and depth at which the disturbance occurs. In shallow, subtidal areas, small rocks that are frequently displaced support fewer species than otherwise. In tropical seas, storms strike coral reefs with such force that waves overturn large corals and force redistribution of huge volumes of matter, destroying old and creating new habitat conditions.

Another form of disturbance to benthic communities is bioturbation (disturbance by activities of living organisms). Bioturbation has major effects at a variety of spatial and temporal scales (Fig. 4.23). Large biota, such as whales and walruses, may cause massive displacements of sediment and biota during benthic feeding activities. At local scales, small infaunal animals redistribute sediments during feeding, burrowing, and other activities that in the aggregate can have major effects on sediment, aeration, and nutrient dispersal (Fig. 4.24).

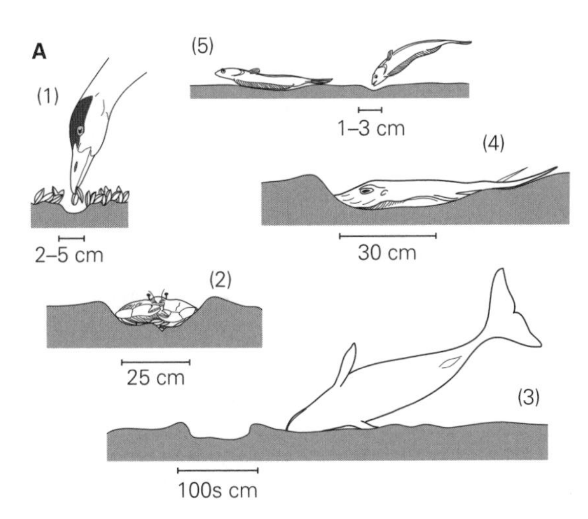

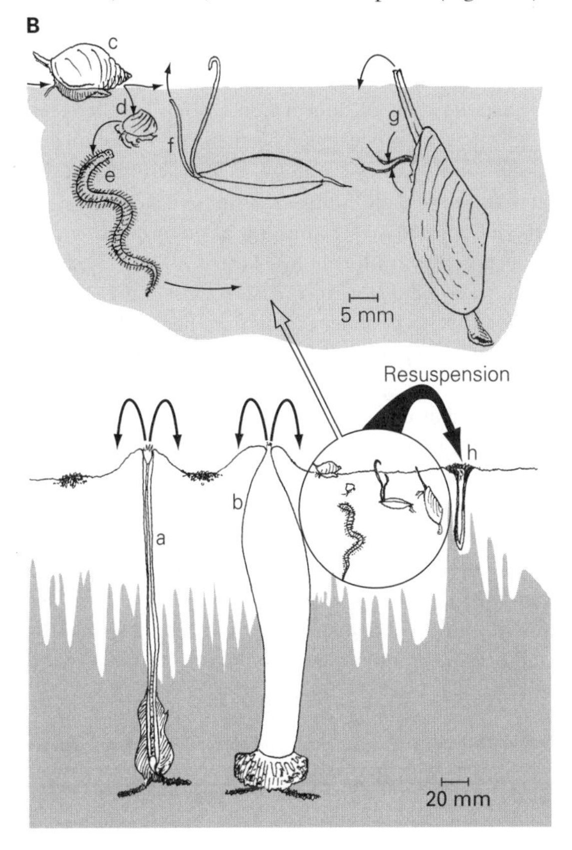

Fig. 4.24 Bioturbation at different scales influences species diversity and feeding type diversity. (A) Benthic patchiness generated by bottom-feeding disturbances: (1) eider duck; (2) crab; (3) gray whale; (4) ray; (5) flatfish. From Hall et al. (1994), with permission. (B) Many deposit-feeders mix and recycle sediment. Arrows indicate routes of sediment ingestion and feces egestion of relatively large sediment organisms: (a) burrowing polychaete worm; (b) sea cucumber; and smaller organisms enlarged: (c) snail; (d) clam; (e) errant polychaete worm; (f) clam; (g) clam; (h) anemone. Conveyor-belt species (a, b) pump reduced sediment to oxidized surface, overturning sediment. Other species (c–g) cycle particles within the oxidized surface. From Rhoads (1974).

Bioturbation can dramatically alter transitions and zonations and can exert major influences on distribution of organisms, and on community structure and function.

4.7 Biogeography and environmental classification

Biogeography is the science of spatial distributions of species. Biogeographers produce maps of biotic assemblages in space and time, and relate them to environmental attributes so that different environments may be compared, in order to promote understanding of how these relationships have come about. Biogeography facilitates understanding of evolutionary and adaptive processes among a suite of different ecosystems, habitats, and communities, and provides conservationists with a basis for selection of protected areas that represent nature's variety. Biogeography also advances science by extrapolation of information among and between environments. Biogeography centers on environmental classification. Just as taxonomy helps scientists sort out evolutionary relationships, environmental classification may be viewed as a taxonomy of environments that reflects ecosystem hierarchy. That is, if biogeographic classification did not exist, each system and each area would have to be independently studied and managed without the benefit of comparisons and previous experiences.

The major problem within the coastal realm is that terrestrial, freshwater, and marine systems are treated differently. Terrestrial provinces have been identified largely on the basis of vegetation under the assumption that plants represent climatic regimes. Watersheds cut across these provinces and are not generally integrated into them. Coastal-marine provinces have been defined both by endemic species and by physiography, and oceanic provinces are based mostly on the physical characteristics of open-ocean water masses in which plankton are tightly coupled to physico-chemical processes (Fig. 3.1, page 59). Thus, the state-of-the-art is that several overlapping and inconsistent classifications exist, each useful in its own right, but not representing the interactions, ecological hierarchies, and dynamics within the coastal realm.

Nevertheless, coastal-realm biogeographic groupings are recognizable. The major task is to identify spatial relationships among components and land–seascapes to facilitate understanding of ecological patterns. First, a classification should be hierarchical (Table 4.6). In one analysis, hierarchical associations among along-the-coast units of island-lagoon marshes reveal that bathymetry and biogeography tend to covary (Fig. 4.25). However, biogeography for sessile species is not applicable to distributions of mobile species such as fishes, which respond to different sets of environmental variables. That is, different groups of species exhibit their own biogeographies, depending on their natural histories. For example, on North America's east coast, shore-to-shelf fish distributions of the warm-temperate Carolinian and cold-temperate Virginian provinces illustrate that species assemblages are over-lapping (Fig. 4.26). For these same fishes, longshore boundaries separating coastal provinces are not discrete; rather, they form gradients, that is, ecotones, indicating that individual species of different assemblages respond differently to environmental properties (Fig. 4.27).

Modern-day biogeography has been based mostly on observed steady states. The existence of gradients among overlapping patterns argues for a new approach. The emergent field of dynamic biogeography shows promise of connecting species assemblages and their distributions to ecological dynamics and change. The application of dynamic biogeography to the coastal realm is presently limited to local and regional studies, but as coastal-realm conditions continue to change,

Table 4.6 Scale properties, units, and objectives of a hierarchical coastal biogeographic classification.

Scale	Ecological unit	Sample objectives
Global	Coastal realm	Recognize land–ocean–atmosphere interactions
Coastal realm	Biogeographic province	Identify species endemics and assemblages that distinguish coastal-realm ecoregions
Biogeographic province	Components: watersheds, estuaries, lagoons, benthos, etc.	Describe component processes that can help explain biogeographic distributions and their dynamics
Components	Land–seascapes: wetlands, biogenic seascapes, water masses, etc.	Understand biotic communities and their relationships *inter alia*

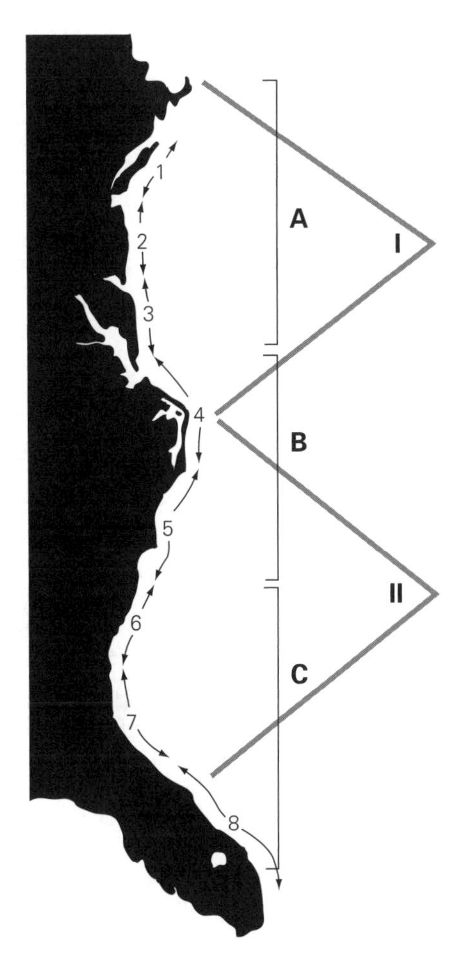

Fig. 4.25 A hierarchical bio-geo-physiographic classification. The biogeographic provinces, Virginian (I) and Carolinian (II), of the eastern United States are derived from coastal species assemblages. Physiographic regions (A, B, C) and smaller subregions (1–8) represent a classification of island-lagoon marsh subsystems that covary with offshore bathymetry. Adapted from Hayden & Dolan (1979); Hayden et al. (1984).

this approach will become ever more essential. Changes in patterns illustrate the complexity of biotic associations, and interpretations depend on integrated knowledge of species' natural histories and ecological processes.

4.8 Conclusion

Charles Darwin based his book, *On the Origin of Species* (1859), on natural history, and concluded that species

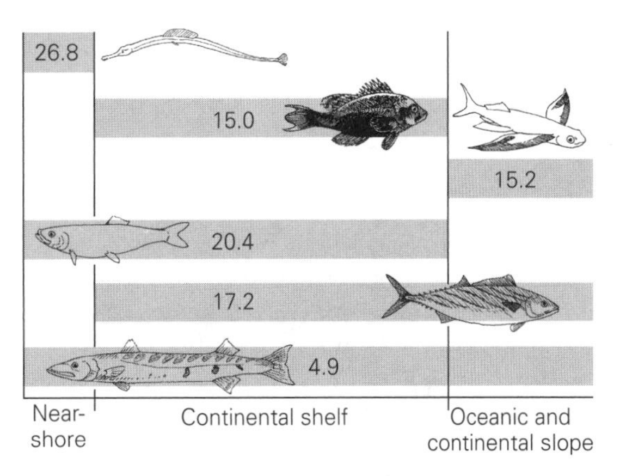

Fig. 4.26 Six overlapping, cross-shelf distribution patterns (represented by shaded areas) are evident for 553 species of fishes of the U.S. east coast's Carolinian and Virginian biogeographic provinces. The numbers are rounded percentages of the numbers of species in each category. Nearshore is defined as waters of approximately 20 m or less in depth. The continental shelf and oceanic waters are separated at the shelf edge. Species typical of these patterns are, from top to bottom: pipefish, sea bass, flying fish, menhaden, albacore, and barracuda. Adapted from Ray (1991).

change by processes of natural selection. At the same time, Alfred Russell Wallace came to the same conclusion. The Theory of Evolution now encompasses a large body of fact from paleontology, genetics, and systematics, which explains how life has emerged to dominate Earth. The eminent geneticist, Theodosius Dobzhansky, aptly stated, in 1973: "Nothing in biology makes sense except in the light of evolution."

The study of natural history illuminates processes that optimize the fitness of species within their habitats. While a species' success can be evaluated only in terms of its own particular environment, it is apparent that adaptive radiations of species have produced ever-increasing biotic and ecological complexity during the past several hundred million years. This has depended, in large part, on the evolution of heterogeneous spatial and temporal patterns, and the anomalies of geography and history.

Conservation is dependent upon natural history more than any other single subject, but must be placed also in an ecosystem context. Natural history reveals evolutionary processes and properties of persistence, resistance, resilience, recovery, and vulnerability to extinction. Population numbers for any species fluctuate from year to year. Some species fluctuate wildly, but many resist impacts of major disturbances through

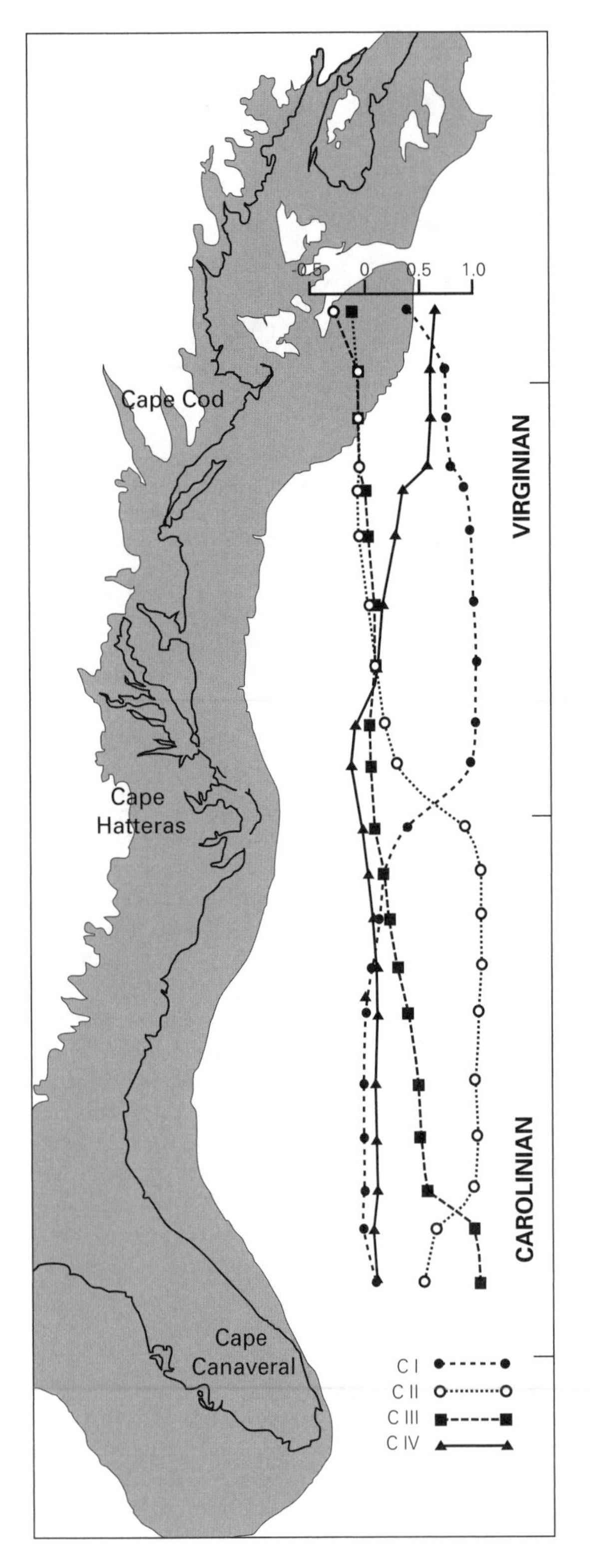

physiological, behavioral, and reproductive adjustments. That is, populations show persistence, resistence, and resilience, as do ecosystems. Patchy environments and heterogeneity affect life histories of organisms in differing ways, and temporal pulses have dramatic impacts on persistence. At any one time, a species may be present in some patches and not others, persisting by means of source–sink interactions and drawing on diverse habitat options. Below minimum thresholds, the species may perish. Conservation depends on identification of these thresholds, as protection of a submarginal environment may limit a species' ability to persist.

Species with higher reproductive potentials presumably have higher potentials for evolutionary change. A population's fitness, measured as the net production of offspring that survive to breed, is critical for its survival. Fitness is a product of competitive abilities, demographic factors, social structure, dispersal, local extinction, symbionts, colonization, and a host of environmental factors. Small populations are at high risk of extinction because they can become inbred and subject to random genetic drift and loss of genetic variability. Theoretically, large populations are expected to persist for longer times. But in either case, long-term persistence of a species is enhanced when a minimum number of interacting populations exists. This is the minimum viable metapopulation, which depends on the optimal distribution of suitable habitats.

Species persistence and extinction rates are uncertain for the vast majority of coastal-realm species. Wide gaps in knowledge exist about how many species there are and what their ranges and habitats are. Even less is known about how species respond to specific kinds of disturbances and environmental change, how fast new species are invading new locations and changing community structures, how species can recover from severe declines, and how feedbacks among species and habitats may lead to ecosystem persistence and resiliency. Species may seem to persist under conditions of envir-

Fig. 4.27 (*left*) Four overlapping, longshore assemblages of fishes result from principal components analysis of ranges of 151 estuary-dependent fish species of the U.S. east coast's Virginian and Carolinian biogeographic provinces. The dominant assemblages are indicated by values above zero: C I, Virginian assemblage; C II, Carolinian assemblage. Weaker assemblages comprise fewer species and largely originate outside the region: C III, tropical assemblage; C IV boreal assemblage. Many species belong to more than one assemblage, as their ranges are broader than either biogeographic province. Adapted from Ray et al. (1997).

onmental change, but as metapopulations are isolated among fragmented land- or seascapes, or as ecosystems shift into altered states, species may move almost imperceptibly, but irrevocably, toward depletion and even extinction. Scores of rare and endangered species may already be among the "living dead." And, for the poorly known coastal realm, species may easily disappear, unrecognized. Even more seriously, as species disappear the natural functions of ecosystems are increasingly compromised.

Section III

Case studies

Introduction to the case studies

With text contributions by Gary L. Hufford

Three case studies highlight differences and similarities among large-scale forces and create a context for understanding regional systems and for conservation. Each case highlights conservation issues in the context of characteristic features of their ecosystems, cultural inheritance, natural history, and ecology. For Chesapeake Bay, restoration of an altered estuarine system is the major conservation issue; a regionally directed program aims to restore valued resources, especially ecologically important oysters and their reefs. For the Bering Sea, the issue is conservation of marine mammals. Two key species highlight problem arenas: the endangered Steller sea lion and its interactions with fisheries, and the walrus, an important resource for Native Americans. For the Bahamas, the major conservation issue is sustainable development of an island nation heavily dependent on tourism and natural resources, such as coral reefs, Nassau groupers, and sea turtles that continue on downward trends.

Although these conservation issues differ, the three case studies illustrate important parallels in biological depletion, emergent phenomena, and desynchronies caused by the removal of key organisms (Table III.1). Additionally, each follows similar time dimensions, marked by four comparable periods (Table III.2): (i) a prehistory of subsistence living that sustained increasing human populations; (ii) a period of western expansion and resource exploitation during the 16th–19th centuries, recorded by apocryphal or traditional knowledge of natural resources and ecosystems; (iii) a period when resource depletions were recognized and calls for conservation emerged; and (iv) a period of dramatic rise of coastal-marine resource exploration and use, and massive ecological change, coinciding with a rapid growth of scientific knowledge and conservation action.

Fundamentally, all three case-study regions are experiencing unprecedented ecological and social changes, in pace, direction, and magnitude, in which resources, habitats, ecosystems, and human societies are continually adjusting. Environmental conditions have strong influences on both species and society. The Bahamas is subtropical, Chesapeake Bay is temperate, and the Bering Sea is boreal to arctic. In each case, day length and climate differently affect ecosystems, species' natural

Table III.1 Cross-regional comparison of major issues.

Issue	Region		
	Chesapeake Bay	**Bering Sea**	**Bahamas**
Conservation goal	Restoration	Marine mammal fisheries conservation management	Sustainable development
Bio-focus	Oysters, fish, waterfowl	Walrus, Steller sea lion	Groupers, sea turtles, coral
Ecosystem type	Temperate estuary	Subarctic sea	Ocean-island system
Primary issues: biological concerns			
Extinction, extirpation	Fringing oyster reef	Steller sea cow	Caribbean monk seal, West Indian manatee
Depleted	Oysters, sturgeon, shad, eel grass, etc.	Steller sea lions, fur seal, whales, fishes	Coral, groupers, conch, lobster, sea turtles
Overabundant	Comb jelly, algal blooms, invasive species	Algal bloom	Marine algae, hutia
Ill-health pandemic	Oyster epidemic, fish lesions, fish kills	Fish shellfish lesions	Diseased corals, sea fans, sea urchins, sea turtles
Toxins, bioaccumulation	PCBs, DDT, PAHs, etc. toxic "hot spots"*	PCBs, DDT, PAHs, etc. in marine mammals*	Ciguatera poisoning
Abnormal behavior	Waterfowl change, fish migration change	Marine mammal strandings, disturbance	Coral bleaching, grouper aggregations
Deteriorating habitats	Oyster beds, wetlands, seagrass, water quality, islands	Sea ice, regime shift?	Coral reef, mangrove, seagrass, beaches, freshwater lens
Secondary issues: human uses and activities			
Extractions	Fisheries, freshwater, sediment (dredging)	Fisheries, marine mammal harvest	Fisheries, freshwater, minerals (aragonite)
Introductions	Exotic species, pollutants	Atmospheric pollutants	Exotic species, pollutants
Physical alterations	Shore stabilization, dredging, urbanization, dams, reservoirs, agriculture, mariculture	Bottom trawling, dredging, port development	Shore stabilization, dredging, harbors, island development, tourist infrastructure
Tertiary issues: ecosystem adjustment			
Social forces	Population, traditional uses, conurbation, globalization	Traditional uses, oil development, tourism, globalization	Population, traditional uses, tourism, globalization
Emergent phenomena	Zoo-epidemics, anoxic bottom, toxic algae, toxic areas, shore erosion	Toxic algae, sea-ice thinning, glacial melt	Invertebrate diseases, algae expansion, shore erosion
Environmental change	Water quality, loss of foundation species, fragmented habitat, toxic sinks	Mammal diversity, climatic-regime shift, food-web change	Herbivore abundance, island, coral reefs, saltwater intrusion

* PCBs, polychlorinated biphenyls; DDT, dichloro-diphenyl-trichloroethane; PAHs, polynuclear aromatic hydrocarbons.

histories, resources, and the timing and duration of events. Understanding how coastal-realm systems differ according to large-scale, top-down variations in day length and climate provides a degree of predictive capability; weather forecasting is one example.

The most obvious effect of latitude is on day length. Day length and solar radiation have major influences on climate, production, life histories, migration, human resource extractions, and other factors (Fig. III.1). At 20°N, the difference in daylight between the winter and summer solstices is 2 hours 34 minutes, while at 65°N the difference is 19 hours and 24 minutes. Common among all regions is

Table III.2 Case studies: phases of change.

Social change	Chesapeake Bay	Bering Sea	Bahamas
Native cultures (before ~ 1500 years ago)	Algonquin nation; settlements, farmers, traders. English explorers	Yupik, Iñupiat, and Chukchi sea hunters; settlements and migratory cultures	Taino Indians enter from south; prosper as roving hunter–gatherers and traders
Western discovery and expansion (~ 1500–1800)	British colony; Indian nation vanquished. U.S. independence	Russian explorations: strong native subsistence cultures	Spanish explorations; Indians vanquished; privateers; northern islands settled; British colony established
Intensified exploitation; beginnings of management and protection (~ 1800–1900s)	Commercial exploitation of sea resources; some resource depletions; resource management begins; vacation centers; railroad	Commercial exploitation of marine resources; severe marine mammal depletions; gold discovered; protection and management begin; subsistence uses continue	Marine resources heavily exploited; artisanal and subsistence uses; salt industry and small plantations; tourism industry begins
Post-World War II accelerated use and management; rise of conservation science (~ 1950–present)	Severe fisheries depletion; oyster reefs degraded; coastal-zone management begins; regional restoration program initiated; sophisticated science; strong infrastructure; strong public conservation interest	Some marine mammal recoveries; very large fisheries developments; marine mammals protected, but subsistence use permitted; U.S.–Russian cooperation begins; science and conservation fragmented	Strong development of island-sea resources; rise of tourism; first land–sea park; Bahamas becomes independent member of British Commonwealth; science and conservation infrastructure marginal; rising conservation interest

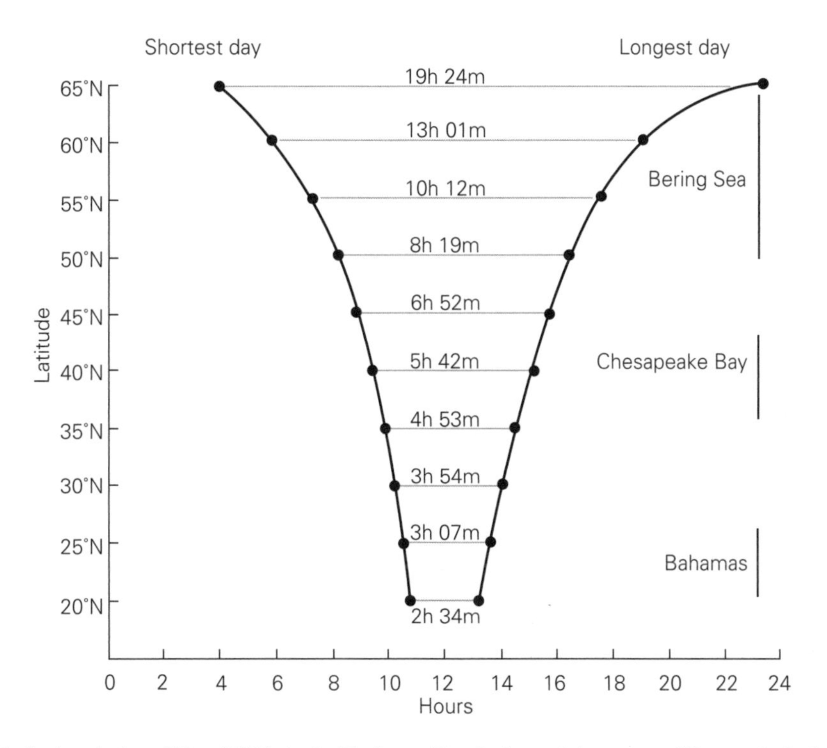

Fig. III.1 Ranges in day lengths from 20° to 65°N latitude. The hours (h) and minutes (m) represent differences in day length between the summer and winter solstices. The bars at the right indicate the latitude ranges for each of the three case studies.

that the longest day (June 21) is longer than the longest night and the shortest day (December 21) is longer than the shortest night, because refraction by the atmosphere causes the upper edge of the sun to appear earlier at dawn and to disappear later at sunset.

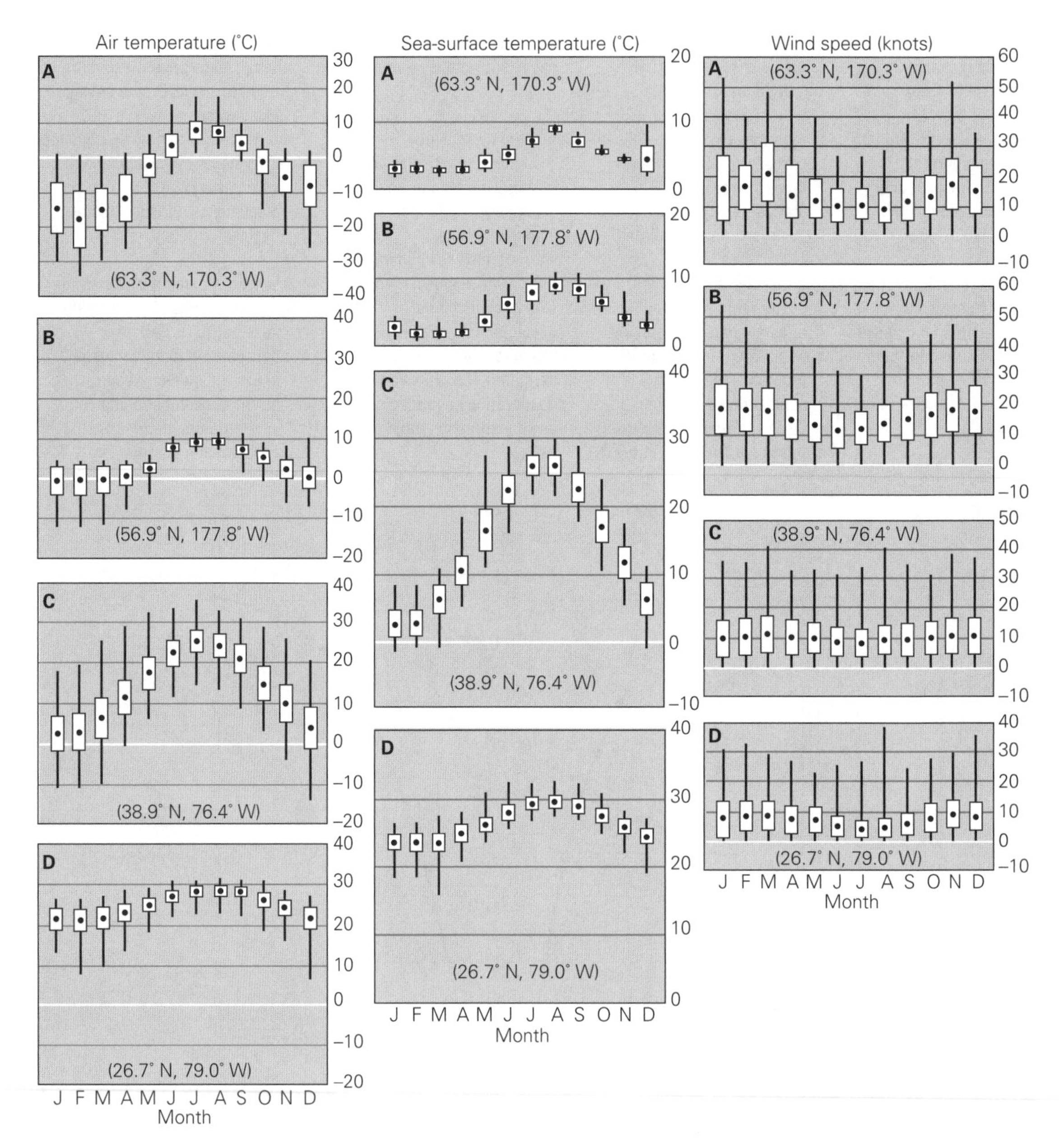

Fig. III.2 A comparison of monthly normalized values for surface air temperatures, sea-surface temperatures, surface wind speeds, and sea-level pressures recorded at NOAA buoy sites in the northern Bering sea for 1981–8 (A), and for the southern Bering Sea (B), Chesapeake Bay (C), and the Bahamas (D) for 1985–93. Vertical bars indicate one standard deviation; vertical lines indicate total range of variation. Buoy latitudes are indicated on the graphs.

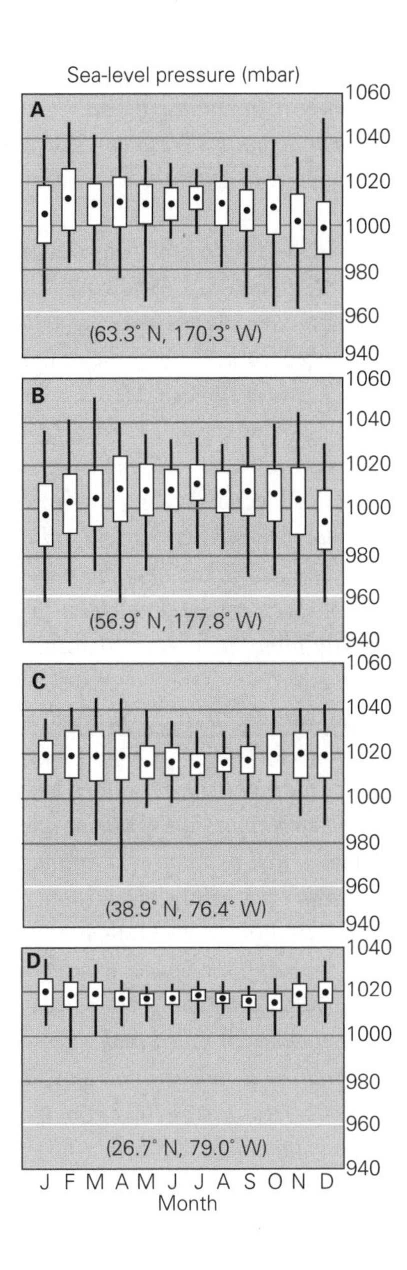

Fig. III.2 Continued

Climate reflects the combined effects of latitude, continentality, air masses, and storms, and is embedded in global atmospheric and ocean circulations. As expected, an increasingly stronger change between seasons occurs from the Bahamas to the Bering Sea. Air-temperature distribution is highest at equatorial and subtropical latitudes, decreases rapidly at middle latitudes, and reaches lowest values near the poles (Fig. III.2, column 1). Air temperatures in the Bahamas are generally 20–30°C year round. Chesapeake Bay air temperatures range annually from below freezing to above 30°C. In the Bering Sea, air temperatures range from well below zero in the winter to nearly 10°C in summer. Also in the Bering Sea, winter temperatures are colder in the northern portion at approximately 63°N than in the southern portion at about 57°N, in part due to the presence of sea ice. Greatest variability in air temperature generally occurs during winter months in all regions.

Sea-surface temperatures (SST) reflect air temperature for each region, with some phase shifts (Fig. III.2, column 2). SSTs result from local heating and atmospheric and oceanic advection. Average values are modulated by transient features (i.e., storms, eddies) embedded in the global circulation. SST variability in the Bahamas is greatest in December through March, in phase with its coldest air temperatures. Chesapeake Bay shows greatest variability in April–May, during a trend toward warmer temperatures. In the Bering Sea, SSTs are relatively cold throughout most of the year. SST variability is greatest in spring and fall, during times of transition between summer and winter. SSTs also lag behind air temperature by about 1 month, because a large extent of the Bering Sea is covered by seasonal sea ice, and the residual effects extend through much of the year.

Surface winds arise directly from the pressure-gradient force generated between relatively high and low pressure (Fig. III.2, column 3). For both the Bahamas and Chesapeake Bay, strong winds occur in August and probably reflect the effects of tropical storms and hurricanes. Winds in the Bering Sea are highly variable during all months, but variability is greatest during winter, when the interaction of cold arctic air and warm Pacific air creates pressure gradients across the region.

Patterns in sea-level pressure reflect the general global atmospheric circulation (Fig. III.2, column 4). Greater variability toward low sea-level pressure is an indication of increased storm-like activity. The Bahamas is influenced by the semi-permanent Bermuda High; the months for greatest variability toward low sea-level pressure are October and February, reflecting the presence of active storms and storm tracks. The greatest variability in Chesapeake Bay occurs in March–April, illustrating the general pattern of a mid-latitude site. The Bering Sea, being located between arctic and temperate air masses, is heavily influenced by the semi-permanent Aleutian Low in winter and the Siberian High in summer, its greatest variability occurring in November and April.

Chapter 5

Chesapeake Bay: estuarine alteration and restoration

As man has uprooted the greatest forests, so can he also annihilate the richest oyster beds . . . The preservation of oyster beds is as much a question of statesmanship as the preservation of forests.

Karl Möbius (1883)

5.1 Introduction

Estuarine systems, worldwide, are undergoing physical alteration, natural resource depletion and ecosystem disruption, with evidence of deteriorating environmental health. As estuaries have progressed toward urbanization, they reflect the benefits of social clustering for taking advantage of resources, and to enhance industry and trade. Estuaries, particularly, also reflect environmental consequences in which change, social constraints, and environmental alterations challenge estuarine resiliency and environmental restoration.

Chesapeake Bay exemplifies this challenge. It is the largest U.S. estuary and one of the largest, most complex, and most productive in the world, and reflects social progress. In a Bay-wide system of long-term self-adjustment, migratory and resident species meet life-history needs as they perform ecosystem roles that benefit human society. Their cumulative roles have yielded important fisheries, transformed chemicals into bioavailable nutrients, and allowed land- and seascapes to recover in response to estuarine change. Social traditions and policies, established under conditions of ecological plentitude during 300 years of social change, have transformed the Bay into an estuarine system in need of restoration. Restoration requires a holistic approach, delineating ecosystem boundaries, and determining roles of species that can return the Bay into a healthy ecosystem. Estuarine fishes, waterfowl, and the Eastern oyster (*Crassostrea virginica*) have played these roles, and their recovery is now a target for restoration.

5.2 Characteristic features of Chesapeake Bay

Chesapeake Bay is a funnel-shaped basin, semi-enclosed by a barrier peninsula (Fig. 5.1). The Bay is fed by numerous freshwater tributaries that drain an extensive, 166 000 km^2 watershed area, which dwarfs the approximately 6500 km^2 Bay surface area (Fig. 5.2). The Bay's surface almost doubles when major tributaries are included. The basin is more than 300 km long, its width varies from about 5 to 30 km, and its mean depth of 8.4 m decreases to 6.5 m if tributaries are included. The middle portion of the Bay and many tributaries have deep channels (some > 50 m) flanked by shallow plains; the southern portion of the Bay contains a broad shoal.

Chesapeake Bay is a coastal-plain, drowned-river estuary. Half of its total water volume comes from freshwater sources, the other half from the Atlantic Ocean. Freshwater, contributed by surface and groundwater, is the Bay's most influential feature. Overland flow contributes more than 90% of the total volume, channeled through watersheds of nine major tributaries (Table 5.1); groundwater accounts for a substantial but unmeasured input. Rivers begin as headwaters at higher elevations, flow through the piedmont, transverse falls and a relatively flat, variable, 24–145 km sedimentary coastal plain, to deliver fluvial products to the Bay. Freshwater input is regionally specific, with half of the Bay's total freshwater entering from the north and west sides. The Susquehanna River alone contributes 50–60% and freshwater marshes are extensive, in contrast to the drier, low-elevation barrier peninsula (Eastern Shore) that adds less than 4% freshwater and where salt marshes are extensive. Atlantic seawater enters from the southeast to interact with freshwater to varying degrees to give the Bay its most distinguishing characteristic – salinity regimes subject to seasonal change.

The Bay is also a partially stratified estuary. The boundary between salt- and freshwater is an important hydrological area, marked by a front, plume, or pycnocline, and is seasonally variable. Fresh- and seawater interactions create varying conditions. As less dense

Fig. 5.1 Place names in the Chesapeake Bay region. The dashed line marks the state boundary. Stars locate state and national capitals; circles locate site names.

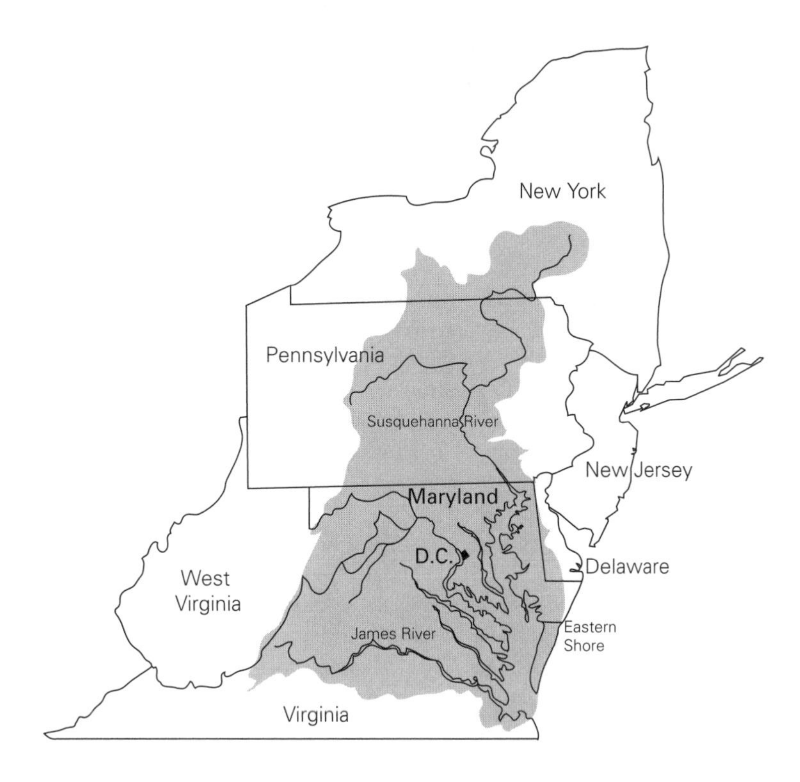

Fig. 5.2 Chesapeake Bay watershed. Freshwater drainage (shaded) crosses several state jurisdictions. From EPA (1983).

Table 5.1 Major tributaries (for map showing location see Fig. 5.1). Total freshwater input to the Bay averages 2300 $m^3 s^{-1}$.

Tributary/drainage basin (Bay location)	Drainage basin area (in square miles)	State jurisdiction(s) [1951–90 flow mean, $m^3 s^{-1}$]	Tributary input and land use
Susquehanna (Upper Bay)	27 000	New York; Pennsylvania; Md [1095]	50–60% freshwater input to Bay. Covers ~ 50% of Bay watershed. Pop. 4 million. For. 66%; Agr. 29%; Dev. 2%
NW Maryland (Northwest)	1 670	Md	Expanding urban belt. Pop. 2.1 million. For. 36%; Agr. 34%; Dev. 17%
Patuxent (West)	957	Md [11]	Contributes ~ 2% total stream flow. Pop. 600 000. For. 42%; Agr. 34%; Dev. 11%
Potomac (West)	14 679	West Va; DC; Md; Va [412]	Second largest freshwater input. Contributes 17% annual inflow. Pop. 5.2 million. For. 57%; Agr. 32%; Dev. 5%
Rappahannock (West)	2 845	Va	Contributes < 4% freshwater input. Low urbanization. Pop. 241 000. For. 57%; Agr. 31%; Dev. 2%
York (West)	3 270	Va	Third smallest input. Pop. 373 000. For. 61%; Agr. 21%; Dev. 2%
James, South (West)	10 432	Va [296]	Third largest watershed. Contributes ~ 14% freshwater to Bay. Pop. 2.5 million. For. 71%; Agr. 17%; Dev. 5%
Eastern Shore (Choptank River is largest river)	5 048 (Choptank = 795 mi^2)	Md; Va	Small freshwater input. Contributes < 1% annual inflow. Pop. 467 265. Agr. 39%; For. 25%; Water 19%; Dev. 2%

Pop., population; For., forest cover; Dev., developed; Agr., agriculture; Va, Virginia; Md, Maryland; DC, Washington, D.C.
Adapted from Boicourt et al. (1999); CBP (2001); EPA (1983); Harding et al. (1999).

freshwater moves down the Bay, and denser seawater moves up, salinity becomes spatially represented as salty (polyhaline), less salty (mesohaline), slightly fresh (brackish), and freshwater zones with boundaries that move with season, tides, climate, and ocean conditions. That is, polyhaline areas can become mesohaline or brackish with spring snow melt and abundant rainfall; in fall, dry conditions create the opposite effect and areas of brackish water may become more saline. Salinity contours (isohalines) identify salinity gradients across the Bay. Due to the Coriolis effect, saltier water moves up the Bay along the eastern shore and fresher water moves down the western shore. Each tributary also has a salinity gradient, which varies with its geometry, watershed coverage, freshwater input, and proximity to tidal influences.

The Bay's hydrological regime, confined in a basin, is subject to intense air–sea–land interactions. Interacting winds and tides create complex conditions. Winds stir the shallow water to alter water masses, increase mixing, and move sediment. Low atmospheric pressure, wind, tides, and basin geometry interact to increase water levels and to deliver forceful surges to shorelines. Under sufficient force, unconsolidated sands, clays, and sediment particles are mobilized, and as one area erodes, another accumulates sediment.

Water movements and shoreline processes differ around the Bay. The benthos and shoreline are continuously modified by stream and overland flows that mobilize sediment in amounts of the order of more than 2000 tons yr^{-1}. Tidal energy creates a progressive wave that travels up the Bay, and within a 24-hour period an entire wavelength is contained when a second wave enters (Fig. 5.3). Semi-diurnal tides, with shifts in ebb and flood influence and direction, force water up and around the Bay in dominant, complex, current-flow patterns. Variable depths, water masses, rough bottom, and boundary conditions alter current amplitudes, turbulence, and velocities. As the progressive wave enters the shallower upper Bay, it is reflected, tidal range increases, and a partial standing-wave results. Lag effects also cause different areas of the Bay to respond differently. For example, on a small scale around slack tide in shallow water, the lag in response-amplitude among measured parameters of salinity, current velocity, and surface elevation varies with time and location (Fig. 5.4), resulting in variable benthic exposures. Lag periods can also be measured in hours, days, weeks, or months, depending on the nature, timing, and duration of the events and spatial parameters.

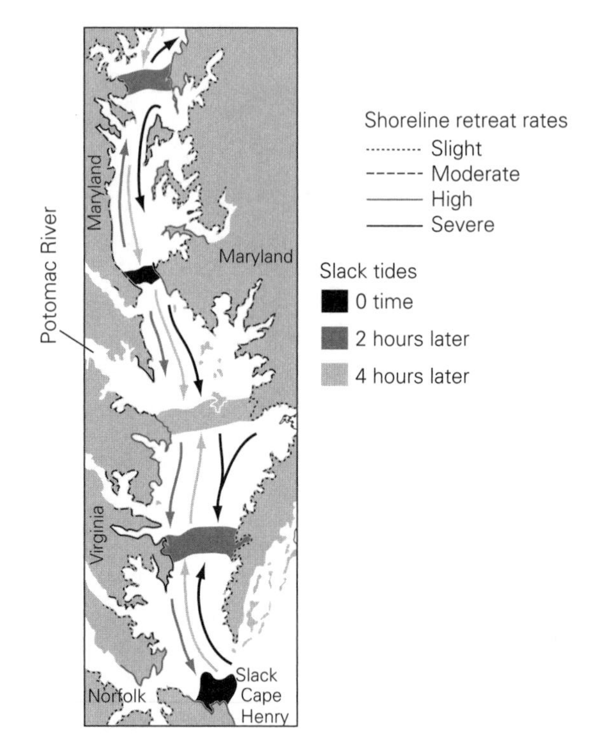

Fig. 5.3 Tide and shoreline retreat. A progressive wave enters the Bay mouth at time 0, subsequently tracked by location of slack tide 2 and 4 hours later; shaded arrows indicate dominant flow at time of slack tide. About every 24 hours another wave enters. Flow speeds are higher at smaller cross-sections, and much reduced in the upper Bay. Complex interactions expose the shoreline to varying degrees of retreat. Compiled from Gross & Gross (1996); Stevenson & Kearney (1996).

The Bay is hydrologically responsive to temporal conditions. Seasons bring changes in air and water temperatures, wind frequency and storms, and duration of light (Figs III.1 & III.2, pages 129–31). In late fall–early spring, extratropical storms ("northeasters") deliver massive amounts of precipitation that dilute seawater, produce erosive waves, mobilize sediment, and increase turbidity. In summer months, hurricanes can strike with great force to alter conditions on a massive scale, in contrast to summer thunderstorms that are only locally significant, short in duration, and erode little. These events can vary in intensity and amplitude and can occur regularly as in tides, irregularly as in storms, or catastrophically in rarer events such as hurricanes. Spring-maximum and neap-minimum tides produce variable current speeds that shift direction and that inundate or expose different locations. Year-to-year variations can also significantly change expected patterns.

These physical processes, interactions, and lag effects all create chaotic stirring and random motion that would predict disorder. Yet, the Bay has sustained a characteristically robust pattern of production, recovery, and biomass yield. Biota capture, store, and process energy and materials through community linkages to estuarine organization, acting collectively to create habitats (marshes, shellfish beds, submerged aquatic vegetation, etc.) that process energy, cycle nutrients, and respond to change. Approximately 295 fishes, 45 crustaceans and mollusks, 29 waterfowl, and 2700 plant species occur in complex associations, their numbers and distribution reflecting environmental conditions, seasons, and physiological domains along a salinity gradient. Highest overall species diversity occurs at the Bay's mouth and the saltier mouths of tributaries and in freshwater, and the lowest diversity occurs in brackish water; however, individual species of taxa (mollusks) exhibit unique patterns of distribution (Fig. 5.5). Estuarine fishes occur from 0 to 30 parts per thousand (ppt), freshwater fishes prefer less than 5 ppt, and fully marine fishes favor more than 30 ppt. Strictly marine or strictly freshwater organisms (stenohaline) dominate higher and lower salinity ranges, respectively. Estuarine fishes also may be permanent year-round residents, or may migrate (Table 4.3, page 105). Dominant benthic residents include blue crabs (*Callinectes sapidus*) and oysters, crabs being scavengers and sessile oysters serving as ecosystem engineers (Fig. 5.6). Both oysters and blue crabs tolerate broad salinity ranges: crabs occur from brackish to fully marine water, depending on sex and age, whereas permanently fixed oysters occur most abundantly in middle salinity ranges (Fig. 5.7). The Bay is also a major habitat for migratory waterfowl that concentrate in October to March to feed. Others remain year round to reproduce and care for young.

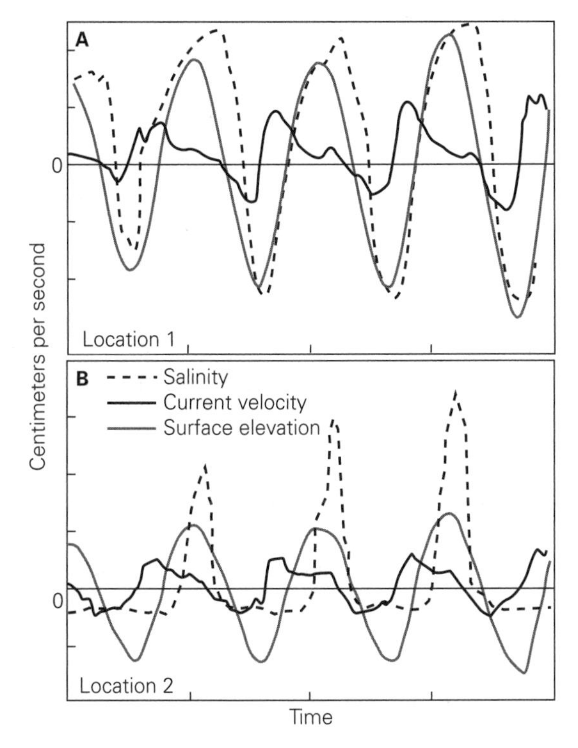

Fig. 5.4 Lag responses expose benthic organisms to asymmetric amplitudes of salinity, current velocity, and bottom surface elevation, observed around slack tide over several tidal cycles and between sites. Adapted from Uncles et al. (1991).

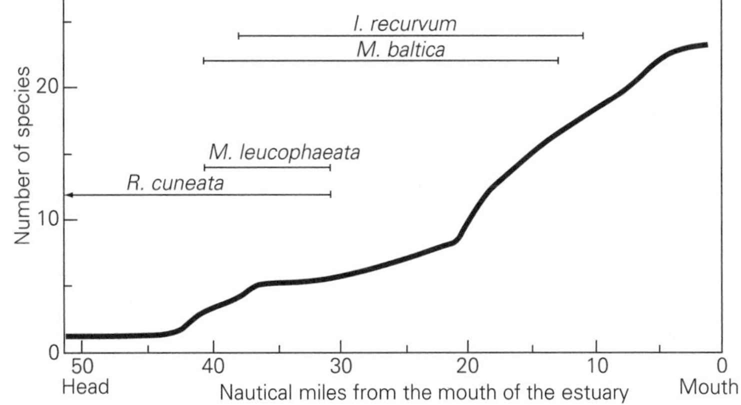

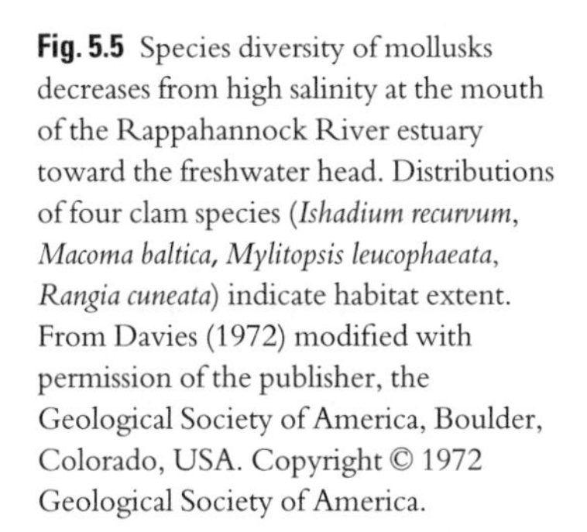

Fig. 5.5 Species diversity of mollusks decreases from high salinity at the mouth of the Rappahannock River estuary toward the freshwater head. Distributions of four clam species (*Ishadium recurvum*, *Macoma baltica*, *Mylitopsis leucophaeata*, *Rangia cuneata*) indicate habitat extent. From Davies (1972) modified with permission of the publisher, the Geological Society of America, Boulder, Colorado, USA. Copyright © 1972 Geological Society of America.

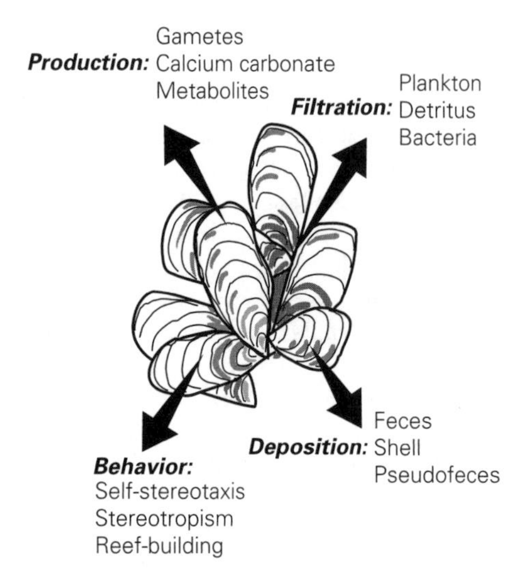

Fig. 5.6 An estuarine engineer. Oysters filter water, release products, deposit materials, and behave (see text for definitions) through specialized adaptations that influence hydrology, biogeochemical pathways, benthic communities, and estuarine function.

5.3 Ecosystem properties under biotic control

Estuarine biota have evolved efficient mechanisms to transform energy into biomass, move chemicals into bioactive compartments, and recover resilient and persistent land–seascape patterns.

5.3.1 Energy conversion

Estuarine species benefit from subsidies provided by wind, tidal currents, circulation, water density and mixing, and pulsed, horizontal and vertical deliveries of materials. Optimal use of these subsidies is apparent in the variety of life-history patterns, sizes, morphologies, and behaviors, as well as the variety of habitats, all energetically linked to biomass production. Survival needs of most estuarine species are met by their undergoing at least three development stages (larval, juvenile, adult), each stage requiring a different diet, habitat, and location. Filter-feeding is a common feeding mechanism, such as for oysters, manhaden, anchovies, and others. Sessile species such as oysters depend on currents for delivery of food and oxygen, removal of waste, and dispersal of offspring; their stationary life subjects them to risks of environmental change, and environmental timing is critical. In contrast, some mobile species meet life-history needs by traveling great distances over the entire mid-Atlantic region (Fig. 5.8). Within the Bay, a species may meet life-history needs by utilizing different areas at different times (Fig. 5.9). Thus, surviving variable estuarine conditions requires specialized adaptations and energy subsidies in which each species' life history requires specific spatial and temporal resources, and all species together collectively contribute to biomass production.

5.3.1.1 Survival, growth, habitat: the oyster example

The oyster is a quintessential estuarine species. It meets variable estuarine conditions by being fecund: one female in one spawning can discharge millions of eggs in warm summer months when the Bay is most biologically active, food is abundant, and risk of storms is reduced. However, the chance for a single egg to become fertilized and mature into a reproducing adult is very small, and timing, opportunity, and luck select winners and losers.

Warm water temperatures of early spring stimulate reproduction. Oysters start life when fertilized eggs enter the zooplankton, and grow for 3 weeks in the water column before they commit to a benthic existence. Eggs the size of a clayey-silt particle (40–50 μm) hatch and then develop into trochophore larvae the size of a silt particle (70–5 μm), then into veligers the size of a fine-sand particle (250–400 μm; Fig. 5.10), as they graze on microalgal cells (< 10 μm). While little is known of their behavior, ecology, and survival in the plankton, evidence indicates that larvae have limited ability to maintain a vertical position within the estuarine current, where they are confronted by hydrological barriers, water viscosity that affects their swimming movements, variations in food supply, disease, and numerous predators. Growth is enhanced in mid-salinity, where benthic structures and slack tide can concentrate oyster larvae with their food particles. Tidal currents may carry larvae to the top of the pycnocline or disperse them widely. The heavier and larger veligers mature into "spat," settle to the bottom, and select a substrate where they commit to a benthic existence by cementation, an irreversible process of electromagnetic adhesion. Spat typically crowd onto substrates, where they grow rapidly and competitively.

Oyster behavior and specialization create conditions for bed and reef formation. Veligers selectively seek oyster shell (self-stereotaxis) and over successive generations a cluster of oysters may vertically accrete

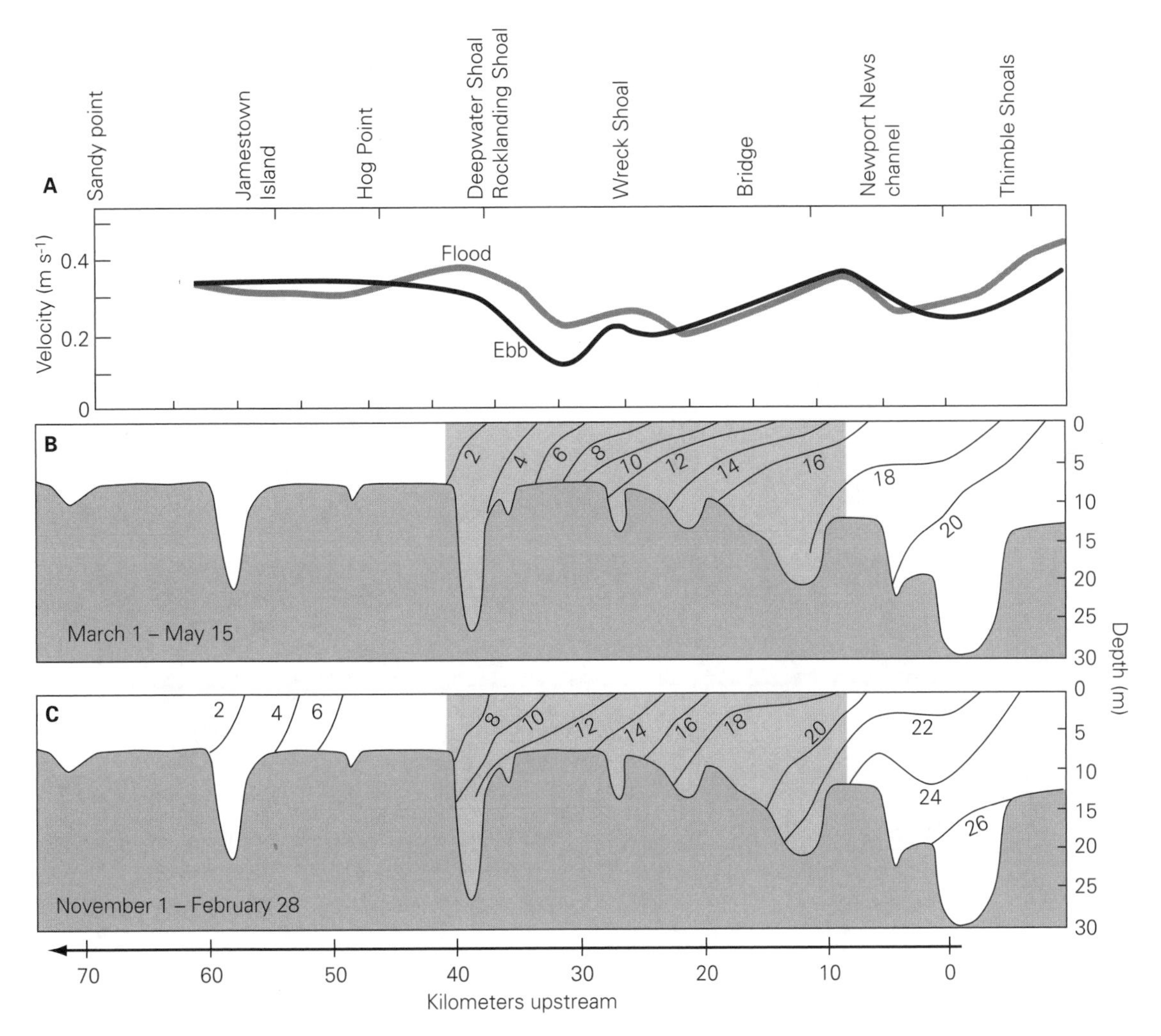

Fig. 5.7 Depth profile, salinity, oyster distribution, and tidal velocities along the James River estuary, at 0 km near mouth to 70 km upstream. Oysters are shown by the shaded area. (A) Flood and ebb maximum velocity averages near bottom, indicating dominance. (B, C) Salinity isohaline averages, 1944–65, in early spring and early winter, respectively. From Nichols (1972), reproduced with permission of the publisher, the Geological Society of America, Boulder, Colorado, USA. Copyright © 1972 Geological Society of America.

(Fig. 5.11). Their growth and orientation are influenced by current-flow direction (stereotropism). Water quality, velocity, circulation, topographic structure, predation, and population density determine survivors. A high degree of specialization is necessary to capture and process food. Currents deliver food particles to actively feeding oysters, which strain the water with specialized, hair-like feeding structures on their gills that generate microcurrents and increase food encounters. An oyster high on the upstream, or channel, side of the bed secures competitive advantages over those downstream, where food abundance diminishes and energetic costs of feeding increase. Many individuals can be smothered under conditions of slow current flow, high turbidity, and sedimentation. Although oysters can tolerate hypoxia for short periods, sustained low oxygen levels can threaten survival. Under optimal water flow, food delivery, sediment-waste removal, and oxygen concentration, tight associations of oysters form a solid ridge along a crest, behind which a bed can form. From

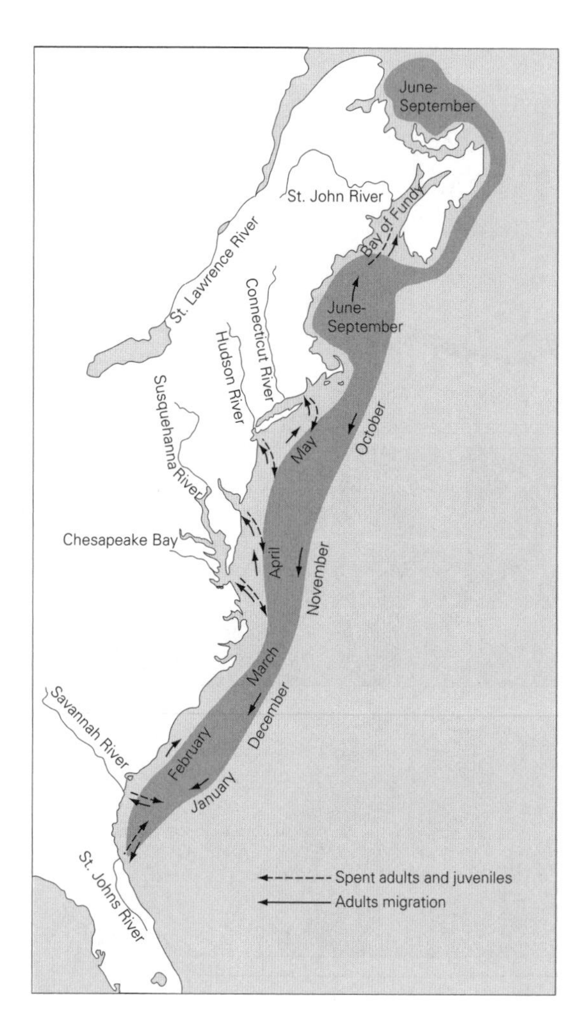

Fig. 5.8 Shad (*Alosa sapidissima*) range over the continental shelf in seasonally directed migratory movements, returning to natal estuaries in spring to spawn. All along their migration route in spring, adults enter rivers to reproduce. Juveniles and some spawned adults exit estuaries in fall.

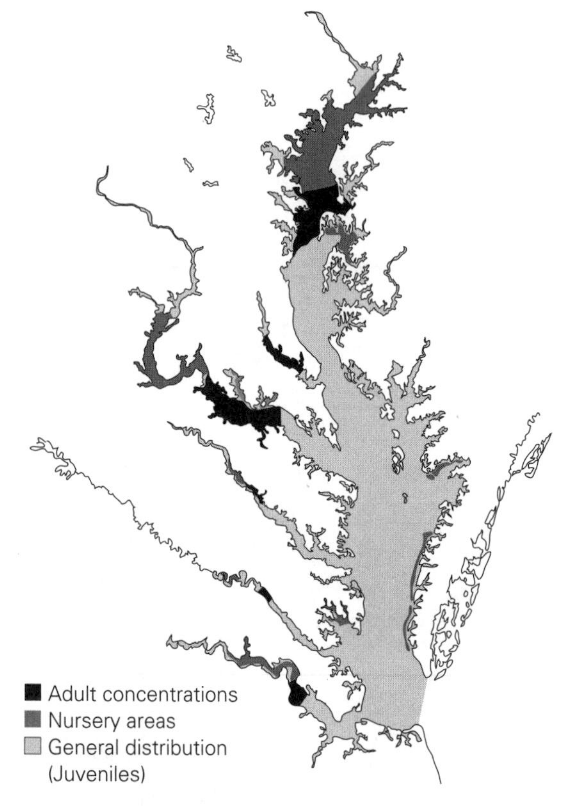

Fig. 5.9 Menhaden (*Brevootia tyrannus*) differentially use parts of the entire Bay and tributaries. From Funderburk et al. (1991).

accumulations of dead shell and feces mixed into sediment, a bed can accrete upward and laterally to form a reef, which intercepts bottom currents, modulates the hydrological regime, increases mixing and particle movements, and modifies the adjacent environment.

Oyster beds form heterotrophic "hot spots" of production. These multidimensional hard structures attract a diverse epizoan community (Fig. 4.1, page 94), where benthic algae, submerged aquatic vegetation (SAV), invertebrates, demersal fishes, and crabs benefit, and to which shorebirds and migratory waterfowl are attracted. During peak biological activity in summer,

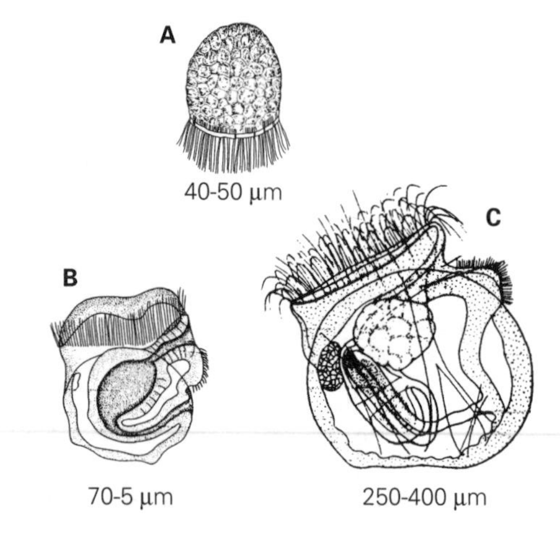

Fig. 5.10 Oyster larvae as microzooplankton in three growth stages: (A) early larva; (B) trochophore; (C) veliger. From Galtsoff (1964).

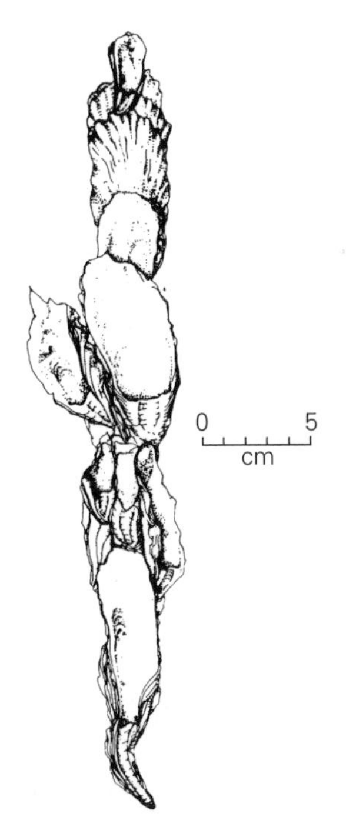

Fig. 5.11 Oysters cluster in vertical growth over successive attachments. From Galtsoff (1964).

beds also attract dozens of species in high abundance, including benthic and demersal fishes. The naked goby (*Gobiosoma bosc*) can occur at a density of more than 40 individuals m^{-2}. Spot (*Leistomus xanthurus*), striped bass (*Morone saxatilis*), and black drum (*Pogonias cromis*) regularly forage on reefs for extended periods of time. Atlantic croaker (*Micropognias undulatus*), hogchoker (*Trinectes maculatus*), and white perch (*Morone americana*) are attracted, and sheepshead minnows (*Cyprinodon variegatus*) feed there at flood tide. And within the oyster's shell, a small parasitic oyster crab (*Pinnotheres ostreum*) secures a specialized habitat. Thus, the bed's complex structure increases species richness and heterotrophic production, which can reach nearly 1500 g C m^{-2} yr^{-1}; community metabolism can be 27 000 kcal m^{-2} yr^{-1}, ranking oyster beds among the highest producers of any benthic community (Bahr, 1974; Bahr & Lanier, 1981).

5.3.1.2 Species and habitat options

Much of the Bay's biomass is reflected in abundant top consumers. Examples include the anadromous American shad (*Alosa sapidissima*), which reach about 75 cm and 5.5 kg, and striped bass (*Morone saxatilis*), which grow to 1.8 m and 57 kg. Their energy efficiency and recruitment success depend on finding suitable feeding and reproduction areas. These fishes require diverse habitats for their spatially extensive life histories. For example, shad use the entire continental shelf to feed (Fig. 5.8), and striped bass migrate over the continental shelf from North Carolina to Massachusetts; both species return to natal tributaries to spawn in fresh to low-salinity rivers and streams. Feeding and reproduction are synchronously timed to temperature, salinity, and hydrology, and tidal currents, channels, and creeks serve as conduits to spawning sites. The first few weeks of these fishes' lives are critical to their survival. Small striped bass larvae prey on zooplankton (copepods, cladocerans, larvae) and grow into juveniles that spend summers in tidal freshwater marshes; they may remain in natal estuaries for up to 2 years. Shad and striped bass transport considerable amounts of ocean-derived metabolic wastes, lipids, nitrogen, and dead bodies to natal streams to benefit other forms of estuarine life.

Species select habitats according to accessibility, availability, phenology, and their specific requirements. Many species, especially juveniles, depend on wetlands, muddy banks, and beds of SAV, where plants grow along a salinity gradient depending on slope, elevation, sediment supply, and nutrients. Tidal marshes form the least erodible shores. Many of the Bay's wetlands are washed by tides, which pump water through networks of meandering channels that absorb tidal energy and facilitate biotic movement (Fig. 5.12). Striped bass, for example, benefit from the meandering flow, seeking tidal freshwater to spawn over sand or mud bottom, where their juveniles find shelter and food in adjacent vegetation. Salt marshes, represented by nine community types in the Bay (Box 5.1), are well adapted to tides: saltgrass (*Spartina alterniflora*) can dissipate more than 50% of the wave energy within the first 2.5 m of marsh. Bivalves that grow along the creek–marsh bank, and the roots and blades of vegetation, stabilize the sediment and dissipate wave energy. Collectively, wetlands support clams, mussels, crabs, vegetation that provide food and shelter for many fishes, crabs, waterfowl, and other wildlife. Abundant larvae and juvenile fishes find protection, shelter, and food in the shallow, nutrient-rich, protected water of wetlands. Alteration in flow patterns and salinity can change the value of wetlands as spawning and nursery grounds for many anadromous species.

Marshes yield exceptionally high net production. Tidal marshes of mostly emergent vegetation trap energy and can reach a high net primary production.

Fig. 5.12 Tidal marsh with natural creek meanders, Eastern Shore, Virginia. Photograph by the authors.

Tidal freshwater wetlands at the heads of tidal reaches produce even greater biomass than salt marshes. This production is consumed directly by wildlife or stored as detritus that is processed by bacteria and fungi to increase nutrient quality. Detritus is a complex reservoir of energy that forms a major food pathway to crabs, shellfish, and fishes (Figs 3.16, page 79 & 4.12, page 102). Most detritus produced by marshes is exported by tidal action to creeks and channels.

Submerged aquatic vegetation (SAV) between marshes and channels provides another habitat option. Fifteen species of SAV inhabit communities in fresh to fully saline water. SAV is adapted to limited water motion, submergence, low nutrient levels, and clear enough water for light intensity to be sufficient for photosynthesis. Most SAV species construct a thick rhizome mat that traps and binds sediments and produces organic detritus that is consumed in large quantities by invertebrates and fishes. SAV forms especially important habitats for developing invertebrates, fishes, and waterfowl. Redhead grass (*Potamogeton perfoliatus*) is a critical food for several diving ducks, and canvasback ducks (*Aythya valisneria*) thrive on wild celery (*Vallisneria americana*). Many animals, such as clams, soft-shell blue crabs, and other waterfowl, historically thrived in abundant SAV beds to feed or seek shelter. These beds are today much reduced.

Oyster beds, seagrass beds, mud flats, and marshes create zoned landscape patterns (Fig. 5.13). In this setting, oyster beds enhance habitat options (Fig. 5.14). Fishes such as tautog (*Tautoga onitis*) and cunner (*Tautogolabrus adspersus*) alternate between reefs and eelgrass and macroalgal beds by day and by night, and seasonally. Mud crabs (*Panopeus* spp.) forage on and around reefs and saltgrass prairies. Blue crabs use oyster beds, seagrass beds, and a variety of other habitats as they move up and down the Bay seasonally in sex-specific migrations. Waterfowl find food and shelter in a mixture of habitats, even during low winter production. Thus, habitat heterogeneity and redundancy enhance diversity by providing options for coexistence in spatially and temporally segregated locations across the Bay.

5.3.1.3 Trophic linkages

Filter-feeding is critical to the Bay's ecological function by clearing the water of particles. What is not consumed by pelagic filter-feeders in the water column falls to benthic filter-feeders (especially mollusks) to support the benthic system. At historical levels, oysters potentially filtered a volume of water in summer equal to the volume of the entire Bay in about 3–6 days; now, depleted oysters would take more than a year. Similarly, the medium-sized filter-feeding menhaden (*Brevoortis tyrannus*) concentrated in the Bay in great numbers to feed on plankton; adults potentially filtered the entire volume of Bay water in less than 2 days. Depletion of oyster filter-feeders represents a fundamental change in the Bay's food-web structure. Evidence suggests that reduced intensity of filtering has resulted in a shift from a food web favorable to benthic species to a pelagic food web more favorable to jellies and microbes. Abundant jellies now thrive in the Bay, and numerous sea nettles (*Chrysaora quinquecirrha*)

Box 5.1 Temperate saltmarsh types of the U.S. east coast

James E. Perry

Marsh types form complex landscapes and are often intermixed. Community names are mostly derived from dominant species.

Saltmarsh cordgrass. Dominated by dense, often monospecific stands of saltmarsh or smooth cordgrass (*Spartina alterniflora*). Occurs from mean sea level to approximately mean high water. Represented by two forms: a tall form, 1.2–2 m high along the water's edge or levees; a short form 0.7 m or less in poorly drained areas behind levees or at slightly higher elevations.

Saltmeadow community. Located landward of the saltmarsh cordgrass community in meso- and polyhaline areas. Also occurs in higher portions of natural levees. Dominant species are saltmeadow hay (*Spartina patens*) and saltgrass (*Distichlis spicata*), which form characteristically dense, low (0.3–0.7 m high), wiry meadows, typically with swirls or cow-licks, and reminiscent of grassland prairies or hay fields. Historically used as a source of cattle fodder or pasture.

Black needlerush community. Interspersed among the saltmeadow community. Common in high marshes of some meso- and oligohaline areas. Black needlerush (*Juncus roemerianus*) nearly always grows in monospecific stands; its dark green (almost black), leafless stem tapers to a sharp point, giving the plant its name.

Tidal saltmarsh community. The only tidal marsh community dominated by woody vascular plants. Occurs landward of saltmeadow and needlerush marshes and usually delineates the tidal wetland/upland ecotone. Dominated by two shrubs: salt bush (*Iva frutescens*) in the lowest physiographic range, and groundsel tree (*Baccharis halimifolia*) in the higher physiographic range. Both shrubs usually reach heights of 1–4 m.

Big cordgrass community. Dominated by big cordgrass (*Spartina cynosuroides*), one of the tallest tidal wetland grasses, reaching 2–4 m high. Mostly occurs slightly above mean high water. Usually forms dense, monospecific stands in oligohaline marshes. Stems are stout, leafy, and have a distinct, coarse-branched flower (seed) head. Leaves have saw-like margins that can lacerate human skin.

Cat-tails. Only one mid-Atlantic cat-tail is common in saline tidal reaches: narrow-leaved cat-tail (*Typha angustifolia*). Usually found in isolated stands in brackish marshes, often near upland margins with freshwater seepage. Broad-leaved cattail (*T. latifolia*), an indicator of high nutrient loads, may also be present.

Reed grass community. A controversial community dominated by invasive reed grass (*Phragmites australis*), a tall, stiff grass up to 4 m tall, with short, wide leaves tapering abruptly to a pointed, purplish, plume-like (feathery) flower head that turns brown in seed. Usually located above mean high water and almost always associated with disturbance, such as placement of dredged sediments, fill, plant die-back, or surface erosion. Usually does not tolerate mesohaline or higher salinity conditions below mean high water.

Pannes. Shallow depressions that form within large saltmarsh cordgrass communities and normally become hypersaline and sparsely vegetated. Usually result from wrack accumulation, which kills plants, or from "eatouts" caused by muskrats or snow geese. Dominated by halophytic, succulent saltworts (*Salicornia virginica, S. europea*, and *S. bigelovii*) 1.5–30 cm tall. Plants may turn dark red by late summer, giving pannes a striking contrast to the yellow-green of surrounding vegetation.

Brackish-water mixed community. A microcosm of all saline-water communities. Occurs in meso- and oligohaline estuarine reaches. No single dominant covers more than 50% of the marsh. High structural and species diversity provides a variety of habitats and wildlife foods.

Sources: EPA (1983); Odum et al. (1984); Perry & Atkinson (1997); Perry et al. (2001); VMRC (1996); Wass & Wright (1969).

consume large quantities of plankton. Production not consumed by macrofauna falls to the muddy benthos for infauna consumption and bacterial decay.

Small floating plants (phytoplankton) sustain zooplankton, invertebrates, and fish. The smallest plankton (0.2–200 μm) constitute the picoplankton, nanoplankton, and microplankton. Among these, blooms of diatoms and dinoflagellates constitute a large floating-plant biomass that produces dissolved organic substances and helps maintain a population of planktobacteria ("microbial loop community"), which in turn are important to phytoplankton and plankton-feeders. Bacterioplankton are the dominant consumers of dissolved organic matter (DOM), transforming it into particulate forms. Numerous tiny microzooplankton predators (rotifers, protozoans, oyster larvae) graze on

Fig. 5.13 Zoned communities along tidal marsh, from exposed oyster bed at low tide, backed by mudflat, brackish marsh, and higher forested land. Photograph by the authors.

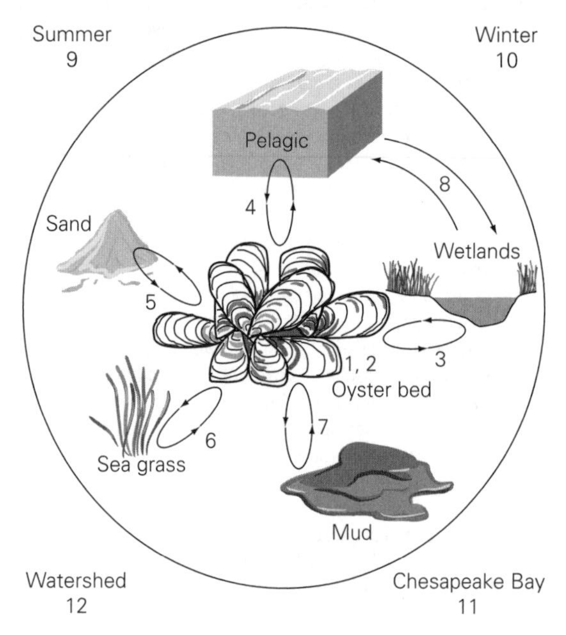

Fig. 5.14 Multiple scales of oysters' role in estuarine complexity/diversity. (1) Oysters as colonizing substrate. (2) Hard reef structure attracting species. (3–7) Interchange with different habitats. (8) Indirect, adding optional choices for species. Direct/indirect support of (9) summer species, (10) winter species. (11) Fringing reef system in estuarine Bay. (12) Regional presence in mid-Atlantic watersheds influencing ocean species.

microphytoplankton (e.g., diatoms) and bacteria. Larger predatory mesozooplankton (copepods, cladocerans) are important food for larvae, juveniles, and adults of many organisms.

The small, filter-feeding bay anchovy (*Anchoa mitchelli*) is an important forager of zooplankton, and its eggs and larvae constitute an important food for other plankton-feeders. Juveniles of filter-feeding herrings (shad, menhaden, etc.) a centimeter in length graze on abundant microplants and larvae, and as the larvae grow, they feed on larger zooplankton. Adult shad feed on larger-sized plankton and small fish. Spot (*Leiostomus xanthurus*) feed on benthic invertebrates, and use nearshore waters of the Bay variously, according to size, age, and food availability. Spot, bay anchovy, and menhaden provide food for striped bass, bluefish (*Pomatomus saltatrix*), weakfish (*Cynoscion regalis*), and flounders, that in turn are consumed by tuna, sharks, ospreys, eagles, and people. Diverse waterfowl and shorebirds feed on fish, invertebrates, and plants, and bottlenose dolphins (*Tursiops truncatus*) enter the Bay to consume a variety of medium-sized fishes. This size-oriented, food-chain coupling strongly influences energy transfer efficiencies and community metabolism, the outcome of which is the complex route to biomass production.

Bay production and consumption vary spatially and temporally. In early spring, with increasing river flow, warmer temperature, and more daylight, plants begin to grow and phytoplankton bloom. Primary production remains high through summer, especially in the mesohaline portion of the Bay, is reduced in fall, and is low in winter (Table 5.2). Fish diversity is greatest in late summer–early fall (August–September). As fall progresses, boreal fishes such as hakes (*Urophycis* spp., *Merluccius bilinearis*) enter the lower Bay to feed, leaving

Table 5.2 Seasonal carbon production in different communities. Mesohaline portion of the Bay.

Community	Seasonal % annual production				Annual production g C m^{-2}
	Summer	Fall	Winter	Spring	
Primary producers	39	20	14	27	347.00
Water-column invertebrates	55	11	6	28	209.00
Macro- and meiobenthos	48	16	4	32	33.70
Nekton	41	25	6	28	0.45
All communities	45	17	11	27	595.50
	% net production, estimated				
Phytoplankton channeled through microbial loop	50	30	28	25	

Modified from Baird & Ulanowicz (1989).

when cold winter temperatures arrive. Low diversity and density of bottom fishes occurs in mid-winter. In late winter–early spring (February–March), the anadromous alewife (*Alosa pseudoharengus*) and American shad (*Alosa sapidissima*) ascend Bay tributaries to spawn. Mature striped bass spawn in April–May, feeding variously on menhaden, spot, and other fishes, according to their abundances. In early summer, juvenile menhaden move into tributaries and remain there until fall, feeding on detritus, benthic diatoms, and zooplankton.

In late fall and winter, adult spot spawn outside the Bay. Spot are most abundant in Chesapeake Bay from spring through fall, favoring muddy areas and reefs in depths of 6–10 m, where they concentrate and where their abundance is correlated with benthic macroinvertebrate decline. Distribution patterns of spot vary with season and age (Fig. 5.15); in spring, their planktivorous larvae, juveniles, and adults are distributed in various age classes throughout the Bay and tributaries. In April–May, spot larvae occur in tidal creeks of more

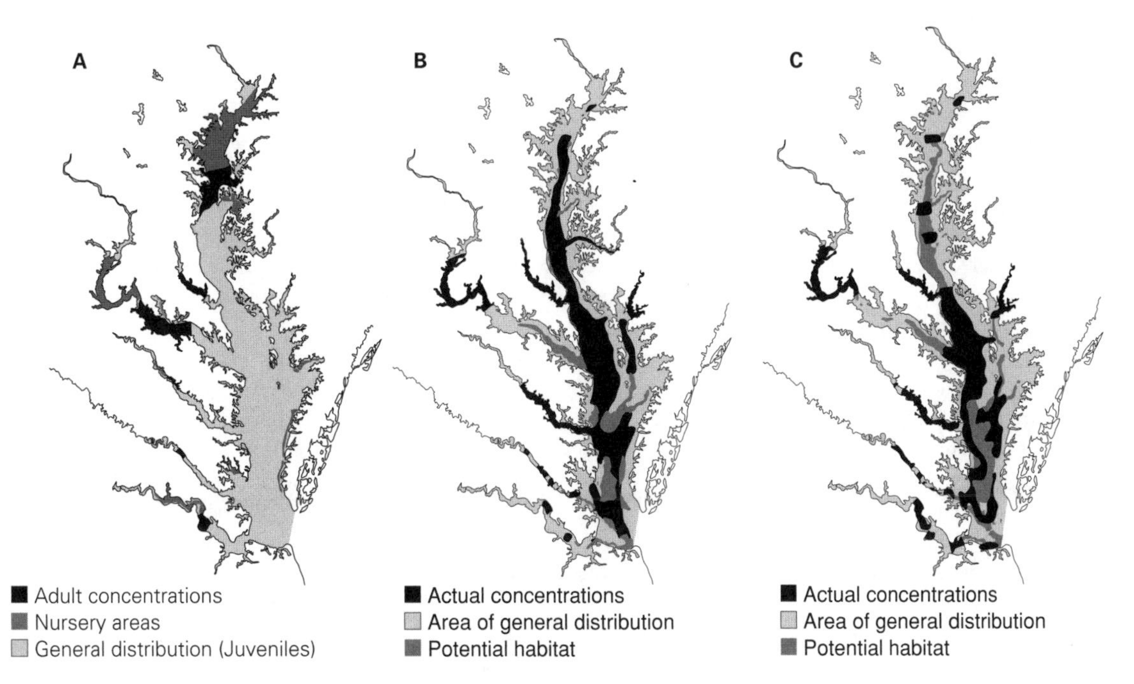

Fig. 5.15 Spot (*Leistomus xanthurus*) meet life-history needs in seasonal shifts of spatial use across the entire Bay: (A) spring; (B) summer; (C) fall. From Homer & Mihursky (1991).

than 5 ppt salinity. Most juveniles, which feed on small benthic invertebrates, leave estuaries in fall. Spot play a key role in estuarine dynamics, regulating benthic community densities, structuring microdistributions, and resuspending surface sediment through feeding bioturbation.

From May through September, bay anchovies concentrate; adults reach peak biomass of more than 23 000 tons in June–July, when spawning egg density reaches 10–1000 m^{-3} and larval densities reach 100 m^{-3}. Anchovy abundance in July shifts from the upper to the lower Bay. A mixed, mostly juvenile fish community of bay anchovies, Atlantic silversides (*Menidia menidia*), striped mullet (*Mugil cephalus*), pinfish (*Lagodon rhomboides*), spot, croaker (*Micropogon undulatus*), and Atlantic menhaden can potentially make up a substantial total fish biomass and standing crop. Thus, the Bay sustains a healthy population of predators at most times of year.

5.3.2 Biogeochemical processing

Estuarine systems are centers of biogeochemical processing, where the physical and chemical environment, boundary conditions, fluxes, and organisms play essential roles in exchanges, transformations, and regeneration of carbon, nitrogen, phosphorus, and other elements. The biota strongly influence estuarine metabolism, regulating the release of chemicals and determining routes taken by bioactive ones. Many estuaries are supersaturated with carbon dioxide and provide a net source to the atmosphere, whereby releases of carbon dioxide and methane can influence climate. Nitrogen and sulfur undergo a number of transformations, and their gases are also released to the atmosphere. However, information on chemical movement through estuarine systems is only grossly understood. How biota trap, release, transform, and move chemicals in estuaries and what roles organisms play in biogeochemical cycles are active fields of research. Where species occur, how they behave, and when they concentrate often give clues to sources of geochemical processing otherwise difficult to detect.

The physical environment sets boundary conditions. Depth, light, water motion, and other variables (degree of saturation of carbonates and oxygen, chemical adsorption at the water–suspension boundary, flocculated aggregates, deposition, etc.) affect chemical transformation. Salinity plays a significant role, since most equilibrium and rate constants in chemical transformations are salinity-dependent. Temperature is critical in rate processes that affect gases (oxygen, carbon dioxide) and the life histories of biota. These physical parameters vary with geomorphology and hydrological behavior, and with fluxes in freshwater and tidal flow, tidal and wind energy, seasonal change, and climate. Sediment plays an important role in sequestering chemicals, whereby tidal dynamics, river discharge, density change, flow patterns, particle dynamics, and the presence and absence of organisms control chemical transport and trapping. Fine-grain sediments are an important reservoir for nutrients, and metals such as mercury associate with particulate organic matter in sediment.

5.3.2.1 Estuaries: large-scale chemical filters

Chesapeake Bay acts like a large-scale filter, processing large amounts of organic matter, inorganic nutrients, and sediment delivered from terrestrial, marine, and atmospheric sources (Fig. 5.16). It performs this role through biogeochemical processes taking place across the watershed Bay system, and regulated most intensely at the dynamic interfaces of land, water, and air.

Residence time is an important measure of a chemical's distribution and transport speed, relative to rates of biogeochemical processes. The different areas of the Bay and tributaries have different residence times, and chemicals move slowly through groundwater, sometimes taking decades before reaching Bay water. Once in the Bay itself, chemicals can be retained for about 1 year before exiting with the estuarine plume, unless consumed by organisms or sequestered in sediment, in which case residence time can be many years. The length of the Bay provides ample time for recycling or burial by sedimentation, and river flow determines rates of flushing. Residence time is one parameter for determining the percent of total nitrogen input that is denitrified.

Chesapeake Bay provides an arena for observing the exchange, trapping, and transformation of chemicals. In intense biogeochemical processes occurring through the system and at discontinuous boundaries, different properties of land, air, water, and biota play essential roles. Large-scale surface discontinuities form sharp transition zones over relatively short distances and under varying environmental conditions, bringing complex chemicals into close proximity. Boundary conditions regulate gas and heat exchange, light penetration, and electrical charge, as well as collecting particles, influencing such important parameters as photosynthetic light, temperature venting, ultraviolet radiation penetra-

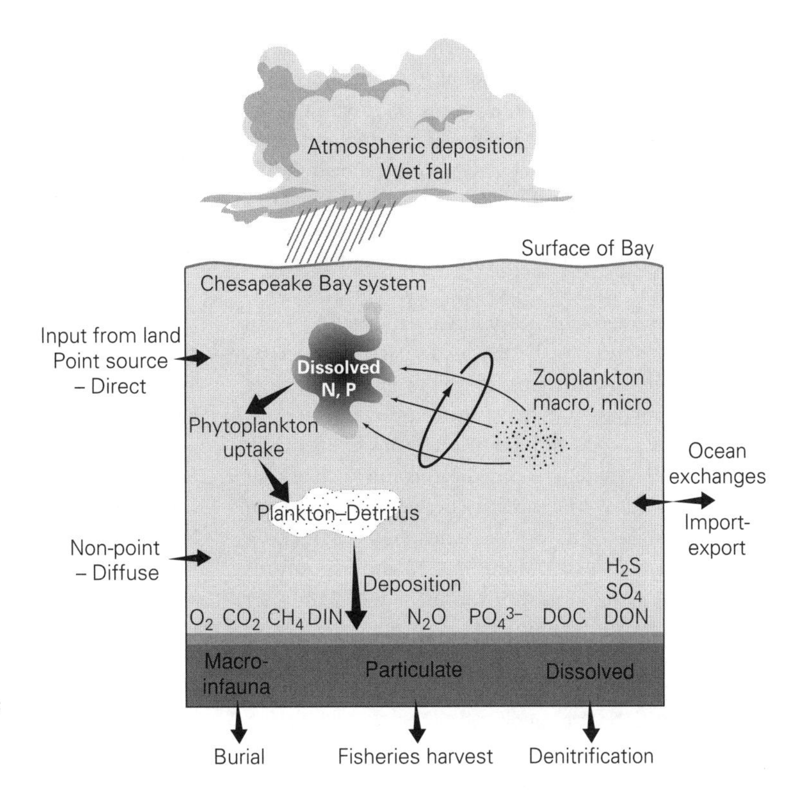

Fig. 5.16 Biogeochemical flow model concept for the Bay. DIN, dissolved inorganic nitrogen; DOC, dissolved organic carbon; DON, dissolved organic nitrogen. Adapted from Boynton et al. (1995).

tion, mechanical disturbances, mixing, and compound transformation. Photochemical processes release labile nitrogen and phosphorus compounds, such as ammonium, amino acids, and phosphate, from dissolved organic matter (DOM). At boundaries, organisms are subject to extreme environmental differences of temperature, salinity, oxygen, and mechanical forces; adapted species may not survive if conditions change.

Nutrients that stimulate primary production, such as nitrogen and phosphorus, come from rivers, groundwater, and aerosols. As saline water is normally limited in nitrogen and phosphorus, a nutrient gradient is created seaward, from high to low concentrations. Nutrient inputs around the Bay vary, resulting in areas of high (eutrophic), medium (mesotrophic), and low (oligotrophic) biological activity. And as some tributaries are typically eutrophic, others may range from being occasionally eutrophic in years of exceptionally high riverine nutrient loading to being mesotrophic during periods of drought. Marsh vegetation performs an important role by filtering sediment, nutrients, and dissolved chemicals flowing into Bay waters.

An area receiving much attention is the mesohaline region. In this mid-salinity (mesohaline) area, currents slow, sedimentation and flocculation are high, and turbidity is at its maximum. The Bay's mesohaline area is dominated by flow from the Susquehanna River and is located north of the Potomac River; its average water residence time is about 42 days. In this area, clays flocculate and high concentrations of suspended sediment settle to the bottom. Organics may reach 80% of the total particulate load, mostly derived from living matter. Bay models suggest that this area can harness an annual production of approximately 600 g C m^{-2}, a value that varies greatly with season (Table 5.2). In summer, phytoplankton reach peak production to account for 33–44% of the Bay's annual carbon production. This important biochemical arena is where brackish marshes are extensive and oysters were historically abundant, and is where anoxic water now prevails in summer.

5.3.2.2 Biota: agents in biogeochemical processes

Organisms have evolved specialized mechanisms to exploit many chemicals and regulate their exchanges at various scales of ecological organization. Minerals play major roles in skeletal formation, with calcium carbonate, calcium phosphate, and silica being principal compounds. Organisms extract chemicals bonded on particulates or dissolved in solution, and remove or accumulate the most soluble forms of iron, calcium, and magnesium, along with pollutants and toxicants, from surface or groundwater. Organisms metabolically transform and excrete dissolved and particulate chemicals in forms available to others in the community. Processes of metabolism release DOM and estuarine gases (carbon dioxide, oxygen, methane, sulfides) through respiration, photosynthesis, and decay. Fecal pellets are a source of DOM, and anaerobic bacteria release sulfide and methane, as well as gaseous forms of nitrogen from processes of denitrification.

The biota process and redistribute tons of chemicals within the Bay system. Rooted plants move chemicals above and below the surface, between aerobic and anaerobic soils, and to the atmosphere. Migratory waterfowl and fish transform, import, and export tons of chemicals. Migratory fish can provide a significant flux of metals. Anadromous fish, in particular, transport and deposit significant amounts of marine-derived organic matter into estuaries. A population of 10 million juvenile Atlantic menhaden, spot, and pinfish in summer months have been reported to defecate a total of about 1 kg of zinc, 56 kg of iron, 1 kg of manganese, and 0.2 kg of copper (Cross et al. 1975). These important trace metals and other chemicals are also redistributed when concentrations of fish graze in areas of high production.

Microbial communities are especially important in biogeochemical activity, through rapidly decomposing a tremendous bulk of organic matter that passes through estuaries. Aerobic *Nitrosomonas* and *Nitrobacter* bacteria are responsible for mineralizing nitrogen into soluble nitrogen compounds for plant growth. Denitrifying bacteria remove excess nitrogen from the water column and transform it into inert nitrogen gas (Fig. 3.18B, page 82). Sulfur bacteria (e.g., *Desulfovibrio*) both oxidize and reduce sulfur, the fourth most abundant element in seawater, and can remove sulfur compounds from seawater to produce toxic hydrogen sulfide. In freshwater, bacteria produce methane. Under conditions of high nutrient concentrations, adequate light, and warm surface waters, blooms of planktonic algae transform nitrogen and phosphorus. Heterotrophic bacterioplankton can salvage a portion of carbon lost from the planktonic food web as dissolved organic carbon (DOC). If nutrient input increases, biological activity also increases, and in excessive amounts nutrients can degrade, toxify, pollute, and alter biological activity.

5.3.2.3 Boundary zones

The intertidal land–sea boundary of variable landscape morphology and hydrology creates favorable conditions for extensive wetland areas surrounding the Bay. Wetlands occur where water saturates the land, especially in periodically flooded, low-lying areas along rivers, streams, lakes, and estuaries. Subsurface water flow often links wetlands to one another or to streams. Groundwater coming to the surface also creates wetlands. An isolated wetland flooded by heavy rainfall forms a pool, and under drought conditions may dry. Wetlands are estimated to cover approximately 3%, or about 5000 km^2, of the Bay watershed, being characterized by dominant species in fresh, brackish, and saltmarsh wetland types. This horizontal and vertically fluctuating boundary also plays important roles in biogeochemistry.

The Bay's extensive intertidal wetlands intercept nutrient flows from land and are exposed to surface-water changes of oscillating tides. These highly variable and chemically active areas are subject to extreme change when alternately exposed to air and submergence, to freshwater leaching, and to periodic tidal pulsing that advects the salinity gradient. Fresh- and saline water ions can alternately affect soil salt content, ionic charges, and hydraulic conductivity. Slightly acidic freshwater delivers humic acids and other fluvial chemicals that reflect the underlying lithology and weathering of soils, and groundwater seepage exerts control on the flux of dissolved substances between intertidal sediments and floodwater. Slightly alkaline seawater delivered by tides contributes sodium (Na^+) and chloride (Cl^-) ions that, along with many other chemicals, change soil oxygen content and pH, which dramatically affects the solubility of many metals. For example, low pH (acidification) leads to dissolution of some metals from sediments. Chemicals in tidal sediments are thus subject to cation exchange processes that are sensitive to salinity, to daily and seasonal temperature change, and to freshwater, oceanic, and climatic pulses.

Intertidal wetlands process sediment and organic matter that upland forests fail to remove, along a salinity gradient from tidal fresh marsh to salt marsh. Freshwater marshes have high plant diversity, brackish marshes are dominated by rushes, and salt marshes are dominated by a few salt-tolerant grasses. These plants control the flux of chemicals, through processes of adsorption, flocculation, precipitation, and sedimentation. Dissolved and oxidized inorganic chemicals pass through tidal-marsh plants, which build root and plant biomass, respire, and export dissolved, particulate-reduced chemical forms. High amounts of organic material, oxidized rapidly by respiratory metabolism, can create anaerobic soils, which vary with tidal mixing and salinity. And within sediments, oxygen content, bacterial activity, and organic matter directly affect the oxygen-reduction potential (redox), which is high under aerobic conditions. Changes in sediment redox conditions alter pathways of sediment mineralization, and high salinity inhibits ammonium production and nitrite-oxidizing bacteria. Saltmarsh sediments, rich in hydrogen sulfide, organic matter, and aerobic and anaerobic bacteria have high oxygen-reduction exchanges, in which the plant–substrate–water interface actively exchanges chemicals through metabolic activities. Molecular oxygen is required for most soil heterotrophic animals and for degradation of some organic compounds (e.g., aromatic hydrocarbons, lignin). However, lack of oxygen does not inhibit anaerobic microbial activity or decomposition of most compounds. Animal activity, such as, fiddler-crab burrows reduce anaerobic conditions by extending a ready source of oxygenated water into the sediment.

Salt-tolerant plants store metals, stabilize shorelines and favor hypersaline conditions, thriving in submerged and waterlogged anoxic sediments. Here, freshwater availability is low, sediment is fine, and summer temperatures are hot. The deep roots of *Spartina alterniflora* support a community of microbes and protozoa, and are well adapted to obtaining nutrients under anaerobic conditions. However, under salinity stress, growth is inhibited. Bordering tidal marshes are shallow, intertidal mudflats and submerged plants. Mudflats trap particulate phosphorus, consume nitrate, and release dissolved phosphate. Subtidal aquatic plants significantly increase oxygen and also influence sediment geochemistry by enhancing regeneration of nutrients, accelerating nitrogen fixation, and increasing diffusion of phosphorus from sediment to water. Benthic algae remove phosphate from seawater and eelgrass beds process organic matter and pump phosphorus between bottom sediments and the water column.

At the air–water interface of the Bay, heat, mass, moisture, momentum, trace gases, and particulates are exchanged. Properties of air and water meet at this interface, which is subject to intense differences in exposure to temperature, oxygen, evaporation, radiation, humidity, etc. At this interface, ultraviolet radiation is absorbed and long-wave radiation is re-emitted. Wind, rain, and light affect the surface, where molecular diffusion and physical exchange processes occur, and winds create conditions for mixing and gas exchange by reducing surface tension. Radiation, turbulent and molecular exchange, and evaporative cooling maintain a heat balance between water and air. In the water, processes of metabolism, decay, and cell breakage release gases into air, and winds and breaking waves release sea salts and chemicals that can be carried great distances in aerosols. And aerosols from land may carry bioactive chemicals (compounds of nitrogen, oxygen, carbon, etc.), as well as pollutants that are deposited with snow, ice, and rain. At this interface, a thin surface film is composed of small plankton (nanoplankton, $\sim 2\ \mu m$), disintegrated parts of organisms, and bacteria. This surface film collects many active substances and floatable materials, such as oils, debris, and silt, which affect water transparency and influence air–water exchange. Surface films are visible as foam lines along tidal fronts, windrows, and regions of water discontinuities observed as fronts.

A third boundary is the estuarine front, which may be observed as a salt wedge, plume, or tidal intrusion that separates water masses at a freshwater-saltwater interface. Fronts can affect sea state, acoustic propagation, and coastal meteorology *via* mists and fog. Generally, a front is dynamically "active" where current velocity drops, pH and ionic composition are altered, chemical changes occur, and nutrients are absorbed and released through differential exchanges. At fronts, hydrocarbonic freshwater ions meet seawater ions at a sharp transitional, but often diffuse, boundary of only a few hours' duration. Fronts are also regions where circulation favors high nutrient concentrations and high phytoplankton biomass that attract herbivore populations, such as juvenile menhaden and blue crab larvae. Metals and other pollutants can also concentrate and become incorporated into the trophic system at fronts. And, along a fresh-to-saline gradient, the different ionic properties of salt- and freshwater create conditions for flocculation involving colloids, which can be carried along a salt wedge. The hydrochemical boundary that occurs at about 2 ppt is the most sensitive salt–fresh transformation area, but transformation of ionic composition is active up to 6 ppt. At this low-salinity area, metals and phosphorus are mostly removed, biotic diversity is low, and chemical and biological reactions are rapid. Transformation is variable to 10 ppt and practically inactive at higher salinity.

The salt–freshwater interface is an important boundary that stratifies the water column in summer. As buoyant, summer-heated freshwater flows seaward over colder, dense, bottom seawater flowing landward, an intense density change separates the water column into a two-layer flow in opposite directions. Stratification at the pycnocline (a region of steep density change characterized by intense mixing) sets up an internal circulation pattern known as a residual, or estuarine, current. Stratification results in an entrainment velocity, in which water and salt are transported in a vertical motion from the lower layer into the upper layer. This vertical circulation is an important feature of a partially mixed estuary like the Bay; it creates an estuarine current that can function as a conveyor belt for nutrients and planktonic organisms, and can retain them within a segment of an estuary (Fig. 5.17). Hence, a phytoplankton bloom occurring in surface flow, can sink and be carried landward in bottom flow. Entrained particles (plankton, sediment) may be deposited at a hydrological "null point," an area of no-net-motion and high turbidity (Fig. 5.18) whereby all but the fine-clay fraction of mineral particles are deposited over a rather short distance. Stratification effectively suppresses vertical exchanges important to bottom oxygen, and after a spring algal bloom and organic mineralization in surface water, respiratory conditions in the bottom layer may create anoxia. Seasonal winds, storms, and precipitation can change mixing energies, destratify, and oxygenate the bottom.

The sediment–water interface is another critical biogeochemical arena, where organic degradation, nutrient recycling, gas exchange, and denitrification take place under shallow-water conditions. Sedimentation from a rain of reworked particles falling from the water column adds a substantial fraction of particulate nitrogen, phosphorus, and silica to the benthic sediment, and over seasonal time scales sediment returns significant amounts of dissolved nutrients back to the overlying water. As benthic conditions can vary, sediment type and vertical profile play key roles, with water depth, current speed, particle size, and deposition rates being important factors. Also, water motion, temperature, adsorption, and biotic processes sort benthic sediment, with grain size and very small changes in current speed or bed roughness affecting particle transport and oxygen availability. If sedimentation is too slow, benthic fauna may lose habitat; if too rapid, plants and animals may be smothered. Well-sorted, mixed particle sizes indicate grain movement and presence of oxygen, with coarser grains free to move independently. Under high oxygen concentrations, phosphorus bound to sediment particles is released slowly, and a heterogeneous benthic community can flourish.

On the benthos, infauna and epifauna play important roles in the Bay system. Their feeding activities influence chemical routes, oxygen, and sediment chemistry. For example, marine animals normally accumulate metals essential to cellular enzyme functions in concentrations 4–5 times greater than surrounding seawater. Oysters construct calcium carbonate shell from calcium and carbonate ions absorbed from water; through filter-feeding, they absorb dissolved chemicals, remove carbon and nitrogen from the water column, and deposit chemicals to the benthos. One hectare of oysters can process as much as about 20 mt of suspended detritus annually, deposit about 500 kg (dry-weight) of fecal pellets per week, and also leave shells when they die. Their fecal pellets contain 77–91% inorganic matter, 4–12% organic carbon, and about 1.0 g kg^{-1} phosphorus. Fecal pellets and shells together create a hard, organic-rich bottom, and alter the textural and chemical properties of sediment. This hard, rough bottom increases turbulent mixing and oxygenation. Metabolizing oyster communities also export ammonium, soluble reactive phosphorus, and dissolved organic carbon to neighboring habitats (Fig. 5.19).

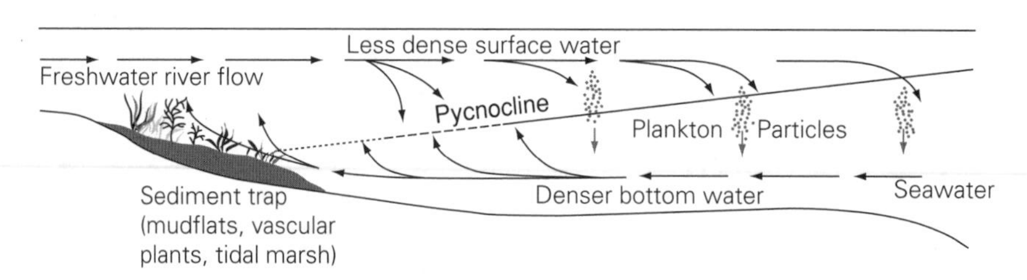

Fig. 5.17 Estuarine current is created by buoyant freshwater flowing seaward over heavier seawater flowing landward, between which there is intense change (pycnocline) and mixing. A residual current is recognized whereby particles can be carried and retained, and deposited at slack. From Correll (1978), with permission.

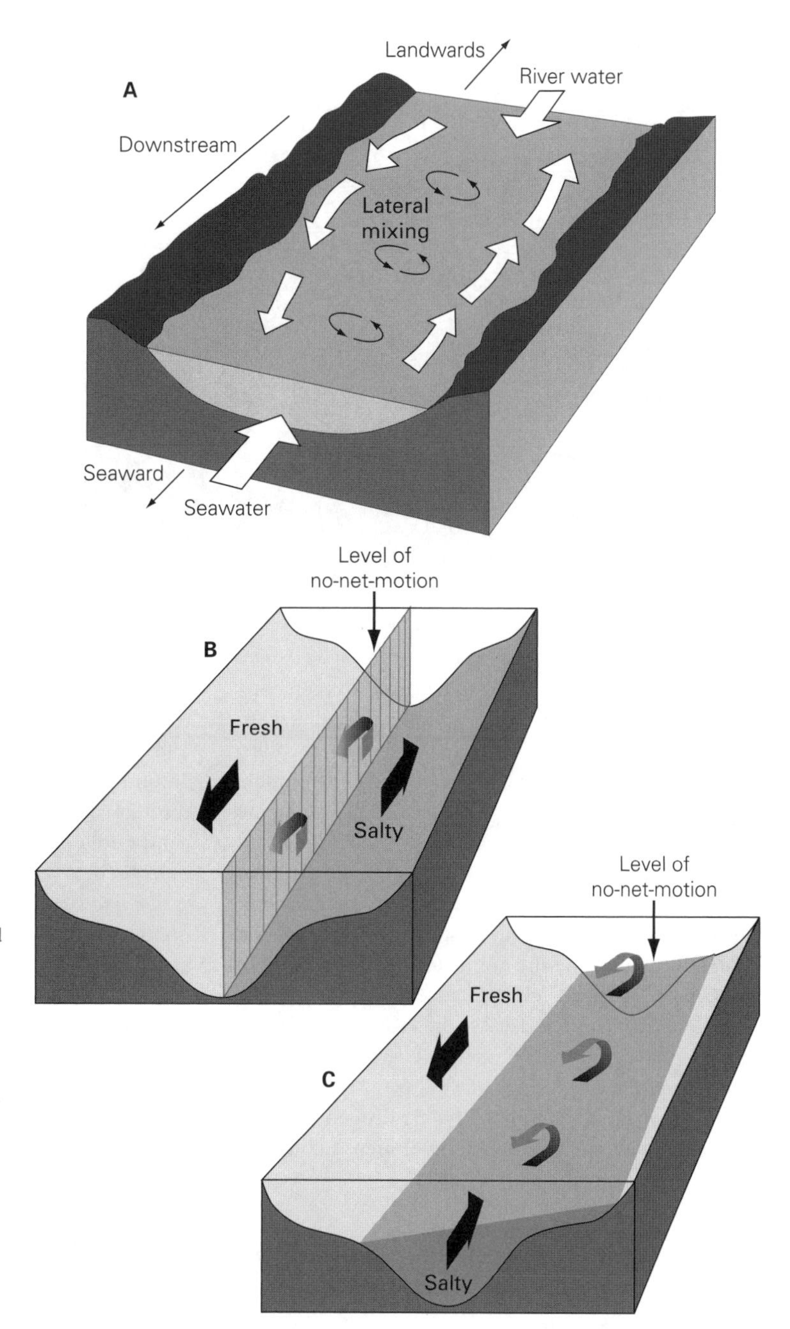

Fig. 5.18 Net, non-tidal residual circulation pattern of a well-mixed, vertically homogeneous estuary (A,B) and a moderately stratified, inhomogeneous estuary (C). Schemata illustrate (A) lateral circulation; (B) vertical two-way circulation with vertical no-net-motion; (C) horizontal two-way density circulation pattern at mid-water interface, with slight horizontal-tilt no-net-motion due to the Earth's rotation. (A) from Bearman (1989) with permission; B–C from Nichols (1972) reproduced with permission of the publisher, the Geological Society of America, Boulder, Colorado, USA. Copyright © 1972 Geological Society of America.

Silts and clays of marshes, mudflats, and benthos create soft sediment and a sink for metals. Biologically produced films and clays (< 2 μm in particle size) result in cohesive, sticky, soft sediment that is not chemically homogeneous with sediment depth. Some metals typically concentrate at the top of the sediment. In interstitial water (pore water, which occupies the space between small sediment particles) important chemical transformations take place depending on oxic or anoxic conditions. Under anoxic conditions the benthos changes and few heterotrophs occur, benthic diversity is reduced, and chemotrophs (bacteria) thrive. If

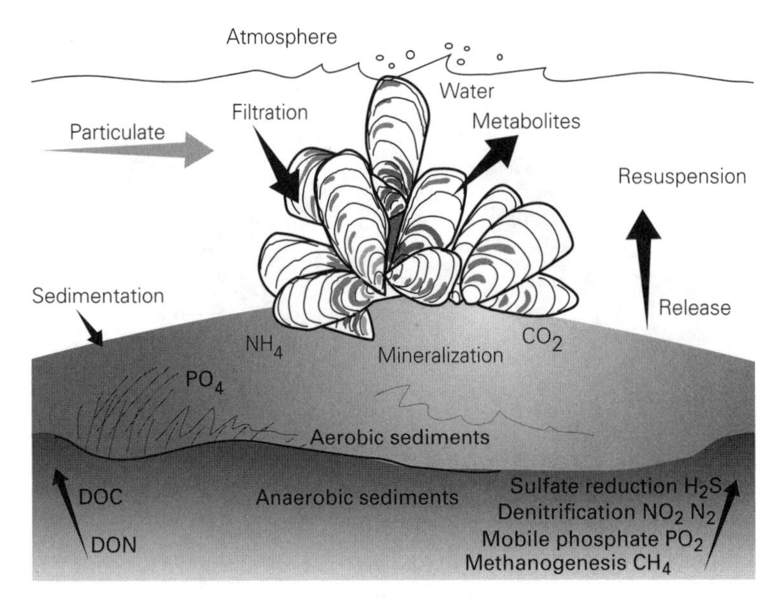

Fig. 5.19 Role of oyster bed in biogeochemical processes. Dense populations remove large quantities of suspended particulate and dissolved organic material, release metabolites, remineralize nutrients in forms useful to phytoplankton, and move carbon, nitrogen, and phosphorus at faster rates than the pelagic food web. DOC, dissolved organic carbon; DON, dissolved organic nitrogen. Modified from Dame (1996).

sediment is shut off from the overlying water by cohesive clay particles and organic matter, oxygen concentration declines in pore water and bacterial activity creates reducing conditions and hydrogen sulfide. With sulfide ions present, metals can form highly insoluble sulfide compounds that have little chance of leaving the sediment. In clays, the relative proportions of cations shift, and dissolved iron may react with sulfide, phosphate, or carbonate to form new compounds. Phosphorus bound to sediment particles is released relatively rapidly, and the dominant transport mode of dissolved components changes to molecular diffusion. Generally, facultative-anaerobic, heterotrophic bacteria carry out the denitrifying process. Gaseous endproducts of nitrogen diffuse through the water into the atmosphere, permanently removing fixed nitrogen from the aquatic ecosystem. Ammonia, hydrogen sulfide, or methane gases can also be released.

5.3.3 The evolving Bay: a continuum of evolutionary change

The present Bay resulted from multi-scale ecological processes embedded in a climatic-geologic evolutionary history. The Bay has been filling with sediment, fluvial inputs, and products of organisms and modified by geomorphological processes, tidal exchanges, storm events, isostatic adjustment, and biotic activities. High sedimentation rates (~ 0.3–1.2 gm cm^2 yr) presently characterize the shallow, fluvial-dominated upper Bay while modest rates (0.1–0.3 gm) characterize the middle Bay, and modest to high rates (0.1–0. 8 gm) characterize the marine-dominated lower Bay (Officer et al. 1984). Land subsidence (sinking) and erosion occurs around the Bay at different rates. And asymmetric tides, through reversals in tidal velocity-dominance, play a central role in net near-bed transport to influence the morphology of small (less than about 20 km in length), well-mixed tidal embayments. Flood stages tend to dominate shallower areas to enhance near-bed transport landward, while ebb tends to dominate deeper areas to enhance near-bed transport seaward. As the Bay contracts in volume, depth, and area, freshwater should move seaward; however, sea-level rise, channel dredging, and blockage of freshwater and sediment by dams are moving seawater landward to flood and erode shores.

5.3.3.1 Top-down structuring

Large-scale climatic, geological, and fluvial interactions have shaped the Bay and small-scale sedimentary, hydrological, and biotic processes have modified it into its present form. The system has been adjusting and coevolving with biota for millions of years. During the late Pleistocene epoch approximately 15 000 years ago, mean sea level was 100–150 m lower than present and the continental shelf was exposed as dry land. The Susquehanna River and its tributaries flowed to the Atlantic Ocean, carving fissures into the landscape, and the James River flowed separately to the ocean. About 10 000 years ago, sea-level rise flooded the wide, sedimentary, coastal plain, penetrating the ancient river valleys and isolating coastal dunes that formed the Eastern Shore. About 6000 years ago, glacial melting slowed; about 2000–3000 years ago sea level approached within approximately 3 m of its present level. As rivers delivered sediment that currents deposited along shores, the Bay enlarged and gradually filled, shifting the shores landward.

On shorter time scales, major storms and weather patterns also bring change. Extreme fluctuations in weather, with alternating periods of flood or drought, deliver high or low freshwater volumes over extended periods of time. Drought can cause saline water to move into freshwater areas. And about 30 times a year, storms ("northeasters" and hurricanes) bring extensive precipitation that radically changes salinity. Storms that occur over days and weeks trigger a sequence of events. Generally, in about 7–26 days, an estuary undergoes an initial response as the storm approaches, then shock with passage of the storm, followed by rebound and recovery (Nichols 1994). Heavy rains can depress salinity for more than a month, and storms can force open new inlets or close others permanently, altering salinity and landforms. Thus, land- and seascape components react, adjusting to change in time-averaged, system-wide responses, and with lag-times of variable durations. Thus, the system adjusts, reorganizing through processes that maintain resiliency in the face of sea-level rise, storm events, and shoreline change.

5.3.3.2 Bottom-up structuring

Bay biota respond to sea-level change and to sedimentary–erosional processes over relatively short time scales. Organisms build marshes, seagrass beds, and oyster reefs that bind sediment, buffer erosional forces, alter current flows, and provide habitat. Yet terrestrial and aquatic plants and animals perform different roles; the terrestrial plant community is inherently longer lasting than animal communities and initiates conditions to which animals respond. On sand dunes, for example, vegetation stabilizes sand that is then colonized by organisms that select appropriate conditions of elevation, depth, salinity, and sediment type according to life history, behavior, intrinsic rates of population increase, and dispersal capabilities. Over time, plant organic matter accumulates, the surface is elevated, moisture content is changed, and the community evolves. Animals may dig burrows that saltwater can penetrate, which selectively kills plants, and initiates new conditions. In this plant-initiated sequence, communities progress across the landscape in pace with physical changes that influence succession.

Aquatic communities present different patterns. Aquatic plant communities, notably phytoplankton, change most rapidly, while animal communities are inherently longer lasting. Also, the presence of biogenic structures in water (e.g., seagrass beds, oyster reefs, clam beds) of variable width and depth creates hydrological roughness that dampens or increases turbulence and subsequently alters hydrological and sedimentation patterns that determine which species may colonize. Filter-feeding benthic animals alter sediment texture and deposit as much as 200 times the annual load of suspended sediment carried in water. In particular, the presence of productive oysterbeds substantially affects water-column clarity and increases sediment cohesiveness and benthic roughness, thereby altering water flow and providing opportunities for colonization.

The persistence of oyster reefs is recorded in geological strata. Oyster-shell basement rock underlies marshes, and oysters occur in the fossil record from early Mesozoic times. Their impressive remains are marked by two parallel bands of dead shells at about 70 and 50 m depths along most of the U.S. mid-Atlantic shelf. Oysters are unique in building hard, stable, persisting benthic structures in an otherwise mobile regime. Oyster reefs historically dominated tidal channels that border marshland (Fig. 5.20). The U.S. Coastal and Geodetic Survey in 1878 sent Lt. F. Winslow to document oyster abundance in the James River and Tangier–Pocomoke sounds. Oysters in Tangier Sound occupied shallow waters around marshy landscapes, concentrating along tidal channels in patterns that reflect channel morphology and meanders (Fig. 5.21), which suggests long-term, time-averaged biophysical processes that balanced natural history, growth, and environmental vagaries. In the James River, the Wreck Shoal oyster bed originated from point-bar topography formed in the late Pleistocene and has grown vertically 1.4 m at the pace of sea-level rise. Despite freshwater and tidal fluctuations, its major inshore portion remained stable for 130 years. As the oysters metabolized, grew, and deposited feces and shell, as observed for 300 years (1550–1855), the bed remained at the tidal surface (Dealteris 1988). Thus, as individual oysters thrived and benthic communities grew, the living reef could expand in co-adjustment with tidal-channel energy and shoreline processes. Some beds, however, may encounter unfavorable conditions; if sediment is excessive, larvae fail to attach, and growth is stifled. The bed then moves toward a process of senescence, decay, and erosion.

Coastal wetlands, with graded degrees of elevation, salinity, and aquatic–terrestrial conditions, respond to rising sea level through evolutionary change. This is made evident by the many beaches described in 1608 that (if they now exist) have moved landward, and by tree stumps of remnant upland forest that are now buried beneath low marshland. A combination of sediment starvation and accelerated rates of relative sea-level rise can cause a marsh to erode or prograde and migrate. In particular, mudflats can be colonized by the salt plant, *Spartina*, whose vegetative growth traps sediment that elevates the mud surface; *Spartina* invades and

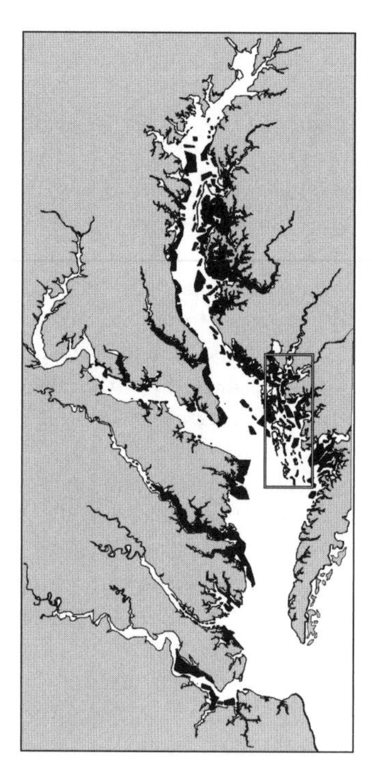

Fig. 5.20 Historical fringing oyster reef (black) of the Bay. Rectangle indicates area of Fig. 5.21. From Funderburk et al. (1991).

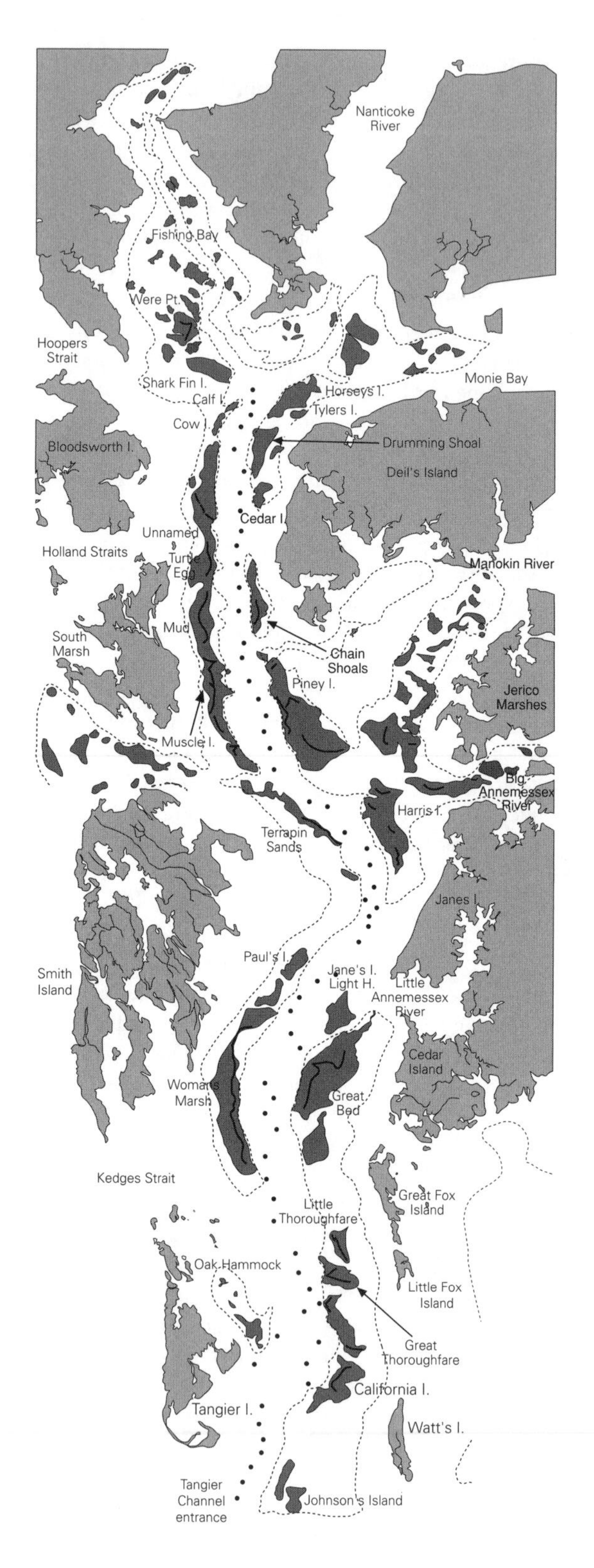

Fig. 5.21 (*right*) Historical fringing oyster reef of Tangier Sound. Oyster abundance followed a pattern of tidal-channel meander (black dots indicate deep channel bottom), creating hard bottom beds (dark shaded area) and concentrating most densely along solid ridge (black line on shaded area). Oysters were relatively sparse outside shaded bed. From McCormick-Ray (1998); Winslow (1882).

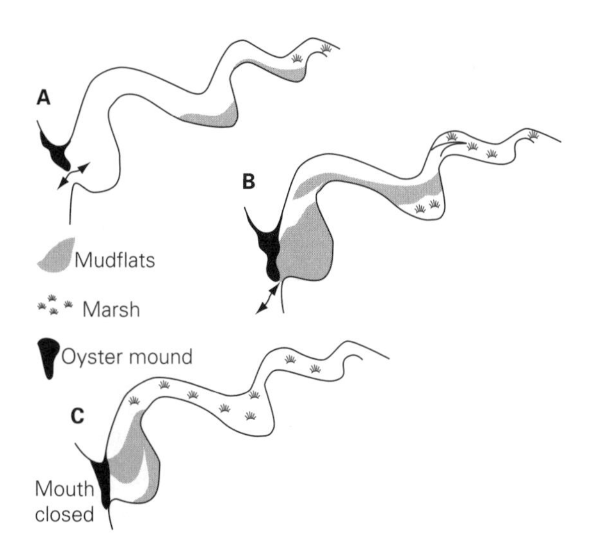

Fig. 5.22 Oysters in marsh evolution: (A) oysters (dark patch) intercept current flow into marsh; (B) sediment is deposited and accumulates (light shaded area); (C) marshland develops. Modified from Hayes & Sexton (1989).

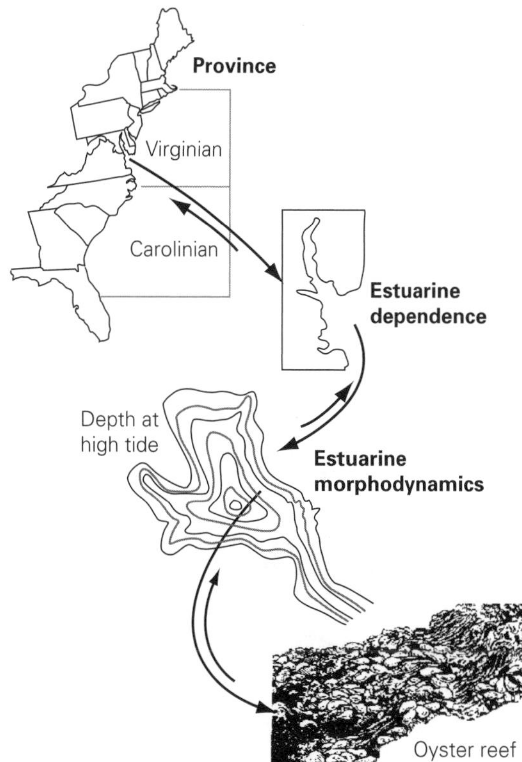

Fig. 5.23 Proposed hierarchical ecological interactions. Oysters interact in top-down/bottom-up feedbacks, and may influence species of the Virginian–Mid-Atlantic Province. From Ray et al. (1997).

the muddy area becomes colonized that with time is converted into intertidal salt marsh. Oysters colonizing favored sites at the mouth of tidal streams may also influence marsh evolution. As stream flow is slowed, sedimentation increases, and a marsh can develop (Fig. 5.22). Dame et al. (1992) suggest that oyster beds may initiate a process of marsh–estuarine evolution over a larger spatio-temporal scale. Furthermore, reef development at local scales can cascade ecologically through Bay-wide hierarchical relationships that can potentially influence regional biota (Fig. 5.23) through physical changes and metapopulation dynamics (Fig. 4.18, page 114).

5.4 Humans: force and magnitude of change

The long-term ecological processes and biotic components that transformed the Bay ecosystem have benefited humans, who have been evolving a social system and modifying the region with increasing intensity (Fig. 5.24). Nomadic Paleo-Indian hunters benefited from the moderating late-Pleistocene (13 000–10 000 BP) climate and ecological diversity that, during the Holocene (10 000–3000 BP), encouraged small family tribes to form more complex societies. They cleared small patches of forest about 4000 years ago, established villages, planted vegetables, and harvested abundant fish and shellfish from creeks and tributaries. Oysters were a staple diet, evident in piled shells in extensive middens, one covering nearly 12 ha. Indians dammed rivers, as did beavers, and they caught fish, grew corn, and introduced tobacco. They produced a surplus of products for local trade. Europeans arrived in the 16th century, first to explore then to colonize and establish settlements, and to create a social system that transformed the Bay and its ecological components.

5.4.1 Inherited riches and initiating policies

The transformation of the Bay was gradual, then intensified. It began under an Indian culture at its high cultural point. Then the Italian Giovanni da Verranzano,

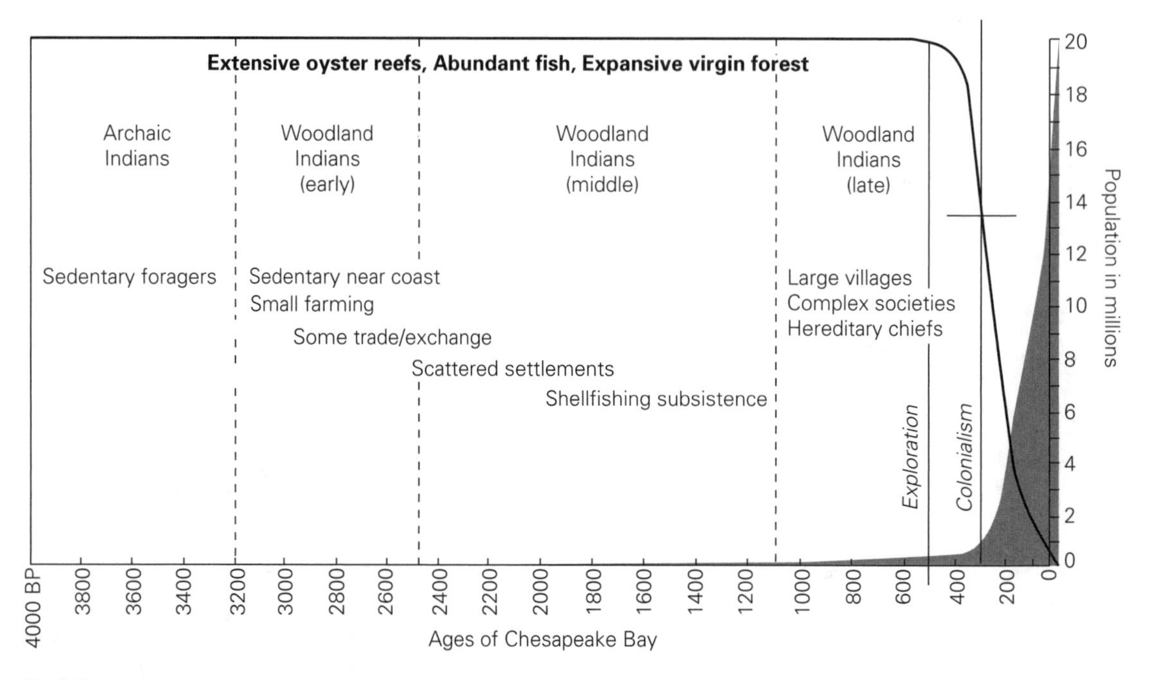

Fig. 5.24 Ages of Chesapeake Bay. Native culture subsisted on natural resources for thousands of years. European culture quickened the pace and altered trends. The 0 line represents the year 2000, after which continued growth is projected.

who explored for France, sailed through Virginia Capes in 1524 to discover a forested landscape, extensive marshes, and abundant natural resources. The English colonizer, John Smith, nearly a century later wrote of natural resource abundance, large fish, flocks of birds, and oysters so plentiful that ships ran aground on extensive beds. A complex Native society had prospered from this abundance, marked by a confederacy of 32 "Kingdomes" of more than 30 provinces, and in excess of 8500 inhabitants in 161 villages scattered across tidewater plains, headed by Chief Powhatan of the Algonquian tribe.

In 1607, George Percy wrote that "upon this plot of ground [Lynnhaven Bay, near Norfolk], we got a good store of mussels and oysters, which lay on the ground as thick as stone . . ." Father White described the Potomac River as ". . . the sweetest and greatest river I have seene so that the Thames is but a little finger to it. There are noe marshes or swampes about it, but solid firm ground, with great variety of woode, not choaked up with under shrubs, but commonly so farre distant from each other as a coach and fower horses may traveale without molestation . . . All is high woods except where the Indians have cleared for corne. It abounds with delicate springs which are our best drinke."

English laws initiated policies of ownership and authority as settlements sprouted around the Bay (Fig. 5.1). Jamestown on the James River became the first English settlement (1607); Kent Island in the north became a fur-trading post (1608). The plentitude of beavers advanced the lucrative fur trade, which moved up the Susquehanna River. Virginia held title to all Chesapeake Bay land until 1632 when King Charles I of England issued a charter to Lord Baltimore establishing the Province of Maryland. At the mouth of the Potomac River, St. Marys (1634) became Maryland's first settlement; Snow Hill (1642) on the Pocomoke River became Eastern Shore's first settlement. Hardwood forests were cut and ships were built, and an abundance of other resources, land, and opportunities became central in initiating a legal system of fairness, dominated by an absent landlord (England) and financed by remote investors. The Navigation Act of 1651 declared that only English-built ships be used in trade. The English Order in Council established that: "no tobacco or other provision of the colonies should thenceforth be carried into any foreign parts until they were first landed in England and the duties paid." The English Crown held all rights to fishing, and ports became lifelines to the outside world, helping to meet

Fig. 5.25 Inverse relationship between oyster size and human population size, 1630s–1720, St. Mary's, Maryland. Adapted from Miller (1986).

England's need for resources and to cover colonial expenses.

In 1682, Norfolk, at the mouth of the James River, became a center for trans-Atlantic trade. In the north, the working harbor at the mouth of the Severn River in 1694 became Maryland's state capital, named Annapolis in honor of Queen Anne. Joppa Town, a prosperous Maryland port on the Gunpowder estuary, was becoming buried by sediment from deforested land, causing Maryland's General Assembly in 1706 to favor the natural harbor of the Patapsco estuary as Maryland's port of entry; this became Baltimore.

England's colonial laws and traditions subsequently formed the basis for federal and state constitutions. The first Continental Congress in 1774 drew from English common law (natural principles of justice) and statute law (acts of will) for developing American law. In the subsequent 80 years of social struggle following the Revolution, the transforming legal system gave a disproportionate share of wealth and power to entrepreneurial and commercial groups that promoted economic growth. Under English common law and public trust doctrine, common practices became U.S. statutory laws, with ownership rights. Traditional privileges preserved the owner's advantage to use resources, placing value on land and property. Rights to tidelands and to lands under navigable waters remained the common heritage of all citizens, whereby fishing, fowling, and navigation fell under sovereign authority. Thus, while land use was governed under private ownership laws, water use and resources fell under public authority, divided between state and federal jurisdictions. Aquatic resources remained public, a commons held in public trust.

In 1668, Philip Calvert of Maryland and Colonel Edmond Scarborough of Virginia signed the Calvert and Scarborough Agreement, which delineated state boundaries and gave Virginia more land. A century later, jurisdictional conflicts over the growing importance of resources and water for trade led President George Washington to settle the Maryland–Virginia boundary and resource disputes by means of the Compact of 1785. This Compact replaced the Calvert and Scarborough Agreement and gave each state equal access to shared Potomac River resources and reciprocal rights to the Pocomoke River.

5.4.2 Social change

Traders, tobacco farmers, and planters turned resources into profit shortly after the first explorers arrived. Furs, timber, marsh hay, and trees were extracted, beavers depleted, and axe and hoe turned wilderness into farmland familiar to English tradition. Farm animals were introduced, and, accidentally, weeds and diseases. Most colonists lived within about 300 m of shore and as population and exploitation grew, some local resources were affected (Fig. 5.25), rebounding when exploitation decreased. Europeans populated the coastal plain up to the rolling piedmont hills at the foot of the Appalachian Mountains. Maryland's 1630 population of 150 and Virginia's 1635 population of 5000 had increased by 1700 to 34 000 and 60 000, respectively. This expansion brought conflict with Indians; despite a peace treaty in 1677, the Native civilization collapsed by 1722. The watershed was being rapidly carved into a mixture of ownership, use, and privileges, although sparing some abundant resources of non-commercial interest. Oysters, crabs, and sheepshead fish (*Archosargus probatocephalus*) remained as "poor-man's food" for subsistence needs. Harvest of oysters required only a small boat and a pair of wooden tongs to pull clumps of them up from the shallow bottom.

Tobacco plantations supported an agrarian economy, and commercial centers grew around port trade. Ships delivered natural products to England and returned with manufactured products. Towns developed at "fall lines" of large tributaries, where rivers abruptly fell into a series of rapids and waterfalls where boat traffic was forced to halt, and dams provided power for gristmills as early as 1646. By 1761, farmers had turned about 20% of the land into farms and Maryland had 120 commercial vessels in operation. All tidewater land and most of the piedmont (above the fall line) were occupied or actively being settled. Baltimore's population of 6755 in 1775 doubled to 13 500 by 1790. The waterways were used for sewage and trash, and sediment accumulated.

By 1800, the Bay watershed population had approached about two million. The Industrial Revolution quickened the pace of development, with Baltimore rivaling New York City and New Orleans as the third largest U.S. exporting center. Roads connected Baltimore to New York, Philadelphia, and Norfolk, ships connected Baltimore and Norfolk to the world, and by the mid-1800s, railroads connected Bay resources to the west. Urban prosperity spread to the countryside, and summer resorts sprang up on the edges of marshes. Maryland's eastern shore became a panorama of prosperous farms, with manicured plantations interspersed with thick forests of walnut, oak, and pine along the water's edge, and attracted vacationers. Towns flourished, with people engaged in orchards, farming, fishing, and oystering.

5.4.3 Emerging resource problems

By the mid-1800s, approximately 40–50% of forested land had been cleared, exposing soil and water to heat, direct sun, rain, and wind. Deep furrows cut by plows exposed native species, changed the microbial community, and subsequently altered abundances of terrestrial species. Freshwater flow, estuarine salinity, and sedimentation patterns were altered and sediment poured into streams and tributaries. The wide bay (Georgetown harbor), in 1751 at the head of the Potomac, by the early 1800s had become filled with silt. By 1831 Baltimore Harbor, its city the hub of industrial growth and commercial activity, required dredging, at a cost of $500 000. Congress had created the U.S. Army Corps of Engineers in 1802, charging it in 1824 with channel dredging and jetty construction, thereby initiating a long history of federal involvement in harbor maintenance that began in Baltimore. The Corps dredged 65 million m^3 of sediment at the mouth of the Bay between 1846 and 1938, where accumulated sediment raised approximately 82.8 km^2 of the bottom almost a meter, causing interference with ship traffic. By the late 1800s, Baltimore's trade and commercial importance was also beset with problems of harbor sewage, pathogens, contaminants, and overenrichment.

Oysters in New England estuaries had been depleted, and entrepreneurs came to the Bay. C.H. Maltby from Connecticut opened an oyster business in Baltimore in the 1830s, and by mid-century, packing oysters for trade had become a major industry, ranking second to whaling as the most profitable U.S. "fishery." The industry in Baltimore involved 45 firms that generated $2 million a year and employed hundreds of vessels and nearly 7000 oystermen. An oyster captain could earn a profit of $2000 a year at a time when most Marylanders earned less than $500 and when Virginians were recovering from Civil War losses. Baltimore's railroad and port extended the oyster trade to the far west and to European markets, making the Bay region the oyster capital of the world. New Englanders introduced new harvesting methods that outraged Bay oystermen as early as 1820 when states outlawed dredging. Dredging broke up oyster clumps, but it also increased an oyster's commercial value, allowing individual oysters to grow round and large. New Englanders also introduced "planting," a systematic procedure that involved "tonging" from small boats, a dredging process that removed oysters, cleaning the bottom for replanting of them.

In the 1870s, following the American Civil War, the waterfront was a tumultuous place. Feuds broke out between neighbors, oyster-dredging laws were repealed, and "the waters were thrown open to every one who would pay the military officials for a permit to oyster . . . ," as reported in the 1887 U.S. National Census. Crisfield on Tangier Sound (Maryland) was the oyster and seafood capital, benefiting from construction of a railroad terminal, and was known for its get-rich-quick schemes. In the oyster-rich area of Pocomoke Sound that bordered Virginia and Maryland, oystermen claimed resources belonging to others and violent disputes erupted, including over state boundaries. The Jenkins–Black Award, in 1877, aimed to settle that boundary dispute, but gave Virginia a larger

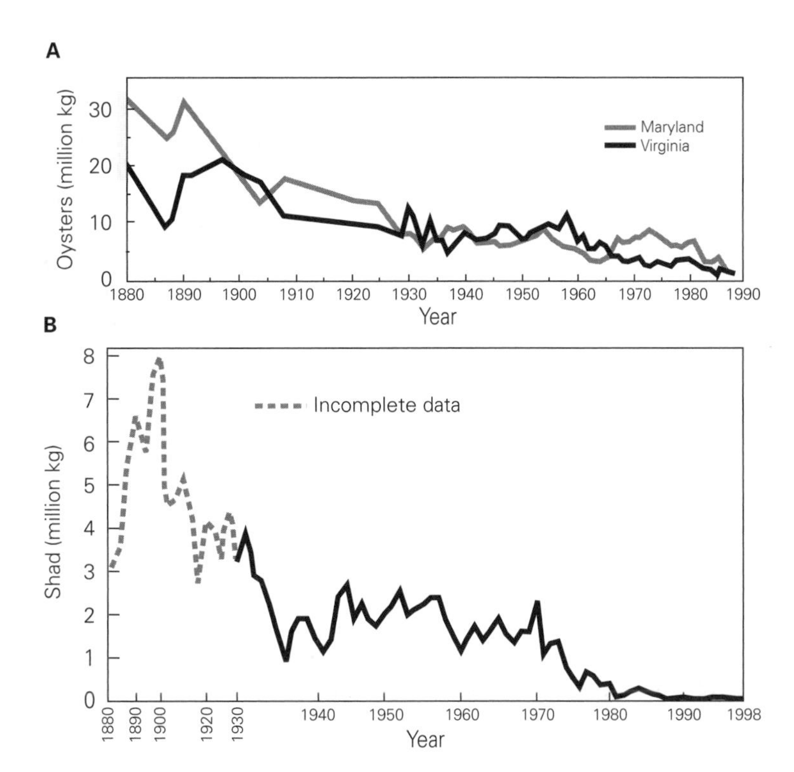

Fig. 5.26 Declining natural resources over 110 years of commercial harvest: (A) oysters; (B) shad (shaded-dashed line reflects change in data collection procedures). From CBP (1999).

share of Tangier and Pocomoke sounds. This decision embittered Maryland watermen and state lines were redrawn shortly thereafter at today's boundary.

Oysters were clearly declining. Calls for conservation and protection echoed through industry, academia, federal offices, state legislators, and were reported in magazines and newspapers. Maryland and Virginia attempted oyster management, establishing commissions and oyster police. However, funding and authority were insufficient. Neither state surveyed the extent or condition of its oyster beds and neither invested in research. Virginia established a Board to handle regulations, advised by Boston and New York businessmen who supplied capital and enterprise, and introduced private leasing for oyster propagation on state-owned bottom. Maryland managed its industry differently, through legislators, and appointed an expert oyster commission, whose advice they failed to follow. Despite passage of many laws, oyster policies of both states vacillated without firm commitments to conservation. Oysters continued to be exploited, reefs destroyed, and their shells valued for lime kilns, fertilizer plants, and road-building. By the 1890s, 900 dredgers continued to harvest relatively scarce oysters (Fig. 5.26A). By 1900, oyster profits were thin and the industry became economically inefficient.

Shad followed a similar decline from overexploitation and habitat loss (Fig. 5.26B). Before European expansion, shad weighing as much as 5.9 kg abundantly migrated into Virginia and Maryland tributaries and up the Susquehanna watershed as far as New York to spawn. Shad wars that began in 1730 on the Susquehanna intensified by the early 1800s as fishing boomed. Mills developed and dams were constructed on rivers important to shad migration and spawning, as state legislators approved construction of canals, large dams, and reservoirs aiming to control floods, produce power, and facilitate traffic. Dam construction in Virginia began in the early 1800s on the James River near Richmond, the state capital, and on the Potomac near Washington, D.C., and elsewhere. The Pennsylvania canal system in the 1830s required feeder dams, and subsequently four hydroelectric dams eliminated all shad runs in Pennsylvania. A Maryland "Act of the Legislature" in 1884 authorized dam construction on the Susquehanna, and in 1927 completed the watershed's largest, the 100 ft.-high Conowingo Dam that affected shad breeding habitat and altered flow patterns of the upper Bay. By the late 1800s, dams obstructed shad migrations in estuaries and rivers along the entire Atlantic coast. From a peak harvest of 3.2 million kg in 1896, that had made Maryland the fourth largest producer, the shad fishery entered a rapid and alarming decline.

Depletion of many other abundant species was evident by 1900. Shorebirds and waterfowl were harvested for market and export; between about 4000 and 5000 were killed each day on the Susquehanna flats alone. North America's most abundant bird, the passenger pigeon (*Ectopistes migratorius*), became extinct by the early 1900s. Other abundant resources, including blue crabs and the diamondback terrapin (*Malaclemys terrapin*), were harvested without regulation.

5.4.4 Twentieth-century management complexity

Baltimore in 1920 had wealth and technology. Its Camden Railroad Station, built in the 1850s, was one of the world's largest. Nearby Bethlehem Steel, established in 1891, expanded shipbuilding facilities. Automobiles, steamships, and railroads sprouted connections to commercial and recreational centers as increasing technological advances moved society from being agrarian to being urbanized. Roads and bridges connected formerly isolated islands and counties to prosperous neighbors, the numbers of dams and reservoirs increased, watershed flows were interrupted, and wetlands were filled, channeled, and dried. As fishes declined, their value increased and fisheries expanded. The need for Bay-wide fisheries management was evident, yet jurisdictional mandates to manage conflicts and resources fragmented the system.

Decline of valued fisheries brought widespread concern and studies. The General Assemblies of Maryland and Virginia in 1948 agreed to establish the Chesapeake–Potomac Study Commission, a group appointed by State governors to re-examine the Compact of 1785 to address conflicts over oysters and fish. Dr. Paul S. Galtsoff of the U.S. Fish and Wildlife Service reported that oyster harvests were a fraction of just 50 years earlier, especially in the upper Bay and the Potomac and York tributaries. An intense study he directed revealed that the formerly high-quality York River oysters were degrading from sulfate waste of a pulp mill, as suspected in 1916 and confirmed in 1939. Galtsoff reported that intensive dredging had depleted oyster beds, and their natural propagation could not keep pace with fishing pressure. By 1945, oyster harvests had declined 50% from a high in 1897, and herring catch had declined 74% from a high in 1908. Yet, the small striped bass fishery expanded and boomed after 1936. Blue crabs were intensely harvested, and menhaden, a fish used for fertilizer, went unevaluated. Fishery laws aimed to maintain individual species at high levels of production and the fishing industry sustained yields through expansion of fishing grounds and use of increasingly efficient gear.

The Potomac River was unique, requiring navigation and dredging to maintain important commercial traffic critical to the nation's capital, Washington, D.C., especially during World War II conflicts. Fishing was important to both Virginia and Maryland, which shared the Potomac. The Corps of Engineers was charged with the dual and conflicting role of maintaining navigation and designating fishing areas for conservation, a mandate that generally favored navigation interests. The Commission determined that fisheries rehabilitation required that both states provide adequate management, and each protect its own oyster stocks by simply abolishing dredging, shelling, and seeding of beds. Also, the management of migratory fish demanded that each state enforce mutually agreed-upon conservation measures. However, when one state took action, the other took a different course and conflicts were not easily resolved, partly because state legislators convened at different times. Thus the Potomac Commission became authorized to protect Potomac fisheries resources.

The twentieth century began with resource depletions. It was a period of industrial expansion and jurisdictional fragmentation. State and national laws favored development, placing land in private ownership for use and development and mandating separate land, wetland, and water resources into smaller units of agency management. Federal authority held control of navigation and commerce in state waterways. Water resources, depending on the species and where they occurred, were managed by either state or federal authority for purposes of productivity and profitability. And the fishing industry, composed of individual organizations that often held widely different views, added layers of complexity onto any goal for protecting Bay resources. Until recently, resource management consisted of regulations to protect public stocks, mandating use of gear, restricting certain uses in certain places, regulating harvest period, amount, and size, and issuing penalties, while lacking sufficient efforts in

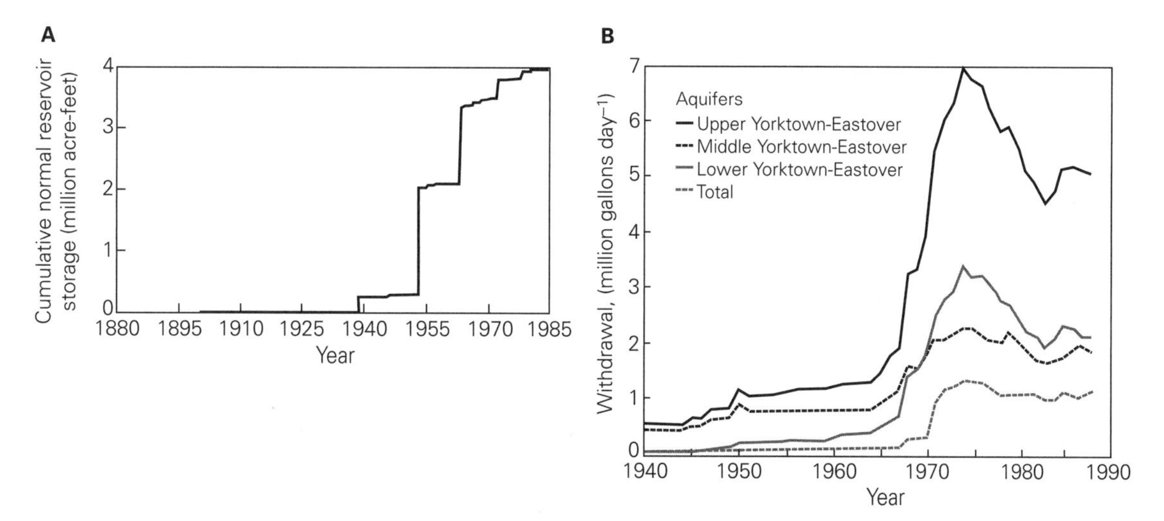

Fig. 5.27 Water use trends in Virginia: (A) reservoir water storage development, 1880–1985 (*National Water Summary*, 1987); (B) water withdrawal, 1940–90, Virginia's eastern shore. From Richardson (1992).

conservation. Furthermore, fisheries research focused on ways to increase production and profitability, and legal restrictions and economic practices interfered with effective management of transboundary estuarine resources. The Commission in 1948 called for a Bay-wide system of coordinated management.

Presently, urban sprawl is spreading rapidly across the watershed. Baltimore's population (in 1994, 703 000) is stretching into suburbs, and Norfolk, with Virginia's second largest population (241 000), is expanding its port area to provide jobs for nearly one-third of the surrounding coastal region. Northern Virginia, with a population growth of 31% since 1975, is becoming a conurbation that straddles Washington, D.C., and Baltimore, and is expanding towards Richmond and Norfolk, as developments throughout the Bay watershed continue to fragment the landscape. Networks of roads connect people to places as vehicles top a million and make the Washington, D.C. area the third most congested in the nation. Roads interrupt water flow (Box 5.2) and vehicles add loads of chemicals to coastal air and water. And each day, 300 more people move into the Bay region – a rate that forecasts nearly 18 million by the year 2020 – most preferring to live along a strip of land within easy access of shores. The rate and concentration of urban growth presents a significant dilemma for sustaining Bay resources, as urban landscapes and associated industries demand energy and freshwater, the Bay's most critical resource (Fig. 5.27), thus requiring more dam and reservoir construction that alters the timing, quantity, and quality of water to the Bay. In 2002, Virginia and Maryland reported a substantial increase in the number of "Impaired Waters."

European colonization and American traditions have left an environmental legacy, turning virgin forest into a herbaceous-plant-dominated system, an estuarine system with depleted resources and blooms of algae and microbes, and a physically restructured watershed of increasing impervious ground and altered tributary flow. As human populations have soared, wetlands have been converted for human uses, nutrient enrichment has increased, and turbidity has worsened the decline of SAV beds. SAV losses have affected waterfowl (Box 5.3). And while some areas of shallow water are expanding due to sedimentation, other areas are eroding, requiring mitigation measures, as in the case of Poplar Island in mid-Bay where dredged sediment is being deposited to counter erosion. The unregulated and hazardous wastes of society have accumulated in streams, open waters, and wetlands, with tens of thousands of organic compounds in commercial use, of which less than 2% have been tested for their effects on the environment (Table 5.3). Pulp and paper mills continue to produce contaminated effluents, including dioxins that affect fish. And in ways not fully understood, the summer hypoxic–anoxic zone is expanding over the benthos.

Box 5.2 Created wetland mitigation: successes or failures?

John E. Anderson

Wetland creation has become increasingly attractive for balancing wetland losses due to construction projects and highway development. The post-World War II revolution in mechanized land development and the real-estate boom produced dramatic changes in growth for America's urban and suburban centers. As cities grew to encompass more land around their borders, ring cities developed and developers sought cheaper lands in rural areas. As a result, the U.S. eastern seaboard began a trend toward becoming a continuous megalopolis, following highway corridors from Massachusetts to Georgia.

Natural landscapes suffer greatly from the swiftness of land development. Where landscape changes were once measured in geological time, human disturbances are almost immediate. Wetlands have become attractive for construction, requiring fill or draining, and are losing their effectiveness as natural buffers for water quality and wildlife habitat. To offset environmental impacts, wetland creation is required. However, wetland creation is often impossible on-site as part of a construction project, as land may not be available; this leads to site selection outside the area immediately affected by the project. Furthermore, wetland creation also must address functions and values and impacts to its watershed. Therefore, because on-site wetland creation is often unrealistic, mitigation banking has been mandated by many agencies to achieve "no net loss."

Regulations for wetland mitigation banking involve permits under the U.S. Clean Water Act, as well as mandates and interpretations by resource agencies to meet the criterion that for every hectare of wetlands lost due to construction, an equal area must be created. The regulatory agencies responsible include the Environmental Protection Agency (federal oversight agency), the U.S. Army Corps of Engineers (permit agency), and a host of local and state agencies. Since 1995, "Federal Guidelines for the Establishment, Use and Operation of Mitigation Banks" has been used by resource agencies as guidance for establishing new wetlands, but does little to offer design and engineering criteria.

Mitigation banks are large-scale created wetlands that are usually engineered to offset construction impacts in the same watershed. They offer many benefits, including long-term management, reductions in the permit review process, consistency of design with what is needed in the watershed, and engineering designs that can accommodate many different wetland types. Problems include habitat fragmentation, poor or uncertain engineering approaches, interbasin transfers of wetland systems, and sacrificing some watersheds for development while others become mitigation areas. A recent example involves the planned construction of the Cohoke Creek Reservoir for Newport News, Virginia. Land acquisition outside the immediate watershed can mitigate impacts of reservoirs planned for construction. The developing agency ultimately bears responsibility for design, demonstrating to regulatory agencies that the mitigated wetland maintains function and value. This includes periodic monitoring and site reporting, a near impossible task for highway departments to undertake, particularly in fast-growing regions, where road and highway construction can overwhelm natural wetland areas and mitigation projects on a grand scale.

Departments of transportation are prime wetland construction agencies. The Virginia Department of Transportation (VDOT) oversees more than 200 wetland mitigation sites scattered across Virginia, with many more planned. Unfortunately, the science of wetland creation has been relatively slow in catching up with construction demands in regions like Chesapeake Bay, and success stories balance the failures. Creation of functional wetlands and acquisition of land in the same watershed can also be expensive; suitable land may have more real-estate value as a parking lot or shopping center than as a tidal marsh.

The practice of wetland creation is not well developed, as it requires establishment of hydrological function and hydric soil development. More often than not, an area is scraped to establish low topographic elevations, and these low areas can become saturated or flooded or involve acid-soil exposure. Often, soils with seed banks must be brought into a site to foster development of the wetland, but establishment of plants on highly acidic soils is nearly impossible without expensive treatment with lime or other buffering agents. In fact, many techniques employed in acid mine drainage are being considered for wetland mitigation.

In successful instances, soil conversion may occur naturally given proper hydrology, and hydrophytes become established. To begin this process, areas are usually planted. But once a wetland is created, whether it will possess the same functions and resource values as the one it replaces is hard to predict. Creating a wetland can be as easy as digging a ditch; many highway ditches contain wetland plants and borrow pits along road corridors are sometimes used for mitigation. But this does not satisfy the function and values of a wetland, as consideration also must be given to overall landscape impacts. The unique landscape position occupied by naturally occurring wetlands has evolved during many years of interacting processes that short-term engineering cannot reproduce.

In the early 1980s, VDOT created the Goose Creek Wetland Bank to offset highway construction in the Elizabeth River watershed, near Norfolk. The 4.5 ha site was originally a borrow pit, then planned as a tidal marsh and planted with cordgrasses. Follow-up investigations using multispectral imagery (Fig. 1A) and field surveys revealed dramatic colonization of endemic, invasive, reed grass. For the portion

Fig. 1 (A) Goose Creek Wetland Mitigation site at Goose Creek, Virginia. Gray areas indicate extent of the invasive *Phragmites australis* in 1997. Stippled areas indicate new areas occupied by the invasive in 1998. White regions indicate common areas unchanged between 1997 and 1999. Vegetation outside the classified regions represents more desirable species. Highway 664 runs north to south (left to right). (B) Franklin Bypass Created Wetland in Virginia. In image, Blackwater River is north; Highway 58 is south.

of the site occupied by cordgrasses, VDOT claims mitigation credit, but reed grass continues to spread vigorously, placing VDOT's mitigation credit in jeopardy. Further, biodiversity was high in the first four years of development, but has declined. Thus, this wetland establishes a poor precedent, as it clearly does not contain the diversity of similar tidal estuarine wetlands.

Franklin Bypass is an example of a created freshwater wetland. It lies near the Blackwater River in Southside, Virginia and is approximately 5 years old. VDOT originally planted a number of native wetland plants, including bald cypress (*Taxodium distichum*). An emergent marsh consisting mostly of bulrushes, cat-tails, willows, water lilies, and pickerel weed soon developed. Recent multispectral imagery (Fig. 1B) indicates that natural marsh zonation has developed. Although part of the area in the higher center of the site dries during droughts, water remains in most of the site due to lower topographic elevation around the margins. The odd shape of the site follows the contours of the adjacent river and the old meander scar. Hydrological connectivity has been established between the river and the site, which benefits hydric soil development. Thus, this site is more consistent with other natural marshes in the watershed, and while plant diversity is average, the marsh offers habitat value as well as water quality and flood control.

Rapid fragmentation of wetlands in the Chesapeake Bay region provides a dramatic example of ecological substitution. With so many wetlands being artificially created, while natural wetlands are destroyed, the region is at risk of losing vital water-quality buffers, habitat, natural recreation opportunities, and flood control mechanisms. Important information regarding ecological diversity and biogeochemistry of natural wetlands in the region is being lost. Recent evaluations of "reference" wetlands may serve as models for measuring future successes in establishing created wetland systems in an ecosystem context.

Sources: Barnard & Mason (1990); Barnard & Priest (1993); Perry & Anderson (1998); Tiner (1984); M. Fitch, VDOT (personal communication, 1998); J. Goldberg, J. Perry & J. Anderson ("Multispectral analysis of Goose Creek wetlands" – unpublished data, 1999).

5.5 Assessing ecosystem condition

The rates and extent of change in Chesapeake Bay reflect centuries of intensifying impact that has transformed the Bay to a condition in which its resiliency is compromised.

Depleted fisheries. The Bay no longer supports economically valuable natural resources at their historical levels of abundance. Striped bass precipitously declined to a low level in 1970 and required federal law that in 1984 authorized conservation and management programs. American shad are depleted throughout their range. In 1980, Maryland placed a moratorium on shad harvest, except in the Potomac River and coastal waters. The largest remaining shad fishery is along the Atlantic coast of Virginia. Oysters reached record lows in 2002, their recovery impeded by disease. Blue crab abundance has been declining in recent years, approaching record lows. Commercial landings of the Atlantic sturgeon (*Acipenser oxyrhynchus*) – a bottom-dwelling anadromous fish that can reach 3 m and 200 kg, and whose

Box 5.3 Waterbirds: changing populations and changing habitats

Michael R. Erwin

Immense flocks of waterfowl and shorebirds in fall in the Chesapeake Bay region have been legendary – so thick they could "darken the skies" (journals of Captain John Smith). Success or failure of the early European colonists' survival depended upon their abilities to harvest the Bay's natural resources, including waterfowl. The most prized resources included oysters and crabs, but also the "King of the Bay" – wintering canvasback ducks (*Aythya valisineria*), Canada geese (*Branta canadensis*), and large shorebirds such as the "curlew" (now known as whimbrel, *Numenius phaeopus*). Waterbirds also played a significant ecological role in the Bay by consuming vast amounts of fishes and shellfish, which almost certainly had a major effect on food-web structure.

As human populations grew during the 18th and 19th centuries, the effects of humans on waterbirds began to mount. The millinery trade (including feathers) of the late 1800s and early 1900s spelled near extirpation of species such as least tern (*Sterna antillarum*), snowy egret (*Egretta thula*), and great egret (*Ardea albus*) in the eastern United States. This led to passage of federal bird protection laws. For other species, market hunting to support demands of many Atlantic seaboard restaurants spelled serious declines for waterfowl and shorebirds, especially in the Bay and along the Maryland and Virginia eastern shores. These, too, were afforded federal and state protection in the early 1900s, following which many species reoccupied former ranges and increased in population.

In the latter half of the 20th century, unprecedented changes occurred, spurred primarily by the exponential growth of the human population in the Bay watershed (see Fig. 5.24). One major impact of human presence was the application of agricultural organochlorine pesticides to control everything from mosquitos to corn borers and boll weevils. The proliferation of these toxic and biologically active substances was harmful to many species of birds, particularly those at the top of the food web, such as birds of prey. Numbers of bald eagles (*Haliaeetus leucocephalus*), ospreys (*Pandion haliaetus*), eastern brown pelicans (*Pelecanus occidentalis*), and double-crested cormorants (*Phalacrocorax auritus*) plummeted in Chesapeake Bay and elsewhere in the United States. Eggshells were so thin that many cracked during incubation. Numbers of many of the pesticide-sensitive species were near a modern-day low in the late 1960s, at the time when Rachel Carson's *Silent Spring* became a best seller. Because most of these chemicals are now banned, depleted bird populations have recovered; now, approximately 400 pairs of bald eagles nest in the Chesapeake Bay and its tributaries! Ospreys nest on nearly every duck blind and channel marker in some river systems. In addition, during the 1990s pelicans and cormorants colonized the Bay for the first time. Reasons for these northern range expansions remain obscure. Pelicans were never recorded as far north as Virginia and Maryland until the late 1980s. One hypothesis is that subtle community changes are occurring in fisheries. The breeding cormorant population in the Poplar Island area, in mid-Bay, grew extremely rapidly during the 1990s, in parallel with growth of this species in the northeast and Great Lakes. It has reached "overabundance" status to the point where controls are being considered by federal agencies, because of suspected effects of these birds on fisheries and aquaculture.

The report for many other species is not so optimistic. As development has intensified in the Bay region, nutrient inputs, sedimentation, animal waste, fish diseases, overharvesting, and anoxia have degraded many aquatic areas and their resources. Declines of submerged aquatic vegetation (SAV) and oyster reefs have had profound effects on some waterbirds. During the 1970s and 1980s, wintering tundra swans (*Olor columbianus*), Canada geese, and greater snow geese (*Chen caerulescens atlantica*) abandoned shallow waters and wetlands and began feeding primarily in agricultural fields on corn and winter wheat, after the harvest when considerable quantities of grain remained on the ground. Recently, numbers of snow geese have reached such levels that they are considered "overabundant" and controls on the breeding grounds in Canada have been implemented. Also dramatic has been the near local extinction from the Bay of the redhead (*Aythya americana*), a diving duck that is an obligate SAV feeder. It numbered nearly 90 000 in the winter in the Bay in the 1950s, but has now moved south to the Carolinas. However, a close relative, the canvasback, has shown a remarkable adaptation by modifying its diet: from being primarily a herbivore (SAV-feeder) it has become largely dependent on the exotic Baltic clam (*Macoma balthica*). Nonetheless, wintering populations declined from more than 200 000 in the early 1950s to about 50 000 by the early 1990s. Changes in the benthos and fisheries in the past 20 years may also help explain declines in some sea ducks. Especially in the 1980s to 1990s, apparent declines have occurred in long-tail duck (old-squaw, (*Clangula hyemalis*) and scoters (three species of *Melanitta*) combined during surveys).

Losses of Bay islands have been well documented during the past 200 years. These losses result from a complex combination of natural and human-induced factors, including erosion, subsistence, and sea-level rise, and have important consequences for nesting waterbirds. The decline of American black ducks (*Anas rubripes*) throughout its Atlantic range prompted a special Joint Venture under the North American Waterfowl Management Plan. Arguments continue about whether losses are due primarily to hunting pressure or losses of breeding or wintering habitat. Island loss has certainly had an effect on nesting least terns, for which numbers of colony sites have been reduced dramatically from the early 1950s to present. Rooftop nesting has become more common in both the Bay and elsewhere along the coast as natural habitats become more disturbed. Surprisingly, for reasons that we do not understand, both black skimmers (*Rynchops niger*) and royal terns (*Sterna maxima*) have colonized islands in the Bay only since the 1980s. Changes in the fisheries, especially of small, non-commercial, forage species (such as anchovies and silversides), account for shifts in population numbers and colonization.

In summary, many changes have occurred, and will continue, among waterbirds in Chesapeake Bay. Some of these are undoubtedly due to human activities, while others are less well understood. What is generally clear is that waterbirds are usually at (or near) the top of their food webs in the Bay and respond strongly to "bottom-up" conditions, such as food supply. Exceptions to this are those species that are either hunted for sport (especially black ducks, geese, and sea ducks) or potentially "controlled" as nuisance species (Canada geese, double-crested cormorant). Without a careful approach, using adaptive management procedures, it will be extremely difficult for scientists and managers to tease apart causal factors for population changes. Probably the single most significant improvement in habitat conditions in the Bay for waterbirds would be the restoration of extensive areas of both SAV and oyster reefs.

Sources: Erwin (2002); Erwin et al. (1993); Funderburk et al. (1991); Perry & Deller (1995).

Table 5.3 Examples and sources of contaminants in the Bay.

Contaminant	Source
Copper	Rivers, especially upper Bay. Diffuse in lower Bay (atmosphere; antifouling paint on boat hulls)
Mercury	Atmosphere: coal burning, incineration; batteries, urban storm water
Tributyltin	Antifouling paint for boats; industrial fungicide; high concentrations near recreational boating
Atrazine	Runoff from farm fields (agricultural herbicide)
PCBs	Sediments (buried), recycled between air and water. Widely spread throughout Bay. Now banned from most uses
PAHs	Urban and industrial areas; urban storm water runoff, atmospheric deposition, oil spills, recreation and industrial boating
Chlordecone (Kepone®)	Accidental spill by plant in James River. Used as stomach poison in baits. Extremely persistent, lipophilic
Dioxins	Domestic, industrial processes (incineration of plastics, industrial processes, combustion of fossil fuels, pulp mills)
Phthalates	Butylbenzyl-phthalate, a plasticizer in vinyl floor tiles, adhesives and synthetic leather; di-*n*-butylphthalate, a plasticizer in food packaging, PVC, some elastomers and insect repellent
Phenol	Coal coking plants, chemical plants, gas works, oil refineries, pesticide plants, and wood preservative and dye manufacturing industries. Produced by municipal wastes, coal liquid and discharges from paper, pulp, and jute mills and paint manufacture
DDT	Very high local levels from ocean dumping. Persistent. Banned in USA 25 years ago

Adapted from CBP Toxic Subcommittee (1992); CBP website (1999); Kline (1998).
PCBs, polychlorinated biphenyls; PAHs, polynuclear aromatic hydrocarbons; DDT, dichloro-diphenyl-trichloroethane.

lineage dates back millions of years – peaked in 1890 and today the species is endangered. The only catadromous Bay species, the American eel (*Anguilla rostrata*), whose 1981 commercial catch was more than 700 000 pounds in both Maryland and Virginia, is declining for unknown reasons. Although Atlantic menhaden have declined sharply, the stock is still considered healthy, and currently supports the most important fishery of the Atlantic coast. Functional roles of these historically abundant species have been much reduced. Mariculture is being attempted to restore commercial production.

Depleted submerged aquatic vegetation (SAV). Vast shallow waters of the Bay historically covered by SAV today lack vegetation or are very sparsely covered. Recovery has occurred in some parts of the Bay from restoration activities and improved water quality.

Changes in waterfowl. Waterfowl abundance varies with the species. Abundance has increased for a few generalist species and decreased for others, and some have been extirpated (Box 5.3).

Extirpated fringing oyster reef. Oyster beds that historically occupied about 40% of tidal Bay bottom (Fig. 5.20) are today, with few exceptions, only silted, remnant beds and grounds for artificial "planting." Bay oysters no longer form natural fringing reefs, but occur in clumps of broken shell, their reef structure obliterated, and ecological function lost.

Wetland loss. Development, fragmentation, and sea-level rise are squeezing wetlands into smaller isolated patches, their functions compromised. The U.S. Fish and Wildlife Service reported that the watershed lost wetlands at a rate of ~3 ha day^{-1} for a net loss of approximately 8000 ha between 1982 and 1989 – an area about half the size of the District of Columbia. Forested wetlands have sustained the greatest loss, primarily due to filling, channeling, and draining. Also, thousands of hectares of wetlands have been destroyed in Virginia as a result of a court ruling (the "Tulloch" decision) that temporarily reopened a legal loophole allowing ditching and draining.

Disease. Unprecedented occurrences of diseases plague many commercial species. Pathogenic organisms in epidemic proportions infect oysters. While oyster diseases are not new, two lethal protozoan diseases, MSX (*Haplosporidium nelsoni*) and Dermo (*Perkinsus marinus*) affect recruitment and may kill up to 80% of immature, 2-year-old oysters. In summer 1992, about 70% of Maryland's oyster beds and 90% of Virginia's beds were infected. And while the cause of the outbreak is unknown, ecological change and oyster cross-regional transfers are suspected. Striped bass have appeared with ulcerative dermatitis and poor nutritional condition, raising concern about the Bay's declining forage base (especially menhaden). Surveys in 1998–9 revealed that half the Bay's striped bass population was infected with bacteria (*Mycobacteria* spp.), formerly known only from aquarium fish. Brown bullhead catfish (*Ameiurus nebulosus*) inhabiting the contaminated Anacostia River near Washington, D.C. exhibit high rates of liver and skin tumors. Shell deformities have appeared on a benthic foraminiferan (*Ammonia parkinsoniana*), an abundant facultative anaerobe in the mesohaline portion of the Bay (Fig. 5.28). Eelgrass (*Zostera marina*) was infected by a fungus in the 1930s (wasting disease) that devastated most beds of its North Atlantic range.

Phytoplankton-algae abundance. Dinoflagellate blooms, causing red, brown, green, and mahogany "tides," appear to be increasing, chlorophyll concentrations are high overall, and diatoms have changed from larger benthic species to very small planktonic forms. Phytoplankton have presently reached excessive concentrations in the mesohaline and tidal–fresh–oligohaline areas in spring and summer, respectively. Large blooms of a microalga (*Prorocentrum minimum*) turned 10 miles of the Choptank River visibly red in May 1998, and a widespread bloom occurred in spring 2000. Blue-green algae (cyanobacteria) are widespread in the upper Bay, small blue-green algae flourish intermittently in the middle portion of the Bay, and diatoms and green algae are increasing in the lower Bay and Virginian estuaries. Large mats of green algae (*Ulva*, etc.) are common in summers and prevent light from reaching SAV.

Nuisance and invasive species. Stinging sea nettles (*Chrysaora quinquecirrha*) and comb jellies (*Mnemiopsis leidyi*) are abundant. Comb jellies have increased more than five-fold, competing with fish for zooplankton prey. Eurasian water milfoil (*Myriophyllum* spp.), a ubiquitous plant invader that arrived in the 1950s, doubled its coverage in one year (1960–1), from about 20 000 to 40 000 ha^{-1}. Today it is no longer so extensive, but the exotic *Hydrilla verticillata* introduced in the early 1980s is threatening native SAV. The invasive common reed (*Phragmites australis*, Box 1.1, page 5) is spreading, replacing other marsh vegetation at a rate of approximately 1–5% annually, altering ecological function by elevating the marsh plain, filling in the microtopographic relief of marsh surface, and sequestering nitrogen. *Phragmites*' dense root structure alters currents and shore-marsh profile, affecting fish movements and free exchanges of detritus. The voracious flathead catfish (*Pylodictis olivaris*) has been introduced from the Mississippi watershed for sport fishing, threatening native freshwater species. And a semi-aquatic rodent, the nutria (*Myocastor coypus*), of South America is locally eradicating wetland plants; measures are now being taken to contain its increase.

Toxic pollutants. The 1999 list of Toxics of Concern for the Bay include DDT, PCBs, PAHs, chlordane, Kepone®, mercury, dieldrin, arsenic, cadmium, zinc, malathion, and more, at levels that may cause toxic impacts to living resources (Table 5.3). The Anacostia River (Washington, D.C.), Baltimore Harbor, and Elizabeth River (Norfolk–Portsmouth industrial complex) are among the nation's most polluted waters. Pollution also originates from the Bay region's high agricultural activity, with application of pesticides in the millions of kilograms a year. Bioaccumulation of organochlorine insecticides by fish and predatory birds has been reported. Mercury occurs in various fish species in some rivers throughout the watershed.

Toxic algae. The toxin-producing algae, *Dinophysis accuminata*, produced a bloom in 2002 of some magnitude in the lower Potomac River. Other toxic species have been reported with at least 12 phytoplankton species of known toxicity identified. Fish kills in late summer 1997 focussed concern on the fish-killing dinoflagellate *Pfiesteria piscicida*, a species that is also toxic to humans. Ulcerative lesions and mass mortalities of Atlantic estuarine fish, particularly menhaden, have been associated with exposure to *Pfiesteria*-like dinoflagellates and their toxins in other areas. Recently, a second species of *Pfiesteria* (*P. shumwayae*) has been identified.

Increased nutrient loading. Human activities have increased Bay-wide nitrogen loading rates six-fold, and phosphorus 17-fold, since early European settlement. Eutrophy is reflected in abundances of phytoplankton and microzooplankton. The upper Bay and major tributaries now receive substantially higher nitrogen loads relative to phosphorus than in the late 1960s and 1970s (Table 5.4). The Chesapeake Bay Foundation estimates about 300 million pounds of nitrogen overloads the Bay's waterways yearly.

Anoxia. Anoxic water was first documented in the central portion of the Bay in 1938, with anoxia–hypoxia becoming increasingly widespread and longer lasting. Dissolved oxygen levels of less than 0.5 ml l^{-1} occurred in about 5 billion m^3 of bottom water by 1980, to dominate the mesohaline area between Baltimore and the Potomac River from May to September. Depleted sediment oxygen is marked by concentrations of the anaerobic foraminiferan *Ammonia parkinsoniana*, which for 30–40 years has dominated the benthic assemblage in parallel with nitrate increase (Fig. 5.28).

Erosion and sedimentation. Sedimentation rates have doubled to quadrupled in shallow waters and upper

Table 5.4 Nitrogen and phosphorus input (kg km^{-2} yr^{-1}) to the upper Bay, and Patuxent and Potomac subestuaries. Values are medians from published sources.

Land use category	Nitrogen load	Phosphorus
Residential		
Low density	543	99
Medium density	963	148
High density	1 803	247
Commercial–industrial	1 630	222
Cropland	889	222
Pasture	494	74
Feedlot, waste storage	289 731	25 935
Forest	247	24.7

From Magnien et al (1992).

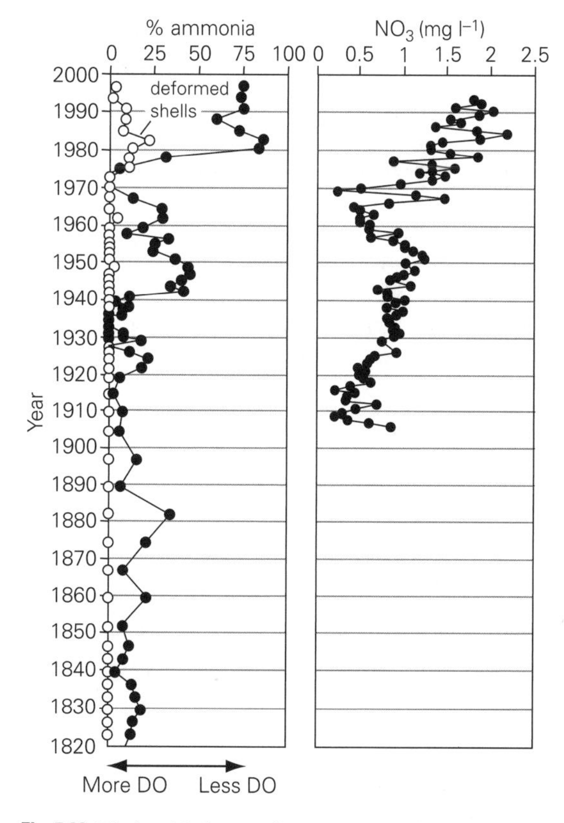

Fig. 5.28 The benthic foraminiferan *Ammonia parkinsoniana* (•), a pollution indicator, increased 75% in sediment, between 1820 and 2000. Shell abnormalities (o) appeared in the mid-1970s paralleling increases in hypoxia and nitrate (NO_3). Unit for nitrate is mg l^{-1}. From Karlsen et al. (2000).

tributaries, more or less following the pattern of land use by humans.

Landscape change. Urban and suburban areas expanded 17% between 1985 and 2000. This trend is expected to continue through at least the next decade, increasing impervious surface cover and loads of pollutants washed from farmland, pavement, and overwhelmed sewage systems. Land clearing parallels increases in floating diatoms and sedimentation rates (Fig. 5.29A), and deposition of total organic carbon (Fig. 5.29B).

Loss of riparian area. Lands bordering streams and tributaries have been significantly utilized and altered. Removal of riparian vegetation, hardening streambanks with concrete, and lawns, riprap, channel realignment, highways, livestock, sediment, urbanization, and recreation have created significant losses.

Blockage of tributary flow. Dams, hydroelectric facilities, highway culverts, and gauging stations interrupt many of the Bay's tributaries. Alteration of natural flow is one of the major causes of changing species populations, including those of resident (e.g., yellow perch, *Perca flavescens*) and anadromous fish (e.g., shad, striped bass). Droughts exacerbate these effects by lowering stream flow and groundwater levels, altering salinity, increasing flooding when rain returns, and increasing pollutant concentrations.

These effects, taken together, indicate ecosystem change. Algal blooms, toxic algae, fish kills, turbid water, oxygen depletion, and toxic contamination indicate a Bay very different from the past. Exposures to pollutants, decreasing natural resource abundance, and altered ecological linkages and land–seascape structure during more than 300 years also clearly indicate a mounting environmental debt. Habitats – wetlands, seagrass beds, water, benthic structures – have been converted to human uses and no longer perform important Bay-wide functions. Community-wide responses, species depletions, reduced water quality and emergent phenomena have added to this loss.

Accounting these losses, identifying change, instituting collective social action, and monitoring recovery involves long-term commitment, documentation, costly human interventions, and sustained public support. Science has played a key role by identifying issues and helping to bring about social and policy changes. Today, the Bay region has sufficient political will,

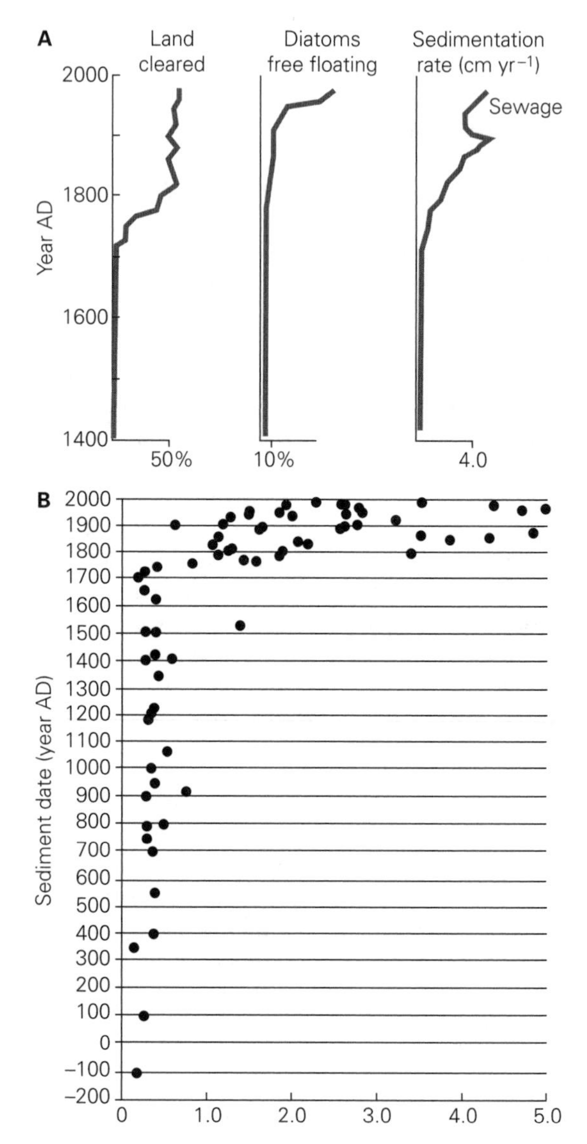

Fig. 5.29 (A) Increases in percentage of free-floating diatoms, sedimentation rates, and nutrients (sewage) correlated to land cleared, as analyzed in stratigraphic profiles of Furnace Bay, Maryland (Brush 1986). (B) Increased sediment deposition of total organic carbon in recent time. From Cooper & Brush (1993).

economic affluence, and social awareness to accomplish restoration. The region also benefits from prominent conservation groups, and a large segment of the public is calling for protection and restoration. Can social policy, channeled through traditional practices, now effect change sufficient to the task of restoring estuarine function and recovering natural resource abundance?

5.6 Restoration: the Chesapeake Bay Program

In the Chesapeake Bay, key attributes have caught public and political attention. Natural resources are valued and human use, with its political history and influence, carries traditional value. Virginia and Maryland are major producers of marine products, which, in spite of depletions, still add hundreds of millions of dollars to state economies. Shipping, tourism, commerce, energy, defense, and recreation are major industries involving dozens of tidewater counties, cities, businesses, and state and federal agencies. These all influence the politics of decision-making, alter state budgets, and shift political authority.

The Chesapeake Bay Program emerged from the landmark signing of the 1983 Chesapeake Bay Agreement that committed policy leaders to protect and restore the Bay ecosystem. This Agreement came about when scientists revealed environmental deterioration, and as a result of the federal National Environmental Policy Act (NEPA, 1969) that established the Environmental Protection Agency (EPA, 1970). Under the Clean Water Act (1972), EPA initiated a Bay study that clearly indicated natural resource and submerged aquatic vegetation losses, nutrient enrichment, and toxic pollution, bringing national attention to focus on estuarine issues.

During the following 20 years, the Program has evolved. The Agreement established the Executive Council (EC) as the policy-making body, represented by state cabinet secretaries, an Implementation Committee (IC), and a Chesapeake Bay Liaison Office. A second, 1987 Agreement renewed former commitments, and amendments in 1992 framed a multijurisdictional approach for restoration and protection. EC membership changed from representation by state cabinet secretaries to state governors (Maryland, Virginia, Pennsylvania), the Mayor of the District of Columbia, and federal EPA administrator, reflecting the growing importance of the Program. The Chesapeake Bay Commission, the legislative body that performs an advisory role to state assemblies and the U.S. Congress, is also included in the EC. The Program has grown in influence and authority, supported by the Federal Estuary Restoration Act of 2000 and a third Agreement, *Chesapeake 2000*, was signed, establishing new priority goals and commitments with a 10-year focus.

5.6.1 Social and political commitment

The Chesapeake Bay Program operates under terms of *Chesapeake 2000*, which stresses ecosystem management of the entire Bay watershed. Hence the governors, EC, and top officials set goals and guide policy through a series of directives, agreements, and amendments that represent citizens and local governments. Sub-Committees on Toxics, Land Growth–Stewardship, Living Resources, and more, carry out Program goals. The Scientific and Technical Advisory Committee (STAC) is an advisory scientific council responsible for suggesting means for restoration and protection. Funding comes from more than 24 federal and state departments and agencies, giving the Program authority. These agencies, together with conservation groups, citizens, and private organizations, undertake important tasks of data management, monitoring, and research.

Chesapeake 2000 commits to:

1 *Living resource protection and restoration.* Restore, enhance and protect finfish, shellfish, and other living resources, their habitats and ecological relationships, to sustain all fisheries and provide for a balanced ecosystem.
2 *Vital habitat protection and restoration.* Preserve, protect and restore those habitats and natural areas that are vital to the survival and diversity of the living resources of the Bay and its rivers.
3 *Water quality protection and restoration.* Achieve and maintain the water quality necessary to support the aquatic living resources of the Bay and its tributaries and to protect human health.
4 *Sound land use.* Develop, promote, and achieve sound land use practices which protect and restore watershed resources and water quality, maintain reduced pollutant loadings for the Bay and its tributaries, and restore and preserve aquatic living resources.
5 *Stewardship and community engagement.* Promote individual stewardship and assist individuals, community-based organizations, businesses, local governments and schools to undertake initiatives to achieve the goals and commitments of this agreement.

These goals are clear, and mechanisms are being tested and evolving, with science, politics, and social and legal factors being woven into adaptive strategies. The challenges are at the forefront of ecological and social theory and practice. Left unresolved, however, is the level of "restoration" to be achieved. The U.S. Committee on Restoration of Aquatic Ecosystems defines restoration as "reestablishment of predisturbance aquatic functions and related physical, chemical, and biological characteristics . . . A holistic process not achieved through the isolated manipulation of individual elements." How the Program will play out, at what point in history that "predisturbance" can be set as a target, and what the Program will be able to restore will provide important lessons for future conservation practice.

5.6.2 Program scope

The Chesapeake Bay Program encompasses the entire estuarine–watershed system, under a Bay-wide banner and partnership agreements. It endorses sustainable development as the ability to meet needs of the present without compromising the ability of future generations to meet their own needs. It also recognizes that: "With more than 15 million people living within the Bay watershed, maintaining a sustainable lifestyle will be necessary to protect the Bay and its resources." Program strategy includes monitoring Bay health, reducing pollution, restoring habitat, managing fisheries, maintaining sustainable development patterns, and using computer models to predict future functions and needs. The partners are engaging local governments and community watershed groups in restoration. Industries, farmers, local community groups, and citizens are working together to reduce the amount of pollutants entering the Bay. Several fishery management councils and Chesapeake Bay Program Fisheries Management Plans guide fisheries, although states continue to manage their resources separately. The Chesapeake Bay Commission, however, announced in 2002 that it would cost $19.1 billion to meet the goals by 2010, at a time when states are facing budget deficits.

Oyster restoration is a major goal, with two important aspects. *Chesapeake 2000* calls for a 10-fold increase in oyster numbers over 1994 by the year 2010. Meeting this goal involves establishing protected areas, whereby states restore both the reef's ecological role and a viable fishery by restricting oyster harvest and designating "Oyster Restoration Areas" – 2000 ha for each state and another 400 ha in the Potomac River. Candidate areas covering more than 10 800 ha have already been designated, and more than 81 ha of oyster habitat have been constructed. The U.S. Congress appropriated $4.15 million in 2001 to boost oyster restoration: $3 million for the U.S. Army Corps of Engineers, $300 000 for EPA, and about $850 000 for the National Oceanic and Atmospheric Administration (NOAA) for construction of oyster reefs, monitoring water quality around reefs, habitat restoration through local-volunteer efforts, and public information. The Chesapeake Bay Foundation, a major conservation group, plans to rear one million disease-tolerant oysters annually to replenish oyster reefs.

The national Sustainable Fisheries Act of 1996 complements existing fisheries management approaches. Under this Act, Bay scientists are developing a Fisheries Ecosystem Plan, calling for fisheries to be ecosystem-based and to include marine protected areas and fishery reserves. This suggests a shift in values, from individual or economic preferences toward collective action, recognizing that the Program helps protect local and state fishing interests.

Chesapeake 2000 is also committed to conserving important lands in the watershed. Specifically, to ". . . permanently preserve from development 20 percent of the land area in the watershed by 2010." A goal of 440 000 ha of protected landscapes, over what already exists, has been set – an area the size of Delaware – at an estimated cost of $1.8 billion. This suggests positive steps in a new wave of conservation for a body of water fast becoming an "urban sea."

5.7 Conclusion

Public awareness, bolstered by scientific information, has culminated in one of the most ambitious and far-reaching programs of its kind. The Chesapeake Bay Program is ecosystem-oriented and science-based, with state, federal, local, and private partnerships carrying out agreed-upon goals. Public education and voluntary compliance are essential. The Program provides a means under which standards of achievement and mandates are to be met through voluntary action, enforcement of laws, and citizen watch-dogs. Yet differences among state and federal agencies, public and private interests, legislative mandates, and traditional uses have created a fragmented approach. As scientists are developing new information and monitoring methods, they query: Are sustainable harvests and the ecological goals of restoration compatible? How successful can establishment of protected areas be in a fragmented urban sea? How big is the Bay-wide problem? What kind, how large, how many, and where should protected areas be placed? Beyond science, what levels of effort, time, and money are required to restore Bay resources? Should restoration seek return to the past or shift toward achievable outcomes of the possible and confront the ecological consequences? Ultimately: What is the carrying capacity for humans in the Chesapeake Bay ecosystem and what level of environmental subsidy in the form of mariculture and waste treatment plants is needed to sustain an expanding urban system?

Chapter 6

Bering Sea: marine mammals in a regional sea

With text contributions by Gary L. Hufford and Thomas R. Loughlin

The many threads of evidence, inquiry, and hypothesis . . . pattern themselves into a tapestry of history – the history of the landscape of Beringia, at one time dominated by sea, at another by a great plain; once covered with forest, then with treeless tundra; once populated by mammoth, horse, and bison, hunted by paleolithic man; now the home of seal, walrus, and polar bear, sought by the world's most skillful sea-mammal hunters.

D.M. Hopkins (1967)

6.1 Introduction

The Bering Sea lies between the North Pacific and the Arctic Ocean (Fig. 6.1). Some of the largest marine mammal, sea-bird, fish, and invertebrate populations on Earth reside there, year-round or seasonally. Native peoples have depended on these resources for thousands of years and subsistence traditions remain strong. Only during the last 200–300 years has industrial exploitation become a major factor. By the 1990s, the Bering Sea accounted for 2–5% of the world's, and more than half of U.S., fishery production. Several populations of fishes, birds, and mammals have fluctuated widely during the past century and a half. Among marine mammals, one species is extinct, some are endangered, and a few have increased. However, pre-exploitation baseline data are questionable, making it unclear whether ecosystem changes may have resulted from the effects of exploitation or have been compounded by effects of cyclic climatic phenomena – or both.

Bering Sea marine mammals inspired the first international treaty wholly dedicated to the conservation of living natural resources – the Fur Seal Treaty of 1911 – which was responsible for recovery of the fur seal and sea otter from near extinction. Until the International

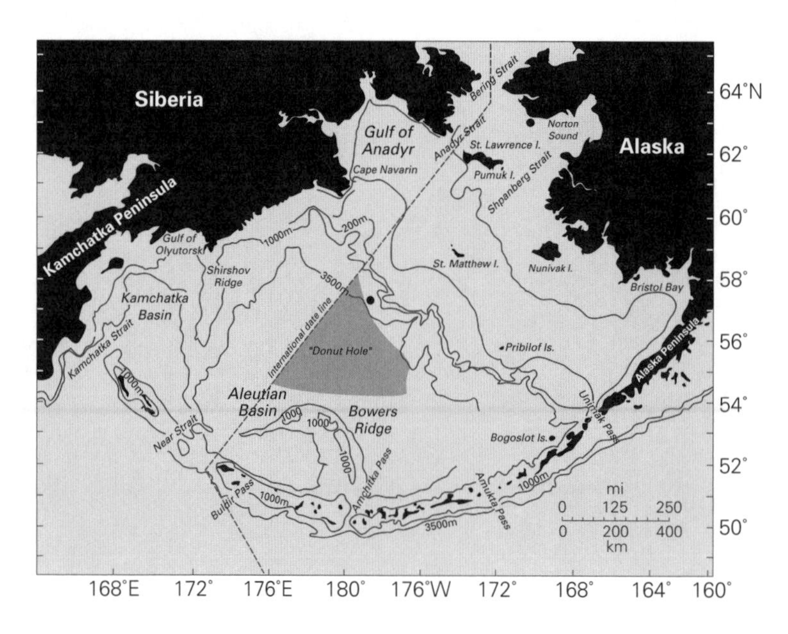

Fig. 6.1 Bering Sea geography and place names. The entire Sea is under U.S. and Russian jurisdiction, with the exception of the international "donut hole." Closed circles (•) indicate weather stations, at 59.9°N, 177.8°W and 63.3°N, 170.3°W; see Fig. III.2 (pages 130–1). Adapted from Loughlin & Ohtani (1999).

Convention for the Regulation of Whaling was agreed in 1931, this Treaty stood alone as an international conservation agreement for marine resources. The U.S. Marine Mammal Protection Act of 1972 was also inspired to large degree by Bering Sea marine mammal conservation (Box 2.4, pages 48–9). In 1973, the U.S. Congress passed the Endangered Species Act, which also affected several Bering Sea marine mammals. In 1976, the Fishery Conservation and Management Act (FCMA: now the Magnuson–Stevens FCMA) extended U.S. marine jurisdiction to 200 miles, and Russia followed the same limit, resulting in two-nation jurisdiction over most of this Sea, with the exception of the "donut hole" (Fig. 6.1).

Two Bering Sea marine mammals illustrate the need to understand interrelationships among climate, oceanography, and human use. The Steller sea lion (*Eumetopias jubatus*) of the North Pacific and southern Bering Sea declined 50–80% from the 1950s to the 1990s; populations have been listed as either "endangered" or "threatened" under the U.S. Endangered Species Act. The Pacific walrus (*Odobenus rosmarus divergens*) is a boreal-arctic, sea-ice-adapted species that has fluctuated in numbers during the past century and a half due to exploitation. Economic and social factors unite these and other Beringian marine mammals into common themes. Conservation and management are undertaken against a backdrop of two contrasting, and unequally positioned, cultural systems of non-common antecedence, one founded on long-evolved beliefs and subsistence needs, and the other on industry and national-to-global economies.

6.2 Physical setting

The Bering Sea covers 2.3 million km², and became partially cut off from the Pacific Ocean by emergence of the 2000 km-long arc of the Aleutian Islands during the Eocene epoch, 80 million years ago. To the north, the 85 km-wide Bering Strait is the only direct link between the Pacific and Arctic oceans.

The Bering Sea is almost equally divided between a deep southern basin and a shallow northern shelf. The basin is ecologically similar to the North Pacific, being almost totally ice-free and geologically complex, and reaches a depth of 3600 m. A steep continental margin, incised by seven of the largest submarine canyons in the world, is transitional from the basin to the shelf. The almost featureless, shallow shelf is one of the world's widest (500–800 km), with a gentle north-to-south slope (0.24 m km^{-1}) down to approximately 200 m; its dominant feature from winter to spring is sea ice, which strongly influences its biology and ecology. The islands of the Pribilofs, Nunivak, St. Matthew, St. Lawrence, and the Diomedes occur on the shelf and are important habitats for waterfowl, seabirds, and pinnipeds.

6.2.1 Oceanography and domains

Net flow from the Pacific Ocean through the Bering Sea is from south to north; the flow through the Bering Strait is approximately 1 $km^3\ s^{-1}$, or about 15% of the global-ocean inflow to the Arctic Ocean. The basin's dominant circulation is counterclockwise, with a mean speed of about 10 cm s^{-1} (Fig. 6.2). Flow reversals through the Bering Strait occur occasionally, apparently due to transient shifts in atmospheric features. That is, the semi-permanent Aleutian low and the arctic high may be replaced by high- and low-pressure systems, respectively, causing a shift in sea-surface slope from north to south (Fig. 6.3).

Differences in temperature, salinity, currents, freshwater runoff, seasonal heating and cooling, sedimentation, sea-ice dynamics, wind stress, and horizontal and vertical mixing interact to subdivide the Bering Sea into domains that influence biotic distributions (Fig. 6.4):

- *Alaska coastal domain*: shore to about 50 m depth; flow northward at 1–5 cm s^{-1}; vertically well mixed and influenced by seasonal discharge of freshwater; summer surface temperatures to 14°C; winter water temperatures near freezing.
- *Midshelf domain*: approximately 50–100 m; little net circulation; surface layer wind-mixed; lower layer (> 30 m) tidally mixed; surface temperatures 3–10°C in summer to near freezing in winter; lower layer 1–9°C.
- *Outer shelf domain*: approximately 100–200 m; dominated by north-northwest current flow at 1–5 cm s^{-1}; surface layer mixed by winds, temperatures to 9–10°C in summer; lower layer tidally mixed, summer temperatures 2–4°C.
- *Gulf of Anadyr domain*: currents generally strong; influenced by the Anadyr River; summer surface layer to 8–9°C; lower layer 0–2°C.
- *Bering deep domain*: depth more than 200 m; greatly influenced by inflow from North Pacific; generally ice-free in winter; surface temperatures usually about 4–6°C.

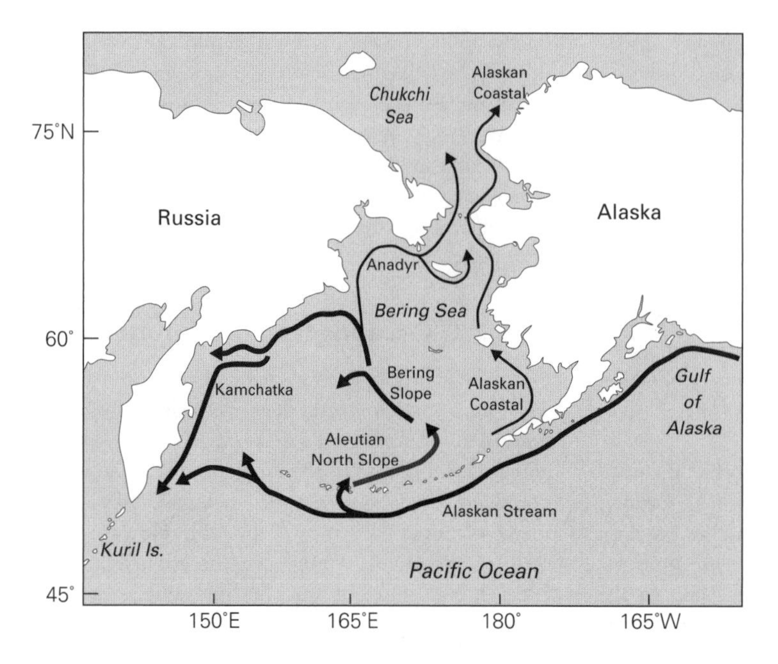

Fig. 6.2 Bering Sea currents. Adapted from Springer et al. (1996).

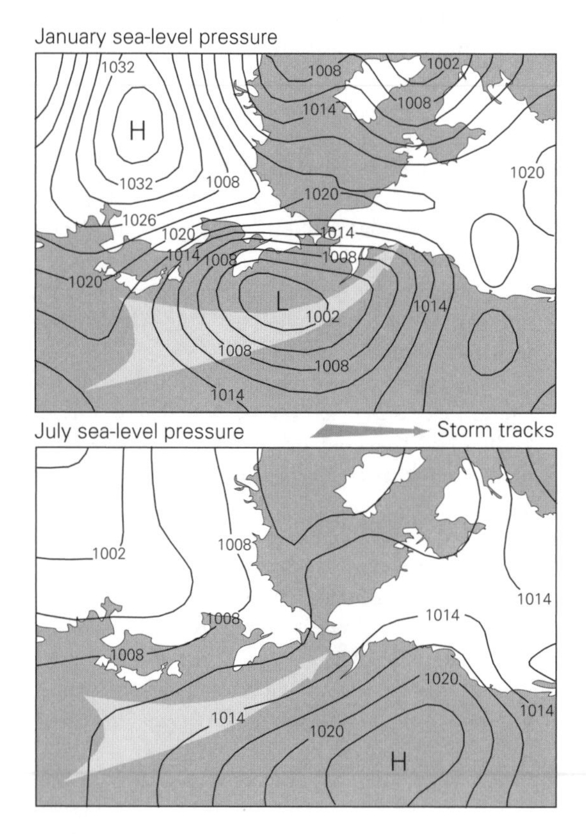

Fig. 6.3 Location of the Aleutian low in winter and its disappearance in summer. Arrows denote storm tracks. See text for explanation.

These domains constitute ecologically distinct areas where physical, chemical, and biotic factors influence reproduction, growth, and survival of marine organisms. Due to different conditions among years and seasons, these domains are highly variable, making predictions of species' occurrences and abundances difficult.

The Bering Sea's landward margins include a wide range of environments, from mountainous shores to terrestrial lowlands. Extensive wetlands support major breeding areas for North American waterfowl, especially on Alaska's Yukon and Kuskokwim deltas and bordering Russia's Anadyr River. Highly seasonal river flows influence the timing of biological events, such as migration and spawning of anadromous fishes. Many rivers support major salmon runs, but salmon enter northern rivers later than southern ones.

6.2.2 Climate and weather variability

The climate of the Bering Sea (Fig. III.2, pages 130–1) is dictated by its latitude, the character of the surrounding land, exchanges among oceanic water masses, the location and intensity of the Aleutian low, movements of atmospheric cyclones and fronts, and global atmospheric circulation, all of which produce strong north–south and east–west variabilities. Additionally, latitude strongly influences day length, which varies from 16 hours in the south to 20 hours in the north in summer to only 3.75 hours in the north and 8 hours in

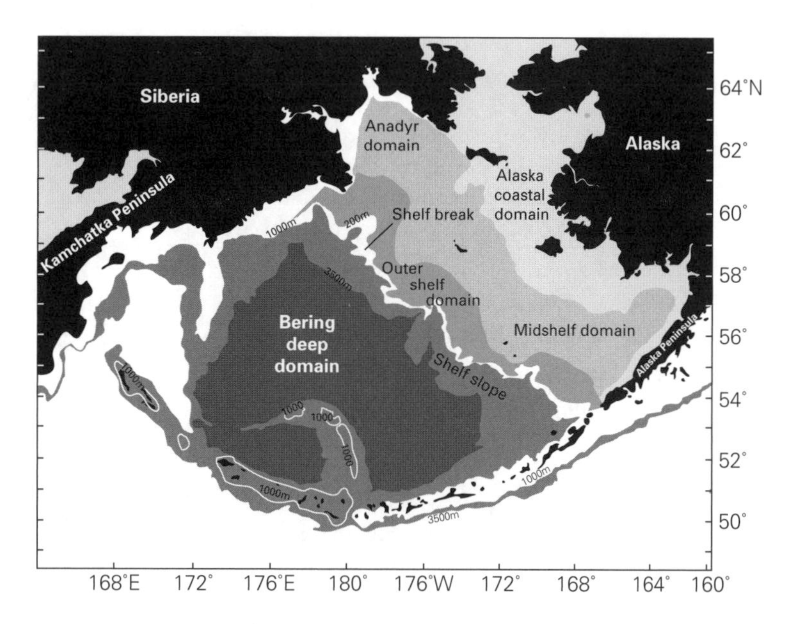

Fig. 6.4 Bering Sea oceanographic domains.

the south in winter (Fig. III.1, page 129). During winter, little radiation is received from the sun in northern areas, which, along with other factors, results in much of the continental shelf being covered by sea ice.

Climatological forcing is greater in winter than in summer by an order of magnitude due to seasonal modulation of the Aleutian low. This low is one of two major low-pressure systems in the high-latitude northern hemisphere and strongly influences the migration of storms through the Bering Sea region (Fig. 6.3). In winter, the Aleutian low is pronounced and normally located near the International Date Line. This low creates a strong pressure gradient that causes winds to come predominantly from the north. Storms are transient features that disrupt this general pattern, with a frequency of about 4–5 per month. The most frequent trajectory of storms is eastward along the Aleutian chain and into the Gulf of Alaska, producing cold outbreaks that are often followed by warm airflow from the North Pacific. In summer, the Aleutian low almost disappears and the east-Pacific high becomes dominant, resulting in weak flow over the region due to lack of a strong pressure gradient. Storms generally move along a general path eastward as in winter, but curve northward into the Bering Sea, with a frequency of only about 2–3 per month.

Long-term observations of Bering Sea weather and climate are limited, making predictions of seasonal, annual, and interdecadal variability difficult. Detailed records of less than 50 years indicate that the Alaskan Peninsula and western Alaska have far less seasonal variability in temperature and precipitation than the Aleutian Islands and Pribilofs. This suggests that a number of climate regimes have yet to be recognized. Interannual and decadal variability of Bering Sea climate appears to depend on positions of the Aleutian low, which responds to variations on a Pacific Ocean and global scale. One interannual climatic shift is El Niño/La Niña, whose spatial manifestations include the Aleutian low and strong variations in temperature. Effects of El Niño include increased air and seawater temperatures in the eastern half of the Bering Sea, but during La Niña, the Aleutian low appears to become less intense and shifts westward, resulting in cooling.

A recently discovered regional climate cycle is the quasi-decadal Pacific Decadal Oscillation (PDO), which is similar to El Niño in the North Pacific, as it includes anomalous cold waters in the central and western North Pacific. First evidence of the PDO came from observations that Alaska salmon catches alternated with salmon catches from the Pacific northwest every 20–30 years, punctuated by abrupt reversals. Assuming that catch variations reflect biological production, it follows that this inverse relationship is related to climate forcing. That is, a cycle of warm waters and mild winters in Alaska occurs when the Pacific northwest experiences cold ocean waters and harsh winters, a situation that reverses abruptly every 20–30 years. Since the 1977 PDO shift from a cold to a warm regime, dramatic shifts in the structure and composition of North Pacific and Bering Sea biotic communities have occurred. Many abundant, short-lived demersal fishes with strong year classes (pollock, Atka mackerel, cod, some flatfishes) that spawned in the mid- to late 1970s have oscillated in abundance, while long-lived flatfishes (arrowtooth flounder, Pacific halibut, yellow-fin sole, rock sole) have remained abundant. Therefore, the ecological responses to such phenomena as El Niño and PDO are of obvious consequence, although not well understood.

6.2.3 Sea ice

In summer, the Bering Sea is typically ice-free and sea ice is far north in the Chukchi Sea. But in winter, sea ice normally covers about 75% of the shelf (Fig. 6.5A). This seasonal ice advance and retreat is the most extensive for any arctic region. The maximum extent of sea ice usually occurs in mid-March, with the average maximum just north of the Pribilofs. However, this is highly variable (Fig. 6.5B, C).

Bering Sea ice formation has been described as a conveyor belt. Sea ice is formed along south-facing coasts in relatively ice-free areas called "polynyas" (Fig. 6.5B-5), which have high rates of heat exchange between the atmosphere and the ocean. Persistent northerly winds blow this newly formed ice southward; the rate of ice movement can be up to 100 km daily. At the shelf break, a thermodynamic balance between ice and surface-water temperatures occurs and the ice melts. However, in and around the Bering Strait area, sea ice is usually transported northward through the Strait by currents. During periods of current reversal, ice from the Chukchi Sea can be transported southward through the Bering Strait and can extend south of St. Lawrence Island.

The dynamics of sea ice are made complex by the interplay of atmospheric and ocean-surface temperatures, atmospheric winds, and oceanic surface currents. Marine life has adapted to this variability, and can even

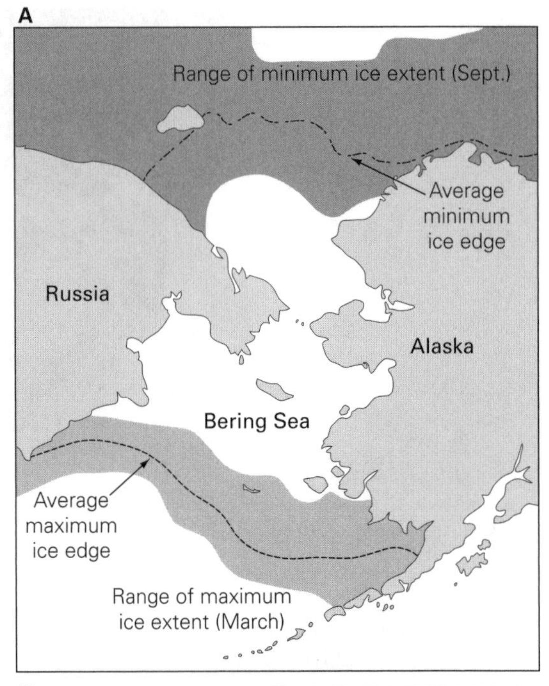

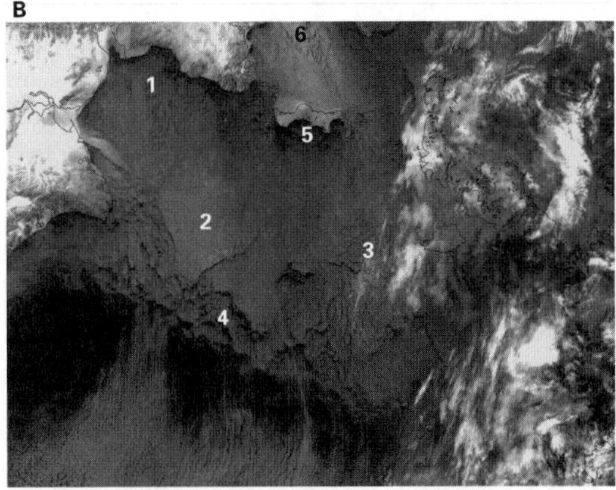

Fig. 6.5 (*right*) (A) Mean, maximum, and minimum extensions of seasonal sea ice (NOAA 1988). (B) NOAA AVHRR (Advanced Very High Resolution Radiometer) infrared image of March 30, 1988: (1) pack ice with leads; (2) broken pack; (3) rounded pack; (4) loose pack; (5) open water and polynyas; (6) continuous ice, which occurs in the north of the Bering Strait region and in the Chukchi Sea (Ray & Hufford 1989). (C) NOAA AVHRR infrared image of March 10, 1999, showing the same ice types as (A) but with shifted location and different range. Since the mid-1990s the ice position for mid-March has shifted north, supporting arguments for global warming. Courtesy of NOAA National Weather Service, Anchorage, Alaska.

take advantage of it. New feeding areas (leads) are continually being opened as the ice moves. Six sea-ice zones are essential habitats for marine mammals and other organisms. For example, the broken pack (Fig. 6.5B-2) is mostly a feature of the northwest-central Bering Sea and is favored by walruses and bowhead whales (*Balaena mysticetus*). Conversely, rounded pack (Fig. 6.5B-3) appears to be inimical to ice-inhabiting marine mammals, as it forms a zone of jumbled pack with little access to open water, especially when outbreaks of sea ice move southward through Shpanberg Strait. Most seals inhabit the loose pack (Fig. 6.5B-4), and polynyas (Fig. 6.5B-5) are favored habitat for seabirds and waterfowl.

Few quantitative studies exist on sea-ice/habitat relationships. A new generation of high-resolution satellite sensors has the potential to provide details on ice type, distribution, and movement, and even to observe groupings of animals in these habitats. Such observations will be critical for conservation if sea ice is affected by climate change, as now appears possible.

6.3 Biotic relationships

Phylum diversity is as high in the Bering Sea as that at lower latitudes, although species diversity is considerably less. Low species diversity is compensated by high biomass production, especially in summer. Reasons for high production are related to physical factors such as upwelling, nutrient input from the land, extensive shelf area, long daylight hours in summer, and the Beringian region's transitional position between the Pacific and Arctic oceans.

6.3.1 Biodiversity

Of the approximately 450 known species of macroinvertebrates and fishes of the Bering Sea, only a small fraction is reasonably well known. Among these are important commercial species, for example king and tanner crabs (*Paralithodes* and *Chionoecetes* spp.), walleye pollock (*Theragra chalcogramma*), and salmon (*Oncorhynchus* spp.). Several commercial fishes are wide-ranging and anadromous. Additionally, hundreds of seabird rookeries occur on cliffs, islands, and coastal wetlands, supporting millions of individuals. Sooty and short-tailed shearwaters (*Puffinus* spp.) are among the Bering Sea's most abundant seabirds; they nest in the southern hemisphere and migrate thousands of kilometers to the Bering Sea to feed during summer. Twenty-five species of marine mammals occur in Beringian habitats: Pinnipedia (sea lions, walruses, and seals), Carnivora (sea otter and polar bear), and Cetacea (whales, dolphins, and porpoises). The majority occurs over the continental shelf and in coastal waters and most are migratory, following sea-ice movements or migrating from warmer waters in spring. A few ascend rivers. Gray whales (*Eschrichtius robustus*) undertake the longest annual migration for any mammal, to and from Baja California lagoons, where they bear young, and the Bering and Chukchi seas, where they feed.

Timing among natural history processes and environmental changes is critical for Bering Sea life. Spring's burst of high primary productivity initiates reproductive cycles of a host of species – for example, among zooplankton and the fishes, birds, and mammals that feed on them – that must be completed in the short season before winter's cold reappears. How various species persist under these changing conditions involves a suite of morphological, physiological, and behavioral adaptations. Abundances of reasonably well-known species can vary dramatically among years, but often with little evidence of cause. Evidence of declines of some crabs, fishes, seabirds, and mammals have been attributed to changes in both food availability and climate. However, a number of other factors may play a role, including environmental change, sea-ice dynamics, atmosphere–ocean interactions, altered food webs, fishing, by-catch, consumption of plastics, oiling, and hunting.

One possible tool for assessing changes in biotic distributions and for clues to environmental relationships is analysis of species assemblages. Analysis of ranges of 86 Bering Sea species of invertebrates and vertebrates reveals six spatially differentiated assemblages (Fig. 6.6A), which are substantiated by features of oceanography, climate, sea ice, and bathymetry. Another set of assemblages may be derived by the same method for marine mammals (Fig. 6.6B); in this case, species include those for which sea ice is important habitat or for which it is mostly avoided. Sea-ice-inhabiting species are the bearded seal, spotted seal, ribbon seal, ringed seal, walrus, and bowhead whale; species that generally avoid sea ice are the Steller sea lion, northern fur seal, Pacific harbor seal, gray whale, and white whale. These analyses demonstrate two levels of biogeography: first, large-scale regional subdivisions and second, how specific variables – in this case, sea ice – may affect distributions among species. In

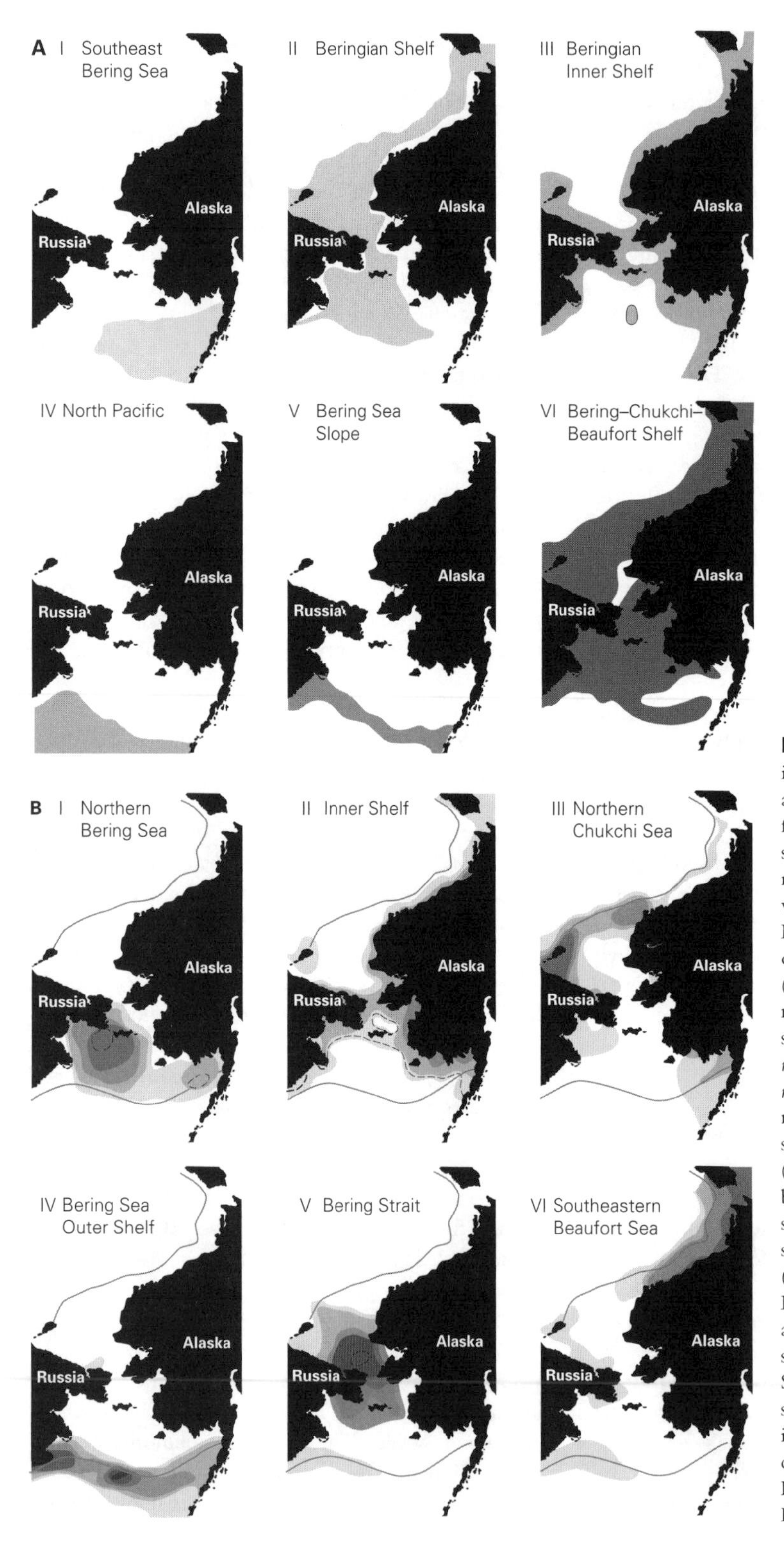

Fig. 6.6 Principal components analysis identifies provinces and species assemblages based on major adult areas from NOAA (1988). (A) Provincial subdivisions, I–VI, are based on individual ranges of 86 species of invertebrates and vertebrates of the Bering, Chukchi, and Beaufort seas shelf: Statistical gradients are omitted. From Ray & Hayden (1993). (B) Assemblages of 11 species of marine mammals: (I) Northern Bering Sea: most strongly represented by walrus (*Odobenus rosmarus*) and bowhead whale (*Balaena mysticetus*) in winter; (II) Inner Shelf: ringed seal (*Phoca hispida*) in winter and spotted seal (*Phoca largha*) in summer; (III) Northern Chukchi Sea: walrus and bearded seal (*Erignathus barbatus*) in summer; (IV) Bering Sea Outer Shelf: spotted seal in winter and fur seal (*Callorhinus ursinus*) in summer; (V) Bering Strait: ribbon seal (*Phoca fasciata*) and gray whale (*Eschrichtius robustus*) in summer; (VI) Southeastern Beaufort Sea: ringed seal and bowhead whale in summer. Dotted lines and shading indicate gradients equivalent to standard deviations. From Ray & Hufford (1989). Data for both analyses are derived from NOAA (1988).

both cases, assemblages prove more useful for interpretation than individual species would suggest because associations among species are implied. For the Bering Sea, distribution of species according to sea-ice type is fundamental. Sea ice is constantly in motion, driven by both wind and current, and any ice type can be readily dispersed or packed in a very short time. Therefore, the percentage sea-ice cover for an area of ocean is variable and probably not very meaningful. On the other hand, ice type and movement can be critical for animals that use specific ice types on which to rest or bear their young. However, few quantitative data are available on the influence of ice distribution and movement on population dynamics.

6.3.2 Production

The Bering Sea's abundance of marine mammals and birds is a striking indication of high production. Beringian warm-blooded (homeothermic) birds and mammals have high energy demands and consume the full spectrum of marine life (Fig. 6.7). The total biomass consumed by these species is difficult to estimate, but probably rivals the total commercial fisheries catch.

Seasonal change in high-latitude seas has major influences on temporal sequences of production. Ambient light levels from late fall to early spring are too low to permit significant phytoplankton growth, and sea ice further limits light in the water column from early winter to late spring. Spring sea-ice melt establishes a density-stratified water column, with fresher, lighter water at the surface. Concurrently, a brief phytoplankton bloom occurs, which can be intensive and follows the retreating ice northward. The physical zonation of the shelf and the nature of atmospheric and oceanographic forcing strongly influence annual primary production, which can be as high as for any regional sea. However, primary production is not distributed evenly, being highest within a variable "green belt" (Fig. 6.8). In some areas, local summer and fall blooms may also occur. Nutrients from upwelling and other factors result in high, but variable, primary production. Overall, high primary production supports high secondary production (zooplankton), which, in turn, supports high seasonal abundances of animals that feed at lower trophic levels, such as squids, smaller fishes, birds, and baleen whales.

Primary production is also affected by a cross-shelf gradient in nutrients, being generally greater in the middle domain than in the coastal domain (125–175 *vs.* 50–100 g C m^{-2} yr^{-1}, respectively). This gradient may break down due to wind-mixing, alterations of nutrient availablity, changes in transport mechanisms, and other factors that can introduce high variances among years. Furthermore, new phenomena may also introduce variability: a coccolithophore (planktonic protozoan) bloom has recently occurred in summer in the southeastern Bering Sea (Fig. 1.1, page 7), for which there is as yet no explanation, nor understanding of its significance to the system.

Productivity of benthic shelf communities also varies across the shelf. In the outer domain, green belt, and oceanic region, organic transfer efficiency from

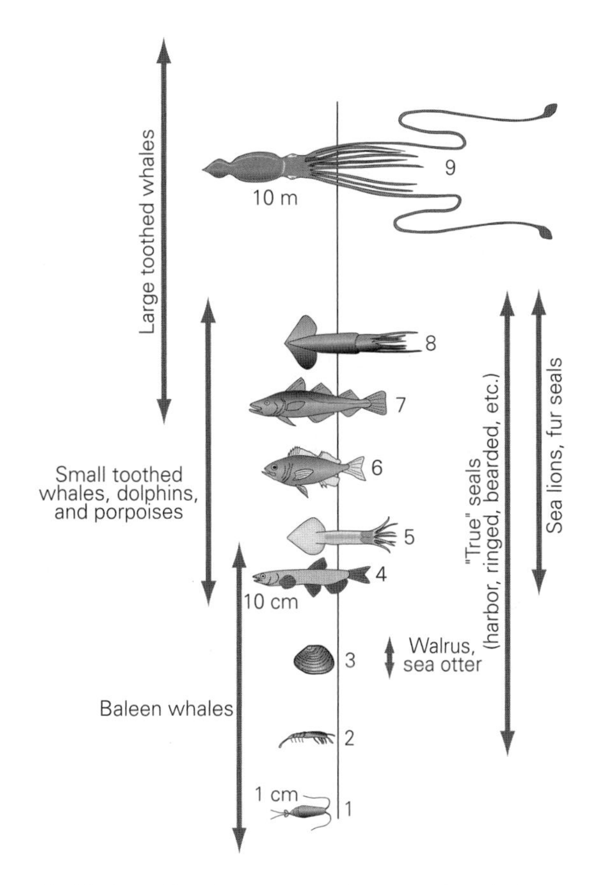

Fig. 6.7 Generalized size ranges of marine mammal prey in the Bering Sea. Prey size is shown to increase on a logarithmic scale from less than 1 cm to more than 10 m. (1) Calanoid copepods; (2) euphausids; (3) benthic invertebrates; (4) shallow- and mid-water fishes; (5) market squid; (6) benthic fishes; (7) demersal cods and their allies (principally juveniles); (8) benthic and mid-water squids; (9) giant squids and small marine mammals. From Loughlin et al. (1999).

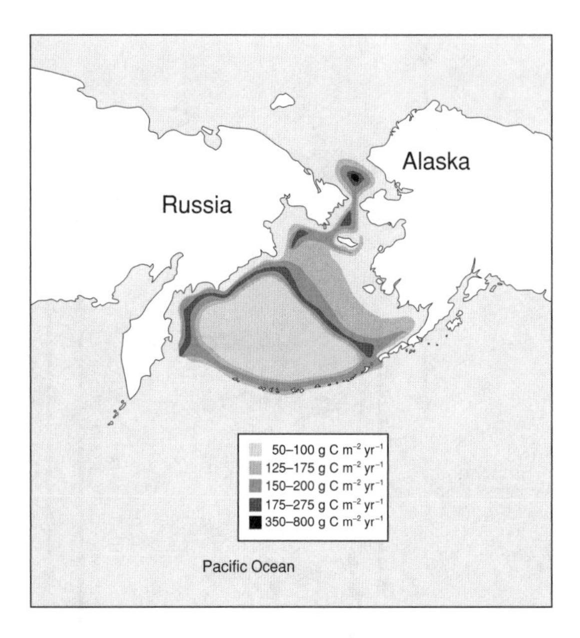

Fig. 6.8 Generalized pattern of primary production in the Bering Sea. The green belt is the region of high productivity at the shelf edge, with northwest and southwest branches. The northward extension reflects high productivity that is augmented by inputs from large rivers. From Springer et al. (1996), with permission.

phytoplankton to zooplankton seems to be about 18–24%, which is about the maximum energy transfer for ectothermic animals. Efficiency drops to around 14% in the middle domain and to 5% in the inner domain, suggesting that primary production is coupled to benthic consumers to a greater extent in these domains than elsewhere. A number of other variables influence benthic production. For example, infaunal biomass is lowest along coastlines due to sea-ice scour and wave and tidal stresses. Infaunal biomass is also regionally variable, being 1–20 g C m^{-2} in the middle domain and rising to 20–40 g C m^{-2} south of St. Lawrence Island, indicating the importance of polynyas for primary production. Infaunal biomass peaks at 50–60 g C m^{-2} in the southern Chukchi Sea, where significant downstream deposition of organic matter occurs at the northern terminus of the green belt; abundant gray whales feed here during summer. The epifauna, in contrast to the infauna, are more difficult to assess, as many species, such as king and tanner crabs, are mobile and are themselves consumers of infauna.

Surprisingly little is known about the feeding habits of many fishes, birds, and mammals, even those of high social value and commercial worth. Demersal species are especially abundant. Walleye pollock alone account for about 50% of the total biomass of demersal fish in the southeastern Bering Sea, although their numbers have varied six-fold during the past four decades. Salmons exhibit similar variability, possibly due to long-term climate oscillations. The nature of this variability, and whether it is caused by shifts in food availability or changes in recruitment, predation, competition, habitat, fishery exploitation, climate, or synergisms among any combination of these, can only be hypothesized.

6.4 Historical exploitation of Beringian marine mammals

Beringian marine mammals have been exploited to meet the subsistence needs of Aleut, Yupik, Iñupiat, and Chukchi peoples for a very long time. The first archeologically documented sites that contain marine mammal bones date from at least 8000 years ago. The principal feature since that time has been long-term continuity (Box 6.1).

The first documented commercial exploitation of Pacific walruses evidently occurred in 1648, when a Russian Cossack crew under Semen Dezhnev destroyed a walrus haulout site at the mouth of the Anadyr River. In 1741, Vitus Bering, a Dane, explored the Bering Sea for the Tsar of Russia. The first Russian commercial trip to Bering Island, where Bering's ship was wrecked and where he died, was in 1743, in search of fur seals (*Callorhinus ursinus*) and sea otters (*Enhydra lutris*). By the 1780s, the era of Russian expansion was in full swing, and the Steller sea cow had become extinct. American Yankees began to pursue whales early in the 1800s. In 1867, the United States purchased Alaska from Russia and marine mammal exploitation increased. Between the 1860s and the 1880s, more than 300 000 fur seals were taken from the Pribilof Islands, bowheads declined to low levels, commercial hunting depleted walruses, and Eskimos sometimes starved as a result. By the early 1900s, sea otters were assumed extinct, few whaling ships remained in the arctic, and fur seals were at their lowest level. In 1911, the Treaty for the Preservation and Protection of Fur Seals and Sea Otters was signed by Britain, Russia, Japan, and the United States, initiating a remarkable recovery for both species.

A new whaling era began in the 1920s, in search of blue, fin, and right whales. Meanwhile, sea otters had been "rediscovered" in 1928 in the Aleutians. By this

Box 6.1 The subsistence era: early prehistory to Euro-American contacts

Igor I. Krupnik

The presence of early humans and the key role the Bering Sea has played in the prehistory of both North America and Siberia had been recognized at least since the late 1800s. However, it was not until the 1930s to 1940s that the time framework and the sequences of local cultures became understood.

Geologists of the late 1800s originated the idea of a "land bridge" connecting northeast Asia and North America. In 1937, Eric Hultén proposed that an ice-free landmass was exposed by lower sea level during glacial epochs in the northern Bering and southern Chukchi seas. He named it "Beringia." Further studies confirmed that both prehistoric faunas and humans used it to enter North America from northern Eurasia. The last time this happened was between 22 000 and 17 000 years ago, when sea levels were about 120 m lower than at present and Beringia was exposed as a broad, grass-covered plain. This plain began to be flooded by post-glacial warming about 15 000 years ago, and Beringia was fully covered by water by 6000 BP, when the Bering Strait again became the key channel for water circulation and for marine mammal migrations between the North Pacific and the Arctic Ocean. The sea-ice and weather regimes that were established have changed little to the present day.

This recent flooding of Beringia destroyed most of the prehistoric records, but recent analysis indicates several human population waves. An early coastal wave presumably originated in northern Japan around 14 000 years ago, and probably brought canoe-using, seafood-exploiting people from the ice-free coasts of the North Pacific to present-day British Columbia and Washington state, and to the tip of South America. A second coastal wave occurred several thousand years later, probably under a harsher environment, and might have brought western Arctic people who already possessed skin boats, sea mammal hunting, and ice-fishing technologies to the northernmost North Pacific. These people most probably were the ancestors of the modern Inuit and Chukchi as well as of the more southern Koryak and Aleut.

Modern archeology offers ample evidence to back a scenario of an established human use of marine resources of the coastal-shelf zone of the North Pacific as early as 10 000 years ago. The earliest evidence within the Bering Sea region proper comes from the Anangula Blade site in the eastern Aleutian Islands, 8750–8250 BP. Its location on a bluff overlooking the water passage that is the key path for migrating whales and seals suggests an economy based on marine resources. Increasingly mild climate and greater availability of abundant marine resources led to more sophisticated full-scale hunting for marine mammals in the ice-free North Pacific and the Aleutians. Along Kodiak Island and the adjacent Pacific coast of the Alaska Peninsula, several sites of about 6000 BP yield remains of harbor seal, porpoise, sea otter, Steller sea lion, waterfowl, albatrosses, salmon, cod, and halibut. On the Asiatic side, evidences of year-round coastal maritime adaptation come from the ice-free zone of northern Hokkaido, where several sites dated from 7000 to 6000 BP yield bones of sea lion, fur seal, dolphin, seal, whale, and fishes. As in North America, hunting was done with barbed harpoons and spears, but there are few clues on the boats and other hunting and fishing equipment involved.

Maritime hunting economies of the northern Bering and southern Chukchi seas did not arrive until about 2500–3000 years later. In the Siberian sector, the earliest evidence comes from Wrangel Island, in the western Chukchi Sea, where excavations have produced midden pits full of fractured bones of walrus, seals, and birds, and artifacts made of walrus ivory, including toggle-head harpoons. This might be the earliest proof of human use of the Pacific walrus – either by offshore hunting with toggle-head harpoons or by killing animals on shore and ice with spears and lances, or even from beached and naturally perished animals.

In Alaska, the earliest known evidences of successful maritime hunting for seals and white whale (belukha, *Delphinapterus leucas*) date from the first prehistoric culture of about 3500–3300 BP, named "Choris," near the present-day town of Kotzebue. A slightly later cultural complex, of about 3300 years ago, from Cape Krusenstern near the village of Kivalina in northwestern Alaska, is known as the "Old Whaling." Certain tools of this culture, such as lances, big harpoon heads, and long-bladed butchering tools, as well as a litter of whale bones, provide the earliest evidences of arctic whaling. Pieces of prehistoric walrus ivory were abundant at the site, but no walrus bones have been found. This suggests that the Old Whaling people either traded for ivory or traveled to other parts of Alaska to hunt walrus. Later prehistoric coastal cultures of western and northwestern Alaska from 2500 to 1500 BP all possessed sophisticated seal-hunting equipment, knives for butchering, scrapers for working hides of marine mammals, and clay and stone oil lamps used for light and heat. Early Alaskan coastal hunters also made extensive use of walrus ivory for hunting tools, house implements, art carvings, and highly decorated burial objects, such as masks. Few walrus bones have been found at these early sites, suggesting limited hunting and the prevalent use of beached animals and/or traded ivory.

These early maritime cultures were succeeded around 2200–1500 BP by people who were presumably direct ancestors of the historic Inuit (Iñupiat and Yupik) and maritime Chukchi in Siberia. These people lived in year-round coastal villages and they possessed the technology and knowledge for effective sea mammal hunting. They used sophisticated toggle-head harpoons, seal-skin floats, killing lances, and skin boats, and their economy was based on successful hunting for walrus, seals, and probably whales. They had dogs, used marine mammal oil for heating and cooking, and were also skilled in storing large supplies of sea mammal meat and blubber in underground ice-cellars, for use through the winter. They built their villages on cliffs and spits and also settled on many offshore islands in the midst of the sea, such as St. Lawrence, and Big and Little Diomede. Their villages were established at the best walrus- and seal-hunting locations. Their extensive use of walrus ivory, baleen, and sea mammal bones for hunting and domestic implements proved beyond any doubt that these people were indeed efficient open-water and sea-ice hunters rather than opportunistic harvesters of beached marine mammals.

With this reliance on year-round sea mammal hunting supplemented by fishing, game and bird hunting, and plant collecting, a year-round maritime economy became firmly established. The only other transition of major importance was the development of communal whaling for bowhead whales from large skin boats, presumably around 1000 years ago. This triggered population growth and building of large villages at sites facing spring ice leads and polynyas along whale migration routes. This, in turn, encouraged stronger social bonds and the emergence of larger social units, such as "tribes" or regional "societies." The sedentary coastal villages housed several hundred year-round residents organized in large kin groups and dozens of skin boat crews engaged in cooperative hunting, transmission of knowledge, communal rituals, and warfare.

The second major technological breakthrough came with European contacts, first with Russians in the 1700s and later with Americans, who introduced iron weapons and, later, firearms, wooden boats with sails, whaling darting- and shoulder-guns, and outboard motors. This new technology made a dramatic impact upon both the efficiency and techniques of Native marine mammal hunting. Nevertheless, neither the foundation of Native reliance upon the key sea mammal species, such as walrus, whales, and seals, nor the core of Native tactics for hunting in ice leads, on ice floes, and in open water, has been altered profoundly. In many ways, the network of Native techniques, knowledge, and skills used in subsistence ("traditional") hunting in the ice-bound coastal and shelf zone of the northern Bering Sea and southern Chukchi Sea is a direct descendant from the indigenous coastal cultures of the region. On the other hand, subsistence hunting for Steller sea lion, walrus, and other marine mammals is much less intensive in the southern Bering Sea.

time, commercial fishing had depleted salmons and other species. Scientists, managers, and Native peoples, recalling their experiences in the 1880s, increased calls for regulation of commercial fisheries. Public demands for conservation followed, and in the late 1960s to early 1970s legislation put in force included the Marine Mammal Protection and Endangered Species acts (Box 2.4), which stopped all commercial exploitation of marine mammals. In 1976, Congress passed the Fishery Conservation and Management Act, declaring a 200-nautical-mile (nmi) territorial sea, which restricted fishing largely to Russia and the United States, except for the international waters of the "donut hole." Bering Sea fisheries exploitation increased greatly, especially by the United States, and by the 1980s Alaska's seafood industry produced 40% of world salmon supply, and walleye pollock (*Theragra chalcogramma*) had become the world's largest single-species fishery. The Pribilof Islands' fur seal population stabilized at about 0.9 million and sea otters were estimated to number approximately 150 000. However, populations of Steller sea lions, harbor seals, and some seabirds have declined. Compensatory increases and decreases also occurred among southern Bering Sea (basin) species, indicating an environmental (mainly climatological) "regime shift" as a possible cause.

This brief history sets the stage for examining biological, ecological, and social aspects of two Bering Sea marine mammals.

6.5 Steller sea lion, *Eumetopias jubatus*

The German physician–theologian, George Wilhelm Steller, first described the Steller sea lion based on a specimen obtained from the western Bering Sea's Commander Islands, while serving as naturalist on Vitus Bering's fateful voyage to Alaska in 1741–2. During this voyage, Steller also described the northern fur seal and the now extinct Steller sea cow.

6.5.1 Life history and vital parameters

The Steller sea lion is the largest member of the Family Otariidae, which includes sea lions and fur seals. Males are 2–3 times larger than females, and can attain 1120 kg in weight and 3.25 m in length. Females average 250 kg and are approximately 2.9 m long. Pups weigh 16–23 kg and are about 1 m long at birth. Both sexes have light buff to reddish-brown fur that is darker to black on the chest and abdomen. Pups are dark chocolate-brown at birth, and molt to lighter brown when about 6 months old. The range extends across the North Pacific Ocean rim, from the Kuril Islands and Okhotsk Sea in Russia to central California. Within this range, sea lions occur at numerous rookeries where they breed and haulouts where they rest (Fig. 6.9A).

In early May, dominant males establish territories on rookery sites traditionally used by females, which assemble near their own natal areas to give birth and to mate (Fig. 6.9B). Males defend the sites, not individual females. Therefore, it is incorrect to use the term "harem" for the assembled females. Territories vary in size depending on animal density. Smaller territories (about the radius of one male length) occur in areas of high abundance, and larger ones (more than two male lengths) at lower densities. Typically, males defend territories using stereotyped posturing and barking behavior and rarely resort to fighting. When fighting does occur, serious injury may result, including to females and pups in the territory. Some females also occur at haulout sites where mating may occasionally take place.

Most females bear a single pup yearly, between about the last week of May and early July. Mating occurs about 9 days after pupping. The embryo is not implanted in the uterus until about October to avoid the simultaneous energetic costs of nursing and pregnancy. Thus, gestation lasts about 9 months (October–June). Females may nurse their pups until they are 4 months to 2 years old, but pups are generally weaned just prior to the next breeding season. Females reach sexual maturity in 3–6 years and may continue to give birth until they are 20 years old or more. Males become sexually mature in 3–7 years, but are not generally physically large enough to establish and maintain

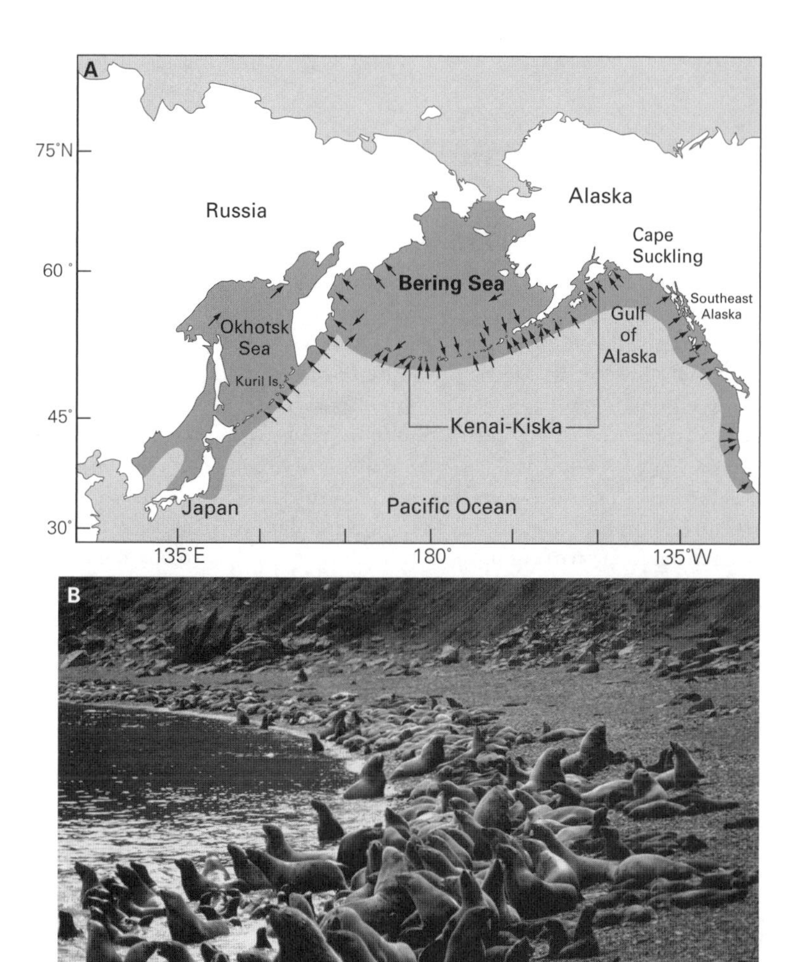

Fig. 6.9 (A) Breeding range of Steller sea lions. The shaded area represents the approximate range of sea lions at sea. Arrows indicate breeding rookeries. Haulout areas are too numerous to be shown. The line separating the eastern and western stocks is at 144°W longitude. (B) Steller sea lions at a rookery in the Bering Sea. Photograph by T.R. Loughlin.

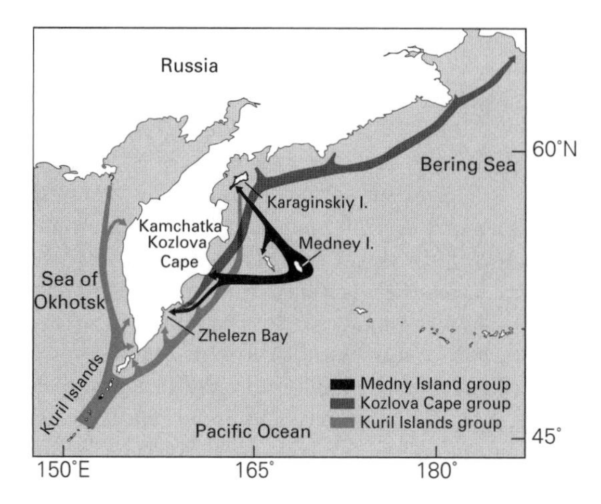

Fig. 6.10 Suggested dispersal of Steller sea lions from birth to 4 years of age from pups marked on Russian rookeries. From Loughlin (1997), with permission from the Society for Marine Mammalogy.

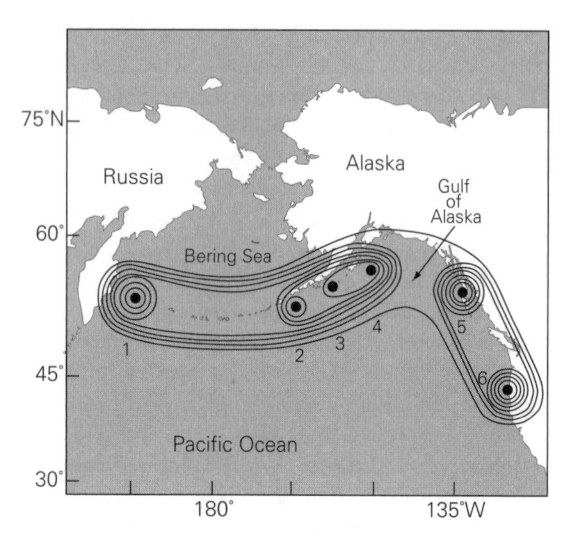

Fig. 6.11 Genetic identity among six grouped localities of Steller sea lions, based on similar mitochondrial DNA haplotypes. The groupings (solid circles) from west to east are: (1) Russia, (2) eastern Aleutian Islands, (3) western Gulf of Alaska, (4) central Gulf of Alaska, (5) southeastern Alaska, and (6) Oregon. Contour lines are drawn at 10% genetic identity and are joined at 30% identity. From Bickham et al. (1996), with permission.

territories until they are 9–11. By age 13–14, they are usually too old and battered to maintain a territory.

6.5.2 Biogeography

Factors that determine Steller sea-lion biogeography include climate and oceanography, avoidance of predators, distribution of prey, reproductive strategy, and movement patterns among sites. Avoidance of terrestrial predation might have been an important factor in determining their present distribution: most rookeries and haulouts are located at sites inaccessible to terrestrial predators. Distribution of prey is also a critical element, and probably determines their dispersal during the non-reproductive season.

Steller sea lions are not known to migrate. However, they disperse widely at times other than the breeding season. Estimated home ranges are about 320 km² for adult females in summer, about 47 600 km² for adult females in winter, and 9200 km² for winter yearlings, all with large variations. Sea-lion pups marked in the Kuril Islands and other locations in Russia disperse various distances throughout the western Pacific Ocean (Fig. 6.10). Some have been sighted near Yokohama, Japan, and in China's Yellow Sea. Pups marked near Kodiak, Alaska, have been sighted near Vancouver, British Columbia. Animals up to about 4 years old tend to disperse farther than adults. As they approach breeding age, they have a propensity to stay in the general vicinity of the breeding islands, and generally return to their island of birth to breed.

Recent analyses of mitochondrial DNA suggest that at least two populations exist (Fig. 6.11). The eastern population includes sea lions east of a line near Prince William Sound, Alaska, including southeast Alaska, Canada, and Washington through California. The western population includes sea lions from Prince William Sound westward through the Aleutian Islands. Additional studies of nuclear DNA, cell proteins, and morphology confirm these results and may allow association of individual sea lions with their birth rookery.

6.5.3 Population abundance, fluctuations, and trends

The western U.S. Steller sea lion population has declined since the 1960s, but reliable survey data are available only from the 1970s. Estimates for this population fell 80% from about 180 000 animals in the late 1960s to about 36 000 animals in 2000. The rate of decline was less from 1990 to 2000 than from the late 1970s to the early 1990s, but still remained high. The rate of decline seems to have increased recently (Fig. 6.12). The Russian portion of the population is

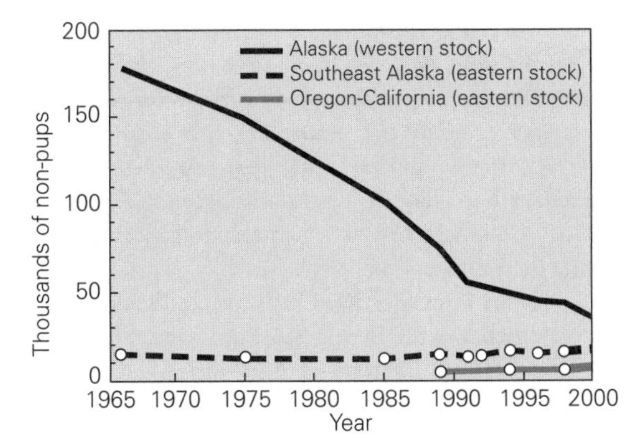

Fig. 6.12 Counts of adult and juvenile Steller sea lions at breeding rookeries and haulout sites in the United States during June and July, 1965–2000, depicting the dramatic decline of the western stock and the low numbers, but stability, of the eastern stocks.

estimated to be about one-third of historic levels. Portions of the eastern population, on the other hand, are either increasing slowly or appear to be stable.

Causes for decline are difficult to determine. Computer modeling and mark–recapture experiments suggest that the likely factor is decreased juvenile survival, but lower reproductive success may also contribute. In some areas where the diet includes numerous prey species, sea-lion numbers have been stable or increasing slightly. In areas where sea lions primarily depend on any one prey item, the population is declining. However, whether population trends are closely associated with diet diversity is equivocal. Other possible causes for decline include disease, pollution, subsistence take, effects of fisheries, and environmental change.

Several diseases have been documented for Steller sea lions, but evidence is not sufficient to demonstrate that disease has played a significant part in decline of the western population. Examination of blood from live Alaskan Steller sea lions revealed antibodies for two bacteria (*Leptospira* and *Chlamydia*), one marine calicivirus (San Miguel sea-lion virus), and a seal herpesvirus, all of which could result in reproductive failure or death. The incidence was low, suggesting that these pathogens played a minor role, if any, in the sea-lion decline. However, additional work remains to be completed before disease can be dismissed as a factor. Pollution has not been considered important in the decline. The magnitude and location of subsistence take by Alaska Natives is also not considered a cause for the observed decline: at most, the present level of take is less than 1% of the western stock and likely consists of older animals not considered vulnerable to population perturbations.

6.5.4 Ecological interactions

The effects of fisheries and probable synergistic effects of environmental change and fisheries are areas of intense research to determine causes for Steller sea-lion population decline.

6.5.4.1 Food, feeding, and energetics

Steller sea lions are opportunistic in diet and eat a variety of fishes and invertebrates. In Alaska, the principal prey are walleye pollock, Pacific cod (*Gadus macrocephalus*), Atka mackerel (*Pleurogrammus monopterygius*), Pacific herring (*Clupea pallasi*), salmon, octopus, squid, flatfishes, sculpins, and a wide variety of other fishes and invertebrates. Prey varies with geography: pollock dominates the diet eastward, but decreases in importance westward where the diet is mostly Atka mackerel and is more varied (Fig. 6.13). During the breeding season, females with pups generally feed at night; territorial males do not eat while on territories. Feeding occurs during all hours of the day once the breeding season ends.

Knowledge of prey choice and foraging ecology is essential for establishing guidelines that segregate feeding from fisheries. Choice of prey is probably influenced by prey biomass, availability, water depth, degree of association with the bottom, reproductive behavior, solitary *vs.* schooling prey behavior, and temporal and spatial distribution patterns. Predation on prey associated with the bottom is a common pinniped strategy, perhaps because the bottom limits the spatial dimension of the predator–prey arena and limits the prey's alternatives for escape. The sea lion's energy costs associated with search and capture are also optimized by schooling behavior of many prey species. Studies of foraging patterns using radiotelemetry have shown that the location, depth, and duration vary by individual, size and age, season, site, and reproductive status. Generally, Steller sea lions are relatively shallow divers, with few foraging dives recorded at depths greater than 250 m. Foraging-trip duration for females in summer is only about 24 hours and averages 17 km in distance, in comparison to winter's approximately 200 hours and 133 km, respectively. This is probably due to the need to conserve energy during lactation. The winter pattern varies considerably, implying that some

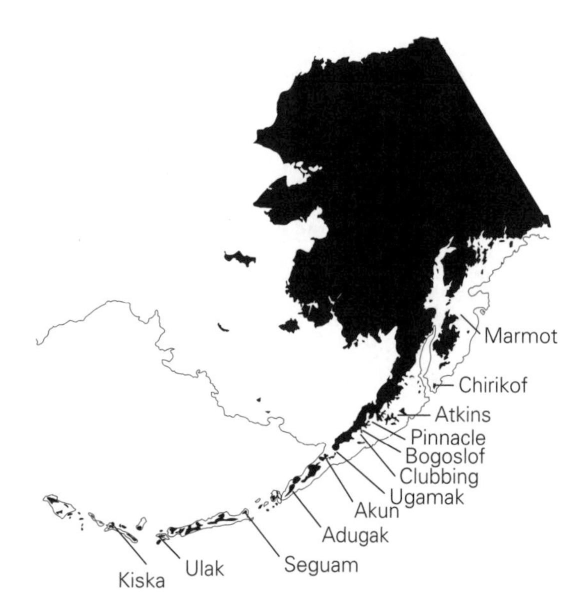

Percent frequency of prey occurrence
P=Pollock; Sa=Salmon; At=Arrowtooth flounder; PC=Pacific cod; Sl=Sandlance; AM=Atka mackerel; H=Herring; C=Cephalopods

Area	P	Sa	At	PC	Sl	AM	H	C
Marmot (64)	69	39	36	5				
Chirikof (74)	69	43	19	9				
Atkins (101)	86	46		6[1]				
Pinnacle (79)	67	71		9	33			
Bogoslof (74)	78	19[2]		1		30		
Clubbing (70)	87	33	26	16				
Ugamak (155)	51	48		3			33	
Akun (58)	36		33	19			55	
Adugak (73)	9[3]	24	73	5				
Seguam (117)		10		3		90		11
Ulak (105)	10		100	1				41
Kiska[4]				16-21		92-95		10-21

1 Sandlance also included
2 Cephalopods and deepsea smelts included
3 Cephalopods and herring included
4 Two sites

Fig. 6.13 Percent frequency of occurrence of prey items collected from Steller sea-lion scats, June–August, 1990–8. Numbers in parenthesis are sample sizes.

females may still be supporting pups. Yearling sea lions in winter exhibit foraging patterns intermediate between summer and winter females in trip distance (30 km average), but shorter in duration (15 hours average).

6.5.4.2 Interactions with other species

Steller sea lions occur at haulouts with California sea lions (*Zalophus californianus*) and northern fur seals, but these species rarely occur together at breeding rookeries. Although physically larger than the other species, Steller sea lions tend to be submissive, moving out of the other animal's way when interactions occur, indicating their non-competitive nature.

Steller sea lions, northern fur seals, and harbor seals occur together in the Bering Sea, but forage at different locations and on different-sized or different species of prey. Fur seals are the most pelagic of the three, commonly foraging more than 320 km from their Pribilof Islands' rookeries. When feeding over the continental shelf, they may dive to the bottom, but when over deep water they feed primarily during the night on squid and small fish, diving up to 50–100 m and following the deep scattering layer (a zone defined by acoustic sensing of organisms). Harbor seals typically feed near shore on bottom fish, in intertidal zones and in and around kelp beds in water up to 100 m deep. Steller sea lions are intermediate between these species. Whether they are potential competitors for food or feeding space is not known. Thus, the three species may partition prey so that little direct competition occurs, or occurs only seasonally.

The possible impact of predation on Steller sea lions is also unknown. Killer whales (*Orcinus orca*) and sharks may occasionally prey on them. The likelihood of shark attack is probably greater for Steller sea lions off the Washington, Oregon, and California coasts than those in waters farther north because of abundance and diversity of sharks southward.

6.5.4.3 Implications of environmental change

Dramatic changes in species numbers and in the structure and composition of biotic communities are to be expected with shifts between cold and warm climate regimes. Sea lions almost certainly have lived through many regime shifts in their 2–3 million years of existence. What may be different about recent declines is the coincident development of extensive fisheries targeting the same prey on which sea lions depend, and climate changes, such as the Pacific decadal oscillation (PDO), which are known to influence species' distributions and, perhaps, abundance. The existence of strong environmental influences could also increase the sensitivity of sea lions to fisheries effects, or to changes in those ecosystems resulting from fisheries, but virtually nothing is known about such interactions.

6.5.5 The conflict arena

The major conflict concerns the effects of fisheries and fishery-related activities on sea lions. Native take does exist to some extent, but not to the extent it once did.

6.5.5.1 *Commercial fisheries*

After World War II, high-seas fisheries developed rapidly throughout the Bering Sea and North Pacific Ocean. Prevalent Bering Sea fisheries included salmon, herring, demersal fish, king crab, and shrimp. Numerous nations were involved. Extensive removals by fisheries sharply reduced the abundance of Pacific Ocean perch (*Sebastes alutus*), other rockfishes (*Sebastes* spp.), and Pacific herring. Following the decline of these species in the late 1960s and early 1970s, other fishes increased in abundance, while crab and shrimp populations also declined.

From the mid-1970s to 1990, the pollock fishery, along with all other groundfish fisheries in Alaska, was transformed by the passage in 1976 of the Fishery Conservation and Management Act (FCMA, Box 2.4, page 48), one of the principal objectives of which was to promote full domestic use of U.S. offshore fisheries. Prior to its passage, most pollock were caught and processed solely by distant-water fleets of foreign nations. Most fishery effort for pollock was located in the eastern Bering Sea and Aleutian Islands regions. Catches of pollock increased quickly from about 100 000 tonnes in 1964 to more than 1.8×10^6 tonnes in 1972. Beginning in 1973, declines in Bering Sea/Aleutian Islands pollock abundance led to reductions in catch quotas, to about 1.0×10^6 tonnes in 1977.

After passage of the MSFCMA, increasing amounts of pollock were allocated to joint-venture fisheries operations (domestic catcher vessels and foreign processing) to encourage development of domestic fishing. Due to declines in the size of the spawning population in the Gulf of Alaska, the joint-venture fishery moved to large spawning assemblages of pollock near Bogoslof Island in the eastern Aleutian Islands basin. Thus, the temporal and spatial distribution of the pollock catch changed as fishery participation changed. Between 1963 and 1997 in the Bering Sea/Aleutian Islands region, the pollock fishery worked increasingly in fall and winter and fished more in areas that in 1993 were designated under the Endangered Species Act as critical habitat for Steller sea lions.

6.5.5.2 *Effects of fisheries*

Steller sea lions may be affected by commercial fishing directly through incidental catch in nets, or by entanglement in derelict debris, or by shooting, or indirectly through competition for prey, disturbance, or disruption of prey schools. The number of sea lions caught in trawl nets was high during the 1960s and 1970s, but is presently at very low levels and is not now considered to be an important component. In some areas, Steller sea lions have been killed deliberately by fishermen, but it is unclear how this may have affected the population.

Commercial fisheries target several of the most important prey eaten by Steller sea lions and remove about 2×10^6 mts of fish from the Bering Sea annually. However, the complexity of ecosystem interactions and limitations of data and fisheries models make it difficult to determine whether fishery removals, directly or indirectly, have influenced sea lions (Fig. 6.14). Pollock are their primary prey, at least in most areas, and have continued to be abundant. Therefore, the sea lion's decline may seem not to be because of prey

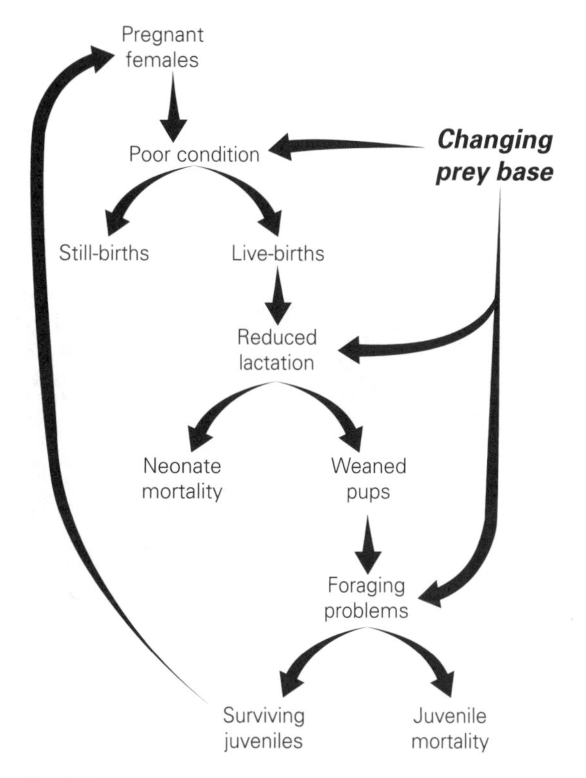

Fig. 6.14 Flow diagram suggesting the possible effects of a declining prey base on Steller sea lions in Alaska. From Loughlin & Merrick (1989).

removal by fisheries. The issue, however, may not be so much the overall abundance of pollock or other food items such as salmons or Atka mackerel, but the availability of prey in time and space when sea lions are feeding. Biologists argue that fisheries cause "localized depletions" when efficient trawlers take fish from small areas where dense fish schools occur, for example near ports, where the fish are most accessible. These are the same, or proximate to, areas where sea lions feed and are where competition between the fisheries and sea lions can be most intense.

The general conclusion is that a combination of factors may be causing the Steller sea lion's decline. First, since the regime shift in the mid-1970s, pollock have become the dominant schooling fish in the Bering Sea. Second, environmental changes may have altered the distributions of prey, such as pollock. Third, alternate prey, such as herring, have been depleted. Fourth, winter fishing activities apparently have reduced availability of pollock at critical times and places for sea lions. Therefore, increased sea-lion mortality in fall and winter might be expected as a result of their need to recover from the rigors of summer reproduction. However, alternate food sources may also be available, making conclusions about a relationship between sea lions and pollock alone untenable.

6.5.5.3 Management responses

The federal government has implemented numerous measures for conservation of Steller sea lions, including prohibitions of shooting, reductions in incidental take by fisheries, trawling restrictions around rookeries, and designation of critical habitat. For example, trawling is not allowed within 20 nmi of all rookeries and major haulouts, nor within foraging areas (Fig. 6.15). A Steller Sea Lion Recovery Plan summarizes these measures, including research on physiology and feeding ecology, monitoring requirements, reassessment of the effectiveness of existing protective measures, and possibilities for full recovery under requirements of the Marine Mammal Protection Act and Endangered Species Act.

Even with these measures now in place, the conservation community has objected to the way that the federal government manages Bering Sea fisheries, especially pollock, and has raised concerns about how conduct of these fisheries may limit sea-lion recovery. In order to reduce the possible impact of fisheries on sea lions, a consortium of conservation groups brought suit against the National Marine Fisheries Service (NMFS) in federal court for violating sections of the Endangered Species Act (ESA) and the National Environmental Policy Act (NEPA). The sequence of events under these acts has been:

- ESA, Section 7, requires all federal "action agencies" to consult with the "expert agency" if actions for which they are responsible may affect endangered or threatened marine mammals or their critical habitat. In the Steller sea lion case, the NMFS Fisheries Management Division (FMD, the "action agency") approves and manages fisheries management plans

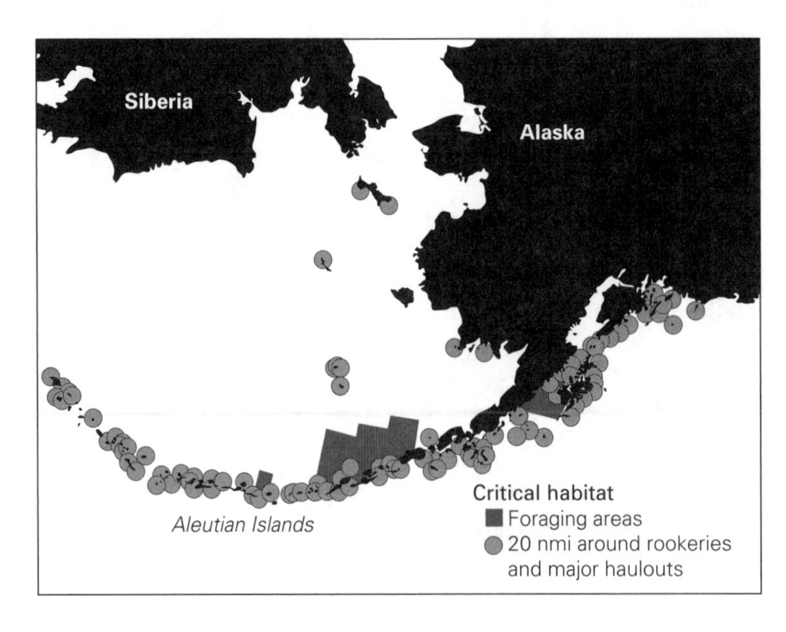

Fig. 6.15 Critical habitat for Steller sea lions in the Bering Sea, Aleutian Islands, and Gulf of Alaska, designated by the federal government in August 1993 as required by the Endangered Species Act. Critical habitat includes a 20 rumi radius around all rookeries and major haulouts, and three aquatic foraging areas.

for Alaskan fisheries. It determined that fisheries may affect Steller sea lions.

- NMFS/FMD did an ESA Section 7 (critical habitat) consultation with the NMFS Protected Species Division (PSD, the "expert agency"), the result of which was a "biological opinion" concluding that the action (fisheries) could jeopardize the continued existence of Steller sea lions or modify their designated critical habitat.
- The action agency is required by the ESA to develop Reasonable and Prudent Alternatives (RPAs) to the proposed action that would not jeopardize the species or affect its critical habitat. In this case, the FMD consulted with the PSD, which found that the pollock fishery would jeopardize the continued existence of Steller sea lions, and proposed RPAs to the action to eliminate jeopardy. These included time and area fishery closures and no-trawl zones around important sea-lion rookery and haulout sites.

In a decision of July 1999, a U.S. District Court judge upheld the jeopardy finding, but found the RPAs lacking. As a result, NMFS was required to revise the RPAs to show that they would, in fact, eliminate the effect of fisheries on Steller sea lions. This case was built on the premise that NMFS allowed the fishery to occur without a thorough Environmental Impact Statement (EIS), as required by NEPA, discussing the possible impact of the fishery on the Bering Sea and Gulf of Alaska ecosystems. The judge agreed with the environmental consortium and required NMFS to develop a new, revised EIS covering the full scope of Alaskan pollock trawl fisheries.

What else can be done? NMFS does not want to impose regulations that might needlessly stifle the fishing industry, yet the government is required to protect and conserve the sea lion. There seems to be little doubt that sea lions and commercial fishing efforts concentrate on the same prey, yet data are not available to conclude that fishing is responsible for the decline. Nevertheless, fishing bears the brunt of responsibility since management of fishing is the parsimonious way to facilitate recovery, as well as being the factor most under human control.

6.5.5.4 Research responses

Questions arise, in such cases as this, about whether observed declines or oscillations in abundance can be attributed to an anthropogenic "cause." In either case, directed research is required. Thus, major questions about the decline of Steller sea-lion populations and the possible implications of fisheries and environmental change are being addressed. For example:

- Are the observed declines due to increased mortality of juveniles?
- Is mortality highest during fall and winter?
- What is the cause of increased mortality – disease, pollution, nutrition, predation, shooting, or some other factor?
- Are Steller sea lions nutritionally stressed, and if true, does fishing contribute?
- What role does environmental variability play in population dynamics and distribution?
- Do prey composition and biodiversity affect population dynamics?
- Is there evidence from the Native community or from archaeological sites suggesting significant past changes in sea-lion abundance?

In order to respond most efficiently, NMFS is working cooperatively with the Alaska Department of Fish and Game, other agencies, and private groups to monitor the Steller sea-lion population in Alaska. Activities include aerial surveys over sea-lion habitat during mid-June to count adults and juveniles; pups are also counted from land during late June and early July. Government scientists also work with physiologists and geneticists at major universities to determine the health, physiological status, population structure, feeding behavior, and vital parameters that provide information on possible causes of mortality.

There are considerable constraints on this research, concerning limited funds, logistics, and priorities. How does one decide which research or monitoring effort is most important and how much emphasis should be placed on it? For example, major research efforts involve Steller sea-lion feeding ecology and movements at sea. Ultimately, the results will be correlated with commercial fishing information to establish the nature and magnitude of overlap between the two. However, should these studies be continued if it cannot be shown that Steller sea lions are nutritionally stressed, as seems possible from available data? And, what other studies can be devised to solve the depletion problem?

6.6 Pacific walrus, *Odobenus rosmarus divergens*

For thousands of years, aboriginal peoples of the north have regarded walruses as having supernatural powers and human attributes, while also utilizing them as a major natural resource for their survival. Many western naturalists, until recently, fancifully perceived walruses as clumsy brutes that fattened on mollusks and crustaceans, which they dug from the muddy bottom with their tusks. Many cultures now perceive the walrus as a symbol of arctic conservation.

6.6.1 Biology and ecology

The walrus is a highly gregarious animal that plays an important ecological role in the Bering Sea's environment, where its feeding activities structure benthic communities and affect recovery times (Fig. 4.23, page 121). Walruses use their tusks for social purposes, defense, and also to aid in hauling out or as an anchor in sea ice (Fig. 6.20), hence the translation of the scientific name: "tooth-walking sea horse."

6.6.1.1 Vital parameters

There are six populations of walruses distributed in a circumpolar pattern (Fig. 6.16). Pacific walruses attain the largest size and bear the largest tusks among these populations. They exhibit a high degree of sexual dimorphism. Males can reach 350 cm in length and 1700 kg in weight; females can reach 300 cm and 1250 kg. Large males have thick, lumpy skin about the neck and shoulders, as protective armor against the tusk jabs of other males. Calves average 115 cm in length and 65 kg at birth. Maximum life span is estimated to be about 40 years. Natural mortality of animals more than a year old is very low, probably only around 1% per year.

Walruses have the lowest reproductive rate among pinnipeds, are slow to mature, have diets largely restricted to benthic invertebrates, and are assumed to be relatively K-selected (Fig. 4.16, page 111). Females mature when they are 4–8 years old and attain full growth by the age of 9–10. Males mature at 9–10 years, but do not usually breed until they are about 15 years old. Mating occurs in winter, from January through March, in sea-ice environments. The sex ratio of the population is about 1 male to 2–3 females, but the ratio of those that breed is about 1 male to 10 females. The maximum rate of reproduction is one calf every 2 years per adult female. The gestation period is about 15 months, but implantation of the embryo in the uterus is delayed 4–5 months until at least June or July. A single calf is born during May to early June; twins are rare. Calves may remain with their mothers for up to 2 years.

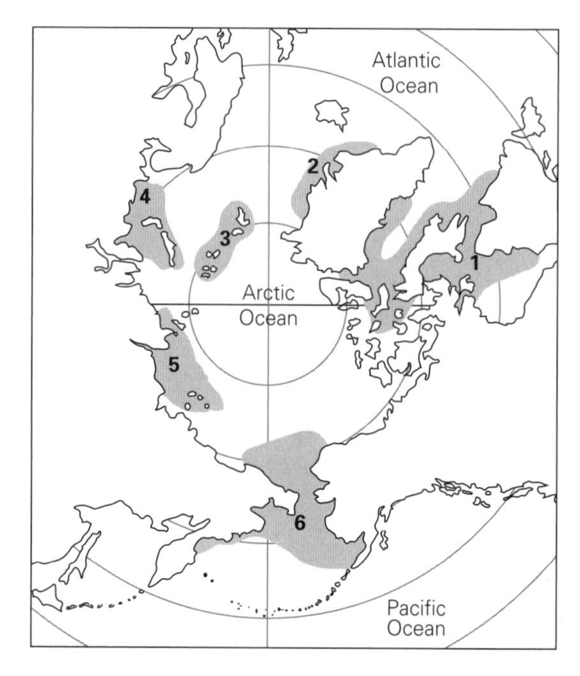

Fig. 6.16 Current world distribution of walruses: (1) Hudson Bay–Davis Strait, (2) eastern Greenland, (3) Svalbard and Franz Josef Land, (4) Kara Sea and Novaya Zemlya, (5) Laptev Sea, and (6) Bering and Chukchi seas. From Fay (1982).

6.6.1.2 Biogeography and habitat

Three centuries ago, Pacific walruses occurred as far south as Unimak Pass, east to southeast Alaska, and west to the tip of the Kamchatka Peninsula. Presently, they are confined to the seasonal sea ice and coastal haulouts of the Bering and Chukchi seas. Male and female distributions differ. In January through April, two subpopulations occur on sea ice of the north-central and southeastern Bering Sea (Fig. 6.17A). Mitochondrial DNA analysis indicates that these groups are part of the same population, but whether a metapopulation structure exists is not known. From April through June, the sea ice disintegrates and moves north, during which time most mature males move to coastal haulouts (Fig. 6.17B), but almost all females with young of the year, juveniles, and a few mature males are scattered widely in the marginal ice of the eastern and western Chukchi Sea (Fig. 6.17C). In October through December, the entire population migrates to the vicinity of the Bering Strait, St. Lawrence, and the Punuk Islands.

The relationship between sea-ice dynamics and walrus distribution is critical but not well known. Winter

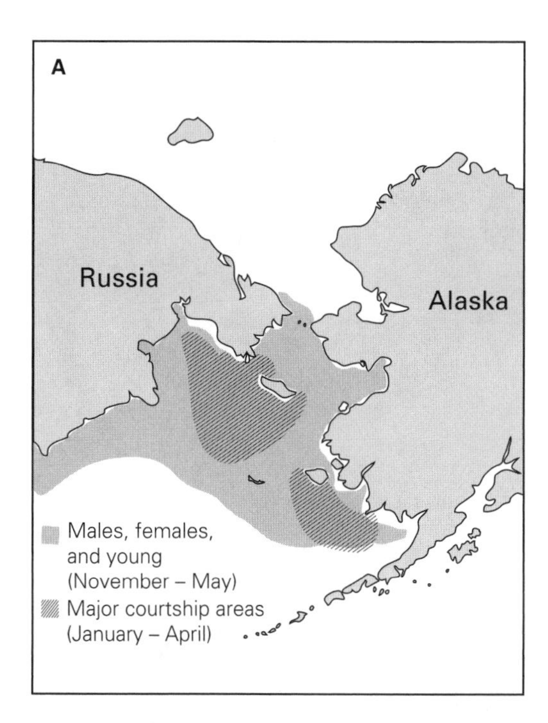

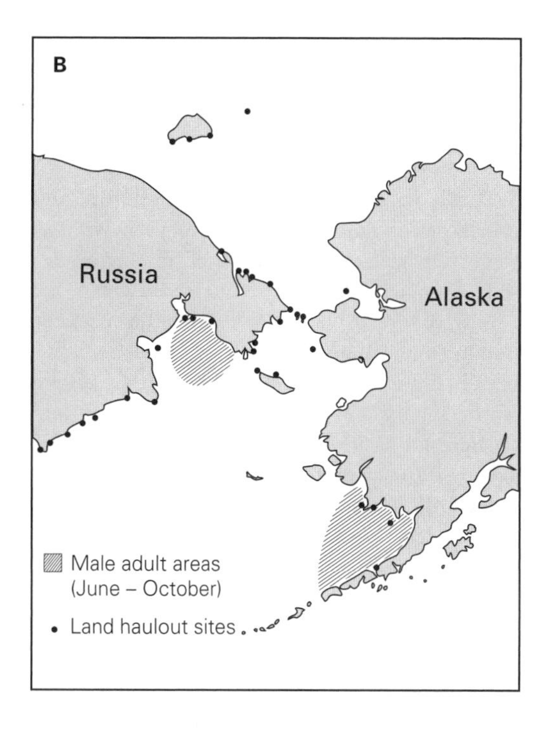

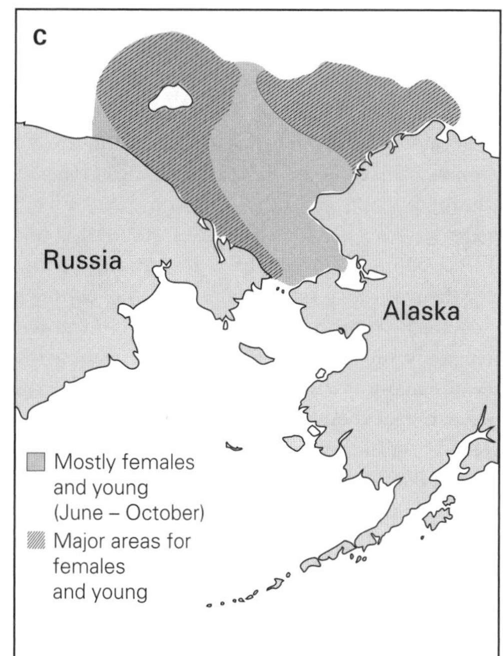

Fig. 6.17 Seasonal distribution of walruses. (A) Distribution in winter, indicating two breeding concentrations. (B) Summer distribution of males. (C) Summer distribution of females, most juveniles, and young of the year. Modified from Fay (1982); NOAA (1988).

distributions, when combined with average distributions of sea-ice types, indicate that the largest, westernmost, concentration favors broken pack, which provides access to water under almost all conditions of wind and current (Fig. 6.18A). Conversely, this seems not to be the case for Bristol Bay walruses whose distribution is not associated with this ice type (Fig. 6.18B). However, in summer, a different situation is evident; sea ice occurs as smaller floes in the marginal ice of the Chukchi Sea, where limited information indicates that hauled-out walruses seem to prefer floes 100–200 m² in area.

6.6.1.3 Behavior

Walruses are gregarious at all times of year. Animals usually rest in contact with one another, with young frequently lying on or in shelter of larger animals, possibly to conserve body heat. On ice in winter, groups usually consist of up to 200 animals, but occasionally number in the thousands. Males in summer congregate in even larger concentrations on terrestrial haulouts. In water, animals also tend to cluster, but more loosely. Reasons for group behavior are not known, but large group size may protect walruses from polar bear or killer whale predation. During feeding, there appears to be considerable synchrony of hauling out and diving. Walrus groups may remain in the water to feed for 2–3 days or more, then haul out to rest for a similar period on previously occupied or neighboring floes. This suggests that groups of walruses follow continually moving ice, possibly to ensure appropriate hauling-out sites. Haulout periodicity may be weather-dependent. Thus, weather influences both feeding periodicity and movements of sea ice, and is a major factor in determining where and when walruses feed.

When hauled out or in water, walruses communicate by means of barks, grunts, and snorts, which presumably serve as threats, greetings, female–calf communication, and other functions. Individual animals can probably identify one another by sound, as well as by smell. Social rank is a function of both tusk and body size, as is evident from the considerable "fencing," posturing, and jabs among both males and females. Wounds may result from these encounters, but serious injury is rare. Walrus mothers are extremely protective of their young. Their tusks are formidable weapons against predators, such as polar bears.

The mating system may be unique among aquatic mammals. During the courtship–mating season, mature males resting on ice are not aggressive toward one another, but underwater they engage in ritualized displays involving tapping and bell-like songs. These songs may or may not serve to attract females, but do appear to establish male acoustic territories (Fig. 6.19). The aggregation of males in an arena is suggestive of a lek, as has been observed for some grouse where females choose among males within a mating arena. However, whether male or female walruses choose sites and/or initiate mate choice is not known.

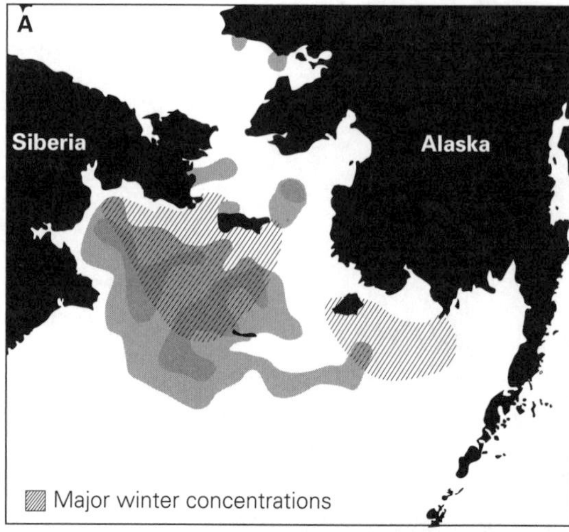

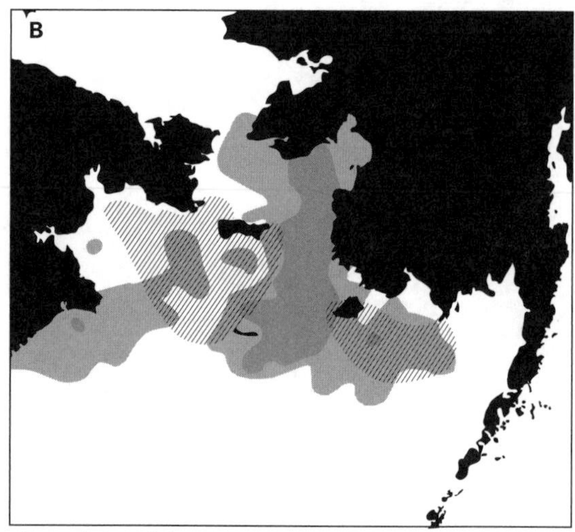

Fig. 6.18 Walrus winter distributions and sea-ice types. The north-central Bering Sea subpopulation appears to co-occur with (A) broken pack, but not with (B) rounded pack (see Fig. 6.5A). The Bristol Bay subpopulation appears not to be associated with either sea-ice type. Distributions of ice types result from statistical analyses of NOAA AVHRR imagery for 10 consecutive years; the darker shading indicates greatest probability of occurrence. Adapted from NOAA (1988); Ray & Hufford (1989).

6.6.1.4 Health and condition

Walruses may have tumors, parasites, hernias, and frostbite, and occasionally become trapped in sea ice. One potentially serious parasite is the nematode worm *Trichinella*, which causes trichinosis. A calicivirus may

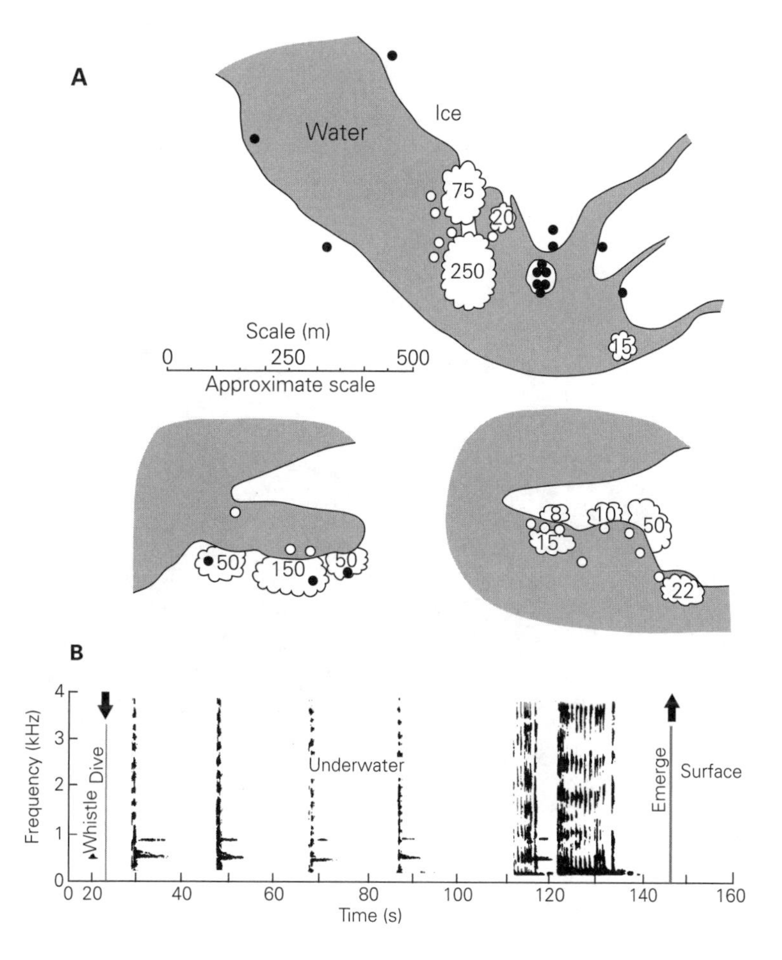

Fig. 6.19 (A) Spatial arrangements of Pacific walruses observed in the north-central Bering Sea in April 1972. Dark areas are water; white areas are sea ice. Wavy lines enclose female groups, with numbers of individuals indicated. Open circles represent bulls active in water; closed circles are inactive bulls on ice. (B) Sonogram of a male "song": a light whistle, emitted at the water's surface, is followed by a series of bell tones, emitted underwater, then a series of knocks, ending with a coda of rapid knocks, followed by surfacing. This sequence may be repeated for hours before a group of females on ice. From Fay et al. (1984).

cause skin lesions and abortion. Pollution is apparently not yet a health problem. However, PCBs (polychlorinated biphenyls), hydrocarbons, pesticides, and heavy metals are present in sufficient quantities to warrant further study. Cadmium does occur in relatively high concentrations in kidneys and liver of Beringian walruses, but its effects on the walruses or on Eskimo health are unknown.

Mass natural mortality of walruses occasionally occurs when walruses rush on- or offshore following disturbance by aircraft or to escape killer whales, polar bears, or hunters. Hauled-out walruses may pile together so vigorously in attempts to escape to water that some may be crushed or severely injured. In the 1950s, approximately 200 partly decomposed walruses washed ashore on the northeast coast of St. Lawrence Island; these animals may have died as a result of an explosion. In 1969–70, about 2000 animals were trapped in heavy sea ice along the Chukchi Peninsula. In 1978, several hundred animals washed ashore on St. Lawrence and the Punuk Islands. Most were emaciated and bore evidence of physical injury from other walruses, the cause of which could not be determined.

6.6.1.5 Food, feeding, and energetics

Walruses eat a wide variety of benthic invertebrates, but favor clams. Feeding-dives usually last less than 10 minutes and are mostly in waters of less than 100 m depth (Fig. 6.20). Although newborn calves can swim, they do not dive far below the surface. On the ice, they require their mother's shelter and warmth for protection against the cold, and depend heavily on nursing for at least 6 months for sufficient energy and to acquire a substantial blubber layer. Walrus milk is about a third fat, which is high for mammals, but relatively low for pinnipeds. As calves gradually wean, they adjust diving and feeding activities until their diving abilities allow them to feed for themselves,

Fig. 6.20 Natural history of the walrus. A male sings before a group of females hauled out on sea ice. One individual feeds on the bottom and another secures itself to the ice with its tusks. Benthic feeding by walruses has widespread effects in benthic communities. Painting by Robert Hynes, with permission of National Geographic Society Image Collection. Ray & Curtsinger, 1979.

possibly during their first 6 months. Calves are usually fully weaned in 2 years.

Food-finding and feeding by walruses are unique. Walruses do not dig with their tusks, nor can they chew with their peg-like cheek teeth. Walruses detect food by moving forward along the bottom while sensing food with their vibrissae and lips. The longer, lateral vibrissae function as detectors and the shorter, central ones are used for finer resolution. Organisms in the sediment are rooted out, much like pigs rooting in soil. The lips manipulate the prey and powerful sucking actions of the tongue extract the meat, which is swallowed whole. An individual walrus requires about 5–7% of its total body weight in food a day, depending on size, age, and reproductive status. Presumably, reproductive males displaying in spring before females feed little or not at all. The number of individual food items consumed per day can range into the thousands when items are the size of worms or small clams.

6.6.1.6 Interactions with other species

Walruses seem not to be prey of any animal other than killer whales, polar bears, and humans. Polar bears are considerably smaller than adult walruses and do not swim as well. However, when walruses on ice detect a polar bear, they flee into the water. Frequently, calves are last to leave the ice, and consequently are exposed to predation.

Walruses apparently compete directly for food with only one other marine mammal, the bearded seal (*Erignathus barbatus*), which inhabits the same areas of sea ice and feeds on many of the same benthic species. Walruses have been known to feed on the dead bodies of whales and occasionally kill and eat seals, especially bearded and ringed seals, as indicated by the occurrence of strips and chunks of seal meat in walrus stomachs. How walruses manage to capture the speedier and more agile seals, or to dissect them, is unknown.

6.6.1.7 Ecosystem interactions

Walruses qualify as a keystone species. They are major consumers of Beringian benthic resources: 200 000 one-tonne walruses require an average of up to 10 000–14 000 mt of food per day (although usually not feeding daily; see Behavior). The yearly total approximates 2.2×10^6 mt, assuming that feeding occurs only every other day on average, equivalent to the weight taken by the entire Bering Sea fishery!

Feeding is influenced by seasonal migration patterns and inter- and intra-annual sea-ice dynamics. Because walrus occurrence is patchy and groups are associated with specific types of sea ice, the effects of walrus-feeding are unevenly distributed. If sea-ice movements were limited by weather or current conditions, walruses would be forced to obtain their food from less than their usual potential habitat, raising the possibility that food supply would become locally depleted, but not regionally depleted. An alternative scenario would occur if sea-ice movements varied widely within or among years, allowing walrus to feed over a large area. If the first scenario were to persist for several years, density-dependent reproductive rate and body condition might be affected; in the second scenario, density-dependence might not be a major factor, allowing the population to increase.

Walrus-feeding has a significant effect on benthic sediment sorting and community structure. Extensive benthic furrows about the width of a walrus's snout, associated with broken clam shell, have been observed directly from submersibles and by scuba-divers (see Fig. 6.20). Side-scan sonar has shown large-scale environmental effects of feeding by walruses and gray whales. The whales are estimated to resuspend approximately 120×10^6 m^3 yr^{-1} of sediment over tens of thousands of square kilometers in the north Bering and south-central Chukchi seas per year – twice the yearly sediment load of the Yukon River! Walrus-feeding may resuspend half that amount. The extent of bioturbation by these two species suggests profound consequences for the structure of the sediment and of benthic communities.

6.6.2 The conservation arena

The walrus population has fluctuated due to past exploitation, but population dynamics and habitat relationships are not well known, resulting in significant uncertainties about abundance and conservation.

6.6.2.1 History of exploitation and management
Walruses have been a staple of Native subsistence for many centuries. Given the state of Native technology prior to the 19th century, a reasonable assumption has been that subsistence hunting had little effect on the walrus population. This assumption leads to the conclusion that a sustainable population baseline ("carrying capacity") might be set at the time when the first commercial hunting began. However, recent archaeological and anthropological studies conclude that subsistence hunting could have depleted walruses somewhat (Box 6.2). Therefore, it seems likely that the walrus population may have been at the lower portion of the range of carrying capacity prior to the period of Russian expansion (Fig. 6.21).

Walruses have been commercially exploited since the mid-1700s. By the late 1700s commercial catch may have removed approximately 1000 animals annually. The total combined kill, including that by Natives, might have been about 5000–7000 animals – a not inconsequential number, comparable to today's subsistence catch (Table 6.1). Commercial catch apparently rose until the U.S. purchase of Alaska from Russia in 1867. Yankee whalers had exploited bowhead whales throughout Beringia since the 1830s, and after the whales were depleted, in mid-century, the whalers intensified walrus hunting. The abundant Pribilof herds were almost exterminated and walruses remained numerous only in Bristol Bay and the northern ice fields. Furthermore, the hunters preferred females, which were more accessible than males and yielded more oil; thus, the hunt not only depleted the productive part of the population, but also caused the death of calves by starvation. The result was that the walrus population was reduced by more than half. By 1880, depletion of whales, scarcity of walruses, and declining prices for oil diminished demand and the whalers continued only a low level of hunting.

Attempts have been made to identify periods of depletion and recovery from the 1880s to World War II. There can be little doubt of significant depletion of Pacific walruses following the American purchase of Alaska, but the walrus population's status until the post-World War II period is a matter of conjecture. For example, during the early 1900s walruses began to occupy former portions of their range, which has been interpreted as evidence for population increase. However, range extension can also be explained by habitat change (e.g., sea-ice changes across years) and redistribution of hunting disturbance. Nor do hunting intensity and catch levels necessarily indicate population status, as hunting effort is also a consequence of changes in market value and other social factors. Most significantly, walrus hunting never completely ceased. Natives continued to depend on walruses for subsistence, and arctic traders, mostly fur hunters, also took walruses from the late 1800s onwards. Therefore, the population probably remained low in numbers; even a low level of hunting

Box 6.2 The walrus in Native marine economies

Igor I. Krupnik

Indigenous coastal hunters have exploited the Pacific walrus population throughout its entire biological range. Within this huge area, the actual role of walrus hunting in Native subsistence economies varied from an episodic or even exotic contribution to being a staple of daily survival. The area with the heaviest human reliance on walrus included residents at the junction of the Bering and Chukchi seas, who may aptly be called the "walrus people" – although they also hunted for a wide spectrum of other marine mammals. However, walrus hunting was at the core of their existence, usually providing 50–90% of the annual caloric intake for most coastal villages, depending upon how well the walrus harvest proceeded each year.

The impact of this early subsistence use on walrus is a matter of hot debate. Assumptions that primitive hunters were not equipped to kill great numbers of walruses, and had no incentive to do so, and that hunting had no greater effect than any other kind of predation have recently been challenged by historical data. While prehistoric hunting equipment might seem "primitive" compared with modern rifles and motorboats, it was nevertheless quite efficient (Fig. 1). Archaeological records and elders' stories document that Native hunting, particularly of large marine mammals, tended to single out juveniles, yearlings, pregnant females, and nursing cows because these animals moved slowly and were easier to catch. Furthermore, the numbers taken were not trivial.

Generally speaking, arctic subsistence hunting stands apart from tropical and temperate-zone models. First, available resources, either marine or terrestrial, are highly seasonal, and those "seasons of availability" are themselves usually very brief. Hence, arctic subsistence hunting was (and still is) crucially dependent upon short periods of abundant game. These few weeks, sometimes days, had to provide hunters, their families, and the entire community with the bulk of food and supplies for long periods. Second, in the arctic resources are extremely difficult to predict from year to year. The Bering Sea subsistence walrus hunter could never be sure what game availability and harvest success awaited in the next season or coming year.

These contingencies highlight the crucial significance of *surplus* catch and *surplus* food storage. During the peak of spring and fall walrus migration, the hunting crews used to kill walruses and to butcher the meat continuously to the point of complete exhaustion – of course, when and if the animals were available. Huge amounts of fresh meat and blubber were laid in storage pits or air-dried on open racks. The objective was to store more than enough meat to last

Fig. 1 Arctic Native people hunted walruses either from one-man kayaks or from much larger skin boats (*umiaks*), with a crew of 5–8 people. Kayaks were used (and still are today in some places) in southern Alaska, the central Canadian arctic, and Greenland. Large skin boats propelled by sails and paddles (today by outboard motors) were used by hunters in the Bering Strait area, on the Chukchi Peninsula, and in northern Alaska. The technology of hunting was similar. This figure illustrates (A) how a hunter in a kayak from Nunivak Island, Alaska, approached a walrus quietly on open water (or floating ice or coastal hauling grounds), (B) fastened a toggle-head harpoon with a skin float attached to the animal, then prepared to kill the animal with a heavier killing lance. These days, hunting is done mainly with rifles, but harpoons with floats and lances are still widely used in many Native communities across the arctic. From Fitzhugh & Kaplan (1982), courtesy of the National Museum of Natural History, Smithsonian Institution.

until the next spring. If the next spring hunt began on time and the new catch was abundant, last year's meat was simply discarded. This attitude and practice appears to account for the extraordinary soil buildup at many ancient coastal sites.

Unfortunately, the earliest statistical data on Native subsistence catch in the Bering Sea go back only to the 1920s, when Native walrus hunting with rifles had already begun, though it was still practiced primarily with harpoons and lances, and in paddle-propelled skin and wooden boats. Based on fragmented catch records, the total walrus subsistence harvest on the Chukchi Peninsula could have been 2000–2500 animals annually, with at least 30–50% additional losses of wounded and unrecovered animals. Native catch on the Alaskan side could have been a half of this, making a total of 4000–5000 animals taken per year, with up to an equal number killed and lost or wounded. This is a fairly conservative estimate, which hardly speaks in favor of a low effect on the walrus population.

The earliest Native catch data available from the Chukchi Peninsula offer poor support for another common assumption: that Native subsistence hunting was done only to facilitate daily needs. Quite to the contrary, reported annual catches fluctuated dramatically from village to village and from coast to coast, often by factors of 2–3 among years. The reasons for these oscillations were primarily natural – local ice and weather conditions, annual position of the ice, and the availability of walrus within easy reach of hunters in small boats. There is little doubt that the very same factors influenced fluctuations in the abundance of walruses and the size of the subsistence catch at Native coastal sites since time immemorial.

A fairly comprehensive picture of the structure of Native economies and of the role played by marine species in human diet may be gained from more detailed records from the Siberian side of the Bering Strait in the early 1900s. Elders' memoirs and early historical accounts may be used to extend this perspective back to the late 1800s. Some Native communities literally lived off walrus and either flourished or just survived from one seasonal walrus hunt to another. When walrus hunting failed on St. Lawrence Island for two consecutive years (1878/9 and 1879/80), more than two-thirds of the island's Yupik population died of starvation, hypothermia, and related hunting accidents triggered by weakness and a desperate search for food. The earliest comparable records from the Alaskan side come from the village of Wales on the Seward Peninsula. Accounts of missionaries and school teachers show that local hunters killed 322 walruses, 32 white whales, more than 80 bearded seals, and 4000–5000 other seals in 1890. For a village population of 539, one would come up with annual meat consumption consisting of 45.7% walrus, 42.6% seals, 8.2% white whale, and 3.5% bearded seal.

This subsistence hunt was divided into four major seasons, referred primarily to the main marine species to be hunted, which fluctuated in accordance to regional geography rather than calendar months. Thus, "winter" was mainly the time for hunting for ringed seals through breathing holes and along the edge of the shore-fast ice, which lasted from December to April, or May–June in more northerly areas. Bearded seals usually start to migrate northward in April and May, and walrus and bowhead whales from April to June. This was prime "spring" time when villages located on the open seacoast were engaged in communal boat hunting in leads, polynyas, and moving ice. During the short "summer" season, from July to August/September, the ice-inhabiting species leave the northern Bering Sea and move far north into the Arctic Ocean, and food was dominated by fishing, bird hunting, collecting birds' eggs, and picking plants and berries in the tundra. Regular large-scale boat hunting resumed in the "fall" (September–November) when walrus, whales, and seals begin their return movement with the advancing, newly formed ice.

According to elders' accounts, game returns in a normal year usually surpassed community needs in "spring" and "fall" seasons, when 10 times as much meat could be obtained as during the rest of the year put together. Thus, every year the prosperity and the survival of the Bering Strait Native communities were critically linked to two brief periods of fervent, almost non-stop hunting that corresponded to the peaks of walrus and whale seasonal migrations. Walrus hunting for meat, blubber, hides for boat covers, and other uses served as the major determinant of the size and the distribution of the human population, its food intake, and the whole spectrum of marine resources it exploited. Meeting human community food needs with any other species would require an unimaginable expenditure of time, labor, and energy. This can be indicated by the following calculations. Five Siberian Yupik communities, with a combined population of 1100 in the 1920s, had a total annual demand for meat and sled-dog food of approximately one billion kilocalories. That volume of food could be supplied by an annual harvest of 1670 "staple" walruses. The actual annual catch during these years oscillated between 1400 and 2200 (average ~ 1500), depending upon weather conditions and success for each year. Thus the average catch covered roughly 90% of human and dog annual needs.

In comparison, the same amount of food could be secured by the harvest of 75 bowhead whales (while the actual annual catch was just 4–8) or of 43 500 small seals (the actual yield was only 4000–8000). Living on fish alone would have been even harder, requiring an annual catch of 1000 tons, as opposed to the actual catch of approximately 50–90 tons, not to mention anything so labor-intensive as living by hunting sea birds. Thus, in spite of the active use of several marine species in Native coastal economies, the role played by each of these was largely determined by how much human effort was required to obtain it. In the Bering Strait area it was always the walrus that was the "core" species in Native economies. It was the most abundant and the most reliable local resource, one that provided the best return per hunting effort in terms of high-quality food, and was easy to secure and store in sufficient amounts.

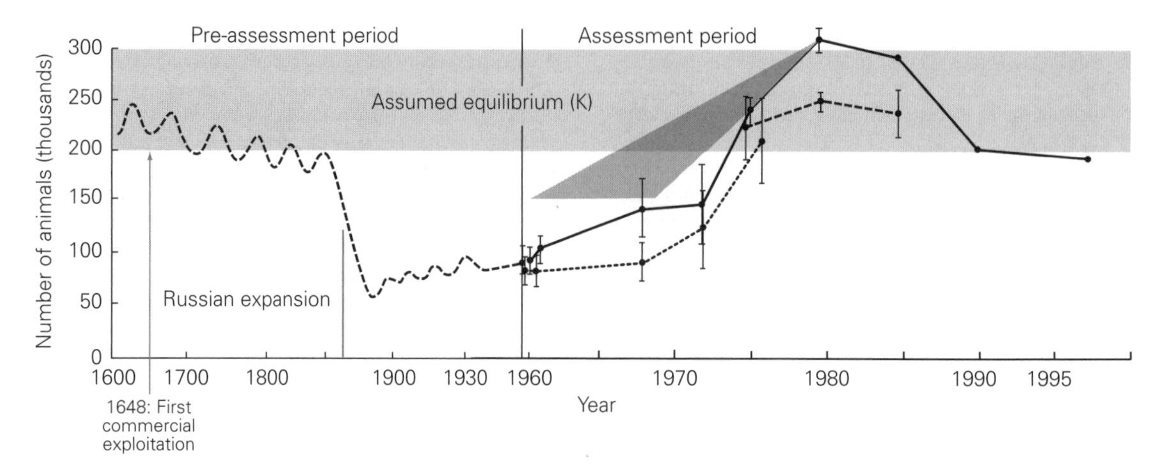

Fig. 6.21 Walrus population assessments from prehistoric times. Native catch is presumed to have allowed the population to fluctuate within the lower portion of carrying capacity (K). Early commercial exploitation depressed the population somewhat, but during the mid-1800s greatly depleted it. From that time to 1960, the population fluctuated, but recovered to an unknown extent. Numbers from 1960 to the mid-1970s are from U.S. aerial surveys (dotted line), and from 1975 through the 1990s from joint U.S.–Russian collaboration (dashed line). Population sizes (solid line) include animals not counted because they were in the water (Table 6.1). The darker gray triangle represents a back calculation of population estimates. See text for explanation. Modified from Fay et al. (1989, 1997).

Table 6.1 Pacific walrus population and total kill.

Year	Adjusted estimate of population size (Mean)	Mean population change (%)	Range of total kill	% population killed
1960	65 500–94 400 (79 950)	NA	8 908	9.4–13.6
1961	75 400–107 700 (91 550)	+15	7 643	7.1–10.1
1968	105 900–159 600 (132 750)	+45	4 094– 6 409	2.5–6.5
1972	97 700–186 200 (141 950)	+7	3 184– 4 902	1.7–2.8
1975	220 300–247 800 (234 050)	+65	4 509– 6 281	1.8–2.9
1980	290 700–310 700 (300 700)	+28	6 617– 9 574	2.1–3.3
1985*	234 020	−22	10 060–17 710	4.2–7.5
1990*	201 039	−15	7 758–12 931	3.9–6.4
1997*	188 316	−6	3 864– 8 621	2.1–4.6

Population assessments were conducted in all years listed, except those marked with an asterisk (*), which represent calculated minimum population sizes. Adjusted estimates of population size include direct visual observations plus extrapolations for walruses not seen because they were presumed underwater. Mean population change represents the increase or decrease from the previous assessment to the year in which the % (rounded) is given. Range of total kill combines U.S. and Russian catch, a 42% correction factor for animals struck and lost, and a proportion of additional calves that die due to loss of their mothers; numbers indicate the kill between the date listed and the previous assessment date, except for 1998 and 1999 which are kills for those years only. % population killed is the total kill divided by the mean adjusted estimate of population times 100. Rates of population change and high ranges of population estimates have led managers to conclude that neither population status nor optimum sustainable population can be known. From 1999 to 2000, total kill has ranged from 4259 to 7757 per year.
Data from Fay et al. (1997); Gorbics et al. (1998); U.S. Fish and Wildlife Service.

could prevent the recovery of a depressed population. Nevertheless, some information seems to indicate that the population increased somewhat from the 1880s to about 1930.

A gradual rise in the walrus population seems to be confirmed by the take of 8000–10 000 walruses during some years from 1930 to World War II. Reduction of hunting during the War presumably allowed the population to increase. Alaska attained statehood in 1959 and in 1960 gained authority for walrus management from the United States Fish and Wildlife Service (FWS). Alaska established a sanctuary for summering male walruses in Bristol Bay and limited subsistence kill to five female walruses per hunter per year; calves could be taken with females, but male subsistence kill was not limited. Additionally, 50 animals per year were allocated for sport hunting. Then, in the mid-1960s, the nature of the subsistence take was altered by an unforeseen change. Until that time, Eskimos used dog teams for transportation and hauling, but in the mid-1960s they adopted snowmobiles. This change shifted the value of walruses away from meat for dogs to tusks as a cash crop for purchasing fuel. Concurrently, the demand for Native handicrafts, including ivory, rose dramatically.

Another significant change was the passage of the U.S. Marine Mammal Protection Act (MMPA) in 1972, which restricted walrus kill to subsistence needs of Native peoples only. The Act also required periodic assessments and non-wasteful hunting methods, and a return of management authority from Alaska to the FWS. An unfortunate consequence of suspension of state management was the loss of the Bristol Bay walrus sanctuary, which had been established by state authority to protect the last remaining handout area within U.S. jurisdiction. However, in 1976, with Federal agreement, Alaska resumed authority for management, and established a kill limit of 3000 animals per year; and in 1979, the state re-instituted the Walrus Islands Game Sanctuary in Bristol Bay. Soon thereafter, Alaska returned authority to the FWS due to "administrative complexities" of the MMPA, which were resolved by amendments to the Act in 1981. Nevertheless, Alaska decided to leave management with the FWS, and recommended that the federal government, state, and newly formed Eskimo Walrus Commission cooperate to develop management plans and procedures. Meanwhile, as jurisdictions shifted, limited attempts at assessment indicated a population increase, although Native take was poorly recorded and assessments were fraught with logistic and methodologic difficulties.

From the mid-1980s, lack of management regulations for walruses and the increasing demand for ivory for the booming cottage industry led to increased harvest rates and waste of the meat. In the early 1990s, the FWS adopted a Pacific Walrus Conservation Plan, intended to reduce waste. Hunters were required to return tusks (tagged for identification), heart, liver, "coak" (brisket), flippers, and "some red meat." The ivory, hides, and penis bone could be sold, but only if they had been transformed into Native handicrafts or clothing. In 1997, the FWS and the Eskimo Walrus Commission established a cooperative agreement, also with Russian counterparts, with two major components: a Marking, Tagging and Reporting Program and a Walrus Harvest Monitoring Project. Together, these are intended to monitor subsistence take, collect biological information, and help control illegal take, trade, and transport.

6.6.2.2 Population assessments

During the 1930s, the Russians made the first attempts to assess the walrus population, in association with their increased catches during that time, but were unable to cover the total range. The first region-wide assessments were conducted by the U.S. FWS in winter 1960, by visual counting from aircraft; the population was estimated at approximately 65 000–95 000 animals (as later adjusted; see Table 6.1). Since 1960, assessments have been conducted at various intervals. The results indicate that the walrus population increased rapidly to about 300 000 by 1980, thereafter declining to around 180 000 by 1997 (Fig. 6.21). These and other data seemed to suggest a carrying capacity between 200 000 and 300 000. From 1951 to 1972, more than 40% of females 7–15 years old were pregnant, but from 1975 to 1985, the age of first pregnancy had risen to 9–15 years, suggesting a rapidly growing population. The proportion of adult females observed with newborn calves was approximately 13 to more than 60% until the mid-1980s, but by 1998–9 had dropped to 5–16%. In addition, more animals were seen in southerly portions of the range during years of increase, and changes in body condition, demographics, mortality, and foods consumed were associated with years of decrease. This combination of indicators is compelling and is consistent with density-dependent population growth and a population near carrying capacity. However, although each indicator may have multiple causes and none can be confirmed without better direct observations of distribution and population sizes.

For purposes of assessing the effects of subsistence hunting, the FWS uses a formula for potential biological removal (PBR), i.e., the maximum number of animals, not including natural mortalities, that may be removed from a marine mammal population while allowing that population to reach or maintain its optimum sustainable population (OSP; see Box 2.4, pages 48–9). PBR is calculated from a minimum population estimate (N_{min}), a maximum rate of increase (R_{max}), and a recovery factor (F_r). A 1988 calculation yielded:

$$N_{min} R_{max} F_r = 188\,316 \times 0.04 \times 1.0 = 7533$$

This result, in conjunction with the most recent population assessment of 1997, led the FWS to conclude that no restrictions on catch are needed because total subsistence catch has usually been less than 7533 animals per year since that time.

Managers emphasize that PBR is a practical management tool, intended as an indicator of risk and not as a replacement for long-term assessments. Nevertheless, the calculation demands examination. First, N_{min} is highly uncertain, particularly for the 1997 assessment used in the calculation above, which the FWS admits was flawed. Second, R_{max} for walruses is even less certain, as exemplified by the variability of pregnancy rates, as well as the few animals observed, relative to the total population. Third, F_r is meant to compensate for uncertainty, that is, to allow for management discretion as consistent with the goals of the Marine Mammal Protection Act to maintain marine mammals within OSP. F_r may be set from 0.1 to 1.0 at the manager's discretion. A value of 1.0 asserts lack of risk, or put another way, that the population is presumed within OSP and capable of recovery. However, this assumption appears to conflict with the low pregnancy rates observed for the apparently declining population, causes for which are unknown. Further, low kill rates, in the 2–3% range (e.g., for 1972 to 1980: Table 6.1), seem to be within the walrus's potential for population growth, but higher removal rates (as from 1985 onwards) seem to be associated with population decline.

Considering the imprecision of counting walruses, managers have recommended more frequent assessment intervals of 1–8 years, depending on the level of precision attained. R_{max} is particularly difficult to address, as it depends both on assessment precision and on measurement of pregnancy rates. Therefore, one could surmise that walrus assessments may reflect variable and imperfect methods, rather than true population numbers. For example, the average annual rate of increase of approximately 9% from 1960 to 1980 seems extraordinarily high, given walruses' low reproductive rate. However, if the population increased at a rate of 3–6%, a back projection from the 1980 number of approximately 300 000 animals yields a population size of about 150 000 between 1960 and 1970 (Fig. 6.21). Which is correct cannot be known.

6.6.2.3 Research implications

At the root of conservation and management are two questions: How many walruses are there? And, how is it possible to determine whether the walrus population is within OSP, as required by the MMPA? The first requires population assessment, the second, environmental assessment. Major questions concern the status and trends of the population, indicators of carrying capacity, and habitat relationships.

At best, walrus population assessments represent highly uncertain estimates, due to judgements involved in visual counts from aircraft, low proportions of the population sampled, and lack of consideration of sea-ice habitat distribution. Commonly, wildlife managers are forced to estimate total numbers, reproductive rate, and demographics extrapolated from sample sizes of less than 1% of total population. Walruses are observable only when on ice or at the water's surface, which raises a number of issues (Fig. 6.22). First, walruses are patchy in occurrence and chance is a factor in observation. Second, most of the walrus population occurs on continually moving sea ice; thus, patch location and day-to-day movements of ice can significantly affect assessment results (Fig. 6.23). Third, assessment methods derive total population number by multiplying animal density by the assumed total habitat area. However, determination of total area requires analysis of hauling-out habitat, namely sea-ice structure and dynamics, which has not yet been incorporated into survey design.

Fourth, because walruses cannot be counted underwater, diving time must be accounted. From observations of individual diving and feeding, walruses appear to spend an average of about 16% of their time at the surface, indicating a correction factor of 6.2 for the proportion of walruses on the surface *vs.* underwater. That is, if 10 walruses are observed at the surface, a total of 62 would be assumed present, 52 of them underwater. However, walruses appear to dive synchronously. That is, if 10 animals are observed at the surface, the total number may be 10 or anything up to 62, or even more. Complicating this is the observation that hauling out of groups also appears to be synchronous: large herds may remain in the water for several days to feed, then emerge for a few days to rest. Consecutive days of observations for walruses on sea ice in winter and on terrestrial haulouts in summer have yielded numbers of hauled-out animals that vary by more than an order of magnitude.

Thus, for walruses, assessment is a formidable problem and few solutions are available. One possibility is

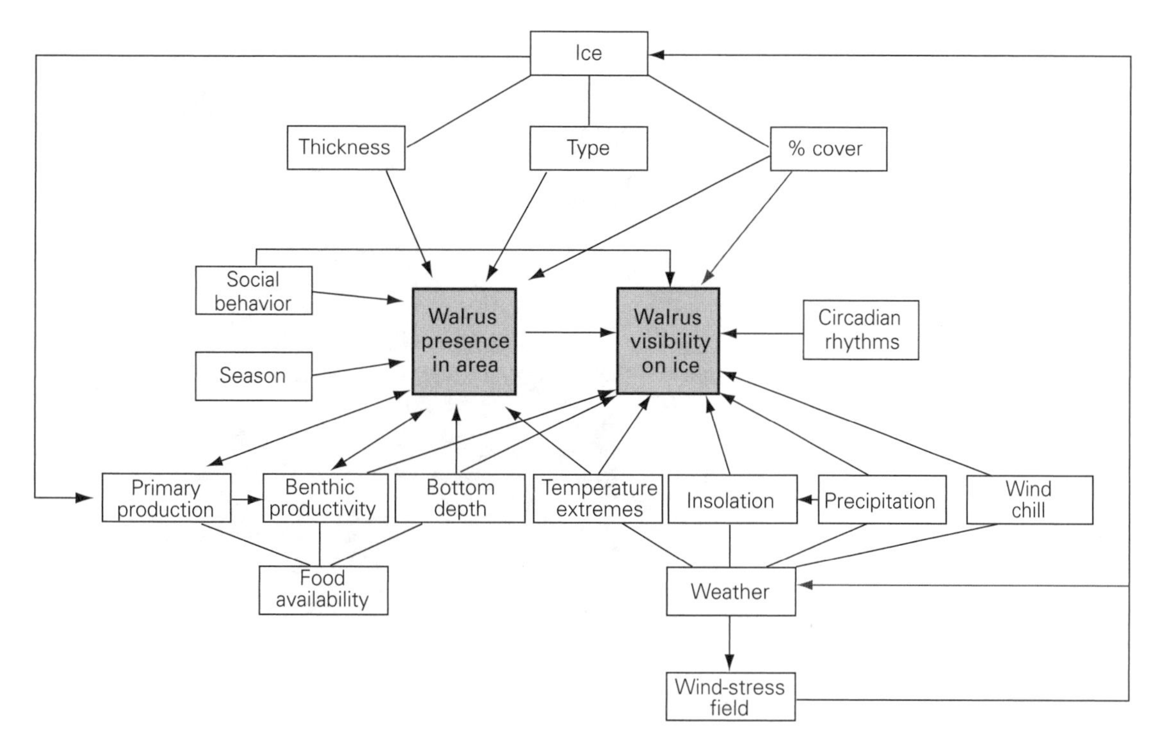

Fig. 6.22 Model of walrus hauling-out dynamics. A number of environmental, behavioral, and physiological factors are involved. No weightings are given to the arrows, due to the paucity of quantitative information available. From Ray & Wartzok (1974).

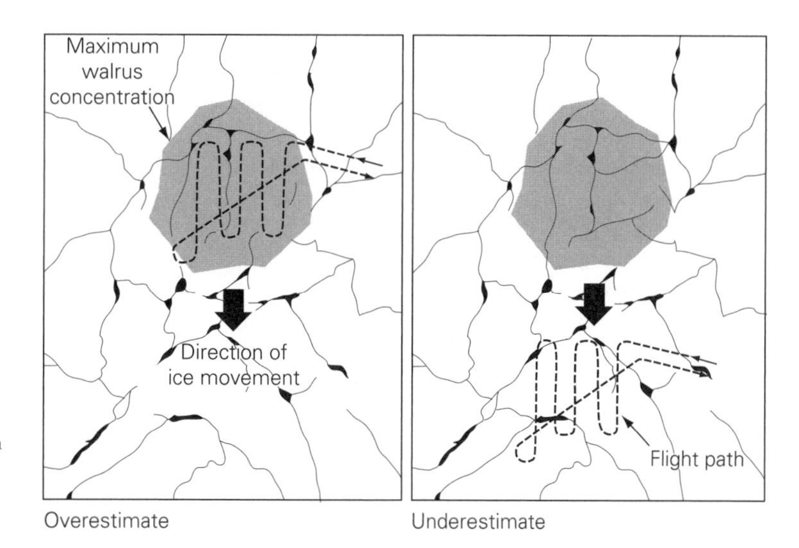

Fig. 6.23 Over- and underestimates of walrus numbers on sea ice may occur unless ice dynamics is taken into account. The shaded area is that of maximum walrus concentration. Grids conducted on different days, but uncoordinated with ice movement, may result in over- or underestimates. From Ray & Wartzok (1974).

thermal infrared sensing to detect groups from aircraft, coordinated with use of aerial photography for counting. In the near future, walruses may be detectable by satellites; this technology is facilitated by the large sizes of groups and very high temperature differences (10–30°C) between animals and ice (Fig. 6.24). This has the advantages of more accurate counting, simultaneous assessment of sea-ice habitat, and achievement of a permanent record for re-analysis, none of which can be provided by visual surveys.

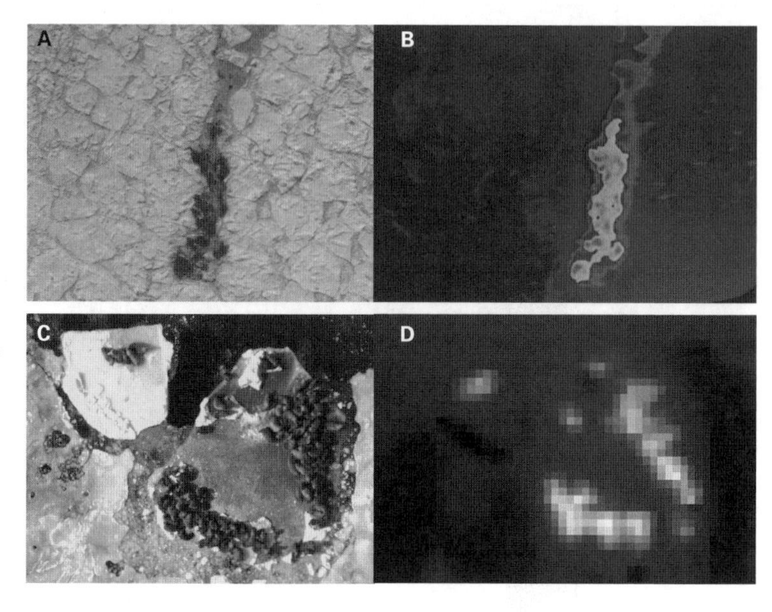

Fig. 6.24 (A) Photographic image of walruses on ice, Chukchi Sea, July 1994; (B) infrared image of the same group; both from an altitude of about 300 m. From Ray & Wartzok (1974). (C) Photographic image of walruses on ice, southwest of St. Lawrence Island, Bering Sea, April 2002, from an altitude of about 500 m, covering an area of about 60 by 63 m; (D) infrared image of the same group from an altitude of about 800 m. Courtesy of the United States Fish and Wildlife Service, Anchorage, Alaska. The animals can be counted with reasonable accuracy in the photographs. Detection of walruses on ice is made possible by a very strong temperature difference of > 20°C between sea ice and walrus skin surface. Detection may thus become possible from high-flying aircraft or satellites. Coordinated detection and enumeration makes possible more accurate assessments than have previously been possible, although this method cannot account for animals in or under the water.

As for estimation of carrying capacity – or determination of optimum sustainable population – benthic food supply is critical. This determination is complicated by the walrus's patchy distribution and because interactions among sea-ice dynamics, weather, and the location and intensity of walrus feeding are virtually unknown. Nor is sufficient information available on fisheries interactions. Studies have estimated that about 20 000 walruses occurring in the Bristol Bay region from May to August might consume as much as 33% of the surf clam biomass (*Spisula polynyma*) and that if commercial claming or dredging were to be conducted in or near areas where walruses feed, walruses and the industry could compete. These studies also concluded that walruses probably could forage elsewhere if the fishery depleted their food, but evidence for this conclusion is also lacking.

6.7 Conclusion

Steller sea lions and walruses respond to environmental changes and human activities in ways that are poorly understood, and only general hypotheses can currently be posed about the resiliency of these species. Changes in abundances of large, slow-growing animals with low reproductive rates lag behind environmental changes. Therefore, assessments reflect adjustments to past conditions, and trends may not become evident until some time after density-dependent responses are initiated.

Marine mammal conservation in the Bering Sea attempts to resolve modern-industrial and traditional, subsistence resource issues within a legal–jurisdictional framework dictated by western European culture. Fortunately, coastal development, pollution, and oil, gas, and mineral exploitation appear not yet to be serious problems in the region. Fortunately, also, almost all of the Bering Sea is under the jurisdiction of only two nations – Russia and the United States – a unique situation for the world's seas. Conservation and research have been facilitated by this fortunate situation (Box 6.3).

At least four effects pertinent to marine mammal environmental interactions may be inferred:

- alteration of food supplies may have precipitated depletions of some populations, while others have increased;
- disruptions of utilized species have been followed by community and ecosystem alterations;
- changes in climate have been correlated with the above effects;
- environmental, technological, and social changes synergize and cascade down to Native American communities and to national and global markets.

Box 6.3 U.S.–Russia Agreement on Cooperation in the Field of Protection of the Environment and Natural Resources

Robert V. Miller

This Agreement originally covered 11 areas of interaction, but Area V – "Protection of Nature and Organization of Preserves" – is the only surviving section, with eight projects including one on marine mammals. Prior to 1972, there had been little joint marine mammal research between the U.S. and the Soviet Union, other than on northern fur seals and a few studies carried out under the academies of science of the two countries. In the late 1960s and early 1970s, the four members of the Fur Seal Commission began occasional exchange of research results and publications on marine mammals. In March 1972, the United States and the Soviet Union began an exchange of information on ice seals and walruses in conjunction with the annual meetings of the Fur Seal Commission. The two countries were also completing a framework for an environmental agreement, which was signed on May 23, 1972, by Presidents Nixon and Podgorny. The two sides met in Moscow in mid-January, 1973, and agreed to a protocol that included cooperative research on whales and pinnipeds. This established a precedent of scientific cooperation that has continued to this day. In 1994, this program was renegotiated as the "Agreement between the Government of the United States of America and the Government of the Russian Federation on Cooperation in the Field of Protection of the Environment and Natural Resources," signed by U.S. Vice President Gore and Russian Prime Minister Chernomyrdin.

Although the Agreement was hampered by lack of funding, some work was carried out during the first year. The scope of research gradually expanded to include dolphins and porpoises of the eastern tropical Pacific and Black Sea, consideration of a conservation agreement on walruses and ice seals of the Bering and Chukchi seas, and parasites of marine mammals. As time went on, communications and project planning improved and the frequency and number of joint activities increased. Between 1993 and 2000, the United States and the Russian Federation (formerly the U.S.S.R.) successfully undertook more than 20 joint cruises, in excess of 50 joint field studies on both cetaceans and pinnipeds, and 25 or more laboratory studies, involving about 300 scientists. This collaborative effort has resulted in direct benefits to Bering Sea marine mammal management, in particular of walruses and Steller sea lions.

For walruses, cooperation had begun in 1972 with a joint assessment between the United States and the Soviet Union, which reinforced the need for standard methods. A series of joint aerial surveys of walrus was initiated in 1975, designed to take place at 5-year intervals. However, major flaws in survey methods were revealed. For example, the unknown numbers of animals in the water *vs.* those on ice makes valid estimates almost impossible to obtain. Nevertheless, this work showed that coordinated surveys could be done, which had potential to yield range-wide estimates of abundance. With respect to Steller sea lions, American scientists had conducted aerial surveys in the 1950s to 1970s that confirmed a marked population decline caused by previous exploitation. By the mid-1980s, U.S. and Russian scientists were undertaking tagging and tracking studies on rookeries in Russia and Alaska, and had begun examining ways of conducting range-wide surveys to estimate total abundance and to determine whether animals were emigrating from east to west. A series of subsequent aerial and shipboard studies have tracked the species' population on a range-wide basis.

These inferences raise questions concerning whether ecological controls operate top-down or bottom-up, whether feedbacks exist between these effects, and the scales at which these interactions occur.

The Bering Sea's resources have been significantly altered during long periods of human exploitation. The ecological consequences of removal or depletion of predatory species can be enormous, and with significant ecological responses. In the southern portion of the Bering Sea, for example, reduction of whales, herring, and some fishes could have made millions of tons of their zooplankton food available every year to feed billions of year-old walleye pollock, and the reduction of fur seals could have dramatically reduced predation on juvenile pollock. These two factors alone could have caused the southern and eastern Bering Sea to readjust and to become pollock-dominated (NRC 1996). However, the question is: Do pollock make a difference to this ecosystem? Two scenarios are possible. If pollock were to decrease, it would be possible that (i) species that feed on pollock might also decrease; (ii) zooplankton on which juvenile pollock feed might increase, leading to increases in planktivorous sea birds, such as auklets; and (iii) piscivorous sea birds that compete with pollock might also increase. On the other hand, if pollock were to increase, (i) their food consumption might cause depletion of food for other species; or (ii) species that prey on pollock might increase (Springer 1992). Therefore, changes in a dominant species can cascade throughout the community food web, possibly also affecting Steller sea lions.

Another example of a potential ecosystem response is the shelf benthos in the northern Bering Sea, which is significantly influenced by walrus and gray whale feeding bioturbation. Both species were severely depleted during the late 1800s and early 1900s, but have since recovered; the walrus population seems now to be near its carrying capacity and the gray whale has been removed from the Endangered Species List in response to complete recovery to pre-exploitation numbers. The earlier depletions of these two species may have had widespread ecological consequences, but whether the benthic communities on which these species now feed are the same as centuries ago cannot be known.

A looming issue for the Beringian ecosystem and its species, as well as its Native people, concerns global climate change. Some seabirds and marine mammals are already being affected by thinning Arctic ice. In spring 2001 Bering Sea ice was thin, and breakup was 3 weeks early. Warming continued into summer, and Point Barrow, Alaska's northernmost point, experienced record high temperatures of 20°C on July 17. Eskimo whalers reported that this was the 10th year in succession that sea ice had broken up early in some areas, making it more difficult for them to hunt bowhead whales because of extensive open water. The year 2002 was similarly warm. Climate models indicate that should this trend continue, the Bering Sea could become ice-free, year-round, in a decade.

The critical conservation question is: Do present management regimes suffice to maintain resources and ecosystem health and, if not, what other actions, agreements, or activities are needed? Current U.S. legislation requires numerical assessments to demonstrate that conservation action is needed; that is, a marine mammal species must be shown to be depleted, endangered, or below OSP in order that regulations can be imposed. A more serious gap concerns lack of an integrated program of research for the Bering Sea region, which is necessary to develop knowledge of biological and ecological relationships. Currently, statements about population trends for most Beringian species and carrying capacity remain subject to interpretations of little-known scenarios on feeding and environmental dynamics.

Chapter 7

The Bahamas: tropical-oceanic island nation

The eastern United States possess very peculiar and interesting plants and animals, the vegetation becoming more luxurient as we go south, but not altering in essential character . . . But if we now cross over the narrow strait, about fifty miles wide, which separates Florida from the Bahamas Islands, we find ourselves in a totally different country . . . such differences and resemblances cannot be due to existing conditions, but must depend upon laws and causes to which mere proximity of position offers no clue.

Alfred Russel Wallace *Island Life* (1880)

7.1 Introduction

The Commonwealth of the Bahamas is the largest tropical archipelagic nation in the Atlantic, equivalent in size to Australia's Great Barrier Reef (Table 7.1 & Fig. 7.1). The climate is subtropical to tropical and day lengths change only slightly from summer to winter (Figs III.1 & III.2, pages 129–31). The archipelago has resulted from 200 million years of tectonic, sedimentary, biochemical, and reef-building processes. Thirty-five major islands, 700 cays (pronounced "keys"), and 2400 rocks are scattered over shallow-water banks cut by deep channels. Bahamian banks constitute a giant carbonate-sediment factory that maintains the islands and reefs at positions near a continuously changing sea level. Many of the Bahamas' characteristic and most valuable species depend upon mangroves, creeks, shallow seagrass beds, and coral reefs close to land. Lucayan Indians first arrived only 1500 years ago and moved from island to island, subsisting on agriculture and marine life. A millennium later, Europeans arrived, beginning an accelerating pace of change.

Today, the Bahamas is relatively prosperous, but highly dependent on tourism and, therefore, on maintaining a healthy environment. The nation has adopted a policy of "sustainable development," which is challenged by a rapidly growing population and limited resources. Many islands and cays lack sufficient water and are uninhabited. Only 22 islands have permanent settlements, although numerous small islands and cays have residences. The capital, Nassau on New Providence Island, is an internationally connected urban-port center. New Providence covers only 1.5% of the total land area of the Bahamas, but invites opportunities and jobs for about two-thirds of the population.

Table 7.1 Characteristics of the Bahamas.

Parameter	Measurement
North–south extent*	720 km (448 mi)
Northwest–southeast extent†	937 km (528 mi)
Greatest width	812 km (505 mi)
Total area‡	325 000 km^2 (124 000 mi^2)
Lands	35 major islands 700 cays 2400 rocks
Total land area§:	13 868 km^2 (4500 mi^2)
Area for 11 largest islands§:	12 455 km^2 = 89.8% of total
Andros	5957 km^2
Abaco Islands and Cays	1666
Great Inagua	1544
Grand Bahama	1373
Eleuthera	518
Long Island	448
Aklins	389
Cat Island	388
Exuma Island and Cays	291
Mayaguana	285
New Providence (including Nassau) and Paradise Island	210
Estimated reef area¶	~ 3150 km^2 (1260 mi^2) ~ 1% of world total ~ 15% of Caribbean

* Measured latitudinally.
† Measured from Little Bahama's northwest bank to Great Inagua's southeast bank.
‡ As included within the archipelagic boundary.
§ From Sealey (2001).
¶ From Spalding et al. (2001).

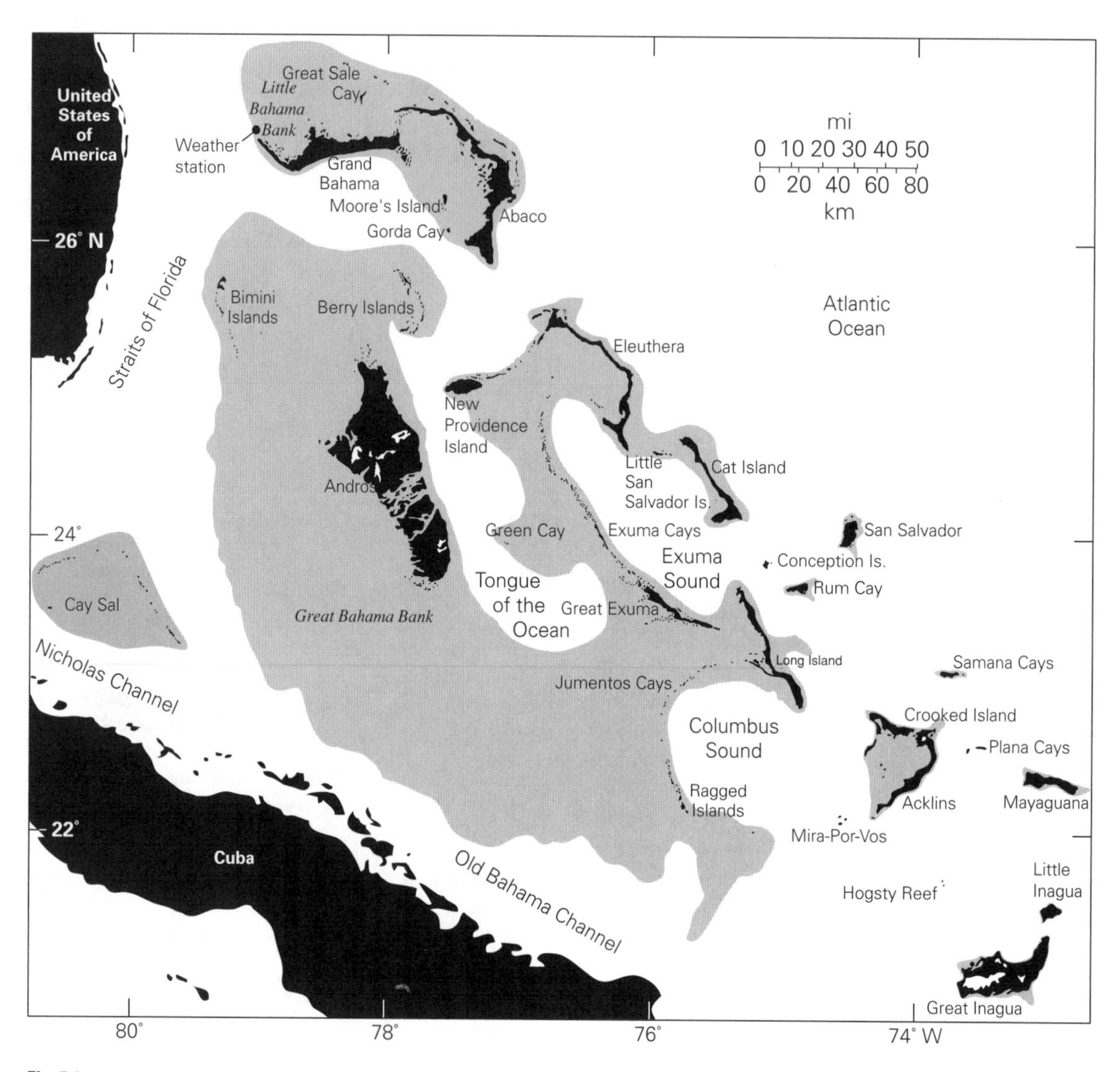

Fig. 7.1 Place names in the Bahamas.

This island nation faces critical environmental issues and is vulnerable to external forces. Although the nation has been a leader among Small Island Developing States (SIDS) in environmental legislation, the ultimate challenge is sustainability of its environment and economy. Biodiversity takes on special significance, due to its capacity to attract tourists, to supply food, and to maintain essential ecological functions.

7.2 Process and pattern

The Bahamas archipelago lies between North America, Cuba, and the Atlantic Ocean, and consists of extensive, mostly submerged, carbonate platforms ("banks"). Most of the land surface is less than 10 m above sea level and the banks are rarely more than 4.6 m deep, except where cut by tidal channels. Tidal range is generally only about 1 m. The archipelago has been evolving for millions of years, and its land–seascapes and life have been integral to that evolutionary process. The accelerating pace of human development sharply contrasts with slower processes of island and bank formation, repair, and regrowth.

7.2.1 Land–seascape formation

The Bahamas platform has resulted from long-term, non-continuous, biophysical and chemical processes and is at an advanced stage of development. The beginnings were in the Triassic period, 200 million years ago, when all continents formed one landmass, called "Pangaea." Rifting between North and South America and Africa and rotational movement of North America away from North Africa resulted in the formation of a small, closed, sediment trap that quickly filled with eroded material and organic matter from the surrounding continents (Fig. 7.2A–D). The weight of these materials caused the basin to subside and carbonates began to be deposited by chemical and organic processes (Fig. 7.2E–G). In early Cretaceous times, more than 100 million years ago, the rate of continental drift increased, which opened the basin to the widening Atlantic Ocean, and the platform became a marginal, continental plateau allowing interchange with Atlantic waters. Improved circulation created favorable conditions for coral–algae growth and calcareous-sand production. Continued deposition of sediment, reef growth, and erosion produced a series of thick, carbonate-capped platforms in the form of atolls rimmed by marginal reefs.

The Bahamas' present habitats have evolved within an oceanic environment during the past 150 000 years. Major changes in deposition and erosion, and changing tectonic, eustatic, and climatic conditions, resulted in

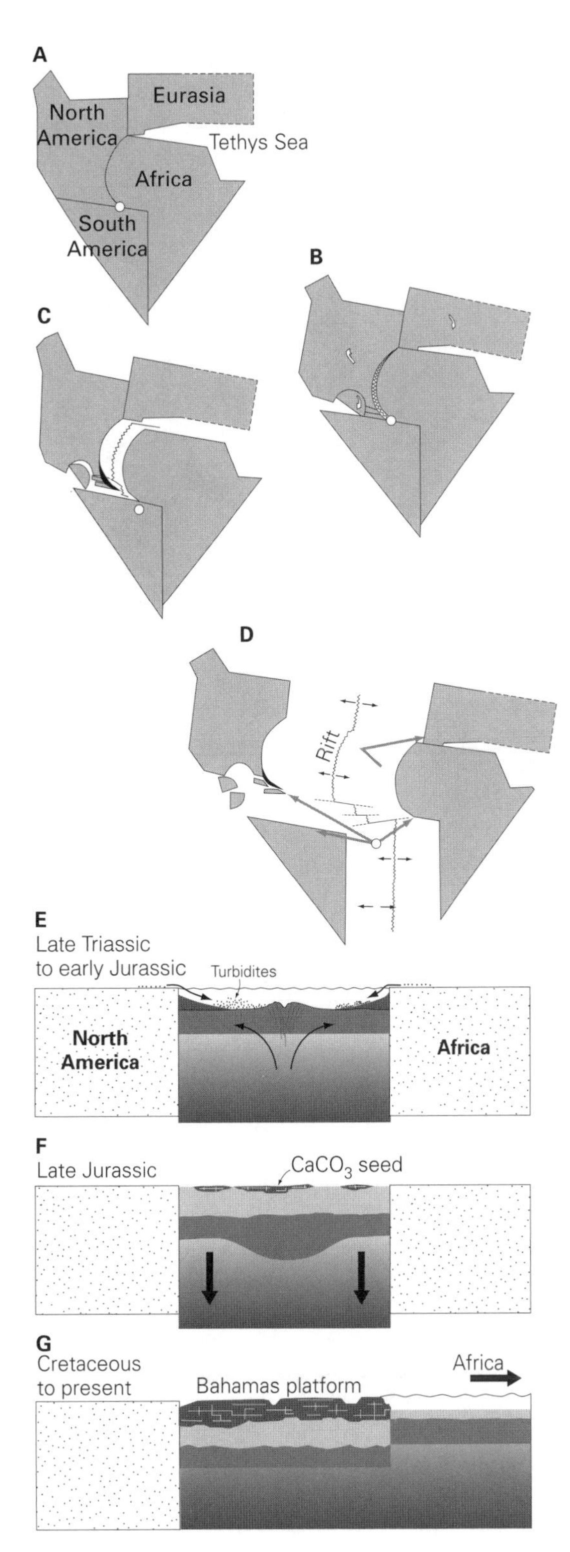

Fig. 7.2 (*right*) *Top*: Evolution of the Bahamas platform by seafloor spreading and plate tectonics: (A) Pangaea in the Permian; (B) rotation of Laurasia (North America and Eurasia joined) around a pivotal Y-junction (O) created a primitive Atlantic Ocean as a crescentic rift; (C) further opening of the North Atlantic; opaque crescent is Blake–Bahamas platform, filled to sea level by sedimentary clastics, then isolated by further seafloor spreading; (D) North Atlantic Ocean of today, showing continents relative to the original Y-junction (O). *Bottom*: Processes of infilling: (E) rifting and infilling by sediment; (F) increased basin subsidence and incipient carbonate deposition; (G) isolation of Bahamas platform by separation of Africa and North America. From Dietz et al. (1970), with permission.

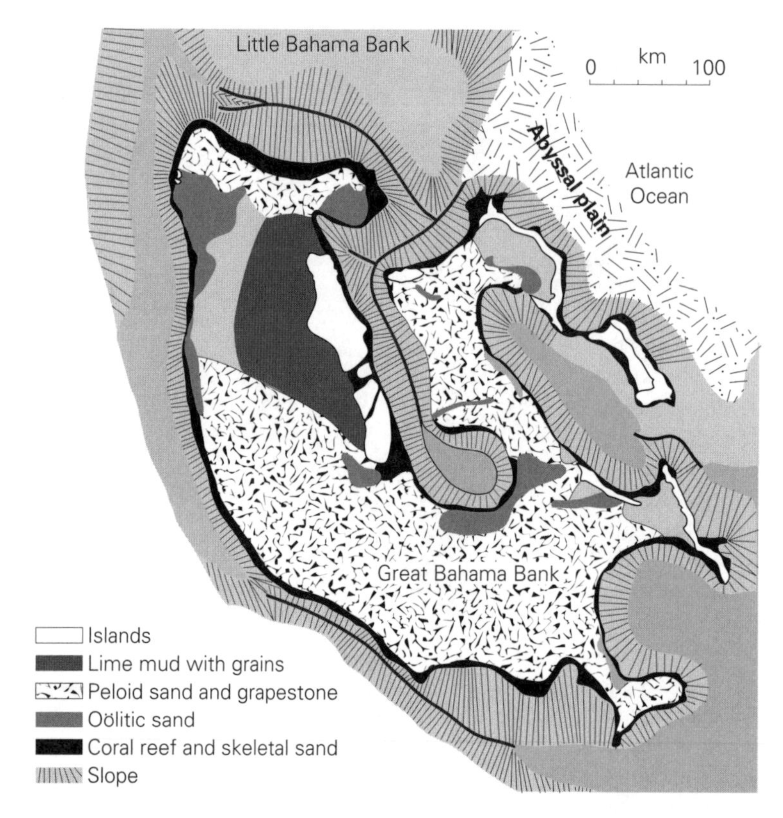

Fig. 7.3 Great Bahama Bank: sediment types, and basins. The Bank is a huge atoll, rimmed by reef structures of ancient origin. Carbonate platforms are surrounded by U-shaped, flat-floored basins, V-shaped canyons, and abyssal plains. See Fig. 7.1 for island orientation, shown in white. Modified from *Marine Geology* 44, W. Schlager & R. N. Ginsburg, Bahama carbonate platforms – The deep and the past, pp. 1–24. Copyright 1981, with permission from Elsevier Science.

the formation of large, submerged banks and smaller, emergent island systems, separated by gullied slopes and canyons that form a drainage system to the abyss. The Great Bahama Bank is the largest of the atoll-like formations (Fig. 7.3). During the Pliocene period, this Bank began to change from a giant reef-rimmed platform to a flat-topped configuration covered with a variety of sands, as a result of organic growth, wind, and circulation. The platform was exposed during the glacial Pleistocene period, when sea levels dropped at least 120 m, subjecting the dry land to solution and erosion. Sea level rose beginning about 10 000 years ago (the Holocene) when biogeochemical and physical processes, including reef formation, provided sufficient upbuilding to maintain the platform at or near rising sea levels. These processes formed an approximately 5 km-thick carbonate cap on the Bahamas platform, which is presently greater than the overlying ocean is deep and, at 1.5 million km^3 in volume, is among the largest carbonate deposits on Earth.

The Bahamas of today remain a vast carbonate factory. Limestone continues to form, as they did millions of years ago, from an interplay of biogeochemical processes, precipitation, evaporation, plant and animal growth, skeletal parts of corals and coralline plants, feces, decay, wind, and erosion. Bahamian land is mostly composed of nearly soil-less karst, a porous carbonate typography formed by solution processes and characterized by sinkholes and caves. Karst is produced when water circulates through rock, serves as a solvent, and forms channels. Water circulation encourages vegetation growth and soil microbial activity, which add acidity and enhance dissolution of rock. When sea levels were low during the Pleistocene, solution processes in the porous limestone expanded into numerous caves. Cave roofs often fell, exposing the cavern at the land surface. Here, water could collect to form a "blue hole" (Fig. 7.4). Many blue holes have sea connections and can be affected by tides; some are almost perfect circles, presumably resulting from tidal circulation that forms a vortex. Blue holes occur on land and on shallow banks (Fig. 7.5). Dean's Blue Hole on Long Island is approximately 215 m deep, the deepest known.

There are no true surface-water streams in the Bahamas, due to the overwhelming occurrence of porous limestone. Rainfall is the sole source of fresh-

Fig. 7.4 Uncle Bill's Blue Hole, north Andros.

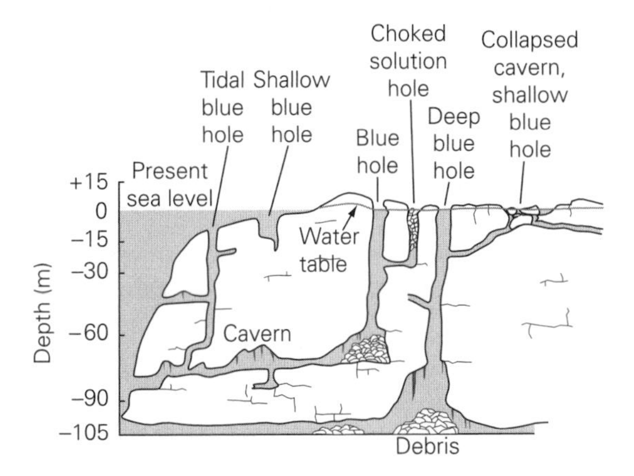

Fig. 7.5 Profile of blue holes on land and below sea surface. From Sealey (1994), with permission.

water and penetrates the karst to collect in lenses that float over denser, saltier water below. Freshwater may exit to the sea at seeps within the intertidal zone or along the reef fringe. Some tidal creeks assume estuary-like conditions when freshwater merges with seawater. Summer is the wet season, when trade winds are laden with moisture, and any disturbance leading to the convective uplift of air can cause rain. The largest islands generate the most heat and create the largest convection currents; thus, the largest island, Andros, collects abundant rainfall. Rainfall decreases from north to south possibly because of the lack of large, warm banks southward and decreased convection. Summer hurricanes and tropical storms generate about one-quarter of Bahamian rainfall, which is deposited over very large areas; high wind speeds generate large storm surges, at times placing the Bahamas at serious risk.

Sediment formation over the banks involves complex biological and physical processes. Small, fine-grained, oval pellets called peloids are formed largely from cemented fecal pellets of small organisms, and usually include carbonates. Oöids are egg-shaped grains, formed of concentric layers of carbonate around nuclei of shell particles or fecal pellets. Oöid formation occurs when the grain is held in suspension by turbulence, and a temperature increase causes a form of calcium carbonate to precipitate as crystalline aragonite directly from seawater that is almost saturated with carbonate. Skeletal sand is a third sediment type, consisting of carbonate particles derived from corals and coralline algae. Other sediments include organic muds, clays, and silts. All of these types are continuously redistributed and sorted by winds, currents, storms, and hurricanes; all can cement together after deposition to form aggregates of various sorts. Muds and oölites predominate on shallow banks near land; grapestone (lumps of cemented oöids) and/or skeletal sands from shells or corals are more abundant on deeper banks. Sediment also accumulates on beaches and forms dunes, or cements into rock from interactions of water and heat. Rock formation is relatively rapid and often incorporates coral skeletons into the rock. These interacting biological, chemical, and geological processes define the Bahamas as a system that depends to a large extent on living organisms for formation of its land- and seascape.

7.2.2 Biodiversity patterns

The Bahamas, with the Turks and Caicos Islands, is surrounded by the Atlantic Ocean to the east, the Straits of Florida on the northwest, and the Old Bahamas and Nicholas channels south and westward (Fig. 7.1). These water bodies form effective ecological barriers. Semi-isolation of the Bahamas archipelago justifies it as a subprovince of the West Indian biogeographic region. Bank and island structure pose the potential for a study of island biogeography. Terrestrial and shallow-water environments fall into three large-scale categories: large island-bank systems (Great and Little Bahama Banks), associated island groups with smaller bank areas (e.g., Cay Sal, Crooked and Aklins Islands), and isolated islands for which bank extent is small (e.g., San Salvador). Each of these groupings may be further subdivided according to (i) sizes and distances among islands, and (ii) arrangements of land–seascapes. This forms the basis for a hierarchical,

Box 7.1 Major Bahamian land- and seascapes

Coral reefs. The Bahamas' most diverse ecosystems. Protect low-lying coastlines from wave action, provide habitats for fish and other sea creatures, and have esthetic value for Bahamians and tourists.

Pine forests. Also known as "pineyards," "pinelands," or "pine barrens." Protect soil and freshwater lenses and provide habitats for many plants and animals, including the endemic Bahama parrot.

Mangrove forests and flats. Occupy about a third of the Bahamas' total land area. Hold sediments in place, minimize flooding and erosion, and help build land. Also provide nursery habitats for many marine animals (including commercial fishery species) and habitat for waterfowl and other fauna. Occur mainly in protected locations on leeward coasts. The Bahamas has about 4286 km^2 of mangrove forest and other wetlands.

Hypersaline ponds. Occur as isolated bays closed off by reefs, mangroves, or sand bars. Microalgae are dominant plants, restricted to mats or dispersed in along the edges. Animals include crabs, insects, brine shrimps, and wading birds. Anaerobic decomposition may release sulfides. Subject to disturbance by heavy rainfall or flooding.

Seagrass beds. Several species, intermixed with benthic algae. Stabilize the sea bed, provide nursery habitats for many commercial fisheries, and are a primary source of food for grazers (invertebrates, fishes, sea turtles, manatees).

Coppice. Comprised of dense, upright, mixed hardwood tree species. Provides habitats for orchids, bromeliads, birds, snakes, iguanas, and crabs. Coastal coppice occasionally floods; on windward coasts receives salt spray and may be shaped and sculpted by wind.

Blue holes. Provide habitats for unique animals. Ocean blue holes host commercially important fishes. Inland blue holes host photosynthesizing cyanobacteria around the edges and several different bacterial species at specific depths and levels of salinity; inland blue holes are also home to several unusual species of fish and invertebrates. Inland blue holes are easily disrupted.

Coastal rock. Occurs close to the sea, and comprises several low-growing and salt-tolerant shrubs, many of which are attractive. Iguanas and other lizards occur in this ecosystem.

Sand strands. Include beaches and strips among coastal rocks. Above the high water mark, plants are typically vines and spreading perennial species and shrubs. Sea oats may occur a little further inland.

Tidal flats and salt marshes. Characterized by saline soils that may be high in silt content. Typically covered with succulent, low-growing plants tolerant of high salinity.

Agroecosystems. Variable. Citrus groves typically comprise blocks of a single citrus species and a single variety, spaced to allow access for field operations. Groves may be irrigated and will certainly be fertilized. Pesticides control many pests, but may also harm benign insects and spiders, as well as birds that help control some of the pests. If inputs of fertilizer, pesticides, and weed control are not provided, the grove will slowly revert to "bush."

Source: Correll & Correll (1982).

biophysical classification of species–environment associations – a task yet to be attempted.

Bahamian environments are subject to constant change from the combined effects of winds, currents, sediment, and living organisms. Land- and seascapes (Box 7.1) occur in rather predictable patterns. The highest land on most islands is windward (usually eastward), where lagoons and reefs also occur. Low-lying wetlands and banks are generally in the lee (westward). North-central islands once supported extensive hardwood and pine forests, and soils and climate are suitable for agriculture. Southern islands are dry and windy, and aerosols are salty, and mostly support stunted tree and scrub communities. Geography, climate, isolation, and the nature of landscapes have provided a rich context for speciation and endemism. Species diversity is relatively low on land, compared with other Caribbean areas. However, endemism is high. An outstanding example is the peanut snail (*Cerion* spp., Fig. 7.6A,B), which takes different forms on individual islands; this species has been called an example of "evolution in progress." Endemic rock iguanas also vary among the islands (Fig. 7.6C,D); all are depleted, but continue to exist in small numbers in scrub and hardwood coppice.

In contrast to land, nearshore marine environments support high biodiversity (Fig. 7.7). These environments tend to become more species-diverse as water depth increases from creeks to reefs. Mangroves grow

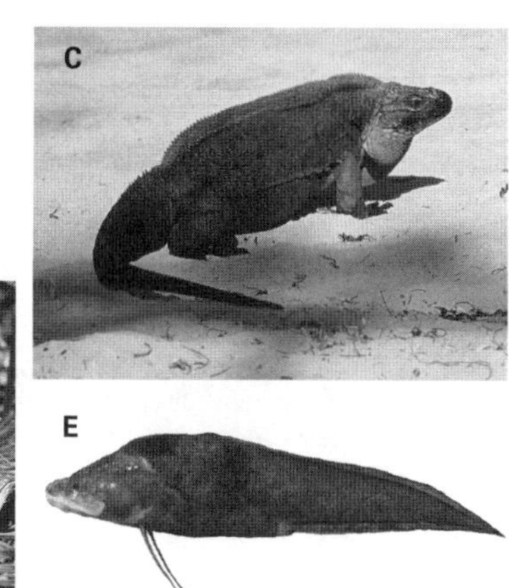

Fig. 7.6 Unique Bahamian fauna: (A, B) peanut shells (*Cerion* spp.) from Great Inagua and Andros, respectively; (C, D) endemic rock iguanas (*Cyclura cychlura* and *C. rileyi*, respectively); (E) a brotulid fish (*Lucifuga spelaeotes*) discovered in a blue hole on New Providence Island in 1968; this individual is from Andros. Photographs by the authors.

Fig. 7.7 The nearshore complex of environments on the eastern side of North Andros: (A) sand- and mudflats; (B) mangrove-lined creek; (C) shallow-water lagoon where ledges and patch reefs occur; (D) reef crest; (E) deep ocean. Vicinity of Staniard Creek, Andros. Photograph by the authors.

most lushly along creeks with adequate circulation. Red mangrove (*Rhizophora mangle*) is the dominant pioneer plant in saltwater ponds, creeks, and bights. Black mangroves (*Avicennia germinans*) are scattered in shallower water. White mangroves (*Laguncularia racemosa*) and button bush (*Conocarpus erectus*) occur landward. Interspersed among these are tidal flats and salt marshes where crabs, insects, brine shrimp, and wading birds can be abundant. The Bahamas' national bird, the West Indian flamingo (*Phoenicopterus ruber*), inhabits shallow, saline wetlands; Inagua's population of more than 50 000 birds is the largest in the Caribbean region. Seagrass and algal beds occur mostly in shallow creeks, lagoons, and bights less than 10 m deep. These habitats provide nursery areas for many commercial fish and are major feeding grounds for green sea turtles (*Chelonia mydas*). Patch reefs surrounded by sand occur in lagoons. Deeper reefs occur at the shelf edge where they form spectacular spur-and-grove formations before descending steeply as a "wall." Coral reefs not only protect low-lying coastlines from wave action, but also provide habitat for sea life and valuable fisheries.

Within shallow-water habitats, fishes occur in a hierarchy of assemblages among mangroves, seagrasses, and coral reefs (Fig. 3.23). Small fishes such as demoiselles (Family Pomacentridae) and butterflyfishes (Family Chaetodontidae) are reef-dwellers. Grunts (Family Haemulidae) depend on shelf habitats that contain seagrasses for nocturnal feeding and reefs for diurnal resting. Snappers (Family Lutjanidae) inhabit both reefs and mangroves, for different purposes. Juveniles of many reef species depend on mangroves and seagrasses, and adults become reef inhabitants. Others such as barracudas (*Sphyraena barracuda*) are ubiquitous. Most important commercial species are members of these shallow-water associations at different life-history stages, for example the Nassau grouper, the Bahamas' most popular and valuable fish, and the queen conch (*Strombus gigas*).

Analyses of Bahamian species diversity and biogeography are imposing tasks. About 95% of the Bahamas' area and most of its habitat diversity are marine, where diversity varies both horizontally and vertically. Caribbean regional biogeography is best illustrated for fishes, the best-known fauna (Box 7.2). The Bahamas is the largest Caribbean insular area, and contains the largest insular fish fauna, of over 600 known species. One approach toward biogeography would be to classify banks and islands into groups according to their sizes,

Box 7.2 Regional diversity among Caribbean fish species

C. Richard Robins

The numbers of species of marine fish known to occur in a region vary greatly from place to place, even in restricted geographic areas. Reasons for such differences are rooted in differences in ecology, climate, geological history, and geography, and their interplay. The Caribbean region's fish fauna is relatively well studied, and because the region is relatively small the transitions between different systems are sharp.

The Caribbean region includes the Bahamas, southern Florida Keys, southern Gulf of Mexico, and the entire Caribbean basin. Almost 1400 fish species occur in coastal waters shallower than 200 m. This represents nearly 10% of the world's marine-fish fauna, but excludes pelagic and deep-water species whose early life-history stages may occur inshore and pelagic fishes that commonly enter shoal waters where steep underwater profiles exist. Distributional patterns of individual species within this region show wide variation, but are of four principal types. Some fish species occur essentially throughout the region and may be termed *ubiquitous*. The *northern continental*, *southern continental*, and *insular* fish faunas cluster geographically, bathymetrically, and ecologically. Effects of freshwater runoff, sedimentation, and seasonal climatic shifts are important environmental factors used to define continental elements. The insular area is marked by: stability of temperature, salinity, and water clarity; lack of runoff of terrigenous sediments; and, where broad shelf areas exist, by calcium carbonate sediments of both biogenic and chemical origin.

In two zones, continental and insular fish faunas are mixed. Large islands with mountainous relief, heavy rainfall, and rivers and bays (especially Cuba, Hispaniola, and Jamaica) have insular species and also provide sufficiently extensive coastal habitats that attract and support continental species. Such islands exhibit an unusually large diversity of fish species. Second, some mainland areas (the Caribbean coast of western Panama and Central America) have mountain ranges that parallel the coast. Drainage basins into the Caribbean Sea in this area are small and carry little sediment to the coast, and then only during the rainy season. This area is avoided by continental species and fits best, in an ecological sense, into the insular type, but with a reduced diversity of species. Within insular areas, the occurrence of fish species is correlated with geological structure, island height, extent of the coastal shelf (especially seagrass and sandflats), surge, and oceanographic features. Size of the land area is far less important than the area of the bottom lying above 200 m. Small Bahamian islands situated on a broad insular shelf have many more fish species than those that have virtually no shelf or grass beds. Grunts (*Haemulon* spp.) are examples of species that depend on extensive shelf habitat.

Any study of fish diversity depends on knowledge of the regional fish fauna. The list of fish species for any well-studied coral-reef area is commonly five times larger than the number of species that can be expected to be caught in a single intensive collection. Approximately 60% of the fish species collected using fish poisons occur in all collections for a given site, and they include nearly all of the species reported by divers. The other 40% shows little consistency from collection to collection, and accounts, in part, for the five-to-one disparity. Some are rare species whose distributional centers may lie deeper on the shelf. However, most are cryptic species living in the reef structure: for example, cusk eels (Ophidiidae), viviparous brotulas (Bythitidae), gobies (Gobiidae), combtooth blennies (Blenniidae), triplefins (Trypterigiidae), clinid "blennies" (Clinidae), and others.

In contrast, visual censuses, even by experienced observers, seldom record these cryptic species, which, when added up for all collections, may be common and comprise the majority of fish species of reef and hard-substrate areas. Little is known of such cryptic species, but reefs with prominent spur-and-groove formations and extensive caves clearly have more of these species than do poorly developed fringing reefs. Therefore, studies of diversity that depend on visual censuses, especially for coral-reef fishes, will detect only a small part of the cryptic fauna and are clearly biased.

Because the Caribbean region is not large, species diversity of the fish fauna for any given island or stretch of mainland coastline is dependent more on its habitat than its geographic location. An exception lies in a comparison of the northern and southern continental sectors, which are geographically defined. The fish fauna of the northern sector was affected by Pleistocene cooling, but that of the southern sector was not. Movements of the Caribbean Plate relative to adjacent plates and to the closure of the Panamic gap may be more relevant to some fish distributions in the southern continental sector. Fishes of the insular sector appear not to be isolated from the mainland by geographic barriers, and the distances from island to island may be too small for geographic barriers to operate. Collections of larval fish show that occurrence of species at a given area may be more indicative of settlement and survival than of dispersal abilities. However, this is equivocal as self-recruitment seems to be common for some species.

The distributions of a few Caribbean fish require special explanation. For example, an endemic Bahamian goby (*Gobiosoma atronasum*) has evidently been limited to Exuma Sound by Pleistocene cooling and resultant isolated refugia. Toadfishes (*Sanopus* and *Opsanus*) lack pelagic larvae and have limited distributions along the Caribbean coast of Central America. Otherwise, endemism among Caribbean fishes seems to be rare and an artifact of collecting effort.

Knowledge of the richness and diversity of Caribbean fish faunas lends insight into the variety of factors that control animal distribution. With knowledge of the life histories of representative species, distributions of Caribbean fish species should be predictable. If so, we will better be able to understand the importance of system changes that occur naturally or anthropogenically.

Sources: Colin (1975); Robins (1971).

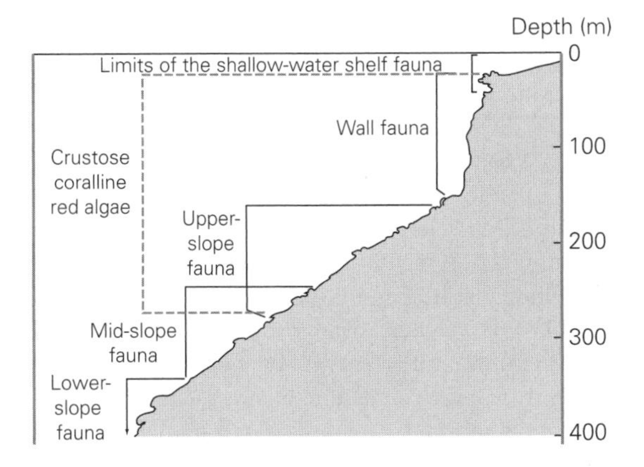

Fig. 7.8 Reef and slope communities extrapolated from research aboard the Johnson-Sea-Link submarines of the Harbor Branch Foundation, 1982–7, at San Salvador, Bahamas. Major faunal changes occur between 100 and 200 m, 200 and 300 m, and 300 and 400 m. Causes for these zonal divisions may be substrate slope, light, and sediment type. These zones probably occur throughout the Bahamas, although the depths may vary with prevailing ocean currents, water clarity, and sediment conditions. Adapted from unpublished work of R.G. Gilmore, C.R. Robins & R.S. Jones, with permission from C. R. Robins.

Fig. 7.9 Stromatolites in Exuma Cays Land-and-Sea Park. Some exceed a meter in height. Photograph by the authors.

on the supposition that larger banks will have many more species than isolated islands with small banks. Bathymetry could serve to define areas according to depth. For example, intertidal areas less than 10 m deep include rocky and sandy shores, tidal creeks, mangroves, ledges, sandy bottoms, seagrasses, and patch reefs. Most patch reefs occur from approximately 10 to 20 m. Reefs deeper than about 20–30 m drop as an almost vertical "wall" to at least 150 m. Although the biota are reasonably well known for the first two depths, comprehensive Bahamas-wide comparisons are not available. Regional comparisons of deep-reef fauna have only recently begun, and the rate of discovery of new species during the past half-century has been high. The blackcap basslet (*Gramma melacara*), one of the commonest fishes of the wall, was first described from a specimen collected in 1958 in the Exumas; the uncommon golden hamlet (*Hypoplectrus gummigutta*) was discovered at the same time and place. More recently, crustose, photosynthetic, red algae have been discovered to approximately 200 m, a depth record for tropical-reef photosynthesis, probably due to the clarity of Bahamian water. On the basis of limited observations from submersibles, fish assemblages for the wall, upper slope, mid-slope, and lower slope to about 400 m appear to be distinct (Fig. 7.8). However, the diversity and productivity of these areas will remain poorly known until sophisticated technologies, such as submersibles, are more extensively utilized.

Some distributions are difficult to explain. Large marine fishes can occur in blue holes and inland lakes, perhaps due to larval transport through cave systems. A brotulid fish (Fig. 7.6E), discovered on New Providence Island in the 1960s, lives in surface freshwaters of blue holes and appears to be widely distributed among several islands; how it could have spread among islands separated by wide seawater expanses remains unknown. And large stromatolites were discovered in the shallow waters of the Exuma Cays Land-and-Sea Park in the 1980s (Fig. 7.9). Stromatolites are among the most ancient life-forms. Formerly, they were known only from Western Australia. A variety of stromatolite types occurs in the Bahamas; all are products of cyanobacterial mats. Why they occur where they do is unclear.

In sum, a host of plants, corals, crustaceans, mollusks, fishes, sea turtles, and others have co-evolved with the continually changing Bahamian system. However, description of the diversity of Bahamian life is far from complete, and biogeography is even less well known. The variety of species–habitat associations that can occur in this large island system is enormous. One major gap concerns the extensive shallow-water banks, which have been presumed to be biologically impoverished. Yet, some bank areas support large vertebrates such as tarpon, bonefish, sea turtles, and extensive sponge populations. Banks can be rich in organic matter, microorganisms, algae, and infauna, perhaps derived from detritus and nutrients carried from wetlands by vigorous tidal currents.

7.3 Natural resource conservation issues

Conservation issues facing the Bahamas vary from local resource issues to national obligations under international agreements, and are linked to social factors such as population growth, freshwater, pollution, and waste management, which have major effects on environmental health and productivity. Fisheries (sport and commercial) require national regulation and management. And endangered species, such as sea turtles, are transboundary issues that require international cooperation. Most issues concern shallow-water environments. Reefs, mangroves, and seagrasses are linked environments that provide many services, extending from production of bank sediments, to protection of the land from erosion, to fisheries and to tourism. Equally critical is the role of biota responsible for the formation of the islands themselves. As much as for any other nation, the Bahamas is a "living system," for which the principal issue is sustaining biodiversity itself.

7.3.1 Coral reefs

Coral reefs are a major conservation priority because of their diversity, productivity, spectacular beauty, and importance to tourism. Modern Bahamian reefs have developed mostly during the Holocene, and are at intermediate stages of development. Coral reefs are presumably the most species-diverse of all coastal-marine environments – the marine equivalent of tropical forests – as they host a rich and complex community of invertebrates and fishes, and are visited by sea turtles, seabirds, and marine mammals. But despite popular attention and evidences of decline, the "coral-reef problem" is complex and requires close examination of ecology, regional seascapes, and an array of socio-economic factors and perceptions (Box 7.3).

Coral reefs are characterized by tightly linked symbiotic relationships, especially between coral polyps and zooxanthellae, which channel the reef's gross production into high community diversity and high energy efficiency, but also result in low net production (Fig. 3.14). In order to photosynthesize, zooxanthellae require clean, clear water with narrow temperature variation (20–30°C). Many reef biota variously inhabit surrounding seascapes and tidal creeks, depending on life histories. Hence, coral reefs are specialized and dependent structures that are exceptionally vulnerable to effects of human activities, such as overfishing, sedimentation, turbidity depletion of surrounding environments, pollution, and warming.

The Bahamas' reefs today are not those of yesteryear when fishes, sea turtles, and seals were extraordinarily abundant. Many reefs are in various states of decline, but for reasons that are not yet clear. Coral reefs thrive worldwide in nutrient-poor waters of subtropical–tropical latitudes. Reef-building (hermatypic) corals live near their upper lethal thermal limit, and elevated water temperatures may be a factor in recent reef decline. In the late 1990s, El Niño events warmed surface waters of the Bahamas sufficiently to cause many corals to "bleach" due to loss of their zooxanthellae. Also, the depletion of large fishes has no doubt taken a toll. These effects are interrelated and possibly synergistic, making principal causes difficult to identify.

Coral-reef conservation involves global, regional, and local aspects. At a global level, climate warming and nutrients from aerosols cause corals to "bleach" and to become diseased. Global transport of atmospheric dust may be implicated in disease (Box 3.1). The Caribbean-wide decimation of formerly dominant species during the past few decades, notably staghorn and elkhorn corals (*Acropora cervicornis* and *A. palmata*, respectively), represents a profound change that

Box 7.3 People and coral reefs

John C. Ogden

The complex and diverse assemblage of organisms that makes up coral-reef ecosystems covers approximately 284 000 km^2 (< 30 m deep) of the global tropical ocean. Should their vertical extent into deeper water be included, the total areal coverage would be greatly expanded. About 90% occurs in the Indo-Pacific and roughly 8% is in the Atlantic.

As the most species-diverse marine ecosystem, coral reefs are often compared to tropical rainforests. Reefs contain far fewer described plant, invertebrate, and vertebrate species than tropical forests, but contain almost twice as many phyla as all terrestrial ecosystems together. This rich diversity and the brilliant display of form and color make coral reefs one of the most attractive environments for the scientific study of origins and maintenance of biodiversity. Economic importance is not less. Fishing on or near coral reefs contributes approximately 10% of the world's present estimated production of 100 million metric tons of marine life, much of it critical to protein-poor, developing countries. Reefs are also important as physical structures, fostering the development of lagoon ecosystems and protecting coasts from the erosion of waves and storms of the open ocean. Thus, reefs represent significant scientific, economic, and ecological resources. Looking to the future, it is not an exaggeration to say that reefs may be the economic hope for many developing nations, through ecotourism and the discovery of new drugs and chemical compounds from their myriad species.

Modern coral reefs are the largest biogenic structures in nature. Coral reefs originated millions of years ago, but reefs of today represent approximately 6000–10 000 years of post-Pleistocene growth, during the most recent period of sea-level rise. They are essentially cemented piles of rock and sand, formed by the limestone skeletons of stony corals and other carbonate-producing organisms (notably algae). The living part of the reef is a thin veneer over the surface. Corals live in all seas, but the hermatypic, reef-building corals are restricted to shallow (< 100 m) waters of approximately 20°C or warmer; waters of more than 30°C can be lethal. Hermatypic corals are distinguished by a symbiotic relationship with photosynthetic algae called zooxanthellae, which live intracellularly in the individual polyps that make up the coral colony and give corals their color. The corals feed by capturing zooplankton using specialized stinging cells on their tentacles, providing their zooxanthellae with nutrients. In return, the coral receives oxygen and photosynthetic products from the algal cells. Therefore, healthy, growing coral reefs require high light, ample oxygen, low nutrient levels, and a narrow range of temperature and salinity in order to survive.

Coral reefs have evolved in association with episodic natural disturbances such as tropical storms, sedimentation from terrestrial runoff, and occasional harmful variations in seasonal seawater temperature. However, with over one-half of the world's population now living in the world's coastal zones, especially the tropics, coral reefs and their associated ecosystems of seagrass meadows and mangrove swamps are increasingly being affected by chronic human disturbances, such as: poor land use practices, including coastal deforestation, sediment runoff, pollution, and diseases; the direct and indirect effects of fishing and aquaculture; destruction of reef habitat by mining, dredging, and coral-collecting; and global climate change. This list is typical of human disturbances in every coastal region of the world, from the tropics to polar regions. An exception is introduction of alien species, which is a major problem in temperate estuaries, but does not yet appear to be a major problem for coral reefs, with the exception of those of the Hawaiian islands.

The distinction between episodic, natural disturbance and chronic human disturbance is fundamental. In the face of relentless coastal development, reefs cannot be expected to continue to support economies based on recreation and tourism without close attention to management schemes such as protected areas and fishery reserves, and coastal land-and-runoff management, all of which can minimize critical human disturbances. Coral bleaching is a special case of disturbance. When corals are subjected to unusually elevated or prolonged periods of seasonal high water temperatures (> 30°C), often in synergy with other stresses such as sedimentation or ultraviolet light, they expel their symbiotic algae and "bleach," turning pale or white. Bleaching is not necessarily fatal, but can kill corals if it persists. While bleaching has been known for a long time, since 1990 bouts of coral bleaching have increased in severity and extent all over the world. This has prompted scientists to propose that corals are the ocean equivalent of the miner's canary in the cage (whose illness or death signaled the presence of poisonous gases), giving an early warning of global warming in the oceans. Alarming recent events indicate that if the canary hasn't died, it may certainly be said to be very ill.

In 1997–8, corals bleached and died across the tropics, associated with the century's most intense and prolonged El Niño. The extent of coral mortality may never be known, but estimates of corals killed are well over 50% for many areas; in some areas virtually all the corals died. This massive and globally coherent event coincides with increased scientific certainty that oceans are warming. Obviously, coral bleaching and global warming cannot be addressed at the same local and regional scale of most other human disturbances to reefs. Unless greenhouse gases are managed by all nations, the future of coral reefs will remain uncertain.

Taken in sum, human disturbances have already extracted a heavy toll on global coral reefs. Recent studies of the status indicate that, by 1992, 10% of the world's coral reefs had been lost; by 2000, 27% were gone; and, by 2050, fully 70% may be lost. Several points emerge from this situation. First, the list of human perturbations to coral reefs is remarkably short and reasonably well known through scientific research. For any one location, disturbances to reefs, with the singular exception of coral bleaching, may be reasonably well identified and reasonable common-sense actions for controlling such disturbances may be suggested and implemented. Second, scientific programs will be necessary to track coral-reef recovery and/or declines and to allow adaptive measures to be implemented. Third, global action on greenhouse gas emissions is needed. Finally, large-scale and conservation areas, far beyond those already in place, are required for integrated protection, human use, and scientific studies of these remarkable structures.

Sources: Birkeland (1997); Spalding et al. (2001).

indicates a region-wide cause, although for reasons that remain unclear. Nationally, the degradation of coral reefs and reduction of reef inhabitants lessens attractions for tourists, reduces habitat quality for commercially important fishes, and exposes adjacent islands to forces of erosion. Locally, reefs are susceptible to pollution, sedimentation, anchoring of boats, excessive use by divers, and coastal development. Additionally, commercial and sport fisheries and tourists have removed significant segments of the reef community, especially fishes, corals, and shells. These effects, taken together, may increase stress on corals, which some scientists suggest might lead to loss of zooxanthellae and increased susceptibility to disease. Furthermore, removal by fishing of herbivorous fishes that feed on algae may result in seaweed overgrowth, which prevents coral recruitment. Thus, changes in the fish community can cascade to the coral community.

Caribbean coral reefs have waxed and waned for many thousands of years, surviving the vicissitudes of climate and sea level, and have recovered each time. High species diversity suggests that species, both animals and plants, provide fundamental roles in reef ecology, and helps explain reef resiliency. Fishes, for example, exhibit a spectacular variety of sizes, colors, shapes, life histories, and ecology. Many fishes depend solely (are obligate) on reef environments, and others are transients. Some form extensive schools, and some are solitary and territorial. Many are conspicuous, but as many are reclusive, particularly small gobies and blennies that hide in the nooks and crannies of the reef's rough and uneven habitat. However, explanations for reef diversity remain hypothetical (Box 7.4).

Reefs and their biota are most valuable to the Bahamas economy as an attractor for tourism. Reef fishes are especially sought for food and pleasure; reef-inhabiting species (e.g., lobsters and groupers) and seagrass dwellers (e.g., conch) supply a large portion of food for the Bahamas, as well as income from tourism. Nevertheless, knowledge of functional relationships among the biota and interdependencies with surrounding seascapes is poor. Presently, coral reefs are globally recognized as a research priority, and many nations, including the Bahamas, have established protected areas for them. The Bahamas forbids collecting of live coral, yet has no overall coral-reef conservation strategy. Further, protected areas that include coral reefs, but fail to include surrounding environments, are insufficient for conservation.

7.3.2 Nassau grouper, *Epinephelus striatus*

The Nassau grouper is the Bahamas' most valuable fish and typifies issues of fisheries and reef conservation. This species is listed as endangered by the International Union for the Conservation of Nature and Natural Resources (IUCN), but has not yet been listed under the Convention on International Trade in Endangered Species of Wild Fauna and Flora (CITES). As a top

Box 7.4 How do so many kinds of coral-reef fishes co-exist?

Mark A. Hixon

Coral-reef fishes compose the most speciose assemblages of vertebrates on Earth. The variety of shapes, sizes, colors, behavior, and ecology exhibited by reef fishes is truly amazing. Reef fishes are dominated by about 30 families, mostly the perciform chaetodontoids (butterflyfish and angelfish families), labroids (damselfish, wrasse, and parrotfish families), gobioids (gobies), and acanthuroids (surgeonfishes). Worldwide, about 30% of the roughly 15 000 described species of marine fishes inhabit coral reefs at some stage of their life cycle. Hundreds of species may coexist on the same reef at one time or another.

A key question for the conservation of coral-reef fishes is: How do so many species coexist? This question is important because conservation requires the identification and protection of natural mechanisms that maintain high species diversity. It is best answered at the level of the ecological guild, which is defined as a group of species that use the same general suite of resources (food, space, etc.) in the same general habitat, such as butterflyfishes that feed on coral polyps inhabiting a reef slope. The central issue is that, as population sizes of species within a community grow to levels where resources are in short supply, one or a few species within each guild should outcompete other species, thereby reducing local species diversity. What prevents such competitive exclusions?

Four hypotheses provide clues to the question of coexistence of reef fishes (Fig. 1). Present information both corroborates and refutes each hypothesis at different reefs, suggesting that all four hypotheses may be valid at some time and place.

A review of the bipartite life cycle of reef fishes is necessary before examining these hypotheses. Many reef

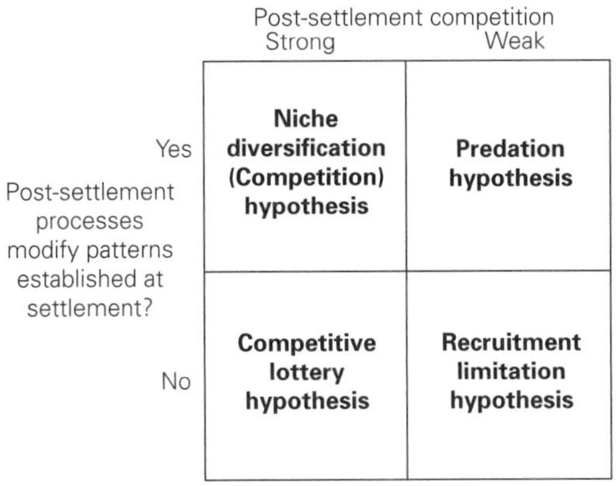

Post-settlement processes modify patterns established at settlement?	Post-settlement competition: Strong	Post-settlement competition: Weak
Yes	**Niche diversification (Competition) hypothesis**	**Predation hypothesis**
No	**Competitive lottery hypothesis**	**Recruitment limitation hypothesis**

Fig. 1 Four hypotheses explaining the coexistence of many species of coral-reef fishes. Modified from G. Jones, in Sale (1991).

fishes (exceptions are gobies, blennies, pipefishes, and a few others) are broadcast spawners, whose gametes and larvae undergo pelagic dispersal, with varying degrees of local retention. Typically, after about a month, late-stage larvae settle in reef or near-reef habitats. Recruitment is the measure of settlement, estimated by counts of newly settled fish. The accuracy by which recruitment actually measures settlement is a major issue in distinguishing among these hypotheses.

The niche diversification and competitive lottery hypotheses both assume that competition is strong among juveniles and adults on the reef, so that coexistence of species is maintained despite the risk of competitive exclusion. The basic idea for the former (sometimes called the "competition hypothesis") is that high overlap in resource use within a guild, combined with competition between the constituent species, selects for lower overlap or diversification of niches. This scenario results in resource partitioning, whereby species within a guild that overlap greatly in diet tend to forage in slightly different microhabitats; alternatively, species that forage in the same location may have slightly different diets. However, a description of resource partitioning provides only a pattern, not the process that caused that pattern.

Some guilds seem to coexist despite an apparent absence of resource partitioning. For example, territorial, herbivorous damselfishes are highly aggressive toward each other, and if all suitable habitat space is occupied by territories, how do such species coexist without niche diversification? The competitive lottery hypothesis (sometimes called the equal chance hypothesis) offers a relatively complex explanation, based on several restrictive assumptions. First, there has to be a strong prior residency effect, whereby a fish that finds a place to live on the reef can successfully defend its territory against all comers. Second, late-stage larvae of all species have to be available to settle in any space that opens on the reef, be it by the death of a territorial fish or by the creation of new habitat space by storms or other disturbances. These larvae are analogous to lottery tickets, in that whichever individual finds the open space first is the winner of that space. Under these conditions, it is proposed that no single species can gain the upper hand in the competition for living space, despite the lack of resource partitioning. In reality, the rate of competitive exclusion may only be slowed rather than prevented, since no two species are truly equal, by definition.

The remaining two hypotheses both assume that competitive exclusion of species is not an issue because some factor keeps population sizes below levels where resources become limiting. Some fish populations have low larval settlement rates, so that living space is not as limiting as the former hypotheses assume. The recruitment limitation hypothesis proposes that low larval supply prevents juvenile and adult populations from reaching levels where substantial competition occurs, in which case post-settlement mortality is density-independent – that is, occurs at a constant proportional rate. Unfortunately, the definition of recruitment limitation has changed through time, so that recruitment is sometimes measured up to months past settlement, and early post-settlement processes are thus ignored. In fact, shortly after settlement, many reef fishes undergo density-dependent mortality in which case mortality rate increases with local population size.

Finally, as an alternative to recruitment limitation, the predation hypothesis suggests that competitive exclusion is prevented by predation rather than low larval supply. In fact, both density-dependent and density-independent predation on newly settled reef fishes, which are typically less than 2 cm long, is usually severe. Many different species of generalized reef fishes and macroinvertebrates – mostly species not normally considered piscivorous – have been found to consume new settlers. There is mounting observational and experimental evidence that such intense predation keeps populations of many reef fishes in check, precludes competitive exclusions, and thereby maintains high local species diversity.

The picture that emerges from the past several decades of research on coral-reef fishes is that a variety of factors maintain high species diversity, and that the relative importance of these factors varies from system to system. This situation indicates the truth of John Muir's admonition that "when we try to pick out anything by itself, we find it hitched to everything else in the universe." Such complexity suggests that the conservation of coral-reef fishes can be best accomplished by preserving entire systems from direct human impact in fully protected marine reserves.

Sources: Polunin & Roberts (1996); Sale (1991, 2002).

Table 7.2 Summary of total recorded landings of marine products in the Bahamas, 1995–9. Later data (2000, 2001) do not indicate any notable changes. Landings do not indicate population sizes or trends.

Product	1995	1996	1997	1998	1999
Lobster (tails)	5 579 059*	5 711 903	5 674 127	5 478 508	6 026 508
	59 825 703†	52 666 139	58 669 158	53 364 247	62 592 798
Snapper (several species)	654 788	751 138	1 655 756	1 721 359	1 908 443
	658 607	1 001 784	2 303 289	2 363 558	2 388 552
Conch (fresh)	1 088 179	1 298 336	1 428 745	1 477 374	1 040 307
	2 106 925	2 715 510	2 942 065	3 651 628	2 619 768
Nassau grouper	788 369	729 719	1 132 264	1 125 817	841 044
	1 613 648	1 699 039	2 477 255	2 674 401	1 999 204
Other grouper	9 085	32 737	167 512	228 235	228 034
	16 216	64 417	365 099	460 581	426 670
Jacks	159 117	200 466	227 626	202 411	175 058
	150 188	226 336	220 602	216 381	184 849
Grunts	17 258	30 230	148 396	198 232	144 441
	12 521	26 129	121 516	155 601	104 916
Stone crab	87 212	55 639	92 801	85 126	109 559
	622 616	394 837	658 967	609 001	680 894
Grouper fillet	133 412	113 548	149 087	108 803	79 534
	394 817	349 051	438 563	327 422	259 594
Lobster (whole)	43 996	359 629	167 069	215 144	51 327
	156 345	1 342 257	677 626	776 233	221 908
Shark	0	9 900	6 013	4 312	3 202
	0	24 471	14 252	10 248	7 223
Green turtle	1 568	3 600	5 328	5 072	2 513
	1 615	3 661	5 923	6 571	4 336
Loggerhead turtle	3 826	2 000	1 690	2 052	744
	3 826	2 310	2 557	3 693	1 454
Other	840 755	848 778	581 004	343 214	307 156
	875 392	876 422	644 148	415 479	337 802
Totals	9 406 524	10 147 623	11 437 418	11 195 659	10 917 910
	66 438 419	61 392 363	69 541 020	65 035 044	71 829 968

* Weight in pounds. † Value in B$ (parity with US$). Data from Bahamas Department of Fisheries.

predator, the Nassau grouper plays an important role in the ecology of reefs. It is a favorite of snorkelers and scuba divers because of its size, inquisitiveness, and photogenic appearance. This fish has a high social value, being named after the Bahamas' capital, Nassau, and is honored on post cards, posters, television, and even in restaurants, where it is consumed as a fish of choice because it has relatively few bones and is easy to eat. Thus, it is not an exaggeration to say that the Nassau grouper has become a symbol of ecological health and social wellbeing of the Bahamas.

The Nassau grouper is exceptionally vulnerable to exploitation. The species, although endangered in most parts of its range, remains among the most sought-after fishes in the Bahamas and accounts for a large portion of total fish landings (Table 7.2). Although it remains relatively common in some areas, it is not as abundant as only two or three decades ago. Insufficient effort is dedicated to study of its ecological relationships and distribution of subpopulations, making its management uncertain.

7.3.2.1 Natural history

The Nassau grouper is a tropical–subtropical species that ranges from Bermuda and South Carolina to the Gulf of Mexico, and throughout the Caribbean to Brazil. It is distributed throughout shallow waters to depths of approximately 150 m. The life cycle is

Fig. 7.10 Life history of the Nassau grouper: (1) adult in bicolor phase; (2) spawning event, with dark-phase female in the lead; (3) newly fertilized, pelagic eggs; (4) approximately 3 mm, pelagic, 5-day-old preflexion larva; (5) 9 mm, postflexion larva; (6) 25 mm, transformed juvenile; (7) newly settled juvenile in inshore algapatch; (8) half-grown juvenile in rocky ledge; (9) subadult in reef. 1–6 are exaggerated in size. Adapted from Sadovy & Eklund (1999).

complex (Fig. 7.10) and ecological limiting factors are not well known. Water clarity and substrate type may be critical, and reef condition may play a part in determining adult abundance. Shallow-water algal flats are important for small juveniles, and nearshore ledges and patch reefs are favored by larger juveniles. Food is not necessarily limiting, as diet is varied for all life-history stages; larvae feed on a variety of zooplankton and juveniles and adults feed on an assortment of crustaceans, mollusks, and other fishes.

Nassau groupers mature at 4–8 years of age and at approximately 40–50 cm in length. They appear to maintain separate sexes throughout life, unlike many sea basses (Family Serannidae) that change sex from female to male as they grow. Almost all individuals are able to spawn at 7 years of age. Older animals continue to reproduce; large fish can produce up to 5–6 million, 1 mm-diameter eggs in a season. The maximum recorded age is 29 years, maximum length is 120 cm, and maximum weight is 23–7 kg.

Spawning occurs in large aggregations. In the Bahamas, at least 23 aggregations apparently once occurred (Fig. 7.11) – almost half of all such sites throughout their Caribbean range. Fishermen have known of these aggregations since at least the turn of the 20th century, but whether all reported aggregations are spawning groups is unclear. Well-documented accounts from past decades have reported as many as 100 000 individuals at a single site. Some aggregations seem consistent in location and numbers from year to year, others are in decline, intermittent, or have apparently disappeared. It is possible that groupers shift aggregation locations from year to year, or within years.

In the Bahamas, Nassau groupers aggregate for about a week during the full moons of November through February, when water temperature is approximately 24–6°C. The bright moon may provide a clue for coordination among individuals, which are scattered thinly in reef environments during most times of year.

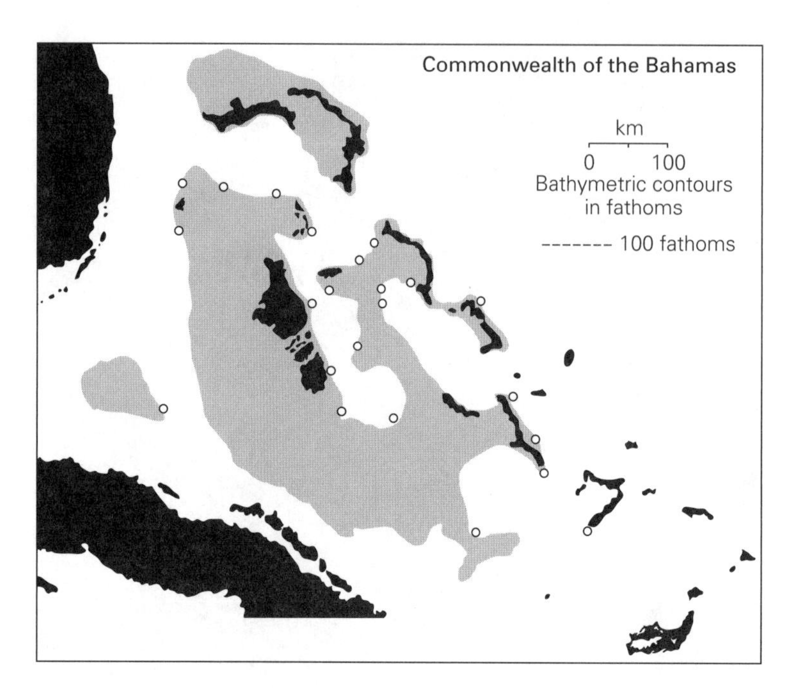

Fig. 7.11 Twenty-three formerly known Bahamian Nassau grouper aggregation sites. The current status of most of these sites is not confirmed. Some are apparently depleted, a few may no longer exist, and some probably remain to be discovered. Adapted from Sadovy & Eklund (1999), after data recorded in 1992 by the Bahamas Department of Fisheries.

To reach a spawning site, groupers swim along the reef in groups ("corridas") of tens to hundreds of individuals, beginning a few weeks prior to spawning. Aggregated individuals exhibit four color phases: barred, white-belly, bicolored, and dark. Spawning occurs at about dusk during a 2–4 day period, when groups of approximately 3–25 individuals, led by a dark-phase, presumed female, rush upwards to near the surface, where eggs and sperm are released. The same individuals may spawn repeatedly during one full-moon period, but whether the same individuals spawn during successive spawning periods is not known.

Aggregation sites exhibit a diversity of features, although few sites have been examined to determine common characteristics. One site, High Cay, Andros, is located at the edge of a wide lagoon on the crest of a steep drop-off to the deep Tongue of the Ocean (Fig. 7.12). According to local people, the aggregation site once extended about 1 nmi, from High Cay south to Long Rock, an area large enough for several thousand spawning fish. The site appears now to be restricted to a hectare-sized area near High Cay. In plan view, the profile of the barrier reef turns south at this area; in cross-section, the profile is that of a promontory. The hard, carbonate-rock bottom of this promontory is uneven and rough, with abundant small coral heads, ledges, very large sponges, and gorgonians. Predominant currents run along the reef from either north or south at less than 1 to more than 3 km h^{-1}. Strong downwelling may also occur. Seas at this location are notoriously rough, especially in winter. Waters appear to be well mixed to a depth of about 150–200 m, where temperature drops approximately 2°C. All of these conditions indicate high turbulence, favorable for fertilization of eggs and possibly for local retention of fertilized eggs and larvae.

The fundamental question is: Why aggregate? The answer lies in the grouper's natural history. Groupers are periodic spawners (see Box 4.2, pages 106–7) that mature at large body size and broadcast large numbers of small eggs that are fertilized in the water column. The grouper's long life allows reproductive effort to be spread over several years. Reproduction at one time and place can risk losing an entire spawning event should environmental conditions be unfavorable. Thus, high survival of larvae in one year may compensate for low survival in others. Aggregation under favorable conditions optimizes fertilization, facilitates dispersal, minimizes egg predation, maximizes chances that larvae can find food, maximizes opportunities for recruits to settle near their point of origin, and reduces natural predation on spawners. Furthermore, the behavior of aggregating fish exhibits considerable social interaction, in which group size may be important. If numbers of

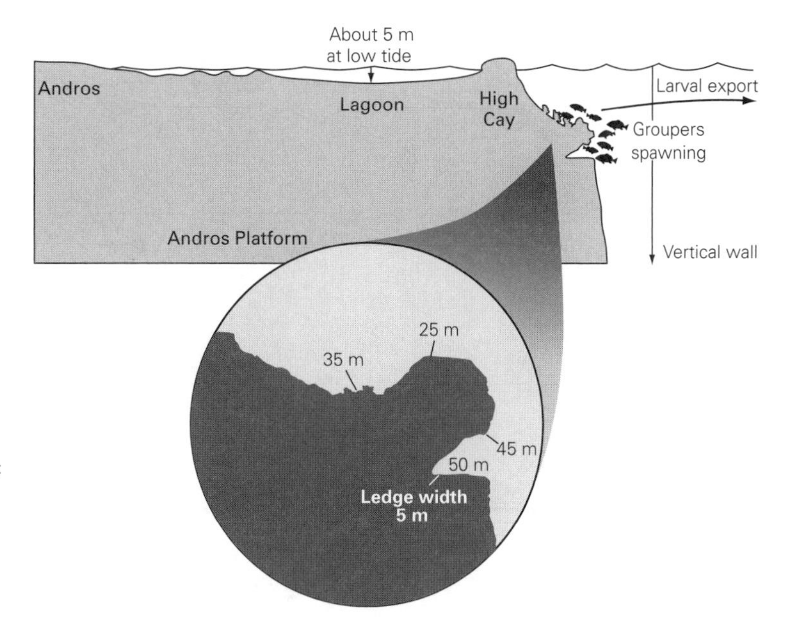

Fig. 7.12 General impression of grouper aggregation site at High Cay, Andros. The distinct promontory is the site of the aggregation. Strong currents and turbulence at this site could strongly affect larval dispersal, and possibly may play a role in self-recruitment of larvae to the same general area where spawning occurred. Illustration is not drawn to scale.

individuals are too small, aggregation may not occur, social behaviors may not be transmitted, and learned memory of migration pathways and aggregation sites may be lost. Conversely, if an aggregating group is too large and too densely packed, individuals may compete for resources, succumb to disease, or even self-pollute their environment. These trade-offs suggest that a critical minimum number of adults must be available to trigger reproductive behavior and that group size and social behavior could be critical to optimize reproductive efficiency.

7.3.2.2 Habitat

For species, such as the Nassau grouper, with complex life histories, a critical mosaic of healthy habitats must be available to assure optimum settlement and growth. Fertilized eggs are subject to local conditions of currents and turbulence, and hatch within a day after fertilization. Larvae 2 weeks old are able to swim, at least weakly, allowing some capability for self-directed movement. Larvae remain in the plankton for about 6–7 weeks and metamorphose into juveniles at approximately 23 mm in length. Mortality is high in pelagic habitats. Whether eggs and larvae drift passively in currents and are dispersed widely, or whether larvae are locally retained and recruited back to the same reproducing population, is unknown. The weight of scientific evidence to date points to self-recruitment, that is, that aggregation sites might be located so as to enhance some degree of local retention of non-mobile eggs and weakly mobile larvae. If so, Nassau groupers in the Bahamas would be structured as a metapopulation, in which subpopulations provide a hedge against depletion.

Sites for larval settlement are also critical. Settlement is influenced by habitat availability, especially macroalgal and seagrass beds. Settlement timing may be correlated with the new moon, when young fishes seek shelter under protection of darkness, and may occur mostly during brief pulses of cross-shelf winds that could assist inshore movements and provide cues for habitat choice. Following settlement, juveniles face trade-offs between predation and growth. Settlement at first favors safe haven from predation, but as the young fish grow they move to larger shelters such as ledges and shallow-water patch reefs where food is more abundant. Only as adults do they inhabit deeper reefs.

The suitability and proximity of all environments necessary for the Nassau grouper during its entire life history represent potential bottlenecks for survival. That is, if any one habitat is lacking, reproductive success and recruitment may be poor. Although the Nassau grouper is considered a "reef fish," this is a misnomer, as creeks, bights, lagoons, and ledges are important nurseries for juveniles. Over much of the Bahamas, creeks and inshore environments are plentifully supplied with plant growth and benthic fauna. Many juvenile fishes that occur in creeks are inhabitants of reefs

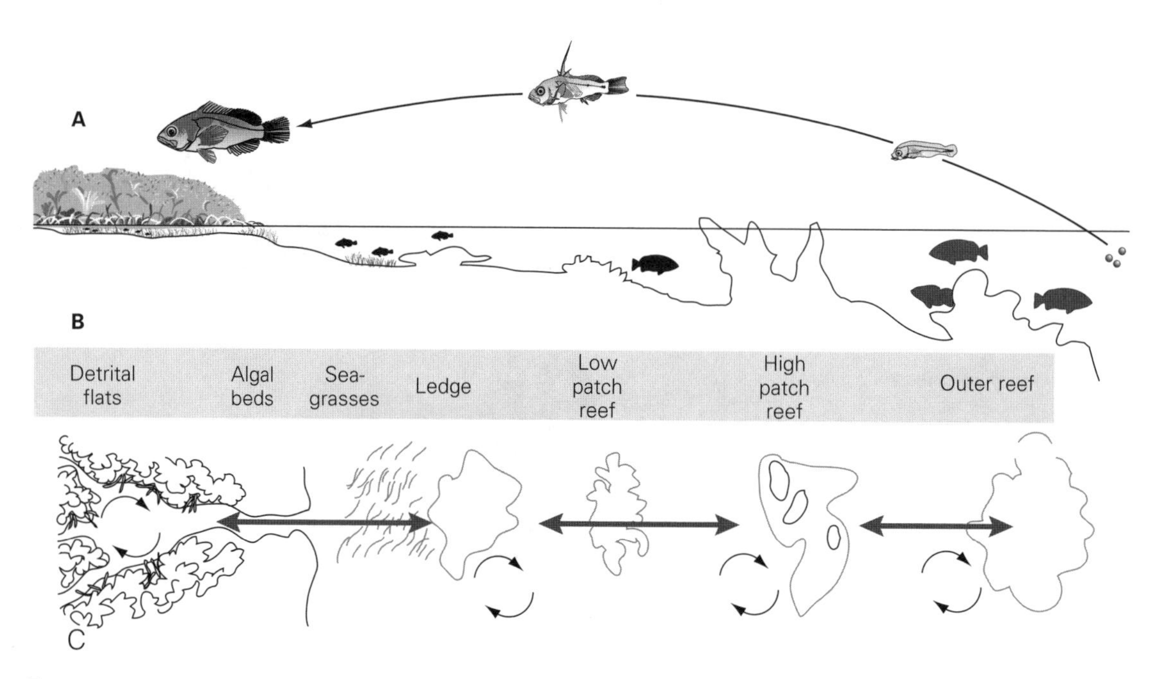

Fig. 7.13 Spatial interrelationships among Nassau grouper habitats may reflect hierarchical food-web relationships, as indicated by the arrows. Profile view (A) and plan view (B) of the Nassau grouper's life history (see Fig. 7.10) associated with habitats. The grouper requires habitats from outer reefs where reproduction occurs, to creeks where juveniles settle, and ledges and patch reefs as they grow.

as adults. Creeks extend into island interiors, where mangroves, sand- and mudflats, and rocky shores are well developed along their borders. The extent to which Bahamian creeks connect biologically and ecologically to reefs is unknown. Research to date indicates that juvenile fishes in creeks feed on invertebrate detritivores (eat dead or decaying plant material and macroalgae), suggesting that creek food webs are detritus-based. Thus, conservation of reef fishes whose juveniles live in creeks requires sustaining detritus-producing mangroves, seagrasses, and benthic algae that sustain an invertebrate food supply. This allows formulation of a hierarchical food-web hypothesis for relationships among creeks, lagoons, and outer reefs relevant to the Nassau grouper. (Fig. 7.13).

7.3.2.3 *Effects of fisheries*

The Nassau grouper is especially vulnerable to overfishing, but is now rare and only incidentally caught in most Caribbean waters. Of 50 known aggregations in the wider Caribbean region, 10 have disappeared and most that remain are overfished, with little sign of relief or recovery. For the Caribbean as a whole, the spawning stock biomass has been reduced to only 1% of assumed pre-exploitation levels. In the Bahamas the Nassau grouper still exists in reasonable numbers, but the same practices that caused declines elsewhere continue to be practiced. Hookah-aided (air transmitted from a surface line to a diver) spearfishing on spawning aggregations is especially deleterious, as it permits repeated dives that disturb natural behaviors and rapidly deplete the population.

The Nassau grouper is long-lived, large-sized, slow-growing, late to mature, curious, easy to catch, tasty, and aggregates to spawn. It exemplifies a top predator that is distributed sparsely during most of the year, and that naturally accumulates large standing stocks of old fish. Thus, fishing is initially highly successful and rewarding. As local populations are reduced, density-dependent reproduction presumably increases, but as fisheries extractions increase, portions of the meta-population may be severely overfished. Fishing generally targets the largest fish; therefore, overfishing affects the average size of fish, the male : female sex ratio, and age structure, all of which act to reduce population growth. Furthermore, heavily fished breeding aggregations have almost always failed within a few years, and may not be re-established.

Optimum population size for Nassau groupers has yet to be determined. Deterministic population-dynamics models that have been developed mainly for temperate or pelagic schooling fishes do not apply to fishes with natural histories similar to the Nassau grouper. Furthermore, fisheries models usually assume equilibrium (density-dependent) conditions in order to estimate sustainable yields. Application of such models to groupers, which forces lumping of the entire metapopulation into a single "stock," omits many aspects of this species' dynamics. Rather, models must also take account of habitat selection and availability, behavior, and interactions among subpopulations.

7.3.2.4 *Implications for conservation*

Since 1998–9, fishing for Nassau groupers at a few aggregation sites has been restricted, including at High Cay. This restriction has been limited to small areas surrounding each site and for short periods before and after the full moons of November through February. A closed season is also being considered. A serious impediment for conservation is lack of information. Among the many questions that challenge management are:

For natural history:

- Are Bahamian Nassau groupers one or many populations?
- How large is the area from which aggregating individuals originate?
- What were the actual sizes of so-called "traditional" (historical) aggregation sites and where were they?
- Do the same individuals return to the same site seasonally or yearly?
- What is the minimum viable population size and optimum sex ratio for reproduction and recruitment?
- What factors influence dispersal and survival?

For habitat:

- What minimum viable habitat area is required to support a minimum population size?
- What optimum mix of habitats and spatial and temporal dimensions is required for population maintenance?
- What suite of characteristics typify aggregation sites?
- To what extent may habitats change over time?
- How threatened are these habitats?
- What human social conditions might permit reserve establishment?

The lack of answers to these questions increases the vulnerability of the Nassau grouper to fishing and hence the probability of depletion of this important species, with consequences for the Bahamian people, the economy, and the reef community.

7.3.3 Green turtle, *Chelonia mydas*

All species of sea turtles are considered to be endangered or threatened, although data for some are deficient. The green turtle has received most attention because of its history of exploitation, its tasty flesh, and the work and writings of one of the most prominent natural historians and scientists of the 20th century, Dr. Archie Carr. Only a few centuries ago, the green turtle was probably one of the most abundant large vertebrates in the world. Andres Bernaldez, writing about Columbus's second voyage in 1494, described Caribbean green turtles as "so numerous that it seemed that the ships would run aground on them and were as if bathing in them." (Cunningham, 1997) This species' life history and wide range place it among species such as tunas, billfishes, and whales that require international conservation. Although its former ecological importance may never become fully known, this large, wide-ranging herbivore no doubt played an important role in the structure and growth of tropical seagrass meadows.

7.3.3.1 *Natural history*

The green turtle attains at least 1.5 m in shell length and up to about 350 kg in weight, a size now rarely recorded. This turtle is semi-gregarious and far-ranging in tropical and subtropical seas. Similarly to the Nassau grouper, it aggregates to lay eggs, but does so on sandy beaches where it is extremely vulnerable to human exploitation, as well as to predation by egg-eating, terrestrial predators.

Research in the Union Creek Reserve on Great Inagua Island has indicated that population regulation is density-dependent – that is, population numbers under natural conditions are regulated by food limitations. This conclusion is supported, first, by the finding that population density and mean annual growth rate are inversely related. Also, an index of condition (ratio of carapace [upper shell] length to weight) is positively correlated with mean annual growth rate, but negatively correlated with population density.

Female green turtles lay more than 100 round, leathery eggs, a bit larger than ping pong balls, in nests that they dig with their hind flippers in the sands of tropical beaches (Fig. 7.14A). Nesting occurs in early to

Fig. 7.14 (A) Green-turtle nesting beach on the lee (western) side of Little Inagua, Bahamas. (B) Beach on southeastern side of Great Inagua, formerly a green-turtle nesting site, but so littered today that turtles may no longer nest there. Photographs by the authors.

mid-summer. Green turtles formerly aggregated to nest in large numbers, but in their presently depleted state they may nest in small groups or even singly. The exception is Costa Rica's Tortuguero Beach, which is the largest nesting area in the Atlantic by an order of magnitude and where green sea turtles appear to be increasing in numbers. Females choose nesting beaches according to conditions of slope, exposure, expanse, vegetation, grain size, and depth. Hatchlings emerge in approximately 60 days, with a carapace length of about 5 cm and weighing about 25 g. They immediately scramble to sea to avoid predation by seabirds, raccoons, and others, where they disappear into a "lost year" – for a time of unknown duration and at a location that has yet to be determined. While in oceanic habitats, they are carnivorous. When the juveniles are about 25–35 cm in carapace length, they appear on shallow-water foraging grounds and become herbivores. They may take as long as 40 years to mature sexually.

Grazing by green turtles can exert profound influences on the structure and growth of seagrass meadows. In the Atlantic, green sea turtles feed primarily on seagrasses, particularly turtle grass (*Thalassia testudinum*). Groups of turtles maintain grazing plots and continually re-crop grazed areas. This initially stimulates seagrasses to grow rapidly, providing a high-quality diet for the turtles and facilitating nutrient recycling within the meadow. When turtle density is high and/or seagrass density is low, grazing plots may merge. Overgrazing causes seagrass growth to decline, and nutrient stores to become depleted, forcing the turtles to migrate to other feeding areas. Thus, a cycle is established, in which seagrasses are initially stimulated, which promotes a dense turtle population, followed by a decline in the food resource, and possible dispersal of the aggregation to new grazing areas.

The Bahamas appears to be a major feeding area for green sea turtles. Carrying capacities of seagrass meadows for green sea turtles vary widely, from about 120 to 4400 kg of green turtles per hectare. Although extrapolations must be handled with caution, this biomass estimate may indicate that 16–586 million green sea turtles may have inhabited the Wider Caribbean Region in pre-Columbian times.

7.3.3.2 Implications for conservation

Late maturity and long life qualify the green turtle as a species "in the slow lane" – that is, recovery from depleted status to reasonable abundance may take many years, even decades. Recovery is problematic, however, because the green turtle's population extends throughout the Caribbean and even into the Atlantic Ocean, where conservation measures cannot be assured (Box 7.5). Therefore, conservation must extend beyond national boundaries if it is to be successful.

Sea-turtle conservation varies widely among Caribbean nations. Regulations in the Bahamas permit green turtles more than 24 inches (~ 61 cm) carapace length to be landed whole for food, except during the nesting season from April through July. No eggs may be taken and no sea-turtle products may be exported; this restriction is required by CITES, to which the Bahamas is a signatory. Nevertheless, from the time of establishment of the Union Creek Reserve in 1961, no further protection has been afforded for sea-turtle habitats, not even seagrass meadows, despite their value for both sea turtles and juveniles of many species. Breeding beaches where sea turtles have the potential to occur in some abundance deserve protection as do their feeding areas. Quality of beaches as breeding habitat is being lessened by the spread of Australian pines (*Casuarina*, Fig. 8.4, page 250), which destroy beach habitat, and by extensive flotsam and jetsam from the sea, which impede turtle nesting (Fig. 7.14B).

Although conservation of green sea turtles is a very high international priority, no comprehensive international conservation strategy exists that responds to their total natural history. Globally, challenges for sea-turtle conservation are compounded by loss of habitat, overexploitation, poaching, boat traffic, disease, pollution, by-catch in fishing nets and trawls, and ingestion by turtles of plastics and other foreign material. Various nations, including the Bahamas, have adopted mechanisms that suit their own particular needs and customs, while also attempting to respond to international conventions. This piece-meal approach lends little confidence for the future of these unique and valuable animals.

Box 7.5 Green turtles in the Caribbean: a shared resource

Alan B. Bolten and Karen A. Bjorndal

More than any other dietary factor, the green turtle supported the opening up of the Caribbean . . . All early activity in the New World tropics – exploration, colonization, buccaneering, and even the maneuverings of naval squadrons – was in some way or degree dependent on turtle.

Archie Carr (1956/1979)

Green turtles (*Chelonia mydas*) are one of five species of Caribbean sea turtles and were once extremely abundant. Because they could be stored alive on ships for months, green turtles were a critical source of food for early explorers and colonists. As a result of unsustainable harvest during the past 500 years, green-turtle populations have been reduced to about 5% of pre-Columbian levels and the species is classified by the IUCN as endangered.

Although thousands of green turtles are still legally killed in the Caribbean each year for food, the value of live sea turtles as an ecotourism resource has been recognized and is now being exploited by many nations. Healthy green-turtle populations are also essential for maintaining the natural productivity, structure, and biodiversity of seagrass ecosystems. Green turtles are the primary herbivore in these ecosystems and, as a result, have major effects on structure and function of seagrass pastures. Therefore, to manage and conserve seagrass ecosystems, green-turtle populations must be protected and allowed to recover to natural levels.

Conservation efforts require the cooperation of many nations in the Greater Caribbean because green turtles are highly migratory throughout their life cycle. Green turtles are long-lived and their life histories span a number of diverse habitats and geographic regions. Female green turtles congregate at specific nesting beaches, or rookeries, where they excavate nests and deposit their eggs. After about 2 months, the eggs hatch, and the hatchlings run down the beach and enter the sea. Carried by currents, these young turtles, measuring only about 5 cm long, enter the open ocean where they remain for several years before returning to shallow coastal areas where they feed on seagrasses. Green turtles require about 40 years to reach sexual maturity. During their long juvenile life, they move among foraging grounds throughout the Caribbean where seagrasses grow. When they reach sexual maturity, they migrate back to their natal beach to reproduce.

Studies in the Bahamas have demonstrated the extent of international movement of green turtles. The Bahamas archipelago is characterized by extensive shallow seas with rich seagrass pastures and diverse coral-reef habitats that provide some of the best sea-turtle foraging areas in the Caribbean. Where do the green turtles we see in Bahamian waters come from? Genetic tags (i.e., mitochondrial DNA sequence patterns) have allowed determination of source rookeries for juvenile green turtles in the Bahamas (Fig. 1A). Rookeries throughout the Atlantic contribute to juvenile feeding populations in the Bahamas. Tagging studies have illustrated where turtles go when they leave Bahamian waters (Fig. 1B). Recently, satellite telemetry has enabled tracking of movement patterns of large juveniles as they leave shallow feeding areas in the Bahamas and disperse to feeding areas in other regions of the Caribbean (Fig. 1C).

The figures demonstrate that green-turtle populations are a shared resource throughout the Greater Caribbean that require coordinated, region-wide management to ensure their survival. Negative effects on nesting aggregations (e.g., harvest of eggs or nesting females or degradation of nesting-beach habitat) in one country will result in changes to foraging populations throughout the region. Likewise, negative effects on foraging aggregations from exploitation or habitat degradation will impact nesting aggregations. Therefore, to be successful, conservation and management plans must be shared by all nations – protection by one country alone will not be sufficient, and overexploitation by any one country may result in loss of sea-turtle populations. The IUCN's Marine Turtle Specialist Group has published a "Global Strategy" as a non-binding guide for conservation. The importance of regional management for sea turtles is recognized in the Inter-American Convention for the Protection and Conservation of Sea Turtles. CITES is a global agreement that has provided significant protection for sea-turtle populations by controlling international trade.

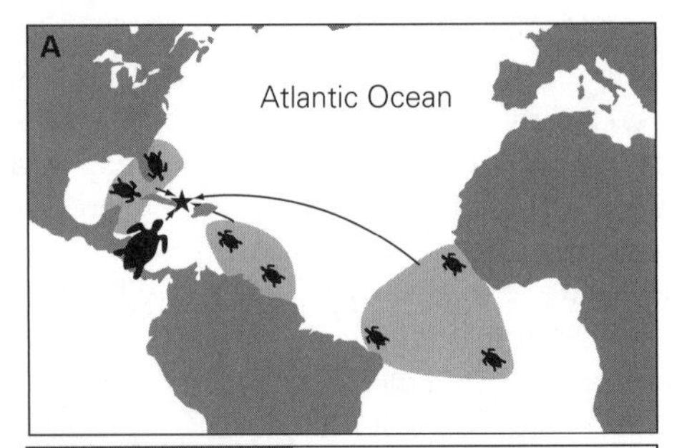

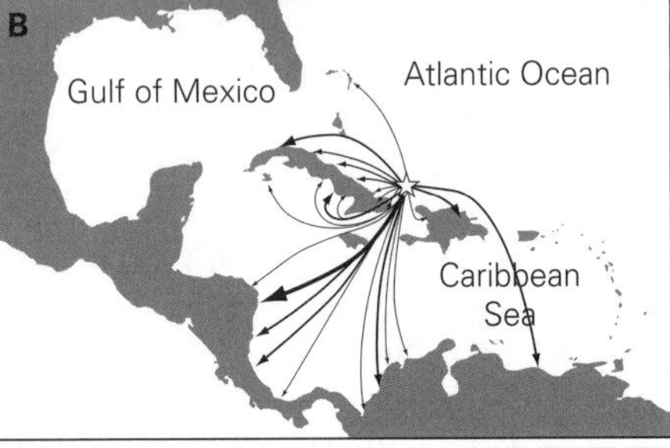

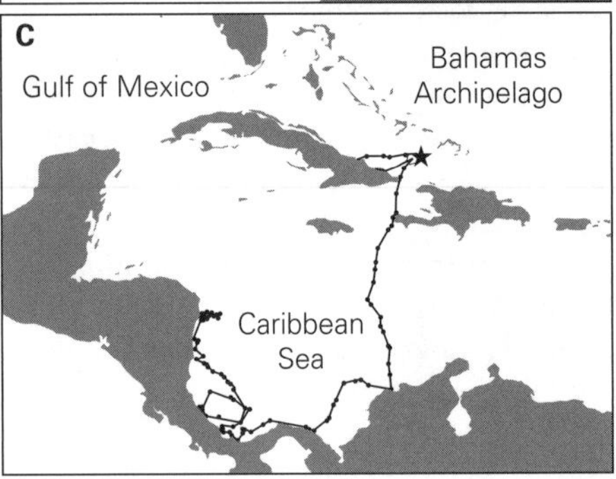

Fig. 1 The green turtle population moves throughout the tropical and subtropical Atlantic Ocean. (A) Mitochondrial DNA sequences demonstrate that green-turtle rookeries throughout the Atlantic contribute to the population of juvenile green turtles at Great Inagua, Bahamas (star); sizes of turtles in the figure are proportional to the contribution of the rookeries or rookery regions (shaded areas).
(B) Destinations of green turtles after they have left foraging grounds at Great Inagua (star), as determined by flipper tags; sizes of arrows are proportional to the number of tag returns.
(C) Satellite telemetry tracks for movements of three turtles from Great Inagua (star) to other feeding grounds in the Caribbean; movement patterns are consistent with the locations for tag returns (B).

Sources: Bjorndal (1995); Carr (1956/1979); Lutz & Musick (1997).

7.4 Roots of conservation in social history

In pre-Columbian times, green turtles abundantly swam over the shallow banks, consuming and structuring extensive seagrass, as did West Indian manatees (*Trichechus manatus*), Caribbean monk seals (*Monachus tropicalis*), groupers, lobsters, conch, and many others. Following European arrival, sea turtles, manatees, and seals were decimated. Green sea turtles now number in the thousands rather than the millions, manatees are rare throughout the Caribbean, and the last monk seal was sighted in 1952 near Jamaica. These losses reflect a history of change to the ecology and economy of the Bahamas' ecosystem.

About 6000 BP, Arawak culture expanded from South America to Caribbean islands and thrived on subsistence methods of extraction. Arawaks expanded northward and developed their own culture, called "Taino." They probably entered the Bahamas around 1500 BP, from Hispañiola. The Taino survived on a slash-and-burn agriculture and a wide variety of terrestrial and marine life. Because most of their protein came from the sea, their villages were close to shore, and they frequently moved when soils and resources were exhausted. Gradually, Taino populations increased and developed a complex society, organized into political groupings that spoke various dialects. The Taino had reached a high level of culture by the 15th century; Spanish records indicate a population of 40 000–60 000.

Columbus landed on the island the Indians called "Guanahani" (San Salvador) in 1492. The Spanish were impressed by Taino agricultural techniques, their use of fibers, and manufacture of canoes, tools, ornaments, and pottery. Columbus's ships sailed throughout the islands, observing natural scenic beauty, noting forests that dotted the landscape and the plentitude of birds. The air was fragrant, the flowers beautiful, and the monk seals, manatees, and sea turtles were there for the taking. Columbus called this place "Bajamar" – islands of a shallow sea – and named the people "Lucayan," after "lukku cairi," the name they gave themselves.

Columbus found no material wealth in the Bahamas, but found gold in Hispañiola. The Spanish returned to the Bahamas to collect Lucayans as slaves to work the mines. By around 1510, the Spanish had systematically removed most of the Taino population. The last to observe a native Bahamian was Juan Ponce de Leon, who sailed in quest of the "Fountain of Youth" and in 1513 encountered only one old Indian. Soon thereafter Lucayans became extinct in the Bahamas – the first part of the New World to become depopulated – and the Lucayans ceased to exist as a people.

To what extent Lucayans affected the Bahamian environment or its resources is unknown. Whatever the impact, the Bahamas appeared to have recovered, even as the Spanish spread their influence throughout the Caribbean, including introductions of a host of domestic and exotic animals and plants that displaced endemic species. The Spanish also required food, and took abundant wildlife. Green sea turtles were especially favored because they were tasty, extremely abundant, easy to catch, and could be stored alive for days or even weeks before being slaughtered. Some historians suggest that were it not for this single source of food, the Caribbean would not have been exploited in so short a period.

The early history of the Bahamas reflects conflicts among the Spanish, French, English, and Dutch for empire and wealth and struggles to control the New World. Rivalry among European nations was intense throughout the 16th century and Caribbean islands became a staging area for European dominance, with the Bahamas at a crossroads. The English gained maritime supremacy after their defeat of the Spanish Armada in 1588. King Charles I in 1629 claimed possession of the Carolinas and Bahamas, issuing a grant to his Attorney General Sir Robert Heath, who never developed his property. France's Cardinal Richlieu similarly granted some Bahamian islands to prominent Frenchmen, in 1633. The French created monopolistic companies, many of which failed. The Dutch also became influential in the Caribbean, and claimed islands in the Antilles. The result was emergence of four powerful European nations with a presence throughout the Caribbean, which created the multinational sea of today.

After the extinction of the Lucayans, the Bahamas remained unoccupied by humans for 130 years. The Bahamas were perceived to have little intrinsic value and were a treacherous place for ships, but still held a strategic geopolitical position. In 1647, a group of English and Bermudan religious refugees called the Eleutheran Adventurers founded the first permanent European settlement on an island they named "Eleuthera" ("freedom"). Living and farming proved difficult and fewer than 200 people endured by 1670. A more successful group of 300 Englishmen occupied another island they named "New Providence," where they founded Nassau on a sheltered harbor. These

Englishmen brought a maritime culture. They gathered conch and caught fish and turtles, rather than farm the land. They had no government, and law was meaningless on the Bahamas' lonely expanse. Buccaneering became a way of life and privateering an honorable calling, as rivalry among European nations intensified throughout the Caribbean.

The Bahamas soon evolved an English culture in the wake of emerging North American influence. The islands became a British Crown Colony in 1717, and Woodes Rogers (a former pirate) was appointed Royal Governor in 1718, bringing law and order as he expelled the buccaneers. Settlers built plantations, mostly for cotton and sisal where soil would allow, and founded salt-producing industries. Slaves were brought from Africa for labor. The Bahamas thrived as the United States approached independence. Britain abolished slavery in 1833, and freed slaves moved throughout the islands and founded small villages. Some practiced a subsistence way of life, in many ways replacing the Natives of centuries before. But as the Bahamian population grew, thin soils became depleted, and plantations failed to compete with U.S. cotton. Plantations were abandoned and the Bahamas became economically depressed.

In the 1860s, the American Civil War brought prosperity when the Bahamas became a haven for Confederate blockade-runners. English and American businessmen dominated Bay Street's business center in Nassau and out-islanders flocked to Nassau for its economic opportunities. Sponging began in 1870 and evolved into a major industry. Shops and warehouses sprouted up along the harbor. Small farms became profitable. However, prosperity was brief. During post-Civil War reconstruction, the economy declined again, and worsened further after a hurricane. Nevertheless, the population continued to rise, from 38 000 in 1870 to 53 000 by 1900.

World War I hit the colony hard as food shortages and cost of living soared. The Bahamas became one of the poorest colonies in the British Empire. Between 1911 and 1921, labor demand in the United States stimulated emigration from the Bahamas, and the population decreased. During the early 1920s, the economy improved again with U.S. prohibition of alcohol. Rum-runners thrived and Nassau's harbor became crowded with shipping as never before. This economic boost gave the Bahamas opportunities to turn to tourism as a source of income, fortuitously at a time when foreign incomes allowed international travel. The government subsidized steamship services, and real estate boomed. The Great Depression slowed growth until World War II, when the Bahamas became a center for flight training and anti-submarine operations. After the War, the Bahamas emerged as a major tourist and international financial center. A new wave of immigration began, characterized chiefly by Caribbean immigrants seeking work and North American and European expatriates leaving their homelands to escape taxation. Tourism became so successful that in 1954 some 100 000 visitors surpassed the Colony's population of approximately 84 850.

The Bahamas achieved self-government in 1964 and on July 10, 1973, the Commonwealth of the Bahamas officially attained full independence as a member state of the British Commonwealth of Nations. Since then, the Bahamas has continued to prosper, although its economy has become increasingly dependent on imports of basic commodities, foreign subsidies and investment, and expertise. Wide gaps in prosperity remain, especially between metropolitan New Providence and the more remote "Family Islands" (as called in the Bahamas). Many Bahamians believe that the history of boom-and-bust could repeat itself if the fickle tourist industry were to fail, overseas employment of Bahamian labor were to fall, foreign investment were reduced, hydrocarbon fuel costs were to soar, or especially if the Bahamas' environment and natural resources were to become further depleted.

7.5 Conservation for sustainability

Sustainability of resources and economic advancement – that is, "sustainable development" – is a high policy priority for the Bahamas. Sustainable development takes account of a broad range of issues, most particularly resource limits (e.g., living natural resources, water supply), growing population and needs for employment, requirements for sustainable tourism, and establishment of protected areas. The Bahamian people have long been aware of these social needs and environmental constraints. The government has passed significant environmental legislation (e.g., the Bahamas National Trust Act of 1959), and has become a signatory to many far-reaching international agreements (Box 7.6). The Bahamas was among the first nations to sign the 1992 Convention on Biodiversity, and hosted the first Conference of the Parties to the Convention in Nassau in 1994.

Box 7.6 Samples of Bahamas legislation and agreements relevant to living resources

Legislation

- *Agriculture and Fisheries Act.* Provides basic legislation for the supervision and development of agriculture and fisheries and protected areas for agriculture and fisheries
- *Bahamas National Trust Act for Places of Historic Interest and Natural Beauty.* Gives the Bahamas National Trust power to acquire and hold lands to conserve natural communities of plants and animals and natural environmental features
- *Fisheries Resources Act.* Makes provision for conservation and management of fishery resources and extends the Exclusive Economic Zone to cover the entire archipelago
- *Plants Protection Act.* No plants may be imported except under the rules of this Act. Expenses for treatment of diseased plants or in removal of plants to be borne by owner
- *Wild Bird Protection Act.* All wild birds in the Bahamas are protected. Game birds allowed to be hunted during open season
- *Wild Animals Protection Act.* No person may capture any wild animal without written authority

International treaties, protocols, and agreements

- *Bonn Convention.* For conservation of migratory species of wild animals
- *Cartagena Convention.* For protection and development of the marine environment of the Wider Caribbean Region
- *Convention on Biological Diversity*
- *Convention on International Trade in Endangered Species of Wild Fauna and Flora*
- *Inter-American Convention for the Protection and Conservation of Sea Turtles*
- *Ramsar Convention.* For wetlands of international importance, especially waterfowl habitat
- *United Nations Convention on the Law of the Sea* and *Agreement on Straddling and Highly Migratory Fish Stocks*
- *United Nations Framework Convention on Climate Change*

As the Bahamian economy continues to grow, development is conflicting with environmental features that sustain tourism and the economy. Therefore, in 1995–7, the Bahamas Environment, Science and Technology Commission in the Office of the Prime Minister accepted UNEP funds to develop a *National Biodiversity Strategy and Action Plan* for comprehensive conservation planning, with an accompanying *Biodiversity Data Management Plan* to carry out the strategy. Bahamian agencies and a U.S. academic team constituted a Task Force that identified and described major policy issues relating to sustainable development. Five areas, in particular, relate to conservation.

7.5.1 Fisheries

The Task Force determined that among the greatest challenges facing Bahamian conservation is sustainable fishing and fish habitat. Most Bahamian resources are marine, as the Bahamas is only marginally suited for agriculture. Bahamian fisheries concentrate on commercial species of high local (subsistence), tourism, and export value, centering on lobsters, conch, and Nassau grouper (Table 7.2). In both weight and value, lobster exceeds all others, and the Bahamas is one of the world's leading exporters. Most commercial, sport, and subsistence species inhabit the narrow band of shallow-water marine environments that surround the islands, including coral reefs. The narrow band of land and shallow water is where development is prominent and also where tourists are attracted.

The conduct of fishing also requires major attention. The way that fisheries data are collected limits its usefulness for determining population status and sustainable levels of catch. Fish data are typically lumped into catch data as "landings"; thus, populations from which species are taken are not known. Additionally, some species (snappers, grunts, jacks) are lumped together and their diverse individual life histories are lost. Furthermore, not all catch is reported. Fisheries data do not include sport, recreational, or subsistence uses; poaching by foreign vessels and violations by domestic fishermen are common. And illegal and/or harmful practices continue, including uses of chemicals, hookah (divers fed air by surface compressors), abandonment of traps, harvesting of protected juveniles, and catches during closed seasons. In the aggregate, these factors make prediction of sustainable levels of catch difficult, if not impossible.

The Department of Fisheries is well aware of these problems. However, fisheries policies that generally encourage overexploitation and overcapitalization fall outside the Department. Regulatory measures (size of harvested fish, establishment of protected areas, seasonal closures, etc.) are clearly inadequate, and are subject to political decisions. Management is handicapped by ineffective surveillance and enforcement over such a

Table 7.3 Freshwater resources of 13 Bahamian islands.

Island	Freshwater-lens acreage	Maximum volume*	Availability†	Population (from census, 2000)
Abaco	116 280	79.1	7 907	10 003
Acklins	15 783	4.36	10 765	405
Andros	338 585	209.92	25 672	8 177
Bimini	395	0.17	75	2 272
Cat Island	149 774	6.80	4 148	1 639
Crooked Island	5 923	1.74	4 223	412
Eleuthera	16 599	8.13	768	10 584
Exumas	6 586	2.90	816	3 556
Grand Bahama	147 884	93.17	2 278	40 898
Great Inagua	3 571	0.86	873	985
Long Island	9 301	2.88	977	2 949
Mayaguana	2 340	0.65	2 083	312
New Providence	17 503	9.63	56	171 500

* Maximum volume is in million imperial gallons per day.
† Availability is in imperial gallons per day per person, not including tourist, industrial, or municipal use.
Water data from Bahamas Water and Sewerage Corporation. Population from Macmillan Caribbean Certificate Atlas.

huge and complex expanse of water, while also seriously lacking financial support, infrastructure, and personnel. Furthermore, many practices are traditional and difficult to change, such as gathering of conch, hunting iguanas and turtles, fishing on spawning aggregations, and harvesting marine life for ornamental purposes.

Available information indicates that most species remain plentiful in at least some parts of their ranges, partly as a consequence of the Bahamas' large size and the remoteness of many areas. Turtles, fish, conch, lobsters, and others have the potential to recover and, if populations do recover, then sustainable use becomes an option, provided adequate measures are taken to monitor the population and to manage them effectively. Many examples of collapsed fisheries elsewhere in the Caribbean indicate a degree of urgency, as many deleterious fishing practices continue in the Bahamas. Especially serious is the wholesale removal of large consumers such as groupers, or key grazers such as green sea turtles, which can have serious ecological consequences. The selective removal of the largest components of the biological community, which began centuries ago with exploitation of sea turtles, seals, and manatees, has no doubt been a cause of fundamental ecosystem adjustments. The extent of these adjustments cannot be known, but continuing resource depletions, especially those imposed by fisheries, can cascade to tourism, subsequently to affect the overall economy and the wellbeing of citizens.

7.5.2 Water

Water affects all life and almost every aspect of environmental quality. Bahamians and tourists demand a healthy environment, for which freshwater is essential. The biodiversity of the Bahamas is also vitally influenced by the distributions of fresh, brackish, and even of hypersaline waters.

Natural freshwater supply depends almost entirely on rainfall that is collected on land in groundwater lenses, and remains the major source. Groundwater has always been easy to exploit in the Bahamas, but is also easy to contaminate due to the porous nature of the rock. Further, groundwater is not evenly distributed (Table 7.3), placing limits on life-style, tourism, and development. Getting water to people involves high infrastructure and transportation costs. New Providence obtains up to 50% of its water supply by tanker transport from the abundant supply on Andros Island. Short supplies of water cause Bahamians and visitors to compete, whereby use by local people may be preempted by resort development. Builders are not required to include cisterns in their development plans, as is required by some other island nations. Furthermore, dredging and channelization for development projects have cut through aquifers, resulting in massive losses of freshwater – for example, as occurred during development of Freeport on Grand Bahama Island. Under such circumstances, freshwater may be artificially

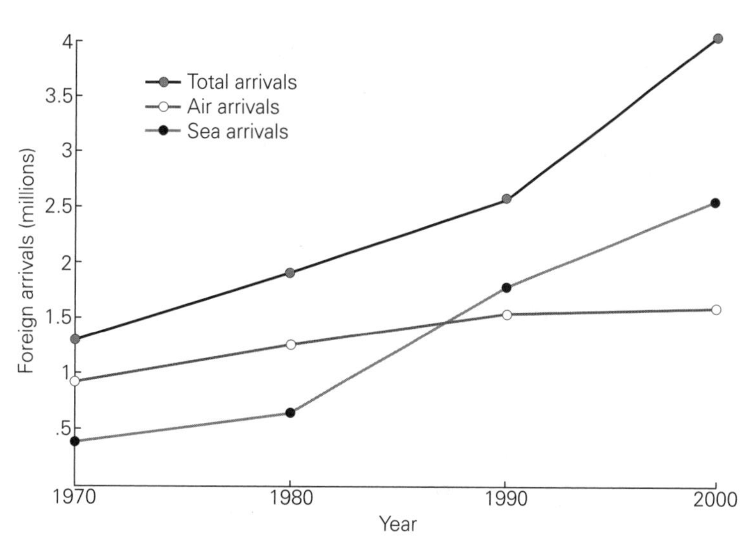

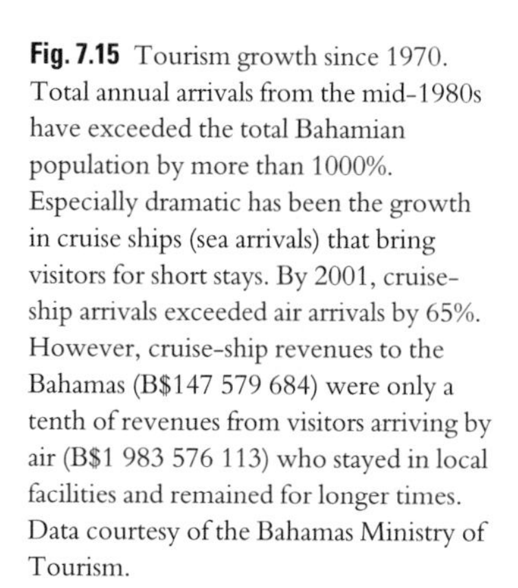
Fig. 7.15 Tourism growth since 1970. Total annual arrivals from the mid-1980s have exceeded the total Bahamian population by more than 1000%. Especially dramatic has been the growth in cruise ships (sea arrivals) that bring visitors for short stays. By 2001, cruise-ship arrivals exceeded air arrivals by 65%. However, cruise-ship revenues to the Bahamas (B$147 579 684) were only a tenth of revenues from visitors arriving by air (B$1 983 576 113) who stayed in local facilities and remained for longer times. Data courtesy of the Bahamas Ministry of Tourism.

obtained by flash desalination and reverse osmosis, which require expensive technology and costly fuel, and produce salt as a by-product that must also be managed.

Destruction and contamination of local water supplies and high costs of technology and energy make the Bahamas vulnerable to shortages, with potentially severe repercussions. Waste and pollution affect almost every sector of society and create conflicts among tourism, fisheries, recreation, development, human health, and esthetics. Porous karst landscapes present problems for waste and sewage disposal. Groundwater under some urbanized areas has become polluted from sewage, domestic wastewater, and urban and tourist facilities. Contaminants include disease agents, nitrates, pesticides, toxic chemicals, and runoff from roads. Overdraw of groundwater lenses can also threaten freshwater supply due to brackish and saltwater intrusion.

Extensive areas of brackish and hypersaline waters are also of economic and biological importance, but are poorly understood in the Bahamas. Brackish water is a potential source of freshwater through desalination, and hypersaline water is a source of salt; both host species of economic and tourism value, such as juvenile fishes and flamingos, respectively.

7.5.3 Tourism

The Bahamas is the tourism leader of the Caribbean. The Bahamian tourism industry has grown remarkably since World War II and now supplies approximately 40% of the gross domestic product and about 70% of gross foreign exchange earnings (Fig. 7.15). The Promotion of Tourism Act of 1964 gave the Ministry of tourism responsibility for marketing, advertising, public relations, sales promotion, research and public awareness, reception services, and hotel licensing. Its stated mission is: "to make it increasingly easier to create, sell, and deliver a satisfying vacation product – satisfying to investors, employees and tourists."

The Ministry of Tourism is concerned about the environmental impacts of tourism and other forms of development, and encourages small, locally owned facilities (Fig. 7.16A). Furthermore, rapid growth of tourist infrastructure has destroyed natural resources and shorelines, contaminated water supplies, and overextended the capacity of waste treatment and disposal. Furthermore, insufficient facilities are in place to allow watercraft to comply with the Environmental Health Services Act of 1987 for the prevention of environmental pollution or contamination from sewage and waste. These impacts have affected tourists' perception of pristine Bahamian environments. Also, overfishing and poaching have depleted once-abundant fishes, and damage to coral reefs by boat anchorages poses problems. Other problems that concern the Ministry are litter and waste, stray dogs and cats, exotic species, and threatened archaeological and cultural resources.

To address these problems, the Ministry of Tourism promotes "greening" of communities, especially tourist facilities. Ecotourism (nature and wildlife watching,

Fig. 7.16 Tourism development. (A) Locally owned Small Hope Bay Lodge, Andros. (B) Castaway Island (formerly Gorda Cay), a small, undeveloped island near Abaco Island, now used by Disney Cruise Lines to bring in more than 2000 visitors per trip. Photographs by the authors.

and the like) is the fastest-growing segment of the travel industry. Yet, locally owned-and-operated vacation facilities and national parks and protected areas important to the tourism industry are seriously underfunded. Rather, marketing is overwhelmingly centered around the mass-market attractions of "sun, sand, sea," foreign-owned hotel complexes, gambling casinos, and cabaret-style entertainment. Tax breaks and marketing assistance are primarily given to resort and casino complexes and the bulk tourist market. Cruise lines and multidestination vacation packages are encouraged. Remote islands have become ports-of-call and some islands have been leased solely for cruise-line use (Fig. 7.16B). Serious environmental costs can result from cruise-line operations. In many ports-of-call of the world, harbor development and dock construction for cruise lines have had devastating local effects on the sea bed, inshore ecosystems, fisheries, and endangered species. Some nations have levied substantial fines, in the order of millions of dollars, for violations of solid-waste and sewage disposal regulations. The multinational cruise-line industry is increasingly being regulated, but enforcement is difficult.

7.5.4 Land use

The Bahamas is especially vulnerable to unplanned development. Until very recently, no environmental impact assessments were required and development was actively sought to build the economy and provide jobs for a growing population. Throughout the Bahamas, marshes, swamps, and wetlands are used as dumpsites, coastal waters have become locally contaminated, particularly in semi-enclosed harbors with heavy boat traffic, and nearshore waters near population centers have become eutrophic and turbid. Vegetation and native wildlife have been affected by introduced species (e.g., wild pigs, donkeys, goats). Even on remote islands, housecats prey on birds and reptiles, and the exotic Australian pine (*Casuarina*) facilitates beach erosion and inhibits native vegetation.

These impacts are recognized, but continue to arise. On the small island of Bimini, dredging and other activities for development of a casino-resort threaten a lagoon that has been approved by government for a future protected area (Fig. 7.17). Serious problems are evident, which have altered the structure and function of this lagoon, and could even threaten the island with erosion due to alterations of natural flow and deposition of sand. The development also has the potential to increase siltation, which can smother coral (Fig. 7.18) and even kill whole reefs.

Crown land is especially pertinent for future development in the Bahamas. Crown "land" includes lands and waters held in trust by the English Crown on behalf of the Bahamian people. Crown-land policy attempts to balance best use with protection of Bahamian rights. Hence, Crown lands have been distributed for many purposes, including development, agriculture, forestry, and national parks. In the mid-1990s, about 70% (~ 1 million hectares) of the total dry land area of the Bahamas remained as Crown land. Of this, approximately 15–20% is under lease for use as agricultural, residential, light industrial, and commercial uses. Forestry and agriculture are promoted, especially in the Family Islands, to allow residents to gain greater income than would be derived from alternative land uses. However, forestry and agriculture account for only about 1–2% of the country's gross domestic product, placing these industries at a disadvantage compared with tourist or

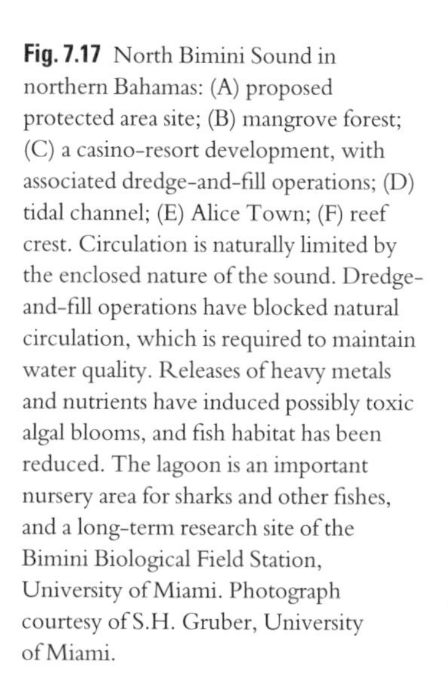
Fig. 7.17 North Bimini Sound in northern Bahamas: (A) proposed protected area site; (B) mangrove forest; (C) a casino-resort development, with associated dredge-and-fill operations; (D) tidal channel; (E) Alice Town; (F) reef crest. Circulation is naturally limited by the enclosed nature of the sound. Dredge-and-fill operations have blocked natural circulation, which is required to maintain water quality. Releases of heavy metals and nutrients have induced possibly toxic algal blooms, and fish habitat has been reduced. The lagoon is an important nursery area for sharks and other fishes, and a long-term research site of the Bimini Biological Field Station, University of Miami. Photograph courtesy of S.H. Gruber, University of Miami.

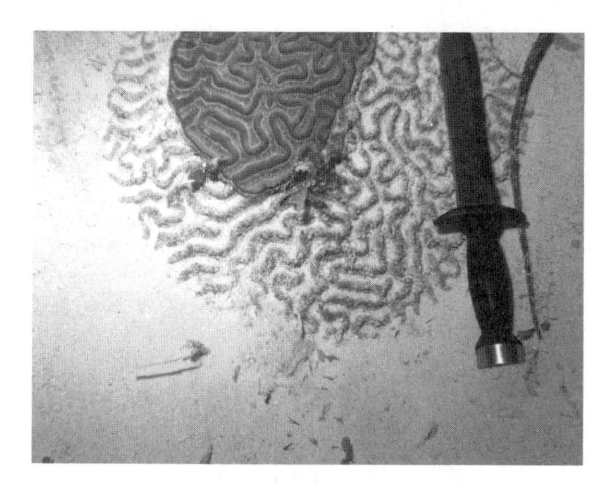
Fig. 7.18 Effect of silt from land clearing on symmetrical brain coral (*Diploria strigosa*). The dead portion was covered with sand resulting from road construction, leaving only the small central portion living. San Salvador, 1974. Photograph by the authors.

other development. Some Crown lands have also been dedicated toward shallow-water aragonite mining. Recently, the Bahamas has undertaken a coastal-zone management policy, in which the Crown lands office will become involved.

7.5.5 Protected areas and reserves

Protected areas serve many needs, including protecting biodiversity and endangered species, fisheries enhancement, water resources, and tourism. The Bahamas was an early leader among island nations in protected-area establishment. The idea for a Bahamian park system began as an inspiration of Colonel Ilia Tolstoy, grandson of Count Leo and a renowned sportsman, conservationist, and innovator. Tolstoy first visited the Bahamas in 1931, and in succeeding years observed "an abundance of wildlife and the clearest water he had seen anywhere." After World War II, he was "shocked by the lack of underwater life." Tolstoy's conversations with the Governor of the Bahamas led to support for a park in the Exuma Cays, on condition that concrete recommendations on resources and boundaries be made to the Bahamian government. That condition was met by a survey in 1958, and in 1959 Parliament passed the Bahamas National Trust (BNT) Act authorizing a protected-area system. In 1964, a 99-year lease to the Trust made the 176-square-mile Exuma Cays Land-and-Sea Park official, as the first tropical land–sea park under one jurisdiction. In the mid-1980s the Park was declared a fisheries "no take" area. Studies have since shown that fish abundance has dramatically

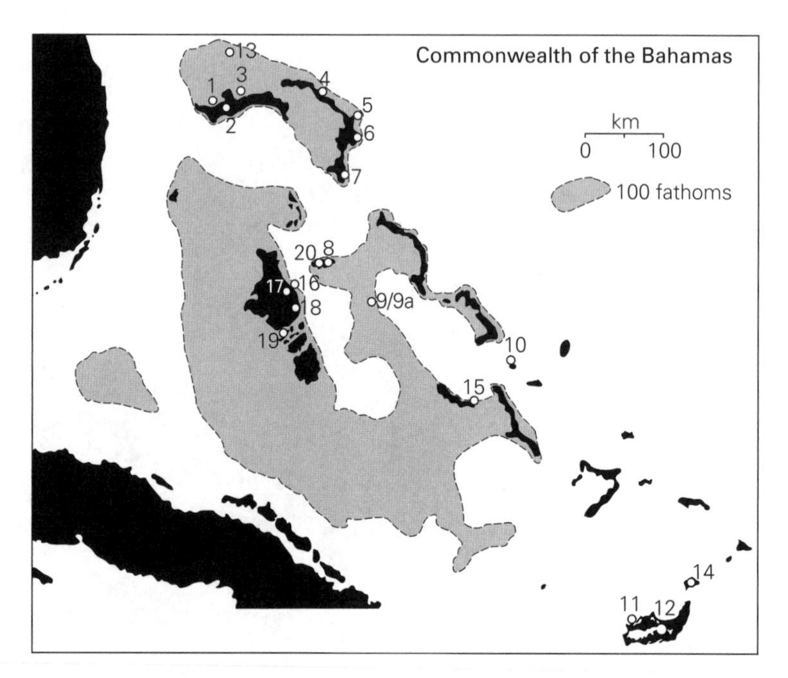

Fig. 7.19 Protected areas managed by the Bahamas National Trust (BNT). *From 1959 to 2001*: (1) Rand Nature Center (1992): Grand Bahama BNT Headquarters, pine forest and coppice, 40.5 ha; (2) Peterson Cay National Park (1971): important tropic bird nesting site, scenic spot, 0.6 ha; (3) Lucayan National Park (1982): site of the discovery of *Speleonectes*, a new order of crustacean; extensive underground cave system, native vegetation, 16.2 ha; (4) Black Sound Cay National Reserve (1988): mangrove, waterfowl, 0.5 ha; (5) Tilloo Cay National Reserve (1990): pristine shoreline, sea-bird nesting site, 4.3 ha; (6) Pelican Cays Land-and-Sea Park (1981): undersea caves, reefs, 840 ha; (7) Abaco National Park (1994): pine forest, coppice, Bahama parrot and other wildlife, 8200 ha; (8) The Retreat (1985): BNT Headquarters, native trees, exotic palm collection, 4.5 ha; (9) Exuma Cays Land-and-Sea Park (1959): cays and reefs, fishery reserve, 45 620 ha (land area includes private inholdings); (9a) Pasture and O'Brien Cays: island inholdings added to Exuma Cays Land-and-Sea Park (2002); (10) Conception Island (1973): a Columbus landfall, migratory birds, green-turtle and sea-bird nesting, 870 ha; (11) Union Creek National Reserve (1965): tidal creek, sea-turtle research, 1980 ha; (12) Inagua National Park (1965): West Indian flamingo and other waterfowl, Bahama parrot, 73 500 ha. *New parks from 2002*: (13) Walker's Cay Marine Area (2002): inshore marine environments, 1540 ha; (14) Little Inagua (2002); island scrub, wetlands, inshore marine areas, 12 550 ha; (15) Moriah Harbour Cay (2002): island environments, inshore marine environments, 5500 ha; (16–19) Andros (2002): collectively 115 830 ha, including: (16) two portions of Andros Barrier Reef, (17) highest concentrations of blue holes, (18) land crab management area, and (19) North Bight mangrove/wetland nursery area. (20) Three areas on New Providence (2002), including: Harrold and Wilson Ponds, waterfowl, freshwater fauna and flora, 101 ha; Bonefish Pond, estuary and creek nursery area, 730 ha; the Primeval Forest, native hardwood forest, 2.8 ha. Data courtesy of the Bahamas National Trust.

increased. The Exuma Park is now a premier tourist and recreational destination.

This Act established the BNT as a non-governmental organization to manage national parks and to advise government on environmental matters and on obligations under international conventions and agreements. By the turn of the 21st century, protected areas included 12 national parks and protected areas totaling 131 077 ha (Fig. 7.19). Only 0.4% of all the Bahamas, 7% of its land, and 0.1% of its sea are included in the system. Although this coverage is small, BNT's protected areas house an impressive array of ecosystems and resources, including: the largest fishery reserve in the Caribbean region (the Exuma Park); one of the longest underwater cave–cavern systems; a world-class marine turtle research project; the world's largest breeding colony of West Indian flamingos; one of the few known wintering habitats of the endangered Kirtland's warbler (*Dendroica kirtlandii*); the only habitats of the Bahama parrot (*Amazona leucocephala bahamensis*) – and much more. In 2002, the government concluded final preparations to create eight new parks, totaling 136 254 ha, potentially doubling the size of the park system, which still would encompass only about 1% of Bahamian territory (Fig. 7.19). A major impediment is that the BNT receives less than 10% of its revenue from government and is forced to raise almost all of its support funds for rangers, logistics, and administration from private sources. These problems challenge further conservation and development of the park system.

Fortunately, the BNT is not the only agency in the Bahamas with concern for, and jurisdiction over, protected areas. The Bahamas Water and Sewerage Corporation has established a conservation area for freshwater on Andros Island and the Department of Agriculture has authority for forest reserves. Additionally, local organizations have influence over environmental matters, including protected areas. For example, the Andros Conservancy and Trust (ANCAT) played a leadership role, together with the BNT and the U.S.-based Nature Conservancy, in recommending and designing the new Andros Island protected areas.

Recently, the government endorsed a policy for 20% of Bahamian territory to be incorporated in a protected-area system, with particular attention to fishery reserves, which would be managed by the Department of Fisheries. At the initiative of the Bahamas Reef Environment Educational Foundation (BREEF), several dozen potential sites were identified throughout the Bahamas as marine reserves. Five sites were selected following studies in the late 1990s, one being the North Bimini lagoon. All five areas are small, and only one has as yet been established. Further, the North Bimini site is threatened by development (Fig. 7.17). Also, questions remain about how large any fishery reserve must be to maintain habitats of sufficient size to sustain a fishery. The lobster is a case in point. Recent research has indicated how optimum reserve sites may be identified for maintenance of a regional population (Box 7.7). The results of this research demonstrate the need for an extensive network of reserves, even if only one species is considered.

Conventions and agreements to which the Bahamas is a signatory commit the nation to establish protected areas of considerable scope. This commitment requires a strategy that would incorporate large areas representative of the total suite of Bahamian environments. The Bahamas government strongly endorses a protected-area system and is a signatory to the Convention on Biological Diversity, which obligates governments to conserve the biodiversity of species and their habitats. The need for a comprehensive protected-area system is made urgent by the lack of comprehensive Bahamian management policies for significant areas of high productivity and biodiversity, such as coral reefs, seagrass beds, and mangroves. The ultimate issue for establishment of a protected-area network for the Bahamas concerns development of a comprehensive, science-based, multiple-use conservation and management system, and an integrative, well-funded authority to manage it. Protected areas are an important part of meeting that goal.

7.6 Conclusion

Humans have inhabited the Bahamas for only about 1500 years. Especially since Columbus's time, human activities have fundamentally altered the living system. A host of social, economic, and environmental problems have now become evident, population growth already being the most challenging (Fig. 7.20). The Bahamas' population is now clearly beyond the carrying capacity of the environment, given the nature of the system and the limitation of resources.

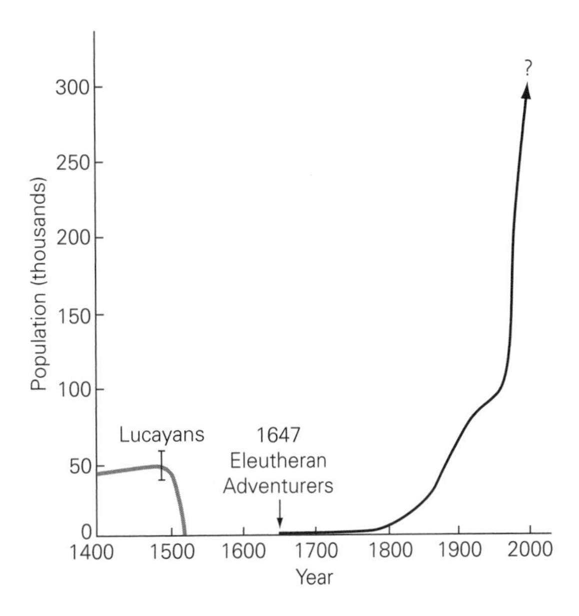

Fig. 7.20 Population growth in the Bahamas. The pre-Columbian Lucayans are estimated to have numbered approximately 40 000–60 000, but were extinct by the early 16th century. The Bahamas remained unoccupied by humans until the mid-17th century, and the population did not again reach 50 000 until the late 19th century, after which the population rapidly grew. Data from Keegan (1997) and from various other historical and Bahamas Government sources.

Box 7.7 Metapopulation dynamics and marine reserves: Caribbean spiny lobster in Exuma Sound, Bahamas

Romuald N. Lipcius, William T. Stockhausen, and David B. Eggleston

One major postulated benefit of no-take marine reserves is that the number and biomass of breeding individuals will increase and subsequently enhance larval production and recruitment to the reserves and exploited areas. The likelihood that this benefit will accrue depends particularly on the manner in which larvae are distributed to reserve and exploited sites through hydrodynamic transport and recruitment processes. This can be conceptually subsumed under the theory of metapopulation dynamics.

Metapopulation dynamics deals with the dynamics of fragmented yet interconnected breeding populations. When populations are fragmented, interbreeding occurs predominantly among individuals within each population. Fragmented populations may be interconnected by dispersive stages, which for many marine species pertains to dispersal *via* oceanic currents. Thus, dispersive stages in species with complex life cycles break the connection at the local scale between reproduction and recruitment; connectivity among subpopulations and metapopulation dynamics are emergent and critical properties of the system.

The degree to which marine reserves enhance recruitment under the influence of metapopulation dynamics is virtually unknown. Caribbean spiny lobsters (*Panulirus argus*) occur commonly in reefs and inshore lagoons ringing Exuma Sound. Therefore, field data on abundance, habitat quality, and hydrodynamic transport patterns for lobsters in the Exuma Cays Land-and-Sea Park no-take reserve and three exploited, potential reserves (Cat Island, Eleuthera, and Lee Stocking Island in Exuma Sound) can serve as a model system for assessing reserve success in enhancing metapopulation recruitment.

The likelihood of success of a reserve at each of these four sites in enhancing recruitment is based on the assumption that together they comprise a semi-closed metapopulation in the Sound. The four sites are assumed to harbor fragmented populations of lobsters because they are separated by either/or: (i) approximately 100 km of shoreline, which is much greater than the average benthic dispersal distance of most juveniles and adults; (ii) extensive shallow banks with little structural relief, food, or shelter, which limits intersite movements; or (iii) the deep, 2000 m basin of Exuma Sound (Fig. 1). Furthermore, major coral reefs, where adults reside and reproduce, are fragmented similarly across the four sites. These populations are, however, interconnected due to larval and post-larval interchange between them.

Therefore, sites other than the Exuma Park provide broad spatial coverage of lobster habitats throughout Exuma Sound. Each of them is of a sufficient size (~ 15% of coastal reefs) to meet the threshold requirement for enhancement of recruitment. All are characterized by coral reefs, patch corals, sand-covered hard-bottom, seagrass meadows, and fields of gorgonians and sponges. Coral reefs crest at approximately 30 m depth before the bottom plunges to 2000 m. Most cays lie on the western edge of Exuma Sound; tidal inlets connect the Sound to the shallow (3–5 m depth) Great Bahama Bank (see Fig. 7.1). Most islands within the Park are uninhabited, though recreational boats are common and poaching by local fishers and visiting boaters is not uncommon.

Theoretically, the effectiveness of the Park and the other three nominal reserve sites can be assessed by estimating

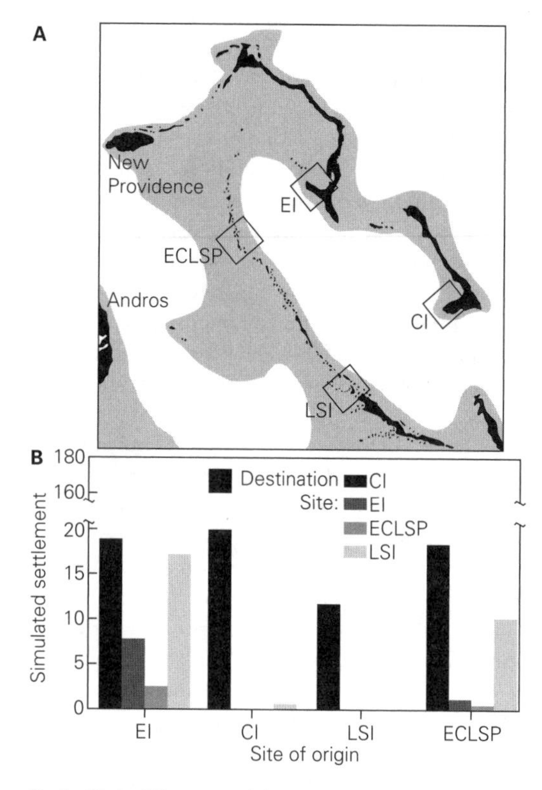

Fig. 1 (A) Caribbean spiny lobster study sites in Exuma Sound: Exuma Cays Land-and-Sea Park (ECLSP), Eleuthera (EI), Cat Island (CI), and Lee Stocking Island (LSI). (B) Planktonic larvae are produced along the coast during the spring reproductive period at each of the four sites in equal numbers. Simulations of larval transport and post-larval settlement assume that larvae enter the offshore region and are transported by advection and diffusion as passive particles. After 180 days in the plankton, larvae metamorphose to post-larvae that actively migrate toward the nearest coast, where they settle. From Lipcius et al. (2001).

the degree to which larvae produced at a reserve may be distributed as recruits among all the sites. This emphasizes the importance of spatial position for reserve success. A circulation model may be used to estimate the degree to which larvae produced at a particular site are transported and redistributed to all the sites. Enhancement of recruitment is deemed optimal when all four sites receive recruits from any one site. The model shows that larvae discharged from the Park and Eleuthera recruited throughout Exuma Sound, with Eleuthera having the most equitable distribution. Larvae released from Lee Stocking and Cat Island recruited back only to those sites. Selective hydrodynamic transport by gyres prevents larvae produced in southern Exuma Sound from reaching northern sites. Hence, a reserve at Eleuthera would be most suitable for enhancing metapopulation recruitment. The Park may function less effectively, though significantly more so than the southern sites.

In assessing criteria useful for metapopulation enhancement, the use of habitat quality or adult density is no more of a guarantee for success than determining reserve sites simply by chance. The only strategy that increases the likelihood of selecting an effective reserve system for augmenting metapopulation recruitment is that which incorporates transport processes. Hence, the designation of a reserve location for exploited marine species requires careful attention to data on metapopulation dynamics and recruitment processes.

Sources: Crowder et al. (2000); Lipcius et al. (1997, 2001); Stockhausen et al. (2000); Stockhausen & Lipcius (2001).

The Bahamas is presently at a crossroads, similar to many small-island, developing nations that seek economic development. Objectives include job creation, tourism, increase of exports, development of business opportunities, and availability of quality products to consumers. The Bahamas government recognizes that tourism, including ecotourism, will remain the principal engine of the economy and that environmental quality and natural resources are key to the nation's future. It follows that the greatest source of security and wellbeing for the people of the Bahamas is environmental quality.

The *Bahamas National Biodiversity Strategy and Action Plan* (BEST, 1997) endorsed a vision for national wellbeing: "A strong nation rooted in a healthy environment." The Bahamas has also adopted many policies that serve to counteract deleterious outcomes, most prominently regarding sustainable development and concerns for social equity and wellbeing, globalization, biodiversity, and global climate change. However, environmental policies and conservation goals continue to conflict with activities and places coveted for development. Thus, the nation and its people continue to become increasingly dependent on influences outside national control.

Section IV

Analysis and synthesis

Diseased sea fan (*Gorgonia ventalina*) encrusted by fire coral (*Millepora alciornis*). Exuma Cays, Bahamas. Photograph by the authors.

The history of human activities in the coastal realm highlights trends that need addressing in order to construct a comprehensive approach for coastal-realm conservation. Evidence continues to mount that when large marine consumers are depleted, or when environments are structurally or functionally altered, ecosystem "health" is compromised. However, without historical knowledge or predictive capacity, conservation is forced into costly restoration and recovery activities and crises management. Among the several major challenges, governance and acquisition of scientific information about the nature of change are most decisive. Thus, coastal-realm conservation depends on placing science and management in the context of systems thinking, with a basis in natural history, and toward the new paradigm of "ecosystem management." The goal is to harness human potential to achieve human-ecosystem sustainability and wellbeing. The question is: How can society gain in the race to conserve both species and entire ecosystems from the inevitable forces of environmental change and the incremental impact of the environmental debt?

Top: Estuary development and alteration also affecting longshore sediment transport. Chesapeake Bay, Maryland, United States. Photograph by the authors. Bottom: A "dead" coral reef. Inagua, Bahamas. Photograph by the authors.

Chapter 8

Coastal-realm change

An analysis of almost any scientific problem leads automatically to a study of its history. The many unresolved issues in evolutionary biology are no exception to this rule. To understand the history of a scientific problem, however, one must appreciate not only the state of factual knowledge but also the zeitgeist of the time.

Ernst Mayr (1991)

8.1 Introduction

The issue facing coastal-realm conservation is not change itself, for change is a characteristic feature of this dynamic environment. However, today changes are occurring with such unprecedented speed, frequency, and magnitude that they are obvious even to a casual observer within a single human generation. From local to global scales, humans are extracting resources, adding novel chemicals and species, altering environments, depleting species, creating discontinuities, changing evolutionary processes, and reordering the coastal-realm system in ways and at rates beyond resource sustainability. Evidence is accumulating that ecosystems are being functionally altered and that their resiliency is being compromised. Further, local changes are cascading regionally to create situations increasingly difficult to manage. Thus, images of healthy coastal ecosystems and natural resource abundance are giving way to realities of stressed environments, emergence of new phenomena, and difficult policy decisions.

Documenting ecosystem change is difficult. Humans have been modifying coastal environments since societies began to cluster there. Therefore, how does one assess change, given the complex dynamics of the coastal realm? How does one restore ecosystems without knowing their historical structure and function? How can a species' status be known without detailed natural history knowledge? In this chapter, we suggest an approach for broadly recognizing change, by providing a reference for environmental health and stress, and by calling attention to human activities as stressors. The discriminating parameter is the time over which human activities create system instabilities.

8.2 Accounting for change

Accounting for change requires identification of a standard, a reference against which to make comparisons. A metaphorical goal, or standard for conservation, centers around ecosystem "health," which is often assumed to be a "natural" condition. This standard is endorsed by much of the public-at-large and is called for in an increasing number of treaties, agreements, and legislation. However, assessing "ecosystem health" can be problematic because of difficulties of measurement. One way to assess health is to identify stressors known to cause a response, and to measure recovery.

8.2.1 Ecosystem health – the reference

Dictionaries almost invariably refer to "health" as a steady-state property of organisms – for humans and other species – that is, wellbeing without disease. This concept evolved from the work of Claude Bernard in the 1880s, who recognized that all organs and tissues of a human body perform functions that help maintain conditions around an equilibrium point that is maintained by homeostatic processes. At each stage in the life history of an organism, adjustment processes maintain a condition of "health" even though individual organs and their relationships change during development stages. That is, functional interactions among organs and tissues constitute an inherited pattern of feedback processes that maintains a recognizable condition of health. Over a span of time, the level that is defended may change, and regulation occurs around shifting set-points in response to stimuli (Mrosovsky 1990). Regulation may be developmental or reactive, cyclical or not. Continuously changing set-points over the course of development create a dynamic stability that allows the individual to persist. Homeostasis of an organism thus involves continuous changes of regulated levels to maintain healthy states. Thus, health is dynamic and implies a steady-state condition under dynamic equilibrium.

However, ecosystems are not organisms. Ecosystems are more open, dynamic, and complex, and are less tractable and not as readily defined or measured. Yet, ecosystems do exhibit self-organizing processes and characteristics of persistence, stability, resilience, and recovery. At any given state, an ecosystem of diverse components may exhibit a persistent pattern that is maintained by processes in a dynamic condition of "health," as defined by scale and choice of condition. Ecosystem "health" may be prescribed by regulation around a shifting set of characteristics or patterns that persist for a specified period of time, as in the case of successional changes. For example, an oyster bed develops through an orderly process of biological accretion and senescence, not unlike set-points in homeostasis of organisms, while undergoing continuous change: (i) an initial phase, as in spawning and colonization; (ii) a growth phase involving clustering and accretion; (iii) a maturation phase of maximum development; and (iv) a senescent phase of disintegration, degeneration, decay, and erosion. Also, a state of dynamic health may be said to exist in high-energy coastal systems, where disturbance dominates, erosion can be prominent, the system oscillates, and fluctuations can help the system return to preexisting conditions. A persistent pattern of the land–seascape may be maintained in different stages of health by small- or large-scale processes, the state or condition depending on the scale and pattern of interest to the observer.

Several parameters have been suggested to define ecosystem health, such as vigor, resilience, persistence, and ecological organization. Vigor may be operationally measured in terms of productivity or throughput of material or energy in the system. Resilience may be observed in terms of a system's ability to maintain its structure and/or behavior under conditions of stress, in which the rate of return (recovery time) to former abundance or quality is important. Persistence is the continuance of some quality or ecosystem parameter through time. System organization may be assessed by comparing the diversity of components, their degree of mutual dependence, and their exchange pathways at different time and space scales. Ecosystem health may also be measured by biodiversity, but whether species-rich environments are more "healthy" than relatively species-poor ones, or whether declines in diversity serve to indicate a decline in "health," are uncertain.

8.2.2 Stress – initiator of change

Stress, as observed by health practitioners, is defined as an internal or external force that causes fatigue, strain, or change in an organism. Selye (1976) challenged the existing medical definition of human health by observing health and disease as conditions during a process of change. He defined stress as the rate at which the body is subject to wear and tear, including nonspecific responses of the body to physiological demands. He argued that the body's reaction to stress is a process of defense to protect the body and to maintain constancy of the "internal milieu." In the presence of a stress condition, a "general adaptive syndrome" describes the sum of morphological, physiological, biochemical, and behavioral changes in three stages of response: an alarm reaction, stage of resistance, and stage of exhaustion. When a stressor is applied, a short-term alarm reaction may involve behavioral or physiological responses. During long-term exposure and resistance, adaptation may involve changes in enzymes, accommodation, or reproduction. Eventually if adaptation fails, an exhaustion stage may lead to death.

Stress is a condition brought on by the summed reactions of the intensity, frequency, and duration of all stressors. Thus, understanding ecosystem health requires understanding stressors and stress conditions that initiate responses (Table 8.1). Ecosystem stress may be measured by responses to suboptimal conditions, environmental change, presence or absence of key or foundation species, declines in diversity, or other factors. An ecosystem stressor is a perturbation that causes stress, and may be foreign to the system or internal when present in excessive amounts. Hence, environmental perturbations (pollution, storm events, physical change, etc.) are external stressors that act on the entire ecosystem. Internal stressors within a biological community may occur as overabundance or crowding, or when limited resources result in keen competition, or when a species suddenly dominates as in cases of pathogens or exotic species. In general, external stressors invoke evolutionary changes; internal stressors act on populations or communities to select winners and losers and to maximize population fitness.

Stressors can be periodic, episodic, or chronic; they can act directly to arouse immediate stress alert, or there may be a lag between the stressor and the response. A

Table 8.1 Examples of ecosystem and community stress factors. Any of these may be viewed as a stressor or a stress, depending on circumstances and scale of interest. Some may also serve as indicators of ecosystem health (Table 8.2).

Ecosystem stressors
Nutrient enrichment (eutrophication)
Hydrological disruptions
Aquaculture
Changes in nutrient transport
Buildup of chlorinated hydrocarbons
Hypoxia and anoxia
Harmful algal blooms
Coastal development and land–seascape alteration
Changes in boundary conditions
Storm events
Climate change and variability
Population and community stressors
Habitat loss
Reduction of species diversity
Exotic and invasive species
Excessive removals of species
Overabundance
Toxic pollution
Depletion and extinction of habitat specialists and keystone and foundation species
Losses of controls on population regulation
Increased disease prevalence and incidence
Changes in mutualistic species interactions

Adapted from Alongi (1998); Rapport et al. (1998); Smodlaka et al. (1999).

proximate response occurs when a species responds quickly to pollution, for example, and its numbers change; a distal response occurs when symptoms develop or change occurs over time. To be proximately stressful, an environmental variable must either increase or decrease from a "normal" state, causing some change in the organism or population. Distal responses, observed as lags over time, occur when a cascade of effects follows an applied stress.

Some stressors may bring benefits or initiate long-term trends. Natural stressors (storm events, bioturbation) may enhance ecosystem performance and maintain ecosystem health and diversity, in which case the perturbed ecosystem returns to an unstressed condition and biota continue to thrive. Natural stressors can also "reset" the system to a former "healthy" condition, as when storm events strike a coral reef, initiating recovery toward high diversity by making space available. Other natural stressors may occur periodically and at many scales to cause major changes in system dynamics, such as successional sequences and system-state changes. Productivity in a water body, for example, can be enhanced with small increases of heat, CO_2, or nutrients when these are otherwise limiting. However, the same stressor applied in excessive amounts can adversely affect survival and health of key organisms to cause an overall depression in productivity, or otherwise depress ecosystem performance. Chronic disturbances that do not mimic the frequency of natural disturbances are especially likely to alter an ecosystem's ability to absorb damage, and can change basic ecosystem properties. In severe cases, the system might ultimately reach a point at which all species and functions are severely compromised, and the system itself may change or collapse without possibility of recovery.

8.2.3 Detecting change

Ability to predict responses to stress at the scale of the ecosystem can be hindered by variability or cycles, lack of baseline data on which to make comparisons, and by what should or could be measured to indicate response. Long-term cycles (centuries, millennia, and longer) recorded only as anecdotal knowledge or during a scientific reconstruction of events or states are difficult to verify, such as medium-scale climate cycles that appear to oscillate over decadal scales. All types of cycles can have profound effects on biota, as they can lead to environmental variability, instabilities, and varying ecosystem expressions, for example multiple steady states, rhythmic phenomena, different spatial patterns, and even to new dynamic behaviors resembling random processes with unpredictable outcomes. The major goal in detecting coastal-realm change, then, is to understand how the variety of biological components of ecosystems are interrelated, how ecosystems respond, and how they may recover and adapt – a most daunting task.

Undertaking a full account of change during the passage of time forces recognition of the "shifting baseline syndrome." Meaningful baselines for assessing health and change are impossible to establish without sufficient accounting of this phenomenon. A widespread example is the depletion of large animals that once played major roles in ecosystems, so-called "ghosts of the past." For decades or even centuries, species loss may go unrecognized, during which time there may be a lag period of system readjustment. If the "ghost" is not recognized, the standard or baseline for analysis may be insufficient for measuring effects of change.

Table 8.2 Indicators of ecosystem health. The top part of the table lists four general goals of indicators and the objectives toward which indicators are directed. The bottom part lists criteria for choice of indicators and their attributes; the first four criteria are general for any indicator that may be chosen and the fifth is species-specific.

Goal	Objective
Intrinsic importance	Valued, depleted, or commercial/sport species; biodiversity; habitat condition
Early warning of stressor and stress conditions	Early exposure to stressor; response rate; lag times of response
Ecological process and function	Rates of production, cycling, feedbacks, decomposition; changes in food webs, biogeochemistry
Conservation relevance	Addresses the "So what" question; species, communities, ecosystems at risk; relevance to adaptive management, regulations, legislation, and agreements
Criteria	**Attributes**
Stress sensitivity	High signal-to-noise ratio; sensitivity to change; false positives minimized
Reliability	High specificity of response; low variability
Complementarity	Multiple indicators possible
Ease of use	Low sampling cost; efficient logistics; ease of laboratory identification; pre-existing database (e.g., fisheries, pollution data on hand)
Species-specific	Ecosystem role (e.g., key, foundation species); clear taxonomy; specialist; well-known natural history; known tolerance levels and resilience; correlated with habitat; limited distribution and mobility at some life stage; easy to find; trends easy to measure

Modified from Hilty & Merenlender (2000); Kelly & Harwell (1989).

One example concerns sea urchins that graze on large kelp forests of coastal Alaska. Sea-urchin populations are held in check by sea otters, but if otters are depleted, as happened during the late 19th century, urchins become so abundant that their grazing can transform the kelp forest into an urchin barren. Kelp forests are also naturally sensitive to warming during large-scale climatic shifts, which complicates this relationship. Because of these biological and climatic influences, conditions naturally alternate along a temporal baseline between lush kelp forests and sea-urchin barrens. In this case, recovery remains possible, as all the elements are still present and abundant. Conversely, in the Caribbean Sea, depletion of green sea turtles has resulted in greatly reduced grazing on seagrasses, and the "shifting baseline syndrome" raises questions about how to determine which condition is "natural" – grazed or ungrazed. The acceptance of a present baseline of ungrazed seagrass beds, without historical information about past sea-turtle grazing, can lead to inappropriate conclusions and predictions about ecosystem health.

Diurnal and tidal oscillations, storm events, seasonal cycles, and interannual and interdecadal oscillations cause recognizable changes in pattern and process. The simplest measurement of ecosystem response to stress is a direct effect on survival, growth rates, and reproduction of an indicator species. Indirect process effects are much more difficult to determine, as they may be mediated by effects on other components in the ecosystem, and/or caused by cascading effects temporally dissociated (measured as lag time) from the initial stressor. Tracking indicator species and environmental variables allows recognition of trends. Changes that result from indirect effects may relate to ecological performance, changes in biodiversity and dominant species, process variability, and rates of change. Measurement of rates of change includes such important parameters as exposure, response, and attributes of species, habitats, or whole systems. Recovery is determined by the duration, timing, magnitude, exposure, and nature of stress. Choice of scale for interpretation is critical.

Various indicators (Table 8.2) may be used to track fluctuations of a quasi-equilibrium state along a temporal continuum. Characterization of long-term fluctuations under circumstances of high variability may be more meaningful than measurement of average equilibrium conditions. Characterization of fluctuations provides a context, a frame of reference or standard, for assessing ecosystem change over time, observed as a trend. The response of an ecosystem is its change of state or dynamics up to the point of maximum deviation from an unstressed condition; that is, the distance from the standard of "health" to an altered stressed condition, or changed state, initiated by a stressor, provides a qualitative and/or quantitative measure of change. Coastal-realm ecosystem change becomes most apparent after long-term, large-scale observation of summed effects. An ecosystem subject to repeated disturbance, or stress, tends to deteriorate in time because of loss of vigor, resiliency, and ability to recover. A meaningful conclusion about change may be described by resilience, that is, measures of recovery rates of key indicators to reference conditions, in which special consideration is given to how rapidly and how effectively response and recovery occur.

8.3 Conservation issues

Extractions, additions, and physical restructuring are human stressors that cause ecological change. The U.S. National Research Council (NRC 1995) has determined that fishing is the top human agent of marine ecosystem alteration, followed by chemical pollution, eutrophication, alteration of physical habitat, invasions of exotic species, and global climate change. These agents of change imply changes of ecosystem state, structure, and function.

8.3.1 Extractions: fishing

Effects of fisheries removals on ecosystems are becoming obvious, given the present state of coastal-marine ecosystems worldwide. In the words of Angel (1991): "It scarcely seems credible that such extensive cases of exploitation have not had major impacts on the structure of the ecosystems concerned, and the energy flow within them." This statement is justified by fishery extractions that are estimated to amount to approximately 8% of the primary production of the ocean, more than 25% for upwelling areas, and roughly 35% for temperate continental-shelf systems (Pauly & Christensen 1995). The extracted production is transported and consumed and waste products are deposited elsewhere. In this process of massive biomass transfer, indigenous flora and fauna are displaced and ecosystem adjustments are required.

8.3.1.1 Fishing: the reference

High historical biomasses of invertebrates, birds, fishes, turtles, and marine mammals are matters of record. As early as the 15th century, great abundances of cod, sea turtles, and marine mammals played a major role in the perception that the ocean's resources were inexhaustible. However, only during the late 19th to early 20th centuries did fisheries science emerge. Scientists discovered that many species are prolific, voracious predators that consume different foods at different life-history stages. Some species grow at different rates and differ slightly in morphology throughout their ranges, suggesting metapopulation structure. These findings make apparent that features of natural history are essential elements for understanding population growth, resilience, and recovery and that complex natural history, community, and ecological relationships surround each resource.

Reproduction and dispersal, food supply and consumption rates, population structure, and community ecology are key factors for recruitment and sustainable fisheries. Three key issues are (Caley et al. 1996): (i) variation of recruitment in space and time; (ii) the extent to which variation in recruitment underlies variation in adult populations; and (iii) the relative importance of recruitment in determining population size and structure. Another prominent factor for recruitment is whether or not metapopulation structure applies. Genetic mixing cannot be assumed from adult movements; species that move considerable distances may exhibit restricted gene flow, whereas those that are relatively sedentary may have unrestrained gene flow (Gold and Richardson, 1999), indicating that the totality of natural history is key to dispersal and mixing.

The extent to which fishes are habitat-dependent is highly variable and difficult to predict, and timing of fishery extractions is critical. For example, fishing that disturbs habitats during post-larval settlement can have a devastating effect on recruitment. Also, the extent of habitat disturbance, such as from trawling, and recovery rate strongly influence recruit survival. This is especially true for species whose habitat requirements are obligate. Additionally, the vast majority of species require a mosaic of habitats during their life histories, and fragmentation of habitat mosaics directly and indirectly affects species' recruitment, biodiversity, and community interactions. Bottom-up factors regulate populations when interactions among trophic levels are altered; top-down regulation occurs when changes are imposed on ecosystem biotic and physical components (habitats). Both types of regulation can simultaneously occur; for example, pelagic fishes living in relatively homogeneous water-column environments are subject to climatic events, which can dissociate developing larvae from their food supply.

In complex communities such as coral reefs, feedbacks among species, habitat, and fishing are important. Reef volume and area, its complexity, vertical relief, texture, spatial arrangement and orientation, and location are important factors for biodiversity and productivity. Changes in species' population size are driven by birth, death, immigration, and emigration of individuals in the population, and which among these factors is habitat- or density-dependent is critical for understanding variations in population size. When some fish species (e.g., grunts, Family Haemulidae) occur at high densities on reefs, their excretions may contribute significant nutrients. Other species that consume coral tissue can adversely affect coral-reef communities. Fishes that graze algae have major effects on succession of algae on reefs. Thus, feeding habits of fish may affect the settlement of larvae on coral. Hence, strong functional relationships exist among fishes and coral reefs. It follows that extractions of fishes can affect reef structure and function and that these effects may cascade to communities throughout adjacent seascapes.

Finally, valued species of fishes, dolphins, whales, and seabirds, as well as non-fished "forage" species (i.e., foods for commercial species), concentrate in the most productive areas of the coastal realm: estuaries, lagoons, reefs, algal and seagrass beds, coastal-ocean benthos, and upwellings, among others. On a regional basis, high fish biomass occurs in geographic units designated as "Large Marine Ecosystems" (LMEs). The 50 LMEs so far described (Fig. 2.1, page 36) presently account for 95% of the total annual yield of marine fisheries. Within these LMEs there is a high internal consumption of recruits by other fishes, as well as external consumption by humans, marine mammals, and birds. By removing significant portions of biomass, fisheries influence these systems, but how and to what extent remains uncertain.

Establishing reference conditions for effects imposed by fishing involves understanding a wide array of interactions, including climate, community ecology, natural history, and the scale of observation. Discriminating natural from human-caused changes and subsequent interactions among them forces hypothetical predictions about anthropogenic effects of fishing on the structure and function of ecosystems. Nonetheless, abundant evidence exists that fishing can affect ecosystems by disrupting associations among life histories or ecological processes, which can lead to sudden changes in system state. Why some biotic communities may be resilient to depletion or loss of a few species while others collapse and change may be explained by the law of collapse (Weinberg 1975), which centers on the degree of interconnectedness and dependency among components that give systems their identity and make them vulnerable to threshold effects. When co-adaptations occur among two or more species, and if any one of the co-adapted species find themselves outside the network of interactions, each member's survival can be impaired.

8.3.1.2 Fishing: the stressors

Fisheries usually proceed in stages (Fig. 8.1). Stevenson (1892) observed the process of oyster exploitation, which is generally applicable to other exploitation patterns: "In every region of the world where the oyster industry has assumed any commercial importance it has passed, or is apparently passing, through the following four stages: first, the natural reefs in their primitive condition and furnishing the entire supply of oysters; second, those reefs somewhat depleted and producing small oysters many of which are transplanted to private grounds and under individual protection permitted to mature; third, the public beds so far depleted that the supply available is very irregular and uncertain and consists almost entirely of small oysters which are transplanted to private areas; fourth, the entire dependence of the industry on areas of ground under individual ownership or protection." This pattern of initial immense harvests at maximum exploitation that concludes with

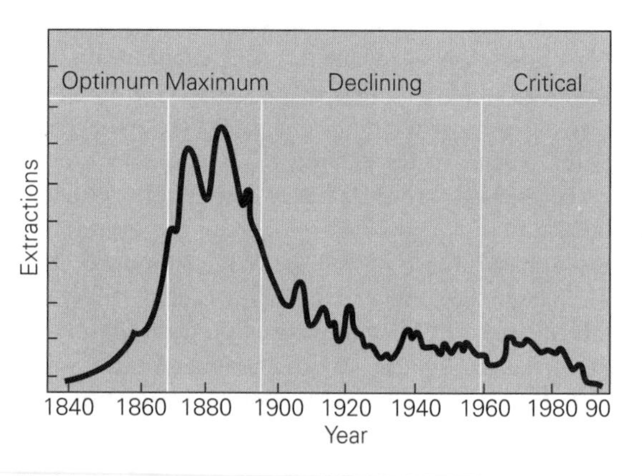

Fig 8.1 Oyster harvest for Chesapeake Bay (1840–1990) illustrates stages in exploitation. Optimum: harvest is on healthy, resilient populations. Maximum: displays fluctuations in yield. Declining: extraction occurs with declining yields. Critical: yields are low and populations require subsidy.

declines to a critical state is typical of fisheries exploitation generally, a situation that continues to the present day.

Human extractions of marine fish have increased four-fold since World War II, reaching more than 100 million mt in 2000. Presently, about one-third of the global fishing industry depends on depleted fish populations. Furthermore, catch figures underestimate true harvest, failing fully to account for commercial extraction and rarely accounting for recreational and artisanal catch or illegal poaching. A pervasive feature is that fisheries are mismatched with the time–space scales of the fished population's dynamics. Species and their habitats are partitioned among different fisheries, agencies, and economic interests; examples include North Atlantic bluefin tuna, walleye pollock, salmons, and many others.

The magnitude and extent of global fisheries strongly suggest that fisheries have become stressors on populations, communities, and coastal-marine ecosystems. Humans have historically selected the largest, tastiest, most abundant, and most accessible species (right whales, oysters, tunas, cods, groupers, etc.). Twenty marine fishes have accounted for almost 40% of the total landed weight; five have dominated global landings since 1950 (Jennings et al. 2001): Atlantic cod (*Gadus morhua*), Peruvian anchovy (*Engraulis ringens*), Alaskan pollock (*Theragra chalcogramma*), Atlantic herring (*Clupea harengus*), and Japanese pilchard (*Sardinops melanostictus*). Humans also prefer many planktivorous filter-feeders (oysters, sardines, herrings, shad, etc.) and detritivores (shrimps, crabs) that are low in the food chain and of greater biomass than larger piscivorous fishes. Fishing operations also destroy habitats. Bottom trawling mobilizes sediment, disrupts benthic communities, and alters benthic systems. Teeth on scallop dredges snag rocks and other sea-bed structures, and hydraulic dredges use jets of water or air to lift dredged material up a pipe and onto the operating vessel or to fluidize the sediment directly in front of the dredge head. Trawling is usually patchily distributed, but can affect large areas when it is intensive. Some North Sea areas are visited more than 400 times per year, rendering extensive changes to benthic habitats. Dynamite fishing, illegally used in poorer regions of the world, indiscriminately destroys structural organisms (corals) and totally destroys local reef habitat (Fig. 8.2).

Fig. 8.2 Rubble and a large crater illustrate the impact of dynamite fishing on coral-reef structure. Kisiti Reef, Kenya, 1969. Photograph by the authors.

Internal community stressors are induced by changes in species abundance and dominance. The removal of large portions of animal populations changes community structure and significantly alters trophic relationships. When populations of top predators (swordfish, tuna) are reduced, prey species (sardines, anchovies), with higher rates of reproduction and growth and lower ages of maturation, increase in abundance and tend to dominate communities. Less competitive species achieve a differential advantage. Thus, removing the largest, most dominant species causes internal changes in community structure, biodiversity, and food-web relationships. Also, seabird and marine mammal feeding areas can be affected when heavily fished areas occur at or near rookeries. When food is demanded to nourish their young, low food supply can decrease recruitment and survival. And, if these important consumers decrease, predation pressure on prey species differentially changes, exerting further stresses on food webs and communities.

Finally, the pace of overexploitation and depletion allows little reprieve for recovery. Genetic and population viability analyses indicate that populations reduced to low levels are susceptible to genetic drift and prone to extinction, and that removal of the largest, oldest, and most reproductively active individuals may alter the genetic structure of the fished population. Genetic drift occurs when allele (multiple forms of a gene) frquencies change from one generation to the next simply due to chance, which is most likely in small populations. Thus, a small population has greater susceptibility to a number of deleterious genetic effects, such as inbreeding and outbreeding depression and loss of evolutionary flexibility. Once a population falls below a certain size, it may become unstable due to failure to find mates or other hindrances to reproduction

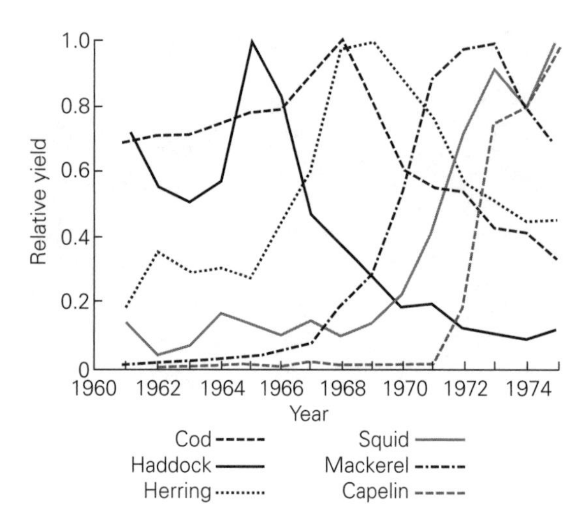

Fig. 8.3 Relative yields of fishes and squid in the northwest Atlantic. Yields are presumed to reflect changes in fished populations. These changes imply change in ecosystem function. From Steele (1991), with permission.

– that is, the Allee effect, or negative density-dependence at small population size.

8.3.1.3 Fishing: the change

Examples of changes caused by fishing are well documented. Removal of gastropods in Chile was followed by an increase in mussels, a principal prey of gastropods. Removal of a few species of urchin-eating fishes by fishers on Kenyan reefs has led to the proliferation of bio-eroding urchins. Overexploitation of Baltic cod stocks has shifted the balance of fish species toward the present dominance of herring (*Clupea harengus*) and sprat (*Sprattus sprattus*). The Pacific sardine (*Sardinops sagax*) fishery that peaked in the 1930s and collapsed during the 1940s and 1950s was followed after a lag of about a decade by an increase of the northern anchovy (*Engraulis mordax*), which then became a major fishery. More than 90% of the baleen-whale biomass of the Southern Ocean was removed by whaling prior to 1920, which has led to an increase in krill-eating crab-eater seals (*Lobodon carcinophagus*) and Adelie penguins (*Pygoscelis adeliae*).

The northwest Atlantic has borne witness to dramatic alterations in fish abundances (Fig. 8.3), implying a change in "functional diversity," that is, "the variety of different responses to environmental change, especially the diverse space and time scales with which organisms react to each other and to the environment" (Steele 1991). These changes appear, at least in part, to be responses to fishing. For example, on Georges Bank excess removal of groundfish and flounder has led to increased abundance of skates (*Raja* spp.), and spiny dogfish (*Mustelis canis*) have since become the major predator. The abundance of heavily fished American lobster (*Homarus americanus*), which prey on urchins, has been correlated with shifts between kelp and urchin-dominated benthos.

Overharvesting has globally affected the entire trophic array of many regional ecosystems, including top oceanic consumers (tunas, billfishes), filter-feeders (oysters, mussels, whales), herbivores (abalones, limpets), anadromous fishes (shad, salmon), and deposit-feeders (clams, crabs, lobsters). Some species have been exploited to the point of endangerment. White abalone (*Haliotis sorenseni*) is a major grazer of kelp in coastal waters of southern California, and has been so intensively exploited that populations have been reduced from 10 000 individuals per hectare 20 years ago to the last few dozen individuals today. This species is now so sparsely distributed that it is unable to reproduce, nor can it fill its former role as a significant grazer on algae. This is one of the few totally marine species that scientists confidently claim is in immediate danger of extinction due to overexploitation; a captive-breeding program is underway.

By-catch can also affect populations and communities. Fishing has severely depleted the formerly abundant barndoor skate (*Dipturus laevis*) of the North Atlantic. Yet, other species have benefited from fisheries. Seagulls have learned to follow fishing boats to take advantage of discarded catch. Some seagulls have become pests that raid garbage disposal sites. These aggressive birds also disrupt and alter the broader seabird community through nest predation and competition.

These changes strongly imply changes of whole ecosystems. Only a few depleted species of fishes have recovered to former abundance, indicating a loss of population or ecosystem resiliency. One of the most striking effects of the selective removal of fishes during the past 50 years' explosive growth of global fishing is the fishing down of aquatic food webs (Pauly et al. 1998, 2000). This phenomenon has occurred as local populations of the largest, most commercially valuable species have been depleted, whereupon fishers move to new locations in search of new populations and new species; as these are depleted, fishing concentrates on smaller, planktivorous species. Hence, food webs shift from those dominated by fishes at high trophic levels to ones dominated by fishes at lower trophic levels. Food and Agriculture Organization (FAO) data show that mean trophic levels of fished species have declined for global-marine fisheries in general, especially in the Mediterranean and the northwest Atlantic. Several factors contribute to this trend. Various "industrial" uses of fish have expanded, for example for fertilizer, aquaculture, animal feed, and pet food (especially for housecats). Also, fishing efficiency has increased. Enormous gill nets, long lines kilometers in length, and fish-finding technology have made possible the exploitation of virtually every population, as well as the amplification of waste.

Extractions by fisheries, accompanied by cyclical climate changes, have long been known to affect community structure and to extend into ecosystem structure and function. Data from early fisheries in the North Atlantic and elsewhere suggest that populations of some species were an order of magnitude greater 100 years ago than at present. However, changes are compounded by factors other than fishing alone. The Russell cycle is a well-documented, multidecadal, reversible shift in oceanographic conditions in the North Sea that is accompanied by a concurrent shift in the fish community. This cycle occurs in response to a North Atlantic climate oscillation and appears to involve a regime shift from one fairly stable configuration of organisms (herring, macroplankton, demersal fishes) to another (pilchard, small plankton). Similarly, in the eastern Pacific, effects of El Niño–southern oscillations (ENSO) events, driven by changes in atmospheric pressure, reverberate through marine food webs to affect fishes and their consumers. Following El Niño events, populations of short-lived species, such as Peruvian anchovies, have crashed. These changes appear to occur naturally, but fishing cannot be excluded as a contributing factor.

8.3.2 Introductions: exotic and invasive species

Exotic species are nonindigenous or alien species and, with invasive species, form a category of biological pollution that results from human activities. Charles Elton (1958), in his classic book on invasions, stated: "In contrast to land and fresh waters the sea seems still almost inviolate." Yet, Elton lists "big changes" that have occurred as a result of digging canals (Panama, Suez), ship transport, and deliberate introductions. Lachner et al. (1970) foresaw future increases in exotic introductions as a result of demands of the aquarium trade and sportfishers, and "accidentally by bilge-water transport." Presently, exotic introductions and invasions are considered to constitute a major feature of environmental change on a global scale. Carlton (1999) noted: "More than 1,000 nearshore plants and animals that are now regarded as naturally cosmopolitan may represent overlooked pre-1800 invasions."

Exotic and invasive species are spreading worldwide. As natural community and ecosystem barriers that maintain ecological structure and organization are weakened, exotic and invasive species can threaten natural diversity and ecosystems. Exotic and native species may proliferate to become invasive and may replace native species, dominate ecosystems, or cause disease, thereby changing the evolutionary pathways that give nature its inventiveness through chance, opportunity, and new options.

8.3.2.1 Exotic and invasive species: the reference

Exotic and invasive species proliferate because of genetics, natural history, dispersal mechanisms, and opportunities that are presented when environmental conditions become favorable. Exotic species may benefit from changes in the home (donor) region or recipient region, and from alterations of dispersal mechanisms. For example, increases in coastal nutrient loading result in large phytoplankton biomass in the home region, which increases the likelihood and frequency of phytoplankton being taken up in ship's ballast water and repeatedly introduced to recipient regions. Emergence of new dispersal mechanisms (faster, larger ships with new destinations) favors increasing frequencies of biological invasions. And changes in recipient regions can alter the region's susceptibility to future invasions, and can increase the potential for pollution-tolerant or generalist (*r*-selected) species to invade elsewhere. Species' invasions are facilitated when new niches become available, as when dominant organisms are removed, or when a change in environmental conditions takes place, populations adjust, and food webs are altered.

Certain biotic qualities contribute to successful biological invasions. It is not possible to characterize all the biological traits of a successful invader; Ehrlich (1986) has posed some generalizations: abundance in the native habitat; facultative feeding habit; short reproductive cycle; high genetic variability; capability of fertilized females to colonize alone; larger size than

Fig. 8.4 Exotic Australian "pine" (*Casuarina*) creates unstable beaches. Pine needles inhibit growth of stabilizing beach plants, leading to erosion and tree fall. Andros, Bahamas, 2001. Photograph by the authors.

Fig. 8.5 Impact of an overabundant endangered species on island vegetation. In 1973, six male and five female Bahamas hutia (*Geocapromys ingrahami*), a native Bahamian rodent, were introduced to a small cay in the Bahamas. A decade later, hutias had overrun the island, stripping it of most non-toxic vegetation, making poisonwood (*Metopium toxiferum*) the dominant plant (Campbell et al. 1991). Bahamas, 1994. Photograph by the authors.

close relatives; natural association with humans; and ability to function well in a wide range of physical–chemical environments. Genetic variability seems especially relevant. As species evolve to maximize fitness, any genetic change that influences fecundity or survivorship alters not only individual fitness but also community fitness. However, why some species are extremely successful invaders and close relatives are not, or why some habitats are colonized while others are not remains a puzzle. Some clues come from general observations, which indicate that exotic plants and animals differ in the ways that they alter native population dynamics and community structure. Exotic plants generally cause major changes in ecosystems by replacing other species (Fig. 8.4), whereas animals cause changes by consumption (Fig. 8.5). Climate change also can increase the probability and extent of biological invasions; for example, species restricted to lower latitudes may become able to colonize higher latitudes, and marginal populations in middle to higher latitudes can increase in abundance and spread. At smaller scales, subtle changes of temperature, oxygen, salinity, flow patterns, and pollution levels change natural community composition by shifting the selective advantage of one organism or population over another.

Exotic species introductions and invasions into marine systems may be leading to profound ecological changes (Carlton & Geller 1993). Distinct biogeographic patterns that took thousands of years to evolve are becoming biologically disordered and homogenized in decades or less. Where close connections have evolved between the biota and its environment, invasive species may become dominant. In disturbed sites throughout the world, where natural biogeographic patterns and barriers break down, invasive species have become especially prominent. Assessment of the consequences of invasions, especially by exotics, must take account of history and changed baselines.

8.3.2.2 Exotic and invasive species: the stressors

Human activities are introducing microbes, pathogens, and exotic, transgenic, and invasive species into coastal-marine ecosystems at unprecedented rates. Exotic species are abruptly and deliberately introduced by activities that benefit humans (mariculture, transgenic species) or enter inadvertently by accident or unintentionally (ship ballast water, spills, wastewater, septic system discharges, sewer overflows, transported by other organisms). Potential pathogens are introduced with discharge of human waste and even sewage treatment fails to destroy most of them. Wastewater treatment systems throughout the world have experienced water-quality or public-health deficiencies; even in the United States, where sewage treatment is common, more than two-thirds of exposed species may be susceptible to pathogenic infection (Patterson et al. 1991). Human-induced changes that alter ecological, biological, chemical, or physical conditions increase the susceptibility of a region to biological invasion and infection. Mariculture exposes susceptible species to

Fig. 8.6 Large oyster rafts for mariculture alter ecological patterns (e.g., sedimentation, flows, production). Filter-feeding oysters deposit tons of excrement and can reduce phytoplankton, with negative consequences for other filter-feeders. Northeast Australia, 1987. Photograph by the authors.

huge quantities of nutrients, pesticides, antibiotics, and potential diseases that enter their environment. Infective species (e.g., protozoa, fungi, bacteria, other parasites) do well under such circumstances. And human stressors acting together with natural events (e.g., El Niño) can amplify conditions favoring pollution-tolerant or opportunistic invaders.

Domestication of species for farming, aquaculture, or the aquarium trade is a form of evolutionary change (Kohane & Parsons 1988). The transfer of animals from one environment into another requires their adjustment to changes in resources, physical and psychological space limitations, suppression of behaviors, and accommodation to proximity of objects formerly associated with danger. In the process of adaptation to a new environment, selection favors individuals that can best accommodate. Selection can be intense, with some genetic trade-offs being made for long-term fitness and survival. So-called "farmed" species are becoming increasingly valued, and have the potential to appropriate much natural productivity. For example, oyster aquaculture has effects on plankton, sedimentation, and other processes (Fig. 8.6). Atlantic salmon may outcompete native salmons or other species. In New Zealand, salmon introductions have led to drastic reduction of galaxiids (small, southern hemisphere, freshwater or anadromous fishes valued for food).

Genetically altered (transgenic) species are intentionally modified to make them more productive, usually under tightly managed conditions, but if these species are introduced into new environments, they can become stressors on natural systems. Major problems occur when genetically altered species escape or are deliberately released. The Atlantic salmon (*Salmo salar*) is a favored aquaculture fish. This species has been genetically engineered and given growth hormones to turn it into a "super salmon" with very high growth rates and early maturation. Genetically altered fish have escaped and now occur wild in the Atlantic and Pacific oceans. Hybridization of these fish with native salmon may lower fitness of wild populations, making them prone to extinction.

8.3.2.3 Exotic and invasive species: the change

Overabundance of species has become a common occurrence and is usually associated with exotic or native species that have become invasive. The exotic green alga *Caulerpa taxifolia* is transforming several inshore communities where it has been introduced, and the native common reed has become invasive under conditions of environmental change (Box 1.1, page 5; Box 5.2, page 162).

A review of marine and estuarine bivalves and gastropods introduced into North America's Atlantic, Gulf, and Pacific coasts revealed a newly established fauna of 36 exotic species (Carlton 1992). Massive inoculation of tons of Japanese oysters (*Crassostrea gigas*) into the Pacific coast of North America occurred for 60 years between 1870 and the 1930s. This has led to hundreds of introductions of associated protists, invertebrates, algae, seagrasses, and perhaps fish. Japanese oysters have also displaced native species in Australia, California, England, and France, despite protective measures. Natural communities have responded, adjusted, and changed, giving rise to regional patterns that are striking. For example, 30 introduced mollusk species are established on the Pacific coast, 19 (63%) of which now occur in San Francisco Bay; the response has been depletions of others.

Two native, invasive coral-reef species stand out. The crown-of-thorns starfish (*Acanthaster planci*) has caused extensive mortality on many Indo-Pacific reefs. Populations of this starfish are presumed to have increased as a result of nutrients introduced by terrestrial runoff into nearshore waters, enhancing the levels of phytoplankton consumed by crown-of-thorns larvae (Birkeland 1982, 1989). The result is a population explosion. *Acanthaster* population explosions have occurred before human habitation, but local population explosions that may have occurred recently as a result of human land uses cannot be ruled out. This starfish is a striking example of a stressor organism; it is a voracious consumer of corals, well known for its ability to reduce coral cover and to change the relative abundances of corals, thereby also altering the coral-dependent biotic community. The ecological effects range from replacement of hard, reef-building corals by soft corals and sponges that do not build reefs, to invasions of algae, and changes in the fish community. Thus, population explosions, followed by rapid declines, may make recovery of corals or attaining equilibrium conditions in the reef impossible.

Another example concerns increase in abundance of the Caribbean native, algae-eating, long-spined sea urchin (*Diadema antillarum*), which probably occurred as a result of overfishing of herbivorous reef fishes. Because the urchin took over the role of major grazer, the effect of overfishing was not accompanied by a change in the coral-to-algae ratio. Only when a rapidly spreading disease caused mass mortality of *Diadema* did a dramatic increase in algae occur, followed by a decline in coral cover and changes in the fish community. The end result has been a sudden "shrinkage" in the entire coral–fish–urchin system (Jackson 1994). What the future effect may be for Caribbean reefs is difficult to predict. *Diadema* appears to be increasing slowly, raising the question of whether Caribbean reefs exhibit multiple stable states (Knowlton 1992). Possibly, effects could also cascade throughout other environments and biota, as reefs widely influence adjacent systems throughout the coastal seascape.

Exotic species, contrary to natives, can introduce entirely new features and mechanisms to recipient environments. The recent discovery of the Indo-Pacific lionfish (*Pterois volitans*) in benthic environments along the Atlantic coast of North America represents the first, apparently successful, introduction of a tropical marine fish to the western North Atlantic (Whitfield et al. 2002). Lionfishes have highly toxic spines that would deter naïve predators; they also feed in a fashion not familiar to Atlantic fishes. Both features could allow for rapid population increase. The potential extent of their predation on small fishes on reefs and wrecks in their new habitat remains to be seen.

Even more striking are disastrous results from introductions of exotic species during routine ballast-water releases, which have affected both biotic communities and human economies. A few examples are: a toxic dinoflagellate (*Gymnodimium catenatum*) that now occurs from Japan to Australian waters; the zebra mussel (*Dreissena polymorpha*) and European shore crab (*Carcinus maenas*) that are flourishing in North America; and large numbers of the comb jelly (*Mnemiopsis leidyi*) in the Black Sea. More marine invasions from ballast-water release and aquaculture activities are expected in the future, and once they arrive, the outlook for dealing with them is bleak.

A serious effect of invasions concerns human health. Ballast water releases large numbers of microbial pathogens throughout the world, including *Vibrio cholerae*, the agent of cholera (Ruiz et al. 2000). Concentrations of bacteria and viruses can exceed those reported for other taxonomic groups by 6–8 orders of magnitude, and the biology of microbes may result in high rates of invasion compared with invertebrates and vertebrates.

Reports of toxic red and brown tides are increasing worldwide. Some diatoms are reported to be toxic, producing a neurotoxin (domoic acid) that causes seizures and death to birds and marine mammals (Mlot 1997); in the 1990s, California pelicans, sea lions, and cormorants died after feeding on anchovies and sardines containing this toxin. A dinoflagellate (*Pfiesteria*), formerly considered rare and innocuous, has become more abundant and lethal to fish populations. Such incidents have had serious impacts on biotic communities and have also brought economic hardships to humans, but their origins remain unknown, and the status of many apparently new species or species forms is uncertain.

8.3.3 Introductions: pollutants

By definition, organic and inorganic pollutants are generally chemically active, with characteristics of persistence, mobility, and short- or long-term adverse biological activity. Key considerations for assessing organic pollutants in the environment include: (i) routes of entry; (ii) reactions of these compounds with the environment; (iii) rates of reactions and movement; and (iv) reservoirs of short-term (days to months) and long-term (years to decades) accumulation (Farrington 1989). However, baseline reference points and ecosystem effects are especially difficult to discern because of the very high number of pollutants entering coastal-marine environments, because pollution has preceded monitoring efforts, and because organisms differ widely in their sensitivities. Persistent inorganic pollutants become a concern when they occur in high amounts, are widely distributed, and take on properties of toxicity. Pollutants become most worrisome when they concentrate in the biota, magnify through food webs, and synergize to cause unexpected consequences.

8.3.3.1 Pollutants: the reference

The route and concentration of pollutants through food webs is complex, with physiology, chemistry, and ecology playing important roles that influence final outcomes. Most aquatic organisms filter water across gills and consume organic matter that may contain soluble or sorb-contaminated particles, making solubility and sorption key factors in determining bioavailability and bioconcentration of pollutants. Many organic chemicals of environmental concern (PCBs, PAHs, etc.: see Chapter 1) have low solubility in water and are hydrophobic and lipophilic. If presented with an opportunity, organic compounds will partition into other substances and undergo sorption and solution in fats. Because almost all particles in aquatic ecosystems have an organic coating – a liquid or gel-like film – hydrophobic and lipophilic compounds tend to sorb onto them. Also, a substantial portion of dissolved organic matter in coastal waters is in the form of humic-like material, which contains colloidal organic matter and has a substantial capacity to sorb hydrophobic and lipophilic compounds.

Plants and animals take up contaminants directly from air, water, or soil in a process referred to as bioconcentration. Pollutants in water and sediment enter aquatic organisms directly through feeding activities or indirectly by sorption. Absorption may occur across the body surface, gut wall, or gill epithelia. Metals are essential to cellular enzyme function in marine animals, and marine animals normally accumulate them in proportional, or nearly proportional, amounts of 4–5 orders of magnitude greater than surrounding seawater. Because of their large surface-to-volume ratios, planktonic organisms can quickly take up metals when concentrations are high. Macroalgae, macroinvertebrates, and fish also absorb metals from water, but much less rapidly than plankton because of their smaller surface-to-volume ratios. PCBs also associate with suspended microparticulates. By processes of adsorption and absorption, phytoplankton can rapidly incorporate PCBs as well as DDT, and once it has accumulated, several species of phytoplankton can metabolize DDT to more toxic and persistent metabolites (DDD, DDE). Responses to toxic pollutants vary considerably among species or organisms.

Bioaccumulation occurs when a contaminant is not broken down, detoxified, or excreted by the organism, and the contaminant body load increases with time. Bacteria can accumulate various compounds to levels 1000 to more than 100 000 times the concentration in water, presumably by direct lipid/water partitioning. Invertebrates exposed to PCBs over relatively long times can have enormous concentrations, typically ranging from 10 000 to 1 000 000 times. Bivalves accumulate very high metal concentrations when other species show little or no accumulation, and also have a high potential for accumulating PAHs. Toxicants can be transformed during the process of bioaccumulation. For example, sulfate-reducing bacteria (*Desulfovibrio*) can make mercury more toxic by converting it to a methylated form (Fowler 1982), under anaerobic, ebb-tide conditions of estuarine marshes (Folsom & Wood 1986). While tissues of many heterotrophic animals can degrade some toxic compounds, many organic compounds (e.g., low-solubility chlorinated and petroleum hydrocarbons) preferentially end up in the lipid components of biota. Physiological mechanisms can detoxify contaminants, as through cellular processes involving metallothioneins (proteins with high affinity for mercury, cadmium, zinc, copper, etc.), or make them more toxic.

Biomagnification is used to describe the generic uptake or transformation of chemicals up the food chain, from plants or plankton to herbivores, small consumers, and larger consumers. Consumers at each level can be affected mildly or lethally, but the final consumer, including humans, can receive the most toxic, or even lethal, dose. Biomagnification is the major reason why persistent environmental contaminants reach high concentrations in top predators, even when levels in air, soil, and water are low. Thus, in the process of biomagnification, trace metals can be mobilized from sediments by algae, detritus particles can bind them, microbes can transform them, macrofauna can assimilate them, and top predators can potentially receive high doses.

8.3.3.2 Pollutants: the stressors

Metals and synthetic organic pollutants are becoming increasingly available to organisms as a result of human activities, most with uncertain consequences. Trace metals and petroleum hydrocarbons are being mobilized and redistributed in toxic proportions. The order of decreasing toxicity of common heavy metals in aquatic organisms is: mercury, cadmium, copper, zinc, nickel, lead, chromium, aluminum, and cobalt (Kennish 1998). Organotins (butyltins), first synthesized in the 1930s, substantially increased in production and usage by 1960s, and are widely used in antifouling paints on boats, fish nets, and lumber preservatives today. Combustion of fossil fuels, waste disposal, and industrial applications (most importantly chloralkali plants, mining operations, incinerators, etc.) release an estimated 10 000 mt of mercury annually to the marine environment (MPB 1998; UNEP 1982). About 196 000 mt of mercury were produced by Latin American silver mines between 1570 and 1900 (Nriagu 1993). Antifouling paint on boats and submerged metal structures equipped with anodes to help prevent corrosion can leach aluminum, copper, and zinc into the water.

Hydrocarbons, in the form of fossil fuels, petroleum, and synthetic halogenated compounds, are ubiquitous products of the post-World War II era. Routine tanker operations, accidents, municipal wastes, and atmospheric fallout contribute substantial amounts to coastal waters. Collisions or spills occur during transport from production centers to refineries. Shipping lanes along approximately 200 m depth contours take tankers relatively close to coastlines and through narrow straits and passages. On average each year, oil and gas production or exploration introduces about 38 000 mt of hydrocarbons into the marine environment worldwide, with 3000 mt to North American waters; transportation activities release 150 000 mt and approximately 9100 mt, respectively. But the vast majority is contributed by individual cars, trucks, boats, non-tank vessels, and runoff from paved urban areas: about 480 000 mt on average worldwide, including 84 000 mt to North American waters (NRC 2002). Aromatic hydrocarbons may account for approximately 1–20% of the total hydrocarbons in crude oil.

Synthetic organic compounds emerged with post-World War II economic and scientific development, and have been applied in increasing quantities since then. About 0.45×10^9 kg were produced annually by the United States just before the War; by 1950, production passed 9×10^9 kg, and by 1985 production reached 101×10^9 kg. Control of pests (mostly insects), plant pathogens, and weeds in the United States alone contributed almost 500×10^6 kg of pesticides (primarily synthetic organic chemicals) and 27×10^6 kg of sulfur and copper sulfate fungicides around two decades ago (Pimentel & Levitan 1986). Presently, humans are speeding up the toxifying processes by synthesizing thousands of new, highly toxic organic compounds with high efficacy, many of which are persistent, have affinity for fat, and may evaporate and travel long distances. Such persistent organic pollutants (POPs) include "hormonally active agents" (HAAs) and "endocrine disrupters" that can cause adverse reproductive and developmental effects in human populations, wildlife, and laboratory animals, although the extent of harm is debated (NRC 1999). Chlorinated hydrocarbons (DDT, pesticides) contain many persistent organochlorines that have been widely banned; however, usage of these compounds remains attractive to developing countries because of their low cost and effectiveness.

8.3.3.3 Pollutants: the change

PCBs, DDT, chlordane, insecticides, herbicides, trace metals, and petroleum hydrocarbons are common coastal-realm contaminants. PCBs have been detected in food packaging and animal feed, soils and sediments, and in industrialized areas like the Hudson River (New York), where very high levels of PCBs occur in sediment. Heaviest loads of hydrocarbons (chlorinated and polynuclear aromatic) occur near urban and industrial developments, and also in varying concentrations in oceans from equatorial to polar regions (Kennish 1998). For example, PCBs have been transported and dispersed by currents from industrializing countries of the Pacific west to Midway Atoll in the northwest Hawaiian islands (Hope et al. 1997).

Overall, there has been a 1000-fold enrichment of tributyltin (TBT) in some surface microlayers and in some surface sediments (Goldberg 1998). Butyltin and TBT contamination is widespread along Thailand's coast, and is associated with large commercial ship and fishing boat hulls (Kan-Atireklap et al. 1997). TBT and its degradation products (Dibutyltin, DBT; Monobutyltin, MBT) have been detected in most Gulf of Maine estuaries, even after the use of TBT on vessels of less than 25 m in length was banned (Larsen et al. 1998). TBT occurs in top predators such as marine mammals, for example bottlenose dolphins in Italian and U.S. waters, porpoises from the Baltic Sea, and seals, Steller sea lions, and a range of cetacean species from Japan and the north Pacific Ocean. TBT, MBT, and DBT have also been found in the livers of porpoises and gray seals stranded or caught around England and Wales (Law et al. 1998), and dead California sea otters have been shown to have high levels of TBT and other butyltin compounds (Henderson 1998).

POPs occur widely in tissues of sea life, in high, often toxic, concentrations. Blue mussels in polluted harbors of southern California concentrate high-molecular-weight aromatic hydrocarbons, including benzo(a)pyrene, a PAH and endocrine disrupter, from creosoted pilings and industrial (especially refinery) effluents, domestic sewage, oil spills, aerial fallout, and bilge water from ships, particularly oil tankers (NRC 1990). Fish from contaminated waters, and larger and older individuals in waters of low pollution, can carry high levels of toxicity. Marine mammals, fish, birds, and invertebrates have high concentrations of PCBs, PCDDs (polychlorinated dibenzo-f-dioxins), and organotins: these pollutants are known to produce epidemiological abnormalities including endocrine disruption, immune responses, reproductive dysfunction, tumors, population declines, and other effects. For many species that concentrate DDT, 80% is converted to DDE, presumably from metabolic breakdown, and is eliminated slowly, making it the major POP in fish, wildlife, and human tissues. In places where DDT is still used as a pesticide, DDE concentrations in human and wildlife tissues can be very high (NRC 1999).

These synthetic pollutants have now spread far into ocean food chains. Polybrominated biphenyls (PBBs) and polybrominated diphenyl ethers (PBDEs), commonly used in electronic equipment, have been detected in sperm whales (*Physeter catodon*), a species that feeds in deep waters off continental shelves. Not unexpectedly, high levels of POPs have been found in seals and dolphins that inhabit the North and Baltic seas, and many cases of reproductive disorders associated with hormonally active agents are being reported. Low calf production, survival, and population decline are reported for belukha whales (*Delphinapterus leucas*) of the St. Lawrence River, possibly due to exposure to PCBs and DDT. Several species of Baltic Sea fish, especially sea trout (*Salmo trutta*), cod (*Gadus morhua*), and Atlantic salmon (*Salmo salar*), have shown reproductive disturbances since the 1970s. A serious but complex reproductive disorder has been described for Atlantic salmon returning from Baltic Sea feeding migrations. And Baltic Sea gray seals (*Halichoerus grypus*) and ringed seals (*Phoca hispida*) have exhibited low fertility rates and occluded uteruses.

Response sensitivity differs greatly among species. Eggs of herring gulls (*Larus argentatus*) hatch and chicks fledge normally when the birds are contaminated with mercury, but mallard ducks (*Anas platyrhynchos*) exhibit reproductive dysfunction. For fish, toxic pollutants primarily target eggs, larvae, and early juveniles stages, with limited effects on adults (Irwin et al. 1997). These examples indicate that impacts can be difficult to interpret. Also, many large oceanic fishes, birds, and mammals, presumably remote from sources of pollution, seem naturally to contain high mercury concentrations, especially the toxic methylated form. It may be "normal" for some species to carry body loads of about 150 $\mu g\ kg^{-1}$ of mercury. Seabirds naturally acquire mercury, and concentrate it mostly in their livers where it can be demethylated, apparently without detrimental effects. Wandering and sooty albatrosses (*Diomedea exulans* and *Phoebetria fusca*) that form breeding colonies on Gough Island in the remote South Atlantic Ocean have high concentrations of up to 271 $\mu g\ kg^{-1}$ mercury in their feathers. The livers of dead Juan Fernández fur seals (*Arctocephalus philippii*), 800 km off the coast of Chile, contained significant concentrations of mercury, cadmium, and lead. And liver, muscle and skin of several species of Canadian marine mammals contained mean levels of methylmercury that generally exceeded the Canadian Federal Consumption Guideline for mercury in fish.

Despite variability of effects, some kinds of pollution especially organics, are known to reduce biodiversity, biomass, and species abundances, depending on concentration (Fig. 8.7). Such changes inevitably cascade through biotic communities and also can affect human health. In the 1950s, the Japanese community of Minamata consumed lethal concentrations of mercury, which resulted in many deaths. Levels of cadmium in kidneys of ringed seals are higher than the critical concentration of 200 $\mu g\ g^{-1}$ wet weight, a level that is associated with kidney damage in mammals. Exposure to high cadmium levels can be lethal for humans, as illustrated by the cadmium-related human disease called "itai itai" that struck a Japanese village following World War II. Chlorinated pesticides, such as Chlordecone (Kepone®), Mirex, DDT, and the plastic PCB, are of most concern for their persistence and toxic consequences. These and other organochlorines

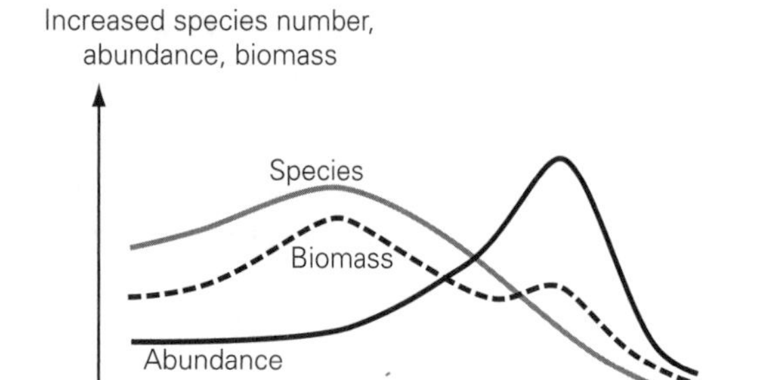

Fig. 8.7 Increased organic input affects biodiversity, biomass, and species abundances differently. Species diversity and biomass are affected first. Eutrophy may cause abundances of a few tolerant species to increase, but extreme eutrophy causes severe declines. Lethal effects occur at high concentrations. From Gray (1993), with permission.

have been widely distributed and are conservative pollutants (not readily degraded by chemical oxidation or bacterial action), and are essentially permanent additions to the marine environment.

8.3.4 Physical alteration

Changes of land–seascape structure have profound implications for a wide variety of conservation issues, from species and habitat protection to ecosystem function and establishment of protected areas. Coastal-realm conservation is necessarily concerned with large biogeographic areas wherein regional scales of interaction take on special significance. Discerning regional patterns and land–seascape character provide useful tools for interpreting physical structure, identifying biotic distributions and disturbances, and understanding processes related to habitat loss, fragmentation, and effects on biodiversity. Physical changes observed at regional scales draw attention to the scope of mechanisms needed for coastal-realm conservation.

Natural land–seascapes are distributed as patches that vary in shape, size, type, heterogeneity, and boundaries. Boundaries define patch units, ecotones describe transitions among patches, and spatial mosaics can indicate how landscape heterogeneity evolves and supports biodiversity. The scale and organization of ecosystems may be revealed by form and pattern, linkages and discontinuities, sources and sinks, persistence and resilience, and degree of apparent randomness and variability.

8.3.4.1 Physical alteration: the reference

Beaches, barriers, estuaries, lagoons, islands, reefs, and the coastal ocean form discrete coastal patterns as a result of biophysical linkages among land, sea, and air. Land–seascapes are continuously subjected to seasonal and climatic cycles, and reworked by exchanges among rivers, tidal flows, storms, and biogenic processes. Land–sea–atmospheric forces interact most intensively on shore and shelf zones, where water, sediment, and organisms interact continuously to give rise to new patterns. Rivers transport water, sediment, and materials from land to coastal waters, and forces of waves, tides, and currents are influenced by basin morphology. Erosive processes release sediment, rework fluvial sediments, expand deltas, and form beaches, dunes, and berms in response to hydrological, geological, and atmospheric conditions. Lagoons and estuaries are long-term sediment sinks formed by fluvial regimes, where the benthos collects products of land and sea, and where biogenic processes build structures that contribute to shoreline defenses and substrate stability, and affect processes of sedimentation, circulation, and erosion.

Global climate change, warming temperatures, and sea-level rise can dramatically affect coasts on longer time and larger space scales. Atmospheric temperature changes correspond to heat transfers and thermal expansion in oceans, and when combined with increased precipitation and melting of polar ice, they initiate sea-level rise and coastal retreat. Increased storm and hurricane activity exacerbates coastal change, and strongly affects gently sloping sedimentary shorelands and low-lying islands (e.g., the Netherlands, Bahamas, etc.). Important factors in mediating these forces of change are the rates of accretionary processes (sedimentary, biogenic) relative to erosion and sea-level change.

In such a naturally dynamic and continuously changing system as the coastal realm, heterogeneous and changing coastal land–seascape patterns appear vulnerable. Yet during the past 10 000 years of the Holocene era, productive environments have been sustained despite wide fluctuations in climate, storms, and biological activity.

8.3.4.2 Physical alteration: the stressors

Human activities are fragmenting coastal space and altering natural restorative processes. Urbanization, industrial expansion, and other coastal developments impede normal shoreline processes important to beach restoration, natural coastal transgressions, and

Fig. 8.8 A "Replenishment Zone" for shoreline development, replacing a mangrove swamp. Grand Cayman Island, Caribbean, 1996. Photograph by the authors.

maintenance of biotic habitat. Elimination of natural breakwaters (e.g., biogenic reefs, beaches, marshes) focuses wave and wind energy onto shores. While erecting beachfront houses, condominiums, hotels, and resorts, developers and property owners demand that coasts stay put, and armor shorelines against erosion. As developments and armoring expand, many shores are *en route* to total stabilization. Yet, shorelines armored against erosion have failed to prevent erosion. Rather, attempts at stabilization often increase erosion problems, requiring further installation not only of bulkheads, but also of nearshore breakwater structures (Fig. 8.8).

Urban patches create areas of land–seascape disturbances where resources (water, food, sediments, minerals) are extracted, materials and energy are imported, wastes are released, and natural flow patterns are altered. Concrete structures and roads replace forests and wetlands, impeding the natural recharge of groundwater. As industry, commerce, and homes extract groundwater, subterranean flow is disrupted, water quality is decreased, and the substrate is weakened, allowing subsidence to occur and saline water to penetrate. Urban patches also need power and water, for which dams are built. Dams modify hydrological and biogeochemcial cycles, interrupt organic flows, change the nutrient balance, and alter oxygen and thermal conditions. As water flow is slowed and sediment is deposited, reservoirs require continual dredging, and coasts are starved of sand. Mining sand and gravel from rivers and beaches for construction material, and modification of inlets, further reduce sediment supply that sustains downstream areas. Construction of jetties, piers, groins, sewage outfalls, and other nonparallel structures on shores interferes with longshore sediment transport that replenishes eroded shores. And extractions of groundwater and oil creates conditions for land subsidence.

A host of other physical stressors are imposed on coastal land–seascape structure. Wetlands drained for development alter hydrological processes important to coastal production. Marinas constructed to serve fishermen and recreationalists alter shore configuration, especially those of semi-enclosed water bodies. Natural bays, estuaries, and harbors converted into ports and industrial centers serve supertankers, cruise ships, freighters, naval traffic, and pleasure craft, thereby increasing pollution, erosion, and sedimentation, eliminating critical habitat, and creating more noise and other disturbances. Bigger ships with deeper hulls and wider girths in confined channels create erosive wakes and require channels to be straightened and deepened by dredging. Altered channel geometry changes current velocities and tidal-prism volumes, with subsequent consequences for natural habitats, basin morphology, turbidity, and saltwater intrusion. Furthermore, dredging requires disposal of enormous amounts of dredged material, some of which is highly toxic, that is commonly deposited along sides of channels bordering shoals, or on marshes and behind dikes, further obstructing natural flows between land and sea.

8.3.4.3 Physical alteration: the change

Worldwide, shorelines are eroding at rates greater than they are accreting. For example, in past years the eastern shore of the White Sea has lost 5 m yr^{-1} and the permafrost lowlands of the western Kara Sea (Russia) have lost 7–8 m yr^{-1} (Hayes 1985). Japan's 33 800 km coastline is frequently attacked by typhoons and tsunamis, and erosion has increased so rapidly during the past 40 years that one-half of Japan's coastline has been armored by human structures. In Holland, sea-level rise and human activities have aggravated shoreline losses, and coastal dikes and seawalls have been constructed to guard against land subsidence. Urban centers with armored shorelines have caused hydrological and other adjustments that extend regionally.

Coastal subsidence has become a major global problem, caused in part by sea-level rise and made worse by human activities. In many locations, removals of groundwater, oil, or natural resources have destabilized substrates and increased subsidence, erosion, and saltwater intrusion. Within the tidal lagoon of the island

Fig. 8.9 Estuarine wetland and salt-marsh development. Chesapeake Bay tributary, 1976. Photograph by the authors.

Fig. 8.10 Marshland educational area and wildlife reserve, "squeezed" between beach development of Atlantic City and urban mainland. New Jersey, 1993. Photograph by the authors.

city of Venice, sea level remained relatively static until 1900. Subsequently, sea level has risen at an alarming rate of 3.5 mm yr^{-1}. While this rate is partly due to global sea-level rise, subsidence from groundwater withdrawal, lagoon dredging, and reclamation have also contributed (Carter 1988). In Bangkok, groundwater pumping has caused the water table to subside more than 10 cm yr^{-1}, and on the Nile delta, where about 1800 humans inhabit every square kilometer of habitable land, natural subsidence is as great as 5 mm yr^{-1} (Milliman 1992). Subsidence from fluid (oil, water) and gas extraction has also been considerable in many other locations (Walker 1990): Tokyo has subsided a total of 4.6 m, Osaka 3 m, Houston–Galveston (Texas) 2.75 m, Lake Maracaibo (Venezuela) 3.9 m, and Long Beach (California) 8.8 m.

Anthropogenic structures, dredging, landfill, and shore armoring affect shores and estuaries worldwide at large and small scales (Fig. 8.9). Dredging and diking between 1870 and 1970 have significantly changed the geometry of the Delaware estuary between New Jersey and Pennsylvania, with subsequent change in tidal hydraulics, estuarine circulation, and sedimentation patterns (Nichols 1988). In New Jersey, structures designed to protect the shore have been placed seaward of development, and bulkheads and seawalls have replaced beaches and dunes in an effort to maintain a fixed shoreline (Pilkey & Dixon 1996). These structures have interfered with sediment availability, requiring persistent, recurrent efforts in beach replenishment. Also in New Jersey, USA, Atlantic City is sited on a coastal barrier backed by a large lagoon and marsh; landward development (villages and farms) is almost continuous, placing a "squeeze" on productive marshes, which will have no place to go should sea-level rise occur. (Fig. 8.10). Presently, seawalls for protection against flooding and erosion, as well as other coastal defenses and developments, occupy large portions of the world's coastline (Table 1.9, page 20).

Biotically rich biogenic reefs, wetlands, and lagoons are especially vulnerable to human activities and are seriously threatened worldwide. Biogenic structures serve multiple purposes by protecting shores and providing productive habitat. Many reefs (formed by corals, algae, mollusks, and worms) are in the process of change, some are dead (page 240), and many have been removed by human activity. Their loss is threatening not only species and communities that depend on them, but also shoreline stability and human habitability. With removal of reefs, landward lagoons are depleted or lost, subjecting shorelines and beaches to destabilizing forces. Removal of wetlands has major effects on coastal-realm biota. And alterations of watersheds have effects that extend throughout the coastal ocean.

In sum, physical alteration of the coastal realm has resulted in widespread change of land–seascape pattern, loss of heterogeneity and habitat, altered biotic communities, and dysfunctional ecosystems. As habitats are fragmented and homogenized (Fig. 8.11), significant losses of production and biodiversity have become apparent. As yet, the magnitude and extent of this loss and its effect on regional biota is poorly understood,

Fig. 8.11 Modified coastal lagoon and barrier beach, reconstructed for human habitation. Miami Beach, Miami, 1978. Photograph by the authors.

mainly because of cumulative, cascading, and synergistic effects.

8.4 The human-dominated coastal realm

Today, human populations, conurbations, and industries increasingly dominate the coastal realm. Humans exert environmental controls through intensified resource extractions, introduction and dispersal of exotics, biochemical transformation, artificial light and noise, and physical changes that fragment the land–seascape, thereby placing novel constraints on species and natural communities. The result is emergence of new physical, biological, and chemical properties, manifested at regional scales. Emergence is expressed as an occurrence of something new at any scale of observation, but is most significant at larger, regional scales, implying ecosystem management at those scales.

O'Neill (2001) has stressed that the "ecosystem concept is a paradigm, an *a priori* intellectual structure, a specific way of looking at nature" and further that ecosystems are "now seen as disequilibrial, open, hierarchical, spatially patterned, and scaled." The hierarchical structure of ecosystems means that each ecosystem operates within a larger set of constraints, which subjects ecosystems to a continuous spectrum of change and variability, governed by physics, chemistry, rate processes, and biotic structures. Also, as ecosystem response is non-linear, a change in one variable can lead to disproportionate change in another. This means that responses that are not dissipated may be magnified up to larger scales. For example, as species are depleted or removed from a system, there may at first be no discernible effect, but with additional removals, resiliency can progressively be weakened until even a minor disturbance can produce an abrupt change. Changes in biota, disturbance frequency, and biophysical structures and functions can result in temporal and spatial discontinuities, with subsequent cascading effects on ecosystem function. When discontinuities occur, the system is temporarily thrown into disequilibrium, requiring an adjustment response that sets linked processes in motion.

8.4.1 Regional seas change

A focus on regional seas exemplifies the magnitude of change induced by the sum of human activities. Presumably, most regional seas exhibited "healthy," quasi-equilibrium conditions for the greater portion of their history, until human activities incrementally induced a continuum of change (Fig. 8.12). These activities began as relatively benign extractions, but as extractions intensified, pollutants were added, and environments altered, fundamental changes occurred, first locally then to whole regions. Further, human activities place chronic stresses on the system, with recovery times being too short for natural resiliency to become effective.

Examples of estuaries, bays, and regional seas that have undergone fundamental changes during the past two centuries are many. The San Francisco Bay has been rapidly and continuously modified by human activity since discovery of gold in 1848. Here, wetlands have been diked and filled, freshwater diverted for irrigation, and many exotic plants and animals introduced. The Asian clam (*Potamocorbula amurensis*) and European green crab (*Carcinus maenas*) have invaded and expanded, changing biotic communities around them. High incidences of physical deformities, reproductive failure, and mortality of fishes and aquatic birds have followed agricultural pollution. Hundreds of square kilometers of saltmarsh habitat have been submerged and eliminated due to subsidence caused by subsurface withdrawal of groundwater, oil, and gas. The Bay is now surrounded by giant industrial parks, a profusion of freeways, power lines, homes, and offices. As a result, Bay surface area and depth have decreased, marshes have mostly disappeared, and valuable natural resources have greatly diminished.

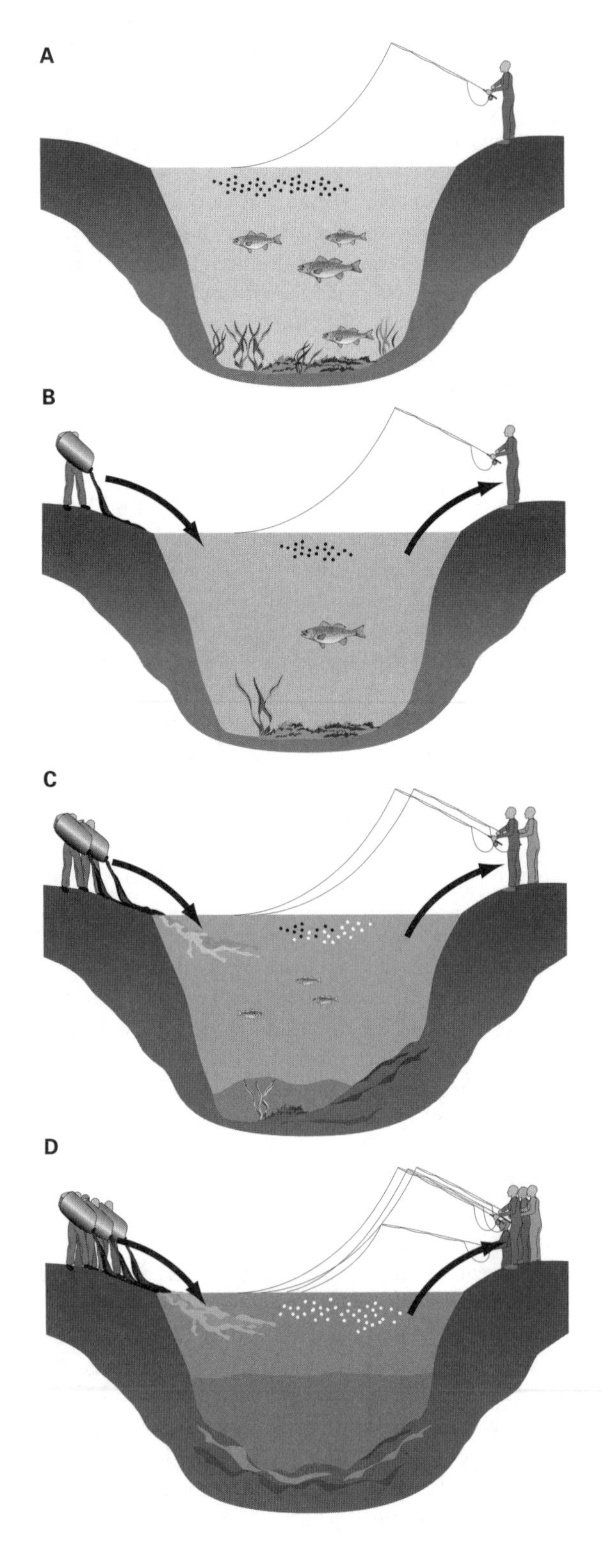

The semi-enclosed, approximately 380 000 km^2 Yellow Sea in the western Pacific is bounded by the Korean peninsula and mainland China, and is influenced from the south by the East China Sea into which the world's fifth largest river (the Yangtze River) drains. The Yellow Sea coast is inhabited by more than 400 million Chinese, and land cultivation is developing at a rate of 3400 $km^2 yr^{-1}$. The coast is undergoing rapid industrialization, which adds heavy metals and other substances to levels that rank this sea among the world's most polluted. Sediments also receive heavy loads of pollutants, from coalburning plants and industrial dumping. The Yellow Sea supports a major fishery, and is one of the most intensely fished areas in the world. Nearly all major fisheries were maximally exploited by the mid-1960s, followed by greatly declining biomass yields. A major shift in species dominance, from longer-lived demersal fish to predominantly short-lived species, notably anchovies, has occurred. The extensive watershed drainage of the Yangtze River, where millions of people dwell, used to deliver large amounts of suspended sediment to the coast, driving the estuary seaward at a rate of 40 m yr^{-1}. Numerous dams have been constructed, interfering with sediment transport, and the Yangtze delta is now retreating. The Yangtze contains a treasury of aquatic living resources, possessing 300 species of fish, of which one-third are endemic. Its annual aquatic production once accounted for 50% of the whole nation's gross output. With construction of the Three Gorges Dam, sedimentation will be further interrupted, many species may become depleted, and the endangered white flag dolphin could become extinct.

Japan's Seto Sea, immediately to the east, is one of the most polluted on Earth. It is more than 423 km long and at least 723 river systems feed into it, creating a large, estuarine sea, where water residence time is 60 years. The Seto Sea supports a major port, trade, and fishing center and is surrounded by highly artificial

Fig. 8.12 (*left*) Generalized stages toward an altered state of a body of water. (A) Low levels of human extractions occur within ecosystem resiliency and recovery rate. (B) Increased fishing and added pollution lead to algal blooms, fewer large fishes, and an altered benthic plant community. (C) Intensive fish extractions and pollution, followed by larger and more frequent phytoplankton blooms and turbidity, lead toward smaller fishes, *r*-selected species, community change, and hypoxia. (D) Toxic pollution and more fish removals lead to altered system function and anoxic bottom water.

shores. It has a long history of fishery production, being one of the first seas to have extensive, large-scale, open-sea mariculture. As a major transportation route between Asia and Japan, the Seto Sea supports a large petrochemical industry. Extensive reclamation has converted 10% of the Sea's shallow water to land for development. Organochlorine pollution has expanded geometrically as industrialized zones have increased, and whole swathes of offshore coastal waters are contaminated by wastewater from coastal disposal outlets. Red tides are common throughout the Sea, with devastating damage to local fisheries. Oil spills also cause serious environmental destruction and damage to fisheries.

The Mediterranean Sea is a highly modified, enclosed sea, bordered by 18 countries. This sea is among the most vulnerable of all seas due to rapidly expanding coastal population, industrialization, and very high seasonal recreation and tourism. If current trends continue, within decades the Mediterranean's coastal region will have nearly 400 million residents, with an additional 250 million or more seasonal visitors. Water exchange with the Atlantic Ocean takes place over roughly 80 years, a sufficiently long residence time to trap pollution and litter. Major tributaries include Spain's Ebro River, with an 83 000 km^2 watershed, France's Rhone River (100 000 km^2), Italy's Po (77 000 km^2), and Egypt's Nile (3.25 million km^2). These watersheds, and others of lesser size, are highly modified by dams, industry, hydroelectric power generation, agriculture, high silt loads, contaminated groundwater, erosion, and toxic algal blooms. As anthropogenic uses continue to grow, water temperature and quality, and circulation patterns are affected. The Sea is severely polluted and nearly one-fifth of all beaches are periodically unsafe for bathing. Fisheries yields are poor and algal blooms are common. Introduced species, such as the alga *Caulerpa*, are causing severe local to regional problems. Hypoxic zones seem to be increasing in extent, and should anthropogenic pollution continue, the Sea may be pushed to conditions of anoxia.

Narrow straits link the Mediterranean to the landlocked Black Sea, which covers 422 000 km^2 and reaches depths of more than 2210 m. Two of Europe's major rivers, the Danube and the Dnieper, flow into it. The Dnieper's 532 000 km^2 watershed is marked by six large dams and 28 large cities. The 796 000 km^2 Danube watershed has 11 large dams, 60 large cities, and 60% of its forest removed. As freshwater input to the Black Sea is limited, there is virtually no flushing, and the Sea is essentially choking from introductions of nutrients and pollutants. Tons of oil enter the Black Sea, along with fertilizers, detergents, waste, and sewage from millions of people. The shoreline is highly engineered, and biotic communities are unevenly distributed. Water takes more than 160 years to flow from the Danube River delta southward through the Black Sea to exit into the Mediterranean. Below its halocline, the Black Sea is naturally anoxic due to weak ventilation of deep layers. Human activities have expanded hypoxic zones, from 3500 km^2 in 1973 to 40 000 km^2 in 1990. A formerly productive fishery is now dominated by small pelagic fishes. The system is marked by large blooms of a zooplankter (*Noctiluca miliaris*) which has low food value for fished species. Also, massive numbers of moon jellies (*Aurelia aurita*) and the exotic American comb jelly (*Mnemiopsis leidyi*) have lead to a nearly total fisheries collapse. Very recently, some signs of fishery recovery have appeared, but the Black Sea remains at high risk.

Two related seas of this region are of special interest. The Sea of Azov and the Caspian Sea provide examples of the potential impact of human activities on semi-enclosed water bodies. The shallow, 37 600 km^2 Sea of Azov lies between the Ukraine and Russia and is linked to the Black Sea through a narrow channel. Freshwater enters from the Don and Kuban rivers and many lesser ones. The 459 000 km^2 Don watershed has two large dams and 22 large cities. The Sea of Azov was once Russia's richest fishing ground and today is highly modified, polluted, and anoxic. During the past three decades, concentrations of phosphorus and inorganic nitrogen have increased four- and six-fold, respectively. The Sea is exposed to accidental and intentional introductions of exotic species, and dikes and breakwaters have been installed to protect seriously eroding shores. The Azov has extensive anoxic areas above its halocline; hypoxia has covered nearly 20 000 km^2 of its shelf, leading to mass mortality of demersal fishes.

The Caspian is the world's largest land-locked sea. This 386 000 km^2, 1000 m-deep salt lake covers an area larger than Japan. Nearly half of its species are endemic, including the Caspian seal (*Phoca caspica*). Once-productive fisheries have been adversely affected by a combination of factors. All the rivers entering it were dammed for power generation and irrigation in the 1940s and 1950s, except the Ural River. The 1.4 million km^2 Volga River watershed, with 14 dams,

61 large cities, and most of its forest removed, contributes a major portion of pollution load. Reduction of the Volga's freshwater input has caused the Caspian's sea level to fall approximately 3 m, to about one-third its original volume. The resulting increase in salinity has affected commercial fishes and a variety of other organisms. Since the 1920s, exotic species have entered the Caspian; this was made worse by the opening of the Don–Volga canal in 1954, when accidental or deliberate introductions added more than 30 new species. Native anadromous species have suffered from dam construction and pollution: the valuable inconnu (*Stenodus leucichthys*) had its 3000 km spawning habitat in the Volga River reduced to 450 km. Catches of salmonids fell from 1500–1900 mt in 1938–9 to 10–20 mt in 1980, and the annual catch of shad (*Caspialosa* spp.) fell from 130×10^3–160×10^3 mt in 1913–16 to 10 mt by the early 1960s. The most important fishery is for three species of sturgeon: the beluga (*Huso huso*), the Russian sturgeon (*Acipenser guldenstadti*), and the stellate sturgeon (*A. stellatus*), all of which are depleted. The greatest biomass in the Caspian is now concentrated in blue-green algae and diatoms.

In northwestern Europe, the 575 000 km^2 North Sea, including the contiguous Baltic Sea, is one of the most intensively used seas in the world. This shallow, northeastern arm of the Atlantic Ocean is the locus of one of Europe's most productive fisheries, where intense dredging has altered benthic communities, and fishery depletion is a major concern. Numerous rivers flow into it; most important are the Elbe and the Rhine-Meuse. The 149 000 km^2 Elbe watershed has 18 large cities and four large dams; the 199 000 km^2 Rhine-Meuse watershed has 68 large cities and six large dams. Additionally, the low-lying Netherlands has reclaimed extensive marine areas behind a line of coastal dunes. During the 1930s, three-fifths of their former sea was reclaimed for farmland, by constructing a dike across the entrance to the Zuiderzee and creating a shallow, freshwater lake, now called the Ijsselmeer. The highly industrialized coastline from Hamburg (Germany) to Brussels (Belgium) is also greatly modified. Thus, both river change and coastal alteration have modified the North Sea region which now accommodates a largely urban and industrialized population, with gigantic chemical and petrochemical complexes, steel mills, metal smelters, engineering and electrical works, cement and power plants, shipbuilding, and other industries. North Sea reserves of petroleum and natural gas are extensive and occasional accidents happen; the largest spill, in 1977 from the Norwegian *Ekofisk* field, released 20 000–30 000 tons of crude oil. However chronic oil pollution from general shipping is more serious than oil spills. Other environmental problems are evident. Nutrient loading has led to eutrophy in some areas, and in 1988, an unprecedented flagellate bloom of *Chrysochromulina polylepis* spread from Danish waters around southern Norway, causing great damage to salmon farms. In 1970, a more than 50% prevalence of a lymphocystis epizootic was reported in flatfish. In 1988, the virus that killed thousands of harbor seals (*Phoca vitulina*) in the North and Baltic seas appears to have been abetted by high levels of PCBs, dioxins, and other organochlorine chemicals discovered in their tissues.

In the adjacent Baltic Sea, pulp–paper mills, mining, metal industries, nutrients from agriculture, atmospheric fallout from western Europe, and sewage plants are major sources of pollution. In deep waters, oxygen deficiency has destroyed benthic communities, killing cod eggs and larvae. In addition, more than 30 exotic species are reported to be abundant.

The Caribbean Sea is included in the subtropical–tropical Wider Caribbean Region, which covers the southern Gulf of Mexico, the Bahamas/Turks and Caicos archipelago, and adjacent Atlantic Ocean waters north to Bermuda and south to Brazil. The Caribbean Sea is mostly low in productivity, is semi-enclosed by an island arc, and is exposed to hurricanes and volcanic activity. Areas of high productivity are restricted to upwellings along the Central and South American coasts, and to coral reefs, mangrove swamps, and seagrass beds found throughout the region. Fisheries production is mostly from shallow waters. The Sea receives freshwater from numerous continental and island watersheds, except for the southeast portion that is affected by several very large river systems, notably the Amazon and Orinoco rivers. The 6.1 million km^2 Amazon River watershed (the largest in the world) and 954 000 km^2 Orinoco play significant roles in circulation patterns and input of materials and nutrients. Neither of these watersheds is developed to the extent of those on other continents. This region is one of recent social and cultural change. "Discovery" by Europeans about 500 years ago initiated dramatic sociocultural, biological, and ecological shifts. The mostly rural/agricultural subsistence life-styles of colonial and independent countries began slowly to change in the late 1800s into the mostly urban/industrial or tourist-based economy of today. In four centuries of

European occupation, many species of birds, mammals, and reptiles disappeared from the West Indies, replaced by horticultural exotics and pests and an overabundance of horses, cattle, pigs, sheep, goats, housecats, and chickens. Today, fisheries are now widely depleted and the region, especially the small-island component (Bermuda, Bahamas, Turks and Caicos, Lesser Antilles), is replete with endangered and depleted species. Since 1980, human populations have grown at an annual rate of approximately 3%, and tourism has grown 9%, with agricultural production and use of fertilizers greatly increasing since the 1960s. Overfishing of reef fish, lobsters, and conch is serious and widespread, threatening life-styles, the economy, and well-being of citizens. Population growth and tourism development have led to widespread impacts on productive inshore environments.

Finally, the Southern Ocean region surrounding the antarctic continent is one of the most productive regions on Earth. The polar continent of Antarctica remains uninhabited, except for scientific and support activities; within the Southern Ocean, the only industries are tourism and fishing. In the Southern Ocean less than a half-century ago, numerous whales, seals, and penguins fed on abundant zooplankton, especially "krill" (*Euphausia superba*), an extremely efficient conduit from primary production to consumers. Whaling greatly reduced whales from the 1880s to the 1980s. As whales were reduced, seals and penguins filled the whale's trophic niche, and the region shifted from a dominance of large whales to a dominance of numerous smaller marine mammals, fishes, and birds. Krill itself is replaced in some years by a pelagic tunicate (*Salpa thompsoni*) as the most abundant filter-feeder, suggesting a potential shift from krill to a jelly, salp-based ecosystem. The effect on birds and mammals is not yet known.

These examples present a snapshot of changes in regional seas and illustrate several features. One feature concerns regime shifts, which are cyclical shifts in ecosystems. Among the best-documented regime shifts occur in the North Atlantic and North Pacific oceans, where biological and ecological changes have been associated with decadal oscillations. These large-scale, climatic events may be natural, *via* such events as El Niño, or partly human-caused, as in global climate change. Cyclical shifts are similar to altered states of systems, the difference being that the latter may be semi-permanent and difficult to reverse. Unfortunately, in practice, the difference between the two may be difficult to detect. A second feature is that regional to local extractions, additions of species, and habitat alteration can synergize to amplify the potential of any one factor for change. A third feature concerns the suddenness of change. Both regime shifts and altered ecosystem states are exceedingly difficult to predict, as change is non-linear in time; that is, change is usually not gradual, but occurs in sudden steps, the major variable being the degree of change that occurs.

A fourth feature is socioeconomic cost. Fisheries depletions, pollution, human health, and ecosystem changes are all shared, transboundary concerns that incur costs that countries are forced to bear collectively, and often can ill afford. The final feature is uncertainty. For example, some water-borne diseases (hepatitis, dysentery, typhoid, cholera) and toxic phytoplankton blooms may appear in cycles associated with climate change and altered quality of coastal waters. A consistent feature is that these changes usually cannot be readily predicted, if at all. Thus, coastal-realm change and the many unknown factors that may be at play illustrate the scope and difficulty of sustainable use, restoration, recovery, and conservation.

8.5 Conclusion

There can be little doubt that the coastal ocean has been enormously altered by human activities, and that changes in spatial and temporal patterns have led to dramatic effects on a variety of linked ecological processes and scales. However, how human-caused and "natural" changes are related can be difficult to tell. Even more problematic is whether altered community and ecosystem states are reversible.

Determinations of health, stressors, stress, and trajectories of change are major challenges for conservation science, as well as for social institutions seeking to integrate governance, management, and ecosystem function toward desirable outcomes. It is likely that many structural and functional changes in ecosystems have resulted from depletion of key species. However, species and communities are not governed only by internal, quasi-equilibrium conditions, but are also under the combined influences of top-down environmental and bottom-up biotic controls. Identification of both sets of controls is critical as conservation moves to protect species and their functioning and to promote recovery of ecosystems toward presumably healthy states. Scientific information and monitoring are required to identify controls together with stressors, and their effects on ecosystem outcomes that would not otherwise become apparent.

Several conclusions emerge. First, a reference is required against which to measure change. Second, all issues exemplified in our case studies are shown to be widespread, making comparisons within a global arena possible. Third, the bulk of coastal-marine conservation issues are regional in scope. Fourth, most coastal-marine systems are highly perturbed, perhaps now existing in altered states very different from historic conditions. Fifth, globally the coastal realm appears to be moving toward major discontinuities between human demands for living resources and the capabilities of ecosystems to produce those resources. Finally, the effects and consequences of human activities on natural populations and ecosystems are best understood in the combined context of natural history, community interactions, ecosystem performance, and society.

Chapter 9

Synthesis

We should respect, cherish, and change with utmost caution this largely unknown natural world because it works as it is and we are totally dependent on its working.

F.H. Bormann (2000)

We must see the larger task – stewardship of all of the species on all of the landscape [and "seascape"] with every activity we undertake as human beings – a task without spatial and temporal boundaries.

J.F. Franklin (1993)

We should not, however, assume that the sea, and especially the arable regions of the continental shelves, can be farmed with the same impunity [as the land]. Indeed, no one knows what risks are run when we disturb this key area of the biosphere.

J.E. Lovelock (1979)

There is a tide in the affairs of men
Which, when taken at the flood, leads on to fortune; . . .
On such a full sea are we now afloat;
And we must take the current when it serves
Or lose our ventures.

William Shakespeare, *Julius Caesar*

9.1 Introduction

Bormann reminds us of the dependence of human societies on nature's resources and services. Franklin tells us that conservation is universal. Lovelock warns that domestication of coastal seas risks unforeseen consequences. And Shakespeare's metaphor symbolizes the urgency we humans face in order to enhance environmental wellbeing. Taken together, these set a prophetic stage for coastal-marine conservation.

The intent of coastal-marine conservation is to maintain ecosystems so as to perpetuate environmental sustainability and human wellbeing. Coastal-realm conservation brings forth issues of traditional approaches of land and water use, and ownership. Lands are governed according to complex jurisdictions and property ownership. Conversely, aquatic areas (lakes, rivers, estuaries, and oceans) have inherited traditions of shared resources, common property, and government authority; personal property is almost non-existent, and responsibility is mostly a government affair. Lack of permanent human occupation has made the seas remote from human experience, having been, until recently, most experienced by fishermen whose livelihoods continue to depend on traditions of hunter–gathering. Even for modern scientists, assessment of resources and detection of environmental change require expensive logistics, specialized instruments, and technical skills uncommon to terrestrial practice.

This book presents the coastal realm as a dynamic system with properties uniquely its own. The foregoing chapters have highlighted issues faced by coastal-realm conservation and have illustrated the exponential growth of agreements and laws, public awareness, knowledge, and management regimes that have slowed but have not reversed trends of degradation and depletion. The conservation challenge is to treat the coastal realm as the altered system that it has become, and to develop innovative, enduring, science-based, and publicly supported approaches that will result in more appropriate mechanisms, better coordination, and renewed social perceptions. The major question is: Can human uses be adjusted to coastal-realm ecosystem dynamics and to the quickening pace of change?

9.2 The rise of coastal-realm conservation

For most of human history, coastal-realm resources have seemed endless. The ocean's vastness reinforced this view until the "marine revolution" (Ray 1970) dramatically accelerated the pace of degradation of the oceans and depletion of resources. The marine revolution is no less important than the previous agricultural (~ 5000–10 000 BP) and industrial (19th century onwards) revolutions, as all have expanded the Earth's carrying capacity for humans. The marine revolution, however, is unique: it marks the period in which humans have been able to explore, exploit, and alter Earth's oceans on massive scales, thus exposing all of Earth to human activities for the first, and possibly last, time.

The history of coastal-marine conservation began early in the 20th century, with scientific recognition of species depletions (e.g., fur seals, oysters, etc.). During the 1920s, offshore oil and minerals heightened national interests in oceans (Belsky 1989). During the 1930s, oceanographic laboratories expanded multidisciplinary knowledge of the oceans. World War II brought realization that the oceans would become an increasing source of living and non-living resources. National and international policies began to emerge, focused primarily on jurisdictions. The next half-century witnessed dramatic changes in exploitation, public attitudes, policy, and knowledge. Conservation moved from a focus on protection of valued species and habitats to a broader context of ecosystem management and environmental policy, accompanied by a search to understand and manage processes of change.

9.2.1 Conservation: 1940s–60s

Following World War II, influential writings of conservationists and scientists set the stage for a global conservation movement. In *Our Plundered Planet*, Fairfield Osborn (1948) became among the first to recognize the extent of environmental degradation of Earth. Later, Osborn's (1953) *Limits of the Earth* queried: "Is the purpose of our civilization really to see how much the earth and the human spirit can sustain?" At about the same time, Aldo Leopold (1950) wrote in *A Sand County Almanac*, perhaps the most influential environmental book of modern times: "Wilderness is the raw material out of which man has hammered the artifact called civilization," to which he also added: "One of the fastest-shrinking categories of wilderness is coastlines . . . No single kind of wilderness is more intimately interwoven with history, and none nearer the point of complete disappearance."

There can be no question that these writings – and many more – were a major inspiration for global conservation. Nevertheless, marine environments remained obscure to most people until writings such as Rachel Carson's (1951) eloquent *The Sea Around Us* and Jacques-Yves Cousteau's *The Silent World* (1953) brought the underwater world to life for the public. The development of scuba (self-contained underwater breathing apparatus) can hardly be overestimated, as it opened a new world to anyone who wished to venture there. For scientists, this invention was no less significant, as it opened new ways for discovery.

During the 1960s, the urgency of conservation dramatically accelerated. Rachel Carson's *Silent Spring* (1962) exposed the politics, economics, and problems of pollution, raising fears among the public for health and wildlife. Her warning was urgent: "Given time – time not in years but millennia – life adjusts, and a balance has been reached. For time is the essential ingredient; but in the modern world there is no time." Concurrently, a "science gap" in ocean research was recognized and became a policy issue. President Kennedy became the first U.S. president to use "oceanography" in a major message to Congress. There soon followed a "virtual blizzard" of hearings and legislation suggesting increased ocean research and development (Belsky 1989). Faced with evidence of pollution and environmental alteration, scientists and conservationists took increasing roles in public policy. The regional nature of coastal-realm conservation also became apparent, one of the first models being the San Francisco Bay Conservation and Development Commission (1965). Meanwhile, influential writings accelerated. Ray Dasmann was among the first to warn of loss of diversity, in *A Different Kind of Country* (1968). And, Dasmann's (1971) *No Further Retreat* focused on development in Florida: ". . . one cannot preserve part of an ecosystem without attending to all of it" – a lesson later to become a guiding principle in ecosystem management and for the biosphere reserve concept, which Dasmann helped innovate.

Only a few global conservation and management groups existed during this period, and even fewer existed internationally. A notable example is the International Whaling Commission, created in 1946 under the International Convention for the Regulation of Whaling to promote the conservation of whale stocks and for the orderly development of the whaling industry. The International Union for the Conservation of Nature and Natural Resources (IUCN) was created in 1948 by scientists and conservationists, at first under the auspices of the United Nations Educational, Scientific, and Cultural Organization (UNESCO), to assist societies worldwide to conserve nature and natural resources. The World Wildlife Fund (WWF) began in 1961, first to support IUCN activities and later to become an independent global conservation force; for both, protecting endangered species and natural areas took center stage, with priority overwhelmingly given to the land. Endangered species in the sea were hardly recognized, except for sea turtles, marine mammals, and a few others. Global recognition of the need for "marine protected areas" began to emerge when IUCN hosted the First World Conference on National Parks in Seattle, Washington, in July 1962. In the keynote talk, U.S. Secretary of the Interior Stewart L. Udall spoke of a "quiet crisis," using a marine metaphor: "The pages of history are littered with wrecks of ships of state which have foundered because natural laws were ignored . . . No feature of the globe has more cultural significance than our great oceans . . . Every sea-touched country has the opportunity to preserve for its people a portion of shoreline with the unique opportunities which they hold for human refreshment and restoration of the soul." At the urging of Peter Scott (later "Sir") the Conference passed a resolution calling for marine protected areas (Box 2.1, page 30).

As global conservation was unfolding, ecosystem science was developing. The stage for ecosystem science was set when the International Council of Scientific Unions (ICSU) organized the International Biological Programme (IBP 1962–72). Technology rapidly advanced; satellites and computers allowed ecosystem-level investigations and analyses. Algal blooms and pollution trails could be detected from space, and monitored by oceanographic instruments. Computer models began to be developed to predict environmental change. The IBP called for the study and conservation of natural ecosystems and human adaptability, thus facilitating a unified approach reminiscent of G.P. Marsh's *Man and Nature* a century before (Marsh 1864).

From the 1960s onwards, pollution, species depletions, and coastal impacts called public attention to the oceans and to the need for marine conservation. Pollution, especially catastrophic oil spills, spread over coastlines. Aquaria and oceanaria displayed charismatic species, especially seals, porpoises, whales, seabirds, and sea turtles, and television spread word of their depletion spurring strong protest among the public of many nations. It soon became obvious to the public that coasts and nearshore oceans were being altered on a massive scale. Yet the oceans also represented to policy-makers a vast area of under-utilized resources that was limited by technology, and ocean information was poor. Politicians reacted by passing laws, creating new agencies, and calling attention to needs for ocean development, advancing knowledge, and attempting to manage valuable resources (oil, marine mammals, fisheries, leisure). In the United States, the President's Commission on Marine Science, Engineering and Resources produced *Our Nation and the Sea*, which led the nation in a new era of ocean affairs to protect national interests in a rising era of federalism, and coined the term "coastal zone" (CMSER 1969).

9.2.2 The environmental era: 1970s–80s

Resource depletion, pollution, and physical alteration initiated an "environmental era" that involved legislative and environmental reform, and public attention. Whaling and sealing, especially, attracted public protest, which led to a complex mix of governance, ethics, economics, marine resource management, and law of the sea. Emotions ran high and management bodies became politicized. As Gerber et al. (2000) observed: "It is here, at the intersection of sentimentality and scientific controversy, that conservation biologists typically face their greatest challenge." In the spirit of achieving common ground, conservationists and international scientists involved in the study and management of whales and whaling gathered in 1972 at the International Conference on the Biology of Whales. The Conference result (*The Whale Problem*, Schevill et al. 1974) noted the depleted state of great whales and summarized the need for information and conservation. A decade later, public awareness became the overwhelming factor in achieving a moratorium on whaling – a dramatic illustration of how public support can achieve a conservation result, whether science-based or not.

The coastal zone was also becoming a conflict arena that needed addressing. A multidisciplinary workshop of scientists, administrators, and stakeholders evaluated coastal-zone issues and produced *The Water's Edge: Critical Problems of the Coastal Zone* (Ketchum 1972).

This group of dedicated scientists concluded that "necessary decisions concerning multiple use of coastal-zone resources require a sound factual base"; the group defined the coastal zone broadly to include continental plains, continental shelves, and the coastal–ocean equivalent to the "coastal realm" in this book. Especially noteworthy was the call for a systems approach: "A most powerful approach developed for attacking complex and otherwise intractable problems is the application of systems analysis . . . Already available are the general conceptual framework; an inventory of management, social, and physical environmental models; some experience in simulation; and fully adequate computer resources which together provide a strong basis for the development of systems analysis approach." The workshop also recognized that decisions made by policy-makers and agencies are vulnerable when confronted by a dynamic ecological, social, economic, and political environment.

In 1976, IUCN initiated the first international program on marine conservation. It established a Marine Steering Committee charged with developing IUCN's 1977/8 Marine Programme, which promised to be the most ambitious program of the Union since its founding. The objectives were emblematic of the times, being aimed toward: global action plans for the protection and rational use of mobile marine animals (e.g., sea turtles, seabirds, whales, seals, and a few other endangered species); regional systems of marine parks and reserves for protection of critical marine habitats; and model management systems for the maintenance of important marine processes (e.g., nutrient flow through watersheds, mangroves, wetland, and fisheries). IUCN's initiative was matched by WWF's Conservation Programme "The Seas Must Live," that included several dozen active and planned projects, overwhelmingly directed toward charismatic species and protected areas, and a few on regional planning. However, at the time public environmental concern for the coasts and oceans was dwarfed by terrestrial conservation. The IUCN–WWF marine program ended in 1978. No equivalent program took its place.

While efforts to develop a global approach to marine conservation languished, the largest oil spill in history, the *Amoco Cadiz* (1978), occurred. This incident, and mounting needs to resolve pollution issues and conflicts about fisheries, mineral exploitation, trade, trans-boundary issues, and rights of passage, resulted in a plethora of national laws and international agreements and conventions, and a rapid proliferation of non-governmental organizations (NGOs). As conflicts increased, public and governmental activities and budgets also increased, and ocean law became a "growth industry," albeit less for conservation and more for purposes of jurisdiction and exploitation (Belsky 1989). Notably, in the early 1980s most nations agreed to the United Nations Convention on Law of the Sea (UNCLOS), which provided a framework for coastal-marine conservation, and has since been amplified by many international agreements and regional programs related to species and environments.

By the mid-1980s, biodiversity and conservation biology emerged on center stage. The central concept of diversity used by most biologists, from the times of early 18th-century naturalists, had been the hierarchy of biological organization, as expressed by the sciences of systematics and evolution. During the 20th century, three levels of biodiversity – genetic, species, and ecosystems – became recognized, and in the early 1980s were widely adopted for conservation purposes (e.g., Norse et al. 1986). Concurrently, the science of conservation biology was born, defined by Soulé (1985) thus: ". . . although crisis oriented, conservation biology is concerned with the long-term viability of whole ecosystems." Soulé's (1986) book, *Conservation Biology: The Science of Scarcity and Diversity*, and new professional societies, notably the Society for Conservation Biology (1987), and their journals legitimized conservation science within academe as a true discipline. Students flocked to new university programs of conservation and environmental studies in response to concerns about continued deterioration of ecosystems and the extinction crisis. Wilson and Peter's (1988) influential book *Biodiversity* signaled a new direction for conservation, but contained only one paper on marine systems (Ray 1988). Despite some attention given to coastal-marine issues (especially protected areas and endangered species) at national and international fora, coastal-realm conservation continued to lag well behind terrestrial concerns among the scientific community, NGOs, governments, and the public.

9.2.3 The post-Rio 1990s

During the 1990s, coastal-marine systems finally achieved preeminence, and the dominant theme of conservation became biodiversity, from genes to ecosystems, most notably at the UN Conference on Environment and Development (UNCED) in 1992 in Rio de Janiero. The Convention on Biological Diversity inspired a number of influential publications and initiatives, and *Agenda 21* (Table 2.7, page 40) gave special recognition to the coastal realm. The World Bank supported a global review of coastal-marine protected areas (Kelleher et al. 1995), and conservation of tropical coral reefs became the marine equivalent of tropical forests due to their beauty and diversity. The global depletion of fisheries inspired recommendations by scientists that at least 20% of the seas be protected. Scientists intensified studies within the coastal realm on land–sea–air interactions, and the scientific and conservation communities raised coastal and marine concerns approximating to the terrestrial. Conservation biology, and community and ecosystem ecology, increasingly influenced conservation practice and policy, joined by a committed and increasingly aware public. And the Law of the Sea treaty firmly placed the coastal realm in the international policy arena.

By the turn of the 21st century, scientists, governments, and conservationists alike recognized that processes that maintain biodiversity cross boundaries. Integrated coastal-ocean management evolved to depend for its success on integrating ecosystem function and biodiversity of regional land–seascapes. At the same time, society inherited elaborate policies and sets of institutions designed for remedying environmental problems one-by-one, and with limited, short-term goals. Gaps became evident between the need for integrated, ecosystem science and management, and institutional capacities to manage across jurisdictional and disciplinary boundaries.

9.3 Present challenges for coastal-realm conservation

We now find ourselves in a period when the need for coastal-realm conservation is widely agreed, richly based on prescient writings of decades earlier and on recent developments in conservation science, management, and policy. Much has been accomplished, many necessary mechanisms are on hand, knowledge is rapidly accumulating, and public concern is growing. Nevertheless, coastal-realm conservation faces complex and value-loaded challenges, with profound ecological and economic consequences.

9.3.1 The challenge of sustainable use

The UN Conference on the Human Environment (Stockholm 1972) declared: "Man has the fundamental right to freedom, equality and adequate conditions of life, in an environment of a quality that permits a life of dignity and well being." This conference brought global recognition to inseparable human–environment interactions and declared that all nations have a responsibility to maintain ecosystems and biodiversity. Subsequently, the World Commission on Environment and Development (Brundtland Commission 1983) conceived "sustainable development" as a goal for blending conservation and growth. However, that growth is exponential and unsustainable without continual input of resources has also been recognized. Thus, "sustainability" has tended to become the goal, explicitly incorporating social, economic, and environmental considerations (Goodland & Daly 1996). Social sustainability is maintenance of human capital, including education, equity, health, nutrition, and other social values. Economic sustainability consists of maintenance of monetary capital. Environmental sustainability means maintaining natural capital: that is, the environment's capacity to process waste, regenerate materials, remain resilient, and persist. These three, if not addressed together, would add to an environmental debt. Sustainability is limited by a total impact equal to the product of population, affluence, and technology. With time, impact increases exponentially, ultimately to limit social and economic growth.

The ultimate challenge for sustainability in the coastal realm is to harmonize two chaotic, non-linear, non-equilibrium systems – society and environment – that operate on very different time and space scales. The global population exceeds six billion people, and each day more than 350 000 new humans enter the world, concentrated mostly in the coastal realm. Daly (1993) estimates that the global economy would have to grow by a factor of seven to give everyone alive today a *per capita* consumption equal to that of the average American. Sustainability of the coastal realm is limited by availability of food, water, space, energy, and resources, with serious consequences for watersheds and the coastal realm. Sustainable agriculture can occur only if ground- or surface water is available, if sufficient fossil fuel energy is available to pump and move it, if monetary resources are available to buy the required technology, and if the soil is suitable for irrigation and fertile enough to support crop growth. Fertilizer and pesticides can increase yields, to a point, but at the same time, wastes and pollutants need to be exported. Thus, intensive farming results in watershed alterations and introductions of pollutants and nutrients into coastal waters.

Fisheries provide the dominant extractive use of the coastal realm. Fisheries production is at an all-time high, but has rarely been sustainable (Pauly et al. 2002). In the early 1950s, fisheries were in a period of extremely rapid growth worldwide. Industrialized fishing continued to expand almost to the 21st century, but has leveled off as most commercial fishes have become depleted. Many fisheries depend on healthy estuarine systems, which face problems of eutrophication delivered by watersheds and altered freshwater flows *via* dam development. Furthermore, fisheries models used by managers to estimate maximum sustainable yields for each fished species assume equilibrium conditions between population size and fishing effort. However, fish population sizes are also driven by environmental variations and community interactions, which until recently fishery models have ignored. Larkin (1977) suggested "an epitaph for the concept of maximum sustainable yield" a quarter-century ago. Yet, simplistic, single-species population models continue to drive many fisheries, despite scientific evidence that these models apply to few species of fish. Furthermore, fishery overcapacity, subsidies to the industry, inadequate accounting of catch, and insufficient natural history and ecological knowledge combine to make projections of fisheries yields highly uncertain. The complexity of fisheries management is highlighted by Ludwig et al. (1993), who suggested four interacting features that lead to fishery overexploitation: (i) wealth or the prospect of wealth generates political and social power that promotes exploitations; (ii) scientific understanding is hampered by lack of controls and replicates, so that each new problem involves new learning; (iii) the complexity of the underlying biological and physical systems precludes a reductionist approach and optimum levels of exploitation must be determined by trial and error; and (iv) high natural variability masks overexploitation until it is severe and often irreversible.

Non-extractive uses, such as tourism and ecotourism, are offered as economic alternatives to extractive or polluting industries, and are most often presumed to be sustainable. However, these too can have unintended and unsustainable consequences, by demanding high energy and water consumption and by production of waste and sewage. Tourism requires considerable investments and infrastructure that few developing countries can afford. When expatriate owners export profits, these nations lose revenue (Harrison 1994). Activities promoted by ecotourism can also be environmentally damaging. Scuba divers in large numbers can injure reefs and tourists can place high demands on local resources. Harassment of animals can also become a serious problem. Whale-watching has become a $1 billion industry in 80 nations, and boats seeking close contact with whales can affect the whale's natural behaviors; this can become especially serious for endangered species, such as North Atlantic right whales, where altered reproductive behavior or injury to only one or a very few animals has the potential to cause a population decline, and even extinction (Fujiwara & Caswell 2001). Swimming with dolphins and fish-feeding by scuba-divers can also affect behavior and nutrition. Thus, the benefits of tourism need to be carefully assessed against risks.

Sustainability, therefore, is a property of the society, culture, and environment, with impacts on human wellbeing. Assessment of wellbeing asks three interrelated questions: How well are the people? How well is the ecosystem? And, how are people and the ecosystem affecting each other? Universally for coastal-realm nations, environmental impact remains high and wellbeing is reduced (Box 9.1).

9.3.2 The challenge for protected areas

For thousands of years, the coastal realm was one large natural refuge within which humans were scattered and depleted resources quickly recovered. When people no longer found sufficient food or resources, they either perished or moved. Today, humans are pervasive, leaving few areas untouched by the human hand, making protected areas imperative.

History shows that protected-area establishment has assumed highest priority when endangered species, degraded environments, and depleted resources were concerned. Today's several thousand existing marine protected areas (MPAs) fall into three general categories: (i) ecosystem representation based on biogeography; (ii) species-specific protection, and (iii) reserves where restoration is encouraged, such as for enhancing fisheries. However, global MPA extent is estimated at only approximately 0.5% of the coastal and open oceans. Furthermore, most MPAs are small, have few restrictions on extractions, are poorly enforced, lack incorporation of region-scale effects into management, and do not include adequate research and monitoring programs – thus being ill-equipped to fulfill their intended goals.

Box 9.1 Wellbeing Assessment

Robert Prescott-Allen

Assessment is an essential part of ecosystem-based management and the integration of socioeconomic and ecological concerns, because it provides the means to:

- measure human and environmental conditions;
- analyze strengths, weaknesses, and options;
- decide policies and strategies to improve conditions;
- keep track of progress and determine the effectiveness of policies and strategies.

Several assessment methods are available. Wellbeing Assessment is recommended for ecosystem-based management because it is the only one that:

1. gives equal weight to people and the ecosystem, considering them together but measuring them separately so that neither is submerged in the other;
2. covers human and environmental conditions comprehensively yet cost-effectively through the selective measurement of the main features of human and ecosystem wellbeing;
3. provides a systematic and transparent way of identifying these features and of choosing high-quality indicators of each feature;
4. combines the indicators into a Human Wellbeing Index, Ecosystem Wellbeing Index, and Wellbeing Index to give a clear picture of human conditions, environmental conditions, sustainability, and major strengths and weaknesses;
5. shows the environmental price of the standard of living by generating a Wellbeing/Stress Index (the ratio of human wellbeing to ecosystem stress).

The core of the method is equal treatment of people and the ecosystem. Human wellbeing is essential because no rational person wants to perpetuate a low standard of living. Ecosystem wellbeing is essential because the ecosystem supports life and makes possible any standard of living. The concept is expressed in the metaphor of the Egg of Wellbeing. The ecosystem surrounds and supports people much as the white of an egg surrounds and supports the yolk. Just as an egg is good only if both the yolk and white are good, so a society is well and sustainable only if both people and the ecosystem are well.

To compare socioeconomic and environmental conditions, the system is divided into two subsystems: *people* (human communities, economies, and artifacts); and *ecosystem* (ecological communities, processes, and resources). Interactions between the two are recorded under the receiving subsystem. Human stresses on the ecosystem (such as pollution and resource depletion), and benefits to it from conservation, are recorded under *ecosystem*. Benefits from the ecosystem to people (from the supply of resources to spiritual comfort), and the human toll of climatic and other environmental stresses, are recorded under *people*.

Each subsystem is divided into five dimensions, providing a common framework for all assessments using the Wellbeing Assessment method. Ecosystem dimensions are land, water, air, species and genes, and resource use. Human dimensions are health and population, wealth, knowledge and culture, community, and equity. The framework allows users to select their own indicators and produce assessments that are tailored to their conditions and needs yet broadly comparable with other wellbeing assessments. The dimensions are designed to group a wide range of topics into a few major categories that are roughly equal in scope and easily communicable to nonspecialists. They are specific enough to ensure that all assessments cover universally important aspects of human and ecosystem wellbeing, yet sufficiently broad and flexible to accommodate concerns that may matter to some societies but not all: any issue regarded as significant for wellbeing and sustainability has a place in one of them.

Because it is impossible to measure human or ecosystem wellbeing directly, assessments must select indicators of their main features. Knowing the essential role of indicators, it is tempting to jump right in and choose them at once. However, it is seldom clear at the start of an assessment how well a given set of indicators represents a desirable combination of human and environmental conditions, what aspects are left out, how much the indicators overlap, or how they relate to each other. Since indicators require the collection and analysis of often large amounts of data, choosing the wrong ones can be a costly mistake. Consequently, it is necessary first to take apart the concepts of human and ecosystem wellbeing to identify the features that need to be measured – and then to unpack each feature to reveal aspects that are both representative and measurable.

Wellbeing Assessment does this by going down a hierarchy of parts and aims, which provides a series of increasingly specific stepping stones from system and goal to indicators and performance criteria (standards of achievement). These steps ensure that the indicators are as representative as possible of the system as a whole and of people's goals for themselves and their environment. Even so, each indicator conveys information only about the particular element it represents. To provide a picture of the entire system and of progress toward the goal of a high level of human and ecosystem wellbeing, the indicators need to be combined into indices. The steps taken to select the indicators are reversed to provide a logical and transparent procedure for combining them, going back up the hierarchy from indicators to system.

Indicators are combined using the Barometer of Sustainability (Fig. 1), a performance scale with two axes, one for human wellbeing, the other for ecosystem wellbeing.

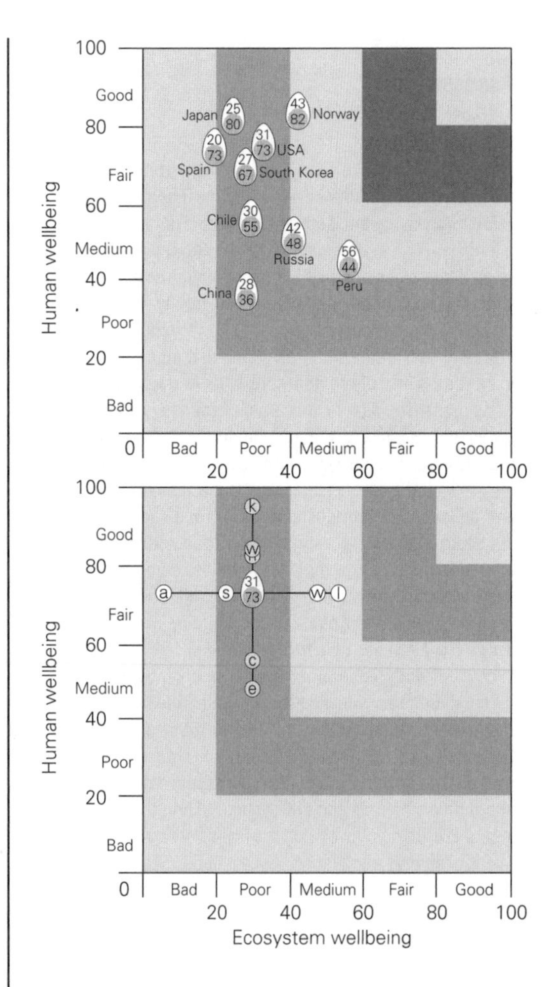

Fig. 1 Indices of wellbeing. *Top*: Barometer of Sustainability providing a bird's eye view of wellbeing of 10 top marine fishing nations (fish catch > 1 million tonnes). The Human Wellbeing Index (HWI) is in the yolk of the egg; the Ecosystem Wellbeing Index (EWI) is in the white. The Wellbeing Index (WI) is the position of the egg – the point where HWI and EWI intersect. Sustainability is the square in the top right corner. The Barometer shows human and environmental conditions, their relationship to each other, and the distance to sustainability. Performance in different years can be plotted on the Barometer to show the direction of change. *Bottom*: Barometer of Sustainability showing the structure of wellbeing for the United States. Circles on the vertical axis are points on the scale of human dimensions: k = knowledge; w = wealth; h (behind w) = health and population; c = community; e = equity. Circles on the horizontal axis are points on the scale of ecosystem dimensions: a = air; s = species and genes; r = resource use (hidden by the egg); w = water; l = land. The Barometer reveals the nation's strengths and weaknesses: progress toward sustainability requires better performance on air (cut carbon dioxide emissions), species and genes (protect wild plant and animal species), resource use (limit energy consumption), equity (shrink the gap between rich and poor), and community (reduce crimes of violence). Courtesy of R. Prescott-Allen.

Each axis is divided into five bands of 20 points each plus a base of 0. Performance criteria – good, fair, medium, poor, and bad levels of performance (corresponding to the five bands) – are defined for each indicator. They set the rate of exchange between the indicator and the scale (the level of performance that is worth a given number of points), enabling indicator measurements to be converted to scores. The scores can then be combined into higher-level indices and ultimately into a Human Wellbeing Index (HWI), an Ecosystem Wellbeing Index (EWI), and a Wellbeing Index (WI).

The Barometers shown here sum 36 human indicators and 51 ecosystem indicators from *The Wellbeing of Nations*, my assessment of the sustainability of 180 countries. The human indicators include healthy life expectancy at birth, gross domestic product per person, school enrollment rates, freedom ratings, and income share between rich and poor and between males and females. The ecosystem indicators include land modification and conversion, inland water quality, carbon dioxide emissions, threatened plant species, the state of major marine fishery species, and fish catching capacity per unit of fish producing area.

By summing a representative set of measurements into clear pictures, wellbeing assessments convert complex inputs into simple outputs, enabling citizens and decision-makers to compare human and environmental conditions, determine sustainability, identify priority actions to improve conditions, and measure the effectiveness of those actions.

Wellbeing assessments can be undertaken for systems of any size, from the Bahamas to Russia, from Chesapeake Bay to the Bering Sea, and from a municipality to a group of nations.

Source: Prescott-Allen (2001).

Until the mid-1980s, only a few marine scientists were involved in MPA establishment. This situation has dramatically changed: a substantial scientific literature now exists on criteria for and design of MPAs. Recent recommendations by the scientific and conservation communities call for a fifth of the global ocean to be dedicated toward a sustainable, global network of reserves for conservation of biodiversity and resources. The challenge is, which 20%! Area selection involves "agonies of choice" (Ray 1999), that is: What are the objectives and how may priorities be allocated? Where should MPAs be located? How many areas are required, and how big should they be? What constitutes "representation"? How can "critical" or "essential" habitat be defined? How may boundaries be designated? How can individual areas be aggregated into a network that reflects ecological ecosystem attributes and processes? And, how can areas dedicated to biodiversity be aligned with those for fisheries or endangered and depleted species? Application of IUCN criteria (IUCN 1976; Table 2.3, p. 31) illustrates several difficulties. First, life resides everywhere on Earth, and every place, land and sea, is critical for some species for some purpose. Clearly, if all taxa were to be considered and all criteria applied (e.g., endemism, endangerment, production, etc.) the entire coastal realm would need to be included, a totally impractical idea. Therefore, decisions about MPA selection inevitably involve value judgments about which species, which environments, or which threats are most important.

Determining priorities also depends on which criteria are given most weight. For example, designation of "hot spots" for threatened, endemic species has been proposed for conservation of biodiversity. This assumes that protection of a few areas of limited size can protect most biodiversity. However, hot spots that apply only for endemic species do not account for wide-ranging species and regional influences. Further, for the majority of coastal-marine life, no one area suffices to cover the suite of most species' habitats or the requirements of its biological community. Even if only one or a very few species are considered, influences can extend far outside their immediate boundaries (see Figs 5.23 [p. 155], 6.20 [p. 194], 7.13 [p. 222], & Boxes 7.5 & 7.7 [pp. 226, 236]). Also, the influences of keystone or foundation species extend over larger areas than their distribution would indicate (Chapter 4). This situation is illustrated by the case studies: oysters (Chapter 5), walruses (Chapter 6), Nassau grouper (Chapter 7), each of which serves a key function in its environment. Sea otters (*Enhydra lutris*) provide another example. With protection from hunting, this keystone species made a rapid recovery during the 1900s, but now faces rapid decline, possibly due to factors outside its extensive MPA (Box 9.2).

A shift in emphasis from hot-spot protection in small, unsustainable areas to adaptive ecosystem management within regional land–seascapes representative of biogeographic regions has been widely suggested, especially since the evolution of the biosphere reserve concept in the mid-1970s. This approach has the potential to take account of regional ecological linkages, mechanisms of dispersal, energy exchange, and biogeochemical processing. MPA design thus involves networks of MPAs as spatial nodes connected by gradients of known attributes, and by natural histories of well-known species (Fig. 9.6).

9.3.3 The challenge of restoration

Restoration is directed toward reestablishing depleted species, habitats, and environmental quality to pre-existing baseline conditions. It requires actions to repair damage and to improve biological and ecological structure and function. Restoration also encompasses suites of remedies that seek to improve ecological benefits, or replenish losses. For example, sewage treatment is a technology for reducing coastal nutrient pollution, and aquaculture is undertaken to restore economic benefits of commercial fishing. Another approach is to reintroduce species where they were extirpated or did not previously exist; yet another is to create habitats, such as wetlands, that have been destroyed. These approaches involve trade-offs requiring consideration.

Scale is an important issue. Most approaches are bottom-up, tactical, local, and issue-by-issue, in which restoration of one resource risks loss of another. Moreover, restoration of a single attribute, such as a small wetland, rarely, if ever, can successfully restore long-term ecosystem function or integrity. Rather, restoration may best be achieved in the context of a regional strategy that incorporates large-scale, top-down ecosystem attributes, linkages, and functions. Our case studies indicate that species-specific and local habitat restoration are insufficient to return whole biotic communities and large regions to former conditions. Conversely, large-scale, top-down approaches involving whole regions and long-term temporal changes can become extraordinarily unwieldy, politically challenging, and uncertain in outcome. Restoration in

Box 9.2 How to design a sea otter reserve

James A. Estes

The Bering Sea/North Pacific Ecosystem (BSNPE) is in a state of spectacular change, characterized in particular by precipitous population declines of various marine birds and mammals. Sea otters are the most recent addition to the list of dwindling species, having declined by 90% or more across large regions of western Alaska since about 1990. In contrast to the situation with most of the region's other declining species, the immediate cause of the sea otter decline is reasonably well understood. But for the moment, let us imagine that we know little more than its pattern, other than to suspect that the cause is in some important way anthropogenic. These conditions define our present understanding for most of the BSNPE's declining species.

Given this state of affairs, how might one design a marine reserve for sea otters? We may begin by defining the goal, which is to protect a sufficiently large segment of habitat from human impacts to provide for a viable sea otter population. One approach is to identify an area capable of supporting an effective population size (N) of 500 individuals. Based on known features of the sea otter's mating system, sex ratio, and demography, this population size translates to an actual population (N) of about 1850 individuals. Armed with this figure and knowing the equilibrium population density of sea otters (D), the area (A) needed to support a viable population is estimated by N/D. Population estimates for sea otters in western Alaska provide values of A ranging from 50 to 200 miles of shoreline, roughly the size of mid- to large-sized islands in the Aleutian archipelago.

Although the reserve size specified by this scenario is large by existing standards for coastal-marine protected areas, in this case it would almost surely fail. The problem is that the ultimate cause of the sea otter decline probably has little or nothing to do with events in the protected area, but rather with those in far removed oceanic waters of the BSNPE. Tagging studies show that the demographic cause of the decline is elevated mortality, not reduced fertility or redistribution. Increased killer whale predation is the apparent reason for elevated mortality. A variety of evidence points to this conclusion, including a highly significant increase in observed attack rate coincident with the decline, and normal survival rates and stable population numbers of sea otters in a lagoonal system that is inaccessible to killer whales. Also, the results of demographic modeling studies show that, if losses due to killer whale predation were solely responsible for the population decline, the expected number of observed attacks is nearly identical to the number actually seen.

Why this has happened is less clear. Perhaps killer whales have hunted otters for years and killer whale numbers are on the rise. Virtually nothing is known about the status of killer whale populations in western Alaska. Or maybe a few killer whales fortuitously discovered sea otters as easy prey. If this latter explanation is correct, it raises the troubling question of why it has happened so recently. The most likely explanation, in my view, is that certain killer whales have turned to consuming sea otters as their preferred prey – harbor seals and Steller sea lions – have declined. The timing of the otter declines, coming on the heels of the pinniped declines, fits this explanation. Even if this scenario is correct to this point, it still leaves the question unanswered of why pinniped populations have declined. Some think it is an influence of the BSNPE ground fishery while others are more inclined to blame temperature regime shifts. Others are beginning to suspect that predation may be important.

This example makes two general points about marine reserve design. The first is that unit ecosystems, even when defined over large areas, are not isolated functional entities, but rather link with adjoining ecosystems. Marine ecologists recognize that linkages of this sort occur through such processes as larval dispersal, physical and biological transport of food and nutrients, and the movement of apex predators. However, the fact that marine reserves must be large enough to capture these diverse linkages remains underappreciated. Although one can argue the details, the take-home message is that marine reserves designed solely around the distribution and apparent resource needs of species targeted for protection are likely to fall short of their intended purpose.

Can marine reserves stem the tide of species decline in the BSNPE? Probably, but only *if* human impacts are instrumental in causing the declines and *if* the reserves are large enough to buffer populations against these impacts. The problem is that the burden of proof traditionally falls on scientists and resource managers to first demonstrate human cause and then to engineer the details of an explicit solution. Time is too short and the system is too complex for that approach to work in most cases. We must cease just thinking about the problem and do something about it. For the BSNPE, we might begin by establishing one or more very large marine protected areas together with the time and resources to demonstrate their effects. These areas must be large because of the demographic characters and ecosystem linkages described above. If overfishing is responsible, either wholly or in part, for the declines of marine birds and mammals in the BSNPE, this approach has the potential to both demonstrate and begin to mitigate those effects. If killer whale predation turns out to be a factor, however, a management dilemma is raised that may be beyond human control.

Sources: Anderson & Piatt (1999); Estes et al. (1998); Ralls et al. (1983).

regions that have acquired a significant environmental debt and are continually undergoing change from large-scale human activities is particularly difficult because the target is always moving, as is the case for Chesapeake Bay.

Restoration of coastal-realm environments is encouraged because many ecologically extinct (no longer functionally important) species survive in sufficient numbers to be restorable, and many ecosystems appear to be in sufficient condition for restoration to succeed. However, intensive and expensive management is almost always required. Furthermore, rarely if ever can restoration replace conservation, which is the most reliable and least costly means of sustaining species and ecosystems. The challenge for restoration lies in determining a reasonable goal, the temporal and spatial dimensions of what is being restored, and the current state relative to past conditions.

9.3.4 The challenge of transboundary issues

Transboundary issues (e.g., air pollution, fisheries, ozone, global warming, trade, etc.) extend to areas beyond control of individuals, laws, jurisdictions, and nations. The coastal realm is a region where sovereignty and globalization play major roles, and where wetlands, migratory waterfowl and fishes, endangered species, pollution, and fluvial inputs to the coastal ocean are major transboundary issues. Responsibility for biological conservation involving transboundary issues usually falls under the United Nations system or on secretariats set up under agreements related to these issues. However, international and institutional mechanisms for resolving transboundary issues are often more symbolic than operational, and remain notoriously weak and difficult.

Urban growth creates a transboundary issue. Cities and urban areas depend on regional to global systems to provide food, clean air, process waste, and many other services that are extended by surface, ocean, and air transport. Thus, urban areas extend their "ecological footprint" to areas many times larger than the areas themselves (Folke et al. 1997). Obvious conflicts result among fisheries, recreation, water supply, pollution, dredging, waste management, and living space, aggravated by ship collisions and accidents, and spills. A similar issue with important biological, economic, social, and political consequences concerns watershed development. An outstanding example is the Paraná watershed: this contains an extensive wetland straddling Brazil, Paraguay, and Argentina in South America (Box 9.3), illustrating consequences of economic stimulation involving watershed development by governments. Removal of wetlands threatens rare species, reduces waterfowl and other valuable species, and alters fluvial inputs to coastal waters. Pollution and sediment from upstream activities in one country can affect the people and economy of other countries downstream. Watershed development often requires translocation of large numbers of people, who may move into crowded coastal cities or seek jobs in other countries, thus expanding urbanization and putting stress on institutional structures and governing systems.

North Pacific salmons (*Oncorhynchus* spp.) also exemplify complex sets of local, national, and international transboundary conflicts that transcend fisheries alone. The conflict arena extends to conflicts over freshwater, dams, forest exploitation, endangered species, and who gets the rights to fish, for example sportsmen, commercial fishers, and/or Native Americans (Box 9.4). These sets of conflicts also plague North Atlantic salmon (*Salmo salar*), which have a similar life cycle to their Pacific cousins. Salmon production is strongly influenced by international commerce, and aquaculture has become especially significant. Native and transgenic Atlantic salmons have been widely introduced into the North Pacific and waters around South America, Australia, and New Zealand for more than a century, where they are farmed. Impacts include replacement of native fishes, introduction of disease, and pollution. The fish-farming issue extends to traditional fishermen, who experience economic hardships when costs of providing wild fish to the market become higher than for aquaculture. To counter this trend, fish vendors often advertize wild-caught fish as being of higher quality.

A further challenge relates to globalization of trade. Multinational corporations (MNCs) have emerged during the past few decades to become a major international economic force, and to play significant roles in environmental affairs by creating transboundary issues of great conservation concern. The top 500 MNCs control 70% of world trade, 80% of foreign investment, and 30% of the world's gross domestic product; collectively, and often individually, they have budgets much larger than many developing countries (Miller 1995). Additionally, MNCs are the largest users of raw materials globally (Elliott 1998): the top 500 MNCs generate more than half the greenhouse gases produced

Box 9.3 The Paraguay–Paraná hidrovía: protecting the Pantanal with lessons from the past

James E. Perry

The Pantanal ("great swamp" in Portuguese) is a 140 000 km^2 alluvial wetland roughly 10 times the size of the remaining Florida Everglades. Located in the upstream basin of the Paraguay River in the heart of South America, the Pantanal stretches across western Brazil and parts of Bolivia and Paraguay. It has one of the most diverse biotas on Earth, with more than 400 species of fish, some 20 species of bats, and more than 650 species of birds and 1700 species of plants. The Pantanal is a key hydrological resource in South America. It sustains flows in the Paraguay River throughout both the wet and dry seasons, which has a major impact on both the ecology and economics of the region between the Pantanal and the Atlantic Ocean.

Threats to the Pantanal

The most profound threat to the Pantanal is the planned development of a 3440 km-long, north–south navigable waterway from Cacares (Brazil) to Nueva Palmira (Uruguay), referred to as the Paraguay–Paraná hidrovía. This waterway would have a substantial impact on the Pantanal's hydrology and ecology. An economic feasibility study showed a positive net return for an engineered system of major channel straightening, dredging, and removal of flow-impeding rock outcrops. Estimated savings in transportation costs were compared with the costs for construction, maintenance, and equipment needed for the project. The report did not, however, consider the large environmental costs of the channel, including changes in water quality and flood amplitude and loss of wetlands. The report also contained a number of mathematical errors, including erroneous additions of simple columns of numbers. Several independent studies focused on the hydrological aspects of the proposed waterway, concluding that the changes would have a substantial impact on the flood regime of the Upper Paraguay River, such as accelerated flood runoff and increased peak discharges in the river, reduced seasonal inundation of the floodplain, and, as a result, changes in the productivity and diversity of the Pantanal. One study found that lowering the level of the Paraguay River by an average of only 25 cm, certainly a realistic estimate, would reduce the flooded area of the Pantanal by 22% at low water.

Lessons from the past

While plans to develop the Paraguay–Paraná hidrovía continue, other countries are trying to undo the massive damage that artificial waterways have done to many of the world's large rivers, such as the Missouri–Mississippi river system, the Everglades/Kissimmee River complex in Florida, the Rhine, Danube, and many others. The United States uses its rivers for commercial navigation and has a network of some 47 000 km of waterways, transporting about 670 million tons of products annually, nearly 17% of its total production. In the early 1990s, the 170 km-long connection between the Rhine and the Danube rivers in Europe was completed, connecting the North Sea with the Black Sea along a continuous waterway of 3500 km. Opponents of the Paraguay–Paraná hidrovía point to the tremendous long-term costs that have resulted from these projects, particularly in terms of what economists call "externalities." These costs (monetary and otherwise) are borne by others than the individuals using the resource and include flooding, deteriorating drinking-water quality, displaced populations, pollution, loss of wetlands and wildlife, and damage to fisheries.

Already hundreds of millions of dollars have been authorized to take farmland out of production and restore a more natural flow of freshwater across the Florida Everglades. An additional expenditure of 3–4 billion U.S. dollars over the next 5–10 years has been proposed to continue this restoration. Today, the Mississippi and Missouri rivers hardly resemble what explorers found two centuries ago. Floodplain wetlands have been reduced by 90% and, in many areas, converted to farmland. Today, these rivers have highly controlled, less dynamic flows (except during floods). Historically, the lower Missouri hydrograph displayed flood pulses in early spring and early summer that coincided with the spawning of many floodplain-dependent fish species. Current hydrographs show a stage increase in early spring that is regulated to maintain navigation through autumn. Ironically, even with dikes and levees for flood protection and navigation, the costs associated with floods have been enormous. In fact, data show that the flood frequency and magnitude has increased following these "improvements." The Great Flood of 1993 alone resulted in U.S. $12 billion worth of damage, much of it incurred when the river attempted to reclaim its floodplain. Based on the Mississippi–Missouri model, changes to the structure and flow pattern of the Paraguay River could lead to a similar disruption of the link between that river and the Pantanal.

Current status

The most significant recent development has been the announcement, in 1998, by Brazil's Federal Environmental Agency (IBAMA) that it has abandoned plans for construction activities along the Brazilian portion of the waterway. Instead, Brazil will now emphasize smaller, non-structural improvements to the Paraguay River. The government of Paraguay, based upon consultation with the U.S. Army Corps of Engineers, is also hesitant to proceed with plans to remove rock obstructions along the Paraguay River. Nevertheless, other countries in the region seem determined to carry out extensive dredging and channel straightening along the course of the river and its tributaries.

Even though the focus of the Paraguay–Paraná hidrovía is on navigational improvements, this waterway may evolve into a multifaceted, economic-development enterprise where a multitude of seemingly independent decisions and projects compete for the same resource. The cumulative effects of small projects can be avoided only if planners, politicians, scientists, and engineers adopt a large-scale, holistic approach. Such a perspective includes consideration of the cumulative impact of all small-scale decisions, and bypasses the pressures of short-term rewards obtained with short-sighted solutions. While it appears that no one is making a deliberate choice to destroy the Pantanal floodplain system, because of the cumulative effect of all these actions the Pantanal of the 21st century may well be as different from its present state as the Everglades and the Mississippi–Missouri rivers are from their pre-1900 condition. Large-scale channelization of the northern Paraguay–Paraná seems to be on hold; but will we witness yet another "tyranny of small decisions?"

Source: Gottgens et al. (2001).

Box 9.4 Pacific salmon: science policy

James H. Pipkin

Historically, more than 10 million salmon and steelhead "trout" (all *Oncorhynchus* species) returned to the Columbia River basin each year to spawn. Now there are less than a million, and most of those originate from hatcheries, not from the wild. Many salmon populations are already extinct and others are headed that way.

Few people would deny that many stocks of wild salmon in the Pacific northwest are in trouble, or that an ecosystem approach must be taken if the problem is to be corrected. However, the application of ecosystem principles to the Pacific salmon issue is enormously complicated and requires the accommodation of a number of important, and often competing, interests. This is true from many standpoints – economic, ecological, social, cultural, institutional, and political. A brief review of those factors, and their application to Pacific salmon, might illuminate both the difficulties inherent in holistic management based on species life-history and habitat requirements and the reasons why it is needed.

First, an example: a juvenile chinook (king) salmon may begin life in a tranquil tributary of the Snake River in Idaho; but after a peaceful beginning, its life is fraught with peril. During its first year, the chinook begins a hazardous journey to the Pacific Ocean, where it will spend the majority of its life before returning to its natal stream. What follows is only a *partial* list of the obstacles it must confront:

- elimination of tree cover and woody debris that give the young salmon a better chance to survive;
- a gauntlet of huge hydropower dams that have altered water temperatures, flow rates, and migratory corridors, and that put the whirling blades of turbines in the young fishes' path;
- voracious pike minnows and other non-native species introduced into reservoirs behind dams, which feed on small salmon;
- water extractions for agriculture that lower river water levels and divert many juveniles into fields, where they perish;
- pollution from industries and fertilizers, pesticides, and herbicides from agricultural and residential properties that affect the salmon's survival;
- siltation from clear-cutting of forests that increases turbidity and raises water temperature;
- increased predation by fish, birds, and marine mammals as young fish approach the ocean and during their years of North Pacific Ocean residence; and
- ocean conditions, such as El Niño, that can dramatically affect survival rates.

If the salmon survives to reach maturity in the ocean, it is again subject to most of the same obstacles as it returns to its place of origin. In addition, all along the return trip salmon risk being caught by commercial, recreational, or tribal fishermen. Finally, if it succeeds in reaching its natal stream, it may find that its gravel spawning beds have been degraded or destroyed by human activity.

This partial recitation of factors affecting chinook survival illustrates why an ecosystem-oriented recovery plan for salmon is so complicated and helps to explain why the expenditure of several billion dollars thus far has failed to produce the desired results.

Severe declines of wild salmon in the Pacific northwest are the product of many factors – some resulting from conscious decisions to sacrifice fish in favor of other economic objectives (such as cheap electric power); some resulting from human actions taken without consideration of their potential impact on salmon (such as habitat degradation through agricultural runoff, clear-cutting of forests, or the building of second homes along rivers); and some even resulting from unintended consequences of efforts designed to *increase* salmon populations (such as the deleterious effects that hatchery-spawned salmon can have on wild salmon).

Because there are so many "culprits," the norm has been for each group to point fingers at the other. Those who benefit from dams say that fishermen are the real problem. Fishermen say that they have sacrificed more than other sectors and that further cuts in their catch will not reverse the real problems, which are rooted in habitat degradation. Indian fishermen point to their treaties (under which they ceded most of their land to the United States), which reserved to them the right to continue to fish in their "usual and accustomed" fishing places. Because of those treaties, they say, all other contributing factors should be addressed before asking them to curtail harvests further. Farmers say that they have relied on irrigation water for decades and cannot give it up without going out of business. Land developers say that their contribution to the problem is minuscule when compared with other factors and that the economic costs of non-development are excessive. (Some would also argue that the right to make maximum economic use of their properties is protected under the Fifth Amendment to the Constitution.)

As the foregoing indicates, many of the arguments about salmon restoration boil down to economics. Obvious questions for consideration include: If salmon survival is to be improved, what is the most economically efficient way to do it? What are the economic trade-offs of one option versus another? The answers to these questions are not always clear. For example, the U.S. Army Corps of Engineers recently estimated that the economic costs to the region of breaching four large dams on the lower Snake River would be $359 million per year, whereas the economic benefits of that action would be $113 million per year, producing an annual net cost of $246 million. Others sharply dispute those numbers.

The economic disagreements often hinge on arguments about the science. For example, disputes about whether salmon produced in hatcheries cause harm to wild salmon (through reducing their genetic diversity or making them more susceptible to disease and predation) result in disputes about whether public expenditures on hatcheries should be continued. Gaps in scientific knowledge about changes in ocean conditions and their impacts on salmon survival cloud economic judgments about what it would cost to produce a desired level of salmon escapement from spawning streams. Disagreements about the effectiveness of collecting and transporting salmon around the dams lead to disagreements about the net costs of breaching those dams, as well as about the likelihood that breaching the dams would lead to salmon recovery.

The arguments about salmon are not limited to economics or science. In the Pacific northwest, salmon are cultural icons. Even to residents who have no intention of ever catching a fish, salmon represent part of their heritage. To many, salmon are like the canary in the coal mine: the serious decline in the abundance of wild salmon is an indication that streams are no longer healthy, that riparian habitat has been degraded, and that desirable recreational opportunities have been lost.

To Indian tribes in the northwest, the salmon has both cultural and spiritual significance. Often the salmon plays a central role in tribal stories and ceremonies. Several years ago, during fishery negotiations between the United States and Canada, a proposal was made to "balance" the economic benefits of salmon produced in the two countries by placing an economic value on all the salmon caught in each country that spawned in waters of the other country and adjusting harvests to equalize those values. For salmon caught in large nets by commercial fishermen and sold to a wholesaler, the valuation task is not difficult to accomplish. But what is the value of a salmon caught by an Indian fisherman for ceremonial or subsistence purposes? According to the tribes, the value is beyond quantification. Such issues frustrate attempts to resolve the diplomatic impasse.

Politically, the salmon situation is also complex. On one level, an elected politician (state or federal) from eastern Washington State whose constituents are primarily farmers who irrigate their fields will have a different perspective on the salmon issue than an elected politician from suburban Seattle whose constituents prize wilderness and sport-fishing opportunities. Both will differ sharply from a politician from the U.S. east coast who questions the large amounts of money being spent to restore salmon in an area that means little to his constituents. The approach of each will make it virtually impossible to find common ground, and in the end the political debate will likely be resolved through seniority and relative political power, rather than by a thoughtful weighing of the merits or a quest for a compromise solution. On another political level, the elected governments of Oregon and Washington are likely to disagree with the government of Alaska (and with tribal governments) about who should bear the pain of harvest sacrifices or habitat restoration. On yet another level, the federal governments of Canada and the United States carry the same arguments, and the same difficulty of achieving consensus, into the international arena.

Jurisdictionally, the problem is equally complicated. For convenience, the major contributing factors are generally described as the "four Hs": harvests, hydropower, habitat, and hatcheries. No single governmental or other entity controls all four Hs. Indeed, no single entity has the power to control fully even *one* H. Take harvests as an example. In general, each state regulates fishing within its borders and off its coast to a distance of three miles. But unfortunately, salmon are not very good at respecting state borders. They swim across state lines, as well as into Canadian waters where harvests are regulated by the Canadian government, and into international waters where their principal protection is international Law of the Sea. Beyond that, tribal governments have treaty rights that affect permissible harvests. The U.S. federal government can step in to regulate salmon harvests only when they affect species "listed" as threatened or endangered under the Endangered Species Act (24 distinct populations of Pacific salmonids are now so listed) or when other federal statutes come into play. Similar dispersions of authority exist with respect to the other Hs.

Other institutions that play a role in the Pacific salmon controversy include the courts, Congress, state legislatures, state and federal fish and wildlife agencies, the Bonneville Power Administration, the Northwest Power Planning Council, the U.S. Army Corps of Engineers, the Environmental Protection Agency, land management agencies such as the U.S. Forest Service and the Bureau of Land Management, harvest management organizations such as the Pacific Fishery Management Council and the U.S./Canada Pacific

Salmon Commission, and tribal governments. Many institutions at different levels (international, federal, tribal, and state) focus primarily on how many fish can be caught, how the catch is to be allocated, and how allocation disputes are to be resolved. Other institutions address isolated issues that bear on salmon recovery – for example, how timber will be harvested on national forest land, or how a hydropower facility will be managed to reduce the impact on fish. No existing institution has the mandate to determine answers to broader ecosystem-based questions, such as how a watershed will be managed. It suffices to say that the existing institutional arrangements are not well suited to development of an ecosystem (or bioregional) perspective on the salmon problem. The U.S. National Research Council (NAS, 1996) has gone so far as to conclude that "the current set of institutional arrangements *contributes to the decline* of salmon and cannot halt that decline."

In sum, it is hard to imagine a more complex problem than what to do about Pacific salmon – in terms of economic implications, social and cultural factors, ecological/scientific considerations, institutional overlaps, or political obstacles. Does that mean we should throw in the towel and schedule funeral services for the remaining wild salmon? Does it suggest that ecosystem-based management is a valid approach only for smaller, more circumscribed issues? For all but the hardened pessimist, the answer to these questions has to be "no." Sadly, the consequences of inaction are now painfully clear. The sharp declines in salmon abundance, the "listings" of salmon populations under the Endangered Species Act, the economic dislocations to commercial fishermen and to the recreational infrastructure of the Pacific northwest, and improvements in scientific knowledge, have dramatically increased the general awareness of the problem. The majority of residents of the region now seem prepared to make economic sacrifices to ensure the survival of remaining salmon species. Hopeful signs are appearing:

- In 1999, the United States and Canada reached agreement on long-term arrangements for ocean salmon fisheries in both countries. The new arrangements put conservation first and, together with harvest restrictions being imposed by the National Marine Fisheries Service under the Endangered Species Act, provide some confidence that harvests will be under control and that attention can be focused on the other Hs.
- Under the Northwest Forest Plan, the federal government has imposed restrictions on timber activities on 24 million acres of federal land in the northwest. In particular, the plan requires leaving buffer strips of intact habitat along rivers and streams for the benefit of salmon and other anadromous fish. Similar standards were implemented for an even larger area on the east side of the Columbia River basin. In addition, forestry regulations designed to help the salmon have been imposed by the states of California, Oregon, and Washington on state-owned and privately owned land.
- The development of salmon recovery plans by the states and tribes demonstrates that the entire region is motivated to find a solution.
- Major changes (such as the breaching of dams) are being seriously analyzed, when most people would have considered them totally unrealistic a few years ago.
- The Endangered Species Act has emerged as a catalyzing force for change, and has given the National Marine Fisheries Service and U.S. Fish and Wildlife Service the power to be central players as the issue unfolds.
- The listing of Puget Sound king salmon in 1999 under the Endangered Species Act has brought home to residents of big cities like Seattle and Tacoma that they are not immune from the sacrifices that will be required.
- The development by nine federal agencies, under the leadership of the National Marine Fisheries Service, of an "All-H" paper, laying out various options that may be pursued for harvests, hydropower, hatcheries, and habitat, is an indication that federal agencies and other state and tribal institutions are beginning to frame the problem in a more holistic manner, despite the fact that decision-making remains highly fragmented. This forms the essence of an "ecosystem" approach, or, as the tribes would put it, "gravel-to-gravel" salmon management.
- Recently, the U.S. Congress significantly increased its funding, both to implement the 1999 U.S./Canada salmon agreement and to provide grants to four states and a large number of tribes for a coastal salmon restoration initiative.

Thus, there are reasons for optimism, but for that optimism to come to fruition will require a huge commitment, by many interested parties, during many years, to fill gaps in scientific information, to coordinate management, to provide funding, and to make and implement hard choices among painful alternatives that have serious, interrelated, ecological, economic, social, and cultural implications. We should not delude ourselves into thinking that the effort to restore wild salmon is anything other than an uphill struggle. The recovery problem is vulnerable to political, social, economic, and ecological pressures. An energy crisis could result in water supplies being used to maximize power production, at the expense of fish passage. Judicial decisions may blur the distinction between wild and hatchery-produced salmon and effectively reduce protections under the Endangered Species Act. Changes of administration, at the national, state, or local level, can produce a new set of economic priorities and dry up money for salmon recovery. Success will no doubt require a willingness on the parts of federal, state, and tribal governments to conclude that current institutional arrangements are not adequate and to establish a new decision-making framework that can address broad issues such as watershed management and fish governance. These tasks are formidable, similar to the daunting obstacles faced by the juvenile salmon making the difficult and challenging journey to the sea. The salmon has no option but to begin the journey and to commit fully to completing its task. For that matter, if we are to bring salmon back from the brink of extinction, neither do we.

annually, and 20 control approximately 90% of global pesticide sales. MNCs often resist strengthening environmental standards in order to reduce costs and increase profits. Their influence on environmental standards and regulations is especially strong in small, developing nations seeking economic growth. International tourism and fisheries development in small-island states with large marine jurisdictions illustrate the global trade problem.

9.3.5 The scientific challenge

Science is a tool for conservation, "a way of knowing" (Mayr 1997; Moore 1984). Science and scientists have helped mobilize the conservation debate and have assisted policy-makers in the elaboration of agreements on many environmental issues. Scientific research has provided important information for conservation, and remains the best way for recognizing change that can otherwise be lost in a chasm of opinion, disagreement, and inaction. At the same time, scientific debate can introduce uncertainty into public and decision-making arenas. Science's self-correcting process often leads to counter-intuitive information not easily amenable to quick-fix solutions, and can counter traditional expectations.

The challenge for science is both internal and external. The internal challenge concerns the acquisition of knowledge itself. Many questions requiring solutions emerge from our case studies, and are reviewed by Hixon et al. (2001). Predicting ecosystem response is an especially formidable scientific task, as it draws on ecosystem and human behaviors that have unpredictable outcomes. Major gaps in knowledge and understanding of the coastal realm add further complexity, involving a mix of highly variable human interactions and antecedent conditions that are often unknown. Furthermore, conservation in the coastal realm requires both specialization and interdisciplinary science. Disciplinary science creates gaps that can be difficult to relate, one to another. For example, marine biology began as a shallow-water tradition, focused mostly on community ecology. Biological oceanography has mostly been concerned with open-ocean production at low trophic levels (phytoplankton), as related to ocean physics and chemistry; large organisms are largely ignored. And fisheries ecology searches for levels of sustainable yields of large, exploitable species, often out of context with fish behavior and ecology. Therefore, these disciplines have evolved separate scientific societies, journals, and departments in universities, with theories and practices that are difficult to relate to one another. Funding for science is also largely disciplinary, and exacerbates these differences. Thus, information gaps exist among scientific specialists, leaving much uncertainty in understanding broader aspects of change and impact.

External challenges fall in the realm of interpreting and applying science to political and social arenas. Many environmental problems are linked to the rise of technology in modern society, with at least two effects on policy. One is growth of knowledge in fields that are not easily accessible to people who lack specialized training in scientific or technical fields. The technical complexity of environmental issues requires expert evaluation, and technical experts now play a central role in environmental decision-making that affects non-technical people. This dependency on technical experts creates a second effect: a threat that policy choices may be placed more in the hands of technical and administrative elitists, and less in those of ordinary citizens (Fiorino 1995). A public perception of "scientific superiority" suggests that scientists have claimed for themselves mandates to "determine the viable management of nature by which the construction of environmental problems is strictly scientific and technical" (Elliott 1998). As a consequence, the social and political connotations of environmental degradation may be perceived as being marginalized. In their contribution to and involvement in decision-making, scientists themselves can become part of the political process, making science susceptible to becoming politicized. Science then becomes a pawn, as politics may rig science to its own end, and agencies can structure science according to their own agendas.

Most scientists recognize these traps. Ecosystem science continues to evolve and scientific research is now an essential factor in understanding the functions and dynamics of ecosystems, the causes of resource loss, and the nature of ecological dysfunction. The challenge for science lies in disciplinary integration and how science may be translated for policy; the challenge for policy-makers and the public lies in understanding the scientific process. Coastal-realm conservation science requires understanding land–sea interactions, and how coastal-realm ecosystems behave despite terrestrial and marine differences (Box 9.5). Scientists are now seeking to develop integrated land–sea process models to guide integrative research, to conserve living resources, and to aid coastal-realm management and policy. Also,

Box 9.5 Terrestrial–marine differences?

Marine–terrestrial differences suggest that oceans may respond to human perturbations in a fundamentally different manner than land. Differences lie mainly in degree and scale and whether pelagic or benthic systems are being compared.

- Fluid media of air and water place different sets of constraints on organisms
- Higher-order (phyletic) diversity is substantially richer in aquatic than in terrestrial environments
- Aquatic portions of marine systems are much more variable and labile than terrestrial portions, whereas benthic portions are as stable
- Geographical ranges of marine species tend to be larger; nearshore organisms can have very restricted distributions, depending on dispersal capabilities
- Marine primary producers are small, mobile organisms (except for macro algae and sea grasses); terrestrial producers tend to be large and sessile
- Marine systems dispersing larvae of many species link distant habitats and are presumed to be more "open". However, spores, pollen, and some arthropods are also transported long distances in air; nevertheless, the atmosphere is a relatively "dead" medium for life
- Most large marine carnivores and grazers have complex life histories, and the largest marine species are much larger than terrestrial counterparts
- Marine predators generally have much higher reproductive outputs (except marine mammals, sea turtles, etc.), which may buffer them from extinction caused by overexploitation, but may render them more vulnerable to threshold effects
- Biomass is hundreds to thousands of times more dilute in pelagic portions of oceans than in most terrestrial systems
- Rates of response to environmental variability is generally greater in marine systems
- Marine species interact trophically with more other species than is the case for land
- The pelagic oceans are on average several to hundreds of times less productive than terrestrial systems; however, benthic and estuarine environments generally exceed terrestrial environments in productivity
- Exploitation of life in the sea is derived largely from unmanipulated populations of species high on the food web, whereas on land most food is derived from domesticated plants and herbivores

Modified from Carr et al. (2003); NRC (1995); Peterson & Estes (2001); Steele (1985).

scientists seek better understanding of the natural histories of species so that relationships between biodiversity and ecosystem function can be established, and resources managed sustainably.

9.3.6 The challenge of gaining political will

Political will is influenced by a wide variety of factors, ranging from social and cultural "spirit of the times," to intellectual and philosophical influences, to geography, climate, history, and political inheritance. Strong political will is essential for maintaining a stable political system, one that survives through crises without internal conflict or warfare. Changes in political will are not simply reactions to "objective" factors, such as economic forces or environmental threat, but also involve conscious manipulation, with adoption of one viewpoint or another that reflects the political climate and cultural tensions of the time. When it comes to the environment, the nature of the interaction between political systems and environment is extremely complex, and gaining political will to conserve is made difficult by political inertia and the *status quo*.

Political will is driven by public perception, and is susceptible to the "fallacy of misplaced concreteness." This fallacy is committed when attention is diverted from the underlying causes that drive change (A.N. Whitehead, in Daly & Cobb 1989). When individual rights to fish (or not) are made complex by rights and privileges granted to other users at the same time (e.g., recreational boaters and tourists), all users' needs can become compromised in a congestion of rules and regulations that satisfy no one and that divert attention from basic issues (Box 9.6). Or, when public attention becomes focused on high-profile depleted species to the neglect of the social issue or environmental dysfunction that causes depletion, political capital can be wrongly spent. Or, when proponents argue that offshore oil and gas exploitation is the least costly way of meeting future energy demands, public attention and political will are diverted from environmental impacts and consumer appetite. And, public debate on global warming risks diverting political will to the costs of taking action, rather than addressing known anthropogenic sources of climate change. In all of these cases, political will reflects time scales of observation, in which time frames of environmental change differ according to the resolution of observation (Fig. 9.1). If action is long delayed a crisis can result; for example, if

Box 9.6 Human factors: conflicts over the Kenai River, Alaska

Gary L. Hufford

The Kenai River, located on the Kenai Peninsula in south-central Alaska, is one of Alaska's most popular rivers for multiple uses: commercial fishing, sport fishing, recreation, native subsistence, and general living. The river, approximately 82 miles long, boasts major runs of all five species of Pacific salmons, in addition to trophy sized rainbow trout and Dolly Varden. At present, the world sport-caught records for king salmon are for Kenai River fish.

Because the Kenai is within easy driving distance of Anchorage, the main population center of Alaska, use of the river has steadily increased with the rise in the Anchorage population and visiting tourists. Demands on the river by these users have produced numerous conflicts. The average commercial catch of Kenai River salmon provides nearly $15 million to commercial fishermen. Sport fishermen not only want to catch one of the five species of salmon, but seek trophy-sized fish. A major fishing guide industry is estimated to bring in more than $20 million to a variety of local businesses. There are also casual boaters and tourists who simply want to drift the river to enjoy its beauty and the abundant wildlife, and those who wish to buy river front property for homes. Not least among the concerns are the local native tribes who wish to take fish for subsistence as they have traditionally for hundreds of years.

The response of federal, state, and local governments, which all have mandated responsibility for portions of the river, has been to manage the Kenai by regulations that don't always take into account the environmental health of the river and its ability to sustain native fish. First, the number of fish that run up the river to spawn is variable and not well known. To manage salmon on the Kenai and ensure sufficient escapement, the state of Alaska puts together a catch management plan that comes under review only every 10 years. The state decided that the catch should be broken into 93% for commercial fishermen, 2% for sport fishing, and 5% for spawning escapement. Decisions on catch are determined each year by predictions made by State Fish and Game; conflicts immediately rise when predicted numbers do not occur. Commercial fishermen then request that the river be shut down to sport fishing to increase their catch, and sport guide fishermen demand that the commercial catch be reduced to ensure a quality experience for their clients. Local sport fishermen demand that guided sport fishing be reduced because it takes too many fish. Meanwhile, the Kenai Native Association demands that natives be able to fish for subsistence. Under federal law, priority is supposed to be given to rural subsistence users. The State of Alaska does not recognize a rural preference and claims that the Kenai, because of its easy access, is "urban," which denies subsistence rights. The court ruled that the federal government take over responsibility for subsistence in October 1999 if the State of Alaska did not modify its law to comply with federal law.

Federal control is now in place with two different agencies trying to manage the fisheries with different agendas. In an attempt to resolve some of these issues, the state has limited commercial openings, closed some days on the river to sport guides, changed the kinds of lures and hooks that can be used, and allowed the natives a token subsistence catch (~ 5000 fish). To control the number of guides, the State of Alaska, in conjunction with federal agencies, requires permits that exceed $2000 per guide boat. To obtain these permits, a guide must demonstrate that he/she has up-to-date safety training and liability insurance, and is fiscally responsible (paid last year's sales tax on all clients). In addition, the state limits the kind of fishing that can be done, and also issues emergency closures and limits the number and size of fish that can be caught and kept.

Changing regulations can make it difficult for guides to provide a quality fishing experience for the sport fishermen. Often dates for a fishing trip are made many months before the trip. It is difficult to tell a potential client not to come because the river is closed, you can't keep fish, or you must use techniques that greatly reduce the possibility of a catch. Also, commercial fishermen don't know how much to invest in gear for the coming season or to plan on how many fish may be available. Homeowners are not sure if they can develop their property and/or must make major modifications to meet property regulations. And last but not least, the traditional native harvest preference on the river is in direct conflict with state law.

Meanwhile, habitat restoration is costing large sums of money and other resources in the lower river. To meet demands of multiple users while maintaining the environmental health of the river is a monumental task when a comprehensive conservation plan remains in flux. Up to now, most plans have only addressed fish catch. But recently, concerns over habitat have caused state and federal governments to issue regulations to protect the shoreline and the river itself. For example, to limit habitat degradation and pollution in and along the Kenai River, the state requires a minimum setback for personal property. However, more than one-third of the homes on the river inside of the present setback were built prior to the regulation and are not subject to a setback. To protect habitat, boat motor size and use have been limited. Motors smaller than 35 horsepower can be used on the lower river and the upper river is closed to motor use. Shore access for portions of the river is also being limited.

A comprehensive conservation plan that accommodates all users of the Kenai River can be accomplished only through a strong scientific understanding of the river environment that is accepted by all users. From that foundation, and recognition of all users' needs, compromises will have to be made by all parties to ensure that the resources of the Kenai River remain healthy and sustainable.

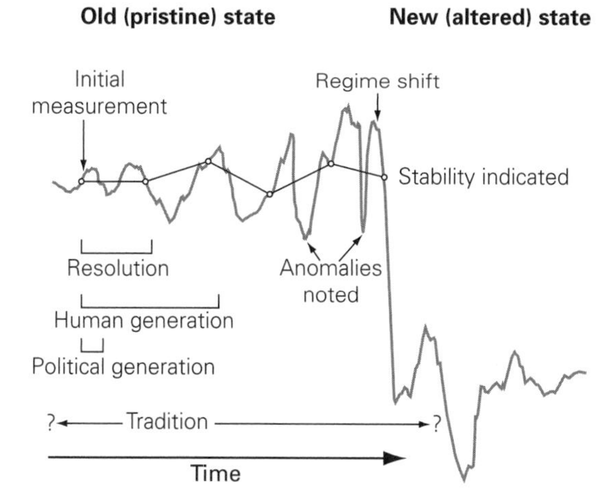

Fig. 9.1 Ecosystems fluctuate over time until a regime change produces a new altered state. Time frames of change differ with observation resolution. If observations (open circles) are too infrequent to recognize natural variability, ecosystems that appear stable may undergo unsuspected anomalous change. Human generation time is sufficient to perceive short-term variability. Political generation time (i.e., the "lifetime" of a politician in office) is short, focused on immediate needs, and insufficient for forecasting. Traditional practices persist for very long times, and are highly resistant to alteration. Thus, when a new altered state occurs, society may be unprepared. Conversely, if observations (monitoring of appropriate indicators) are frequent, large-scale, and long-term, increasing magnitudes of variation might warn of a bifurcation point and regime shift.

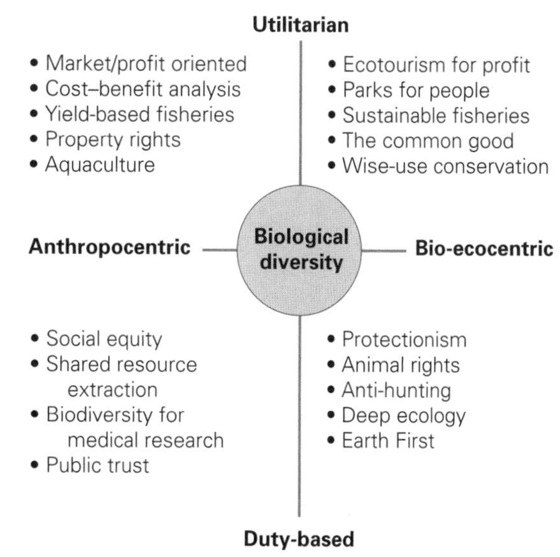

Fig. 9.2 Perceptions about values (e.g., biological diversity) are conceived in a decision matrix on two axes: from utilitarian to duty-based and from anthropocentric to bio- or ecocentric. Holders of various positions are exemplified in four segments, suggesting discrepancies for biodiversity conservation. Concept adapted from Beatley (1994).

the ecosystem flips at some bifurcation point and a regime shift occurs, political will may be unprepared for needed change in management approaches.

Political will reflects the values of a society and can become compromised when individual perceptions widely differ and when individuals place themselves in differing positions with respect to a particular issue (Fig. 9.2). For example, if an endangered species stands in the way of a marina, which position would a wildlife conservationist or a yachtsman adopt? Or, if an economically important shipping or cruise-line industry crosses a whale migration path or valuable fishing ground, what alternative is acceptable? Or, if nutrient pollution to coastal waters results from lawn or farm fertilizers, what are the options and how are life-styles affected?

The influence of political will transcends national boundaries to determine the outcome of international negotiations. During the early 1970s, the environment was put on national, international and legislative agendas, reflecting public concern. Two decades later, the United Nations Conference on Environment and Development (UNCED, 1992) placed conservation, including the coastal realm, directly into the international political arena, reflecting global interest. The summit produced *Agenda 21*, but ended without agreement on its primary goal – a political statement indicating *how* UNCED objectives might be met and a schedule for suggested actions. And rather than focusing on shared environmental concerns and the need to consider environment and people together for protecting our planetary ecosystem, Principle 1 of UNCED's *Declaration* stated: "Human beings are at the center of concerns for sustainable development" (Table 2.7).

In the coastal realm, where issues are especially difficult for politicians to resolve, implementation of treaties and national programs remain unfulfilled, and political will lags behind scientific understanding and public perception. Assessments of UNCED have brought into question whether it really provided a set of principles to shape international action on environment and development, whether it forged global partnerships, and whether it provided the basis for a global environmental ethic (Elliott 1998). Kullenberg (1999) observed: "Rather than achieving the desired integration and holistic approach agreed upon at UNCED in 1992, subsequent actions appear to have increased global fragmentation. National institutional structures cannot respond in an inter-sectoral way. International organizations have a similar sectoral structure, as does the management of the regional bodies . . . Thus, coordination is not only to be sought among the intergovernmental organizations, but perhaps also primarily at the national level."

9.3.7 The challenge of governance

The coastal realm is embedded in complex social, economic, historical, political, and geopolitical factors that fragment decision-making, conservation, and management. As protecting rights of citizens almost always precedes environmental degradation, governmental and legal systems are tilted toward economic and social gain with lesser concern for the environment. The result is that the coastal realm is divided into domains of jurisdictions and uses, where laws and legislation separate land and sea, and differing values lead to multiple-use conflicts. A case in point concerns conflicts among oil and gas exploitation, fisheries, and recreation, which are managed separately and offer quick pay-offs, but can have long-term environmental and social consequences. Thus, governance for the coastal realm is ill-equipped to handle widespread environmental problems involving the high rates of environmental change and large scales of human impacts.

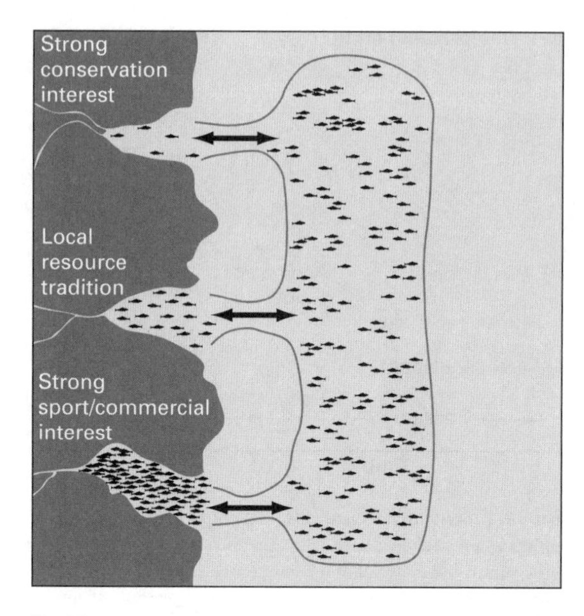

Fig. 9.3 In this model, a valued, continental-shelf fish species is sustained by recruitment from three estuaries. A metapopulation model suggests that the population should be managed as a unit. However, governance may vary with location. If the species is relatively rare in the north, estuarine governance may favor strong protection. In a middle estuary, the species is common and governance may favor traditional use. In a southern estuary, the species is abundant and governance would favor commercial harvest. Thus, the species' conservation depends on agreement among diverse stakeholders under different circumstances.

Among coastal nations and in international negotiations, problems of integration of issues involving fisheries, pollution, and multiple-use protected areas are common. Under UNCLOS, nation states have a general obligation to protect and preserve marine resources, an obligation that Belsky (1989) interprets to extend to all states "as a statement of customary international law"; that is, if a nation rejects a comprehensive, ecosystem approach it could be in violation of international law. Further, in many nations international law is part of their internal law; thus, there can be a duty under domestic law to apply an integrative ecosystem approach. Nevertheless, as coastal areas grow more crowded, management agencies continue to issue site-by-site or permit-by-permit procedures that allow urban expansion, resource loss, and environmental change to slip by. This situation is exacerbated by strong resistance among nations and the public to comprehensive policy. The regional, long-term scale of management that is required to redress these shortcomings is beginning to be addressed in special cases (e.g., watersheds and some commercial fisheries), but is far from becoming common practice.

The primary challenge for governance lies in recognizing the coastal realm as a unit of management rather than a disconnected patchwork of bureaucratic directives with unequal powers and conflicting goals. Missing are the long-term forecasts that anticipate crisis (Box 9.7). For example, when scientific evidence reveals sets of regional fishery metapopulations, and stakeholders within the region have diverse interests, fisheries management can become fragmented (Fig. 9.3). Or, when dams obstruct freshwater flow, management of estuarine and coastal-ocean fishes is affected. Integrated coastal-zone management (ICZM) is an attempt to deal with fragmented issues by negotiating conflicts, setting priorities, and mitigating adverse impacts. Applying ICZM, however, usually involves reaching

Box 9.7 Inventing foresight

Frank M. Potter

An ability and a willingness to engage in serious national foresight suggests Mark Twain's remark about the weather: "Everyone talks about it, but nobody does anything about it" (a statement that may once have been accurate, although it is clearly no longer the case). Books have been written about the subject, conferences held, workshops convened, and occasional tentative experiments have been undertaken.

What a forecast process offers is a way to manage problems before they have surged out of control and can only be treated as urgent (and, almost certainly, expensive) crises. The concept is not complex, although the consequences would indeed be far-reaching. Essentially, it would require the Chief Executive in a given country to establish a process under which the various government agencies would be required to identify the long-range implications of programs in which they were engaged, or which they propose to establish. Programs that fail to meet foresight requirements would be put on hold until they complied.

This process would be not be easy, simple, or without controversy. On the other hand, the process would significantly improve the quality of executive decisions, allowing the review of government policies already in place as well as those presented for approval, and enabling the identification and selection of options before those options have been eliminated by the passage of time – all with relatively trivial investments of time, effort, and funds. Those who have examined the concept have generally supported it. Why has it only rarely progressed beyond the conceptual stage? To begin with, the process threatens many bureaucrats involved in the process of government, as it tends to establish accountability. Also, it creates the prospect of an unconfined and uncontrolled policy review capability that has the potential of revealing obstacles to plans that might better be left undiscovered until later on.

Foresight is, first and foremost, a process; it should never be confused with the products that it may produce. The results of the process ought to be a better understanding of the effects of policy initiatives that have been, or may be, undertaken in both government and the private sector. Some private companies have developed substantial foresight programs, and their affairs have prospered. Royal Dutch Shell has had this capability for years, and has used it as a critical part of the corporate decision-making process. The majority of governments, on the other hand, have never done so, and we have paid a substantial price for that omission.

Had governments put a foresight process into place, they might have been able to anticipate and take steps to mitigate some of the consequences of the energy crises of the 1970s – massive economic disruption, unemployment, trade balance problems, gasoline lines, and transportation and utility disruptions. Similar disruptions occur with fisheries or coastal alterations. It would appear to make sense to anticipate some of these events through the wise use of foresight, and to develop options for dealing with them. The time to fix the roof is when the sun is shining, not when it has started to rain.

A host of environmental challenges confront us today, including resource conservation and use, environmental quality, disrupted ecosystems, shortfalls of water supply, evolution of resistant disease-causing organisms, trade, economic wellbeing, and social equity. But only a few countries have actually tried to create such a process and even there, it has not survived for long. Foresight offices were established in Sweden and New Zealand, but were later terminated. Interestingly enough, in both cases the reasons were similar: they were too successful. Foresight offices were set up to advise the Prime Ministers of both countries and performed, according to reports, well indeed – so well that opposition parties in the legislatures felt at such a disadvantage that they took steps to withhold funding for further foresight efforts.

It would, of course, be possible to institute foresight capabilities at other than national levels. State and local governments could benefit from rigorous foresight analyses and, as mentioned, it has been used quite successfully in the corporate world. Foresight is nothing more than a tool; it can be used or it can be misused. If used wisely, it should produce benefits far outreaching its inconvenience and cost.

consensus among a host of stakeholders that hold different values, or focusing on one or a few specific issues that can satisfy the lowest common denominator. A single unenthusiastic individual or nation can stymie agreement on progressive goals, even when at stake is the health of the environment for protecting the common good.

9.4 Strategies for coastal-realm conservation

In this final section, we examine our environmental debt and suggest means for expanding conservation to regional proportions under the guidance of adaptive ecosystem management. An overarching goal for conservation is development of strategies to sustain societies while also conserving natural resources. This goal requires placing human needs in the context of ecosystem diversity, vigor, persistence, and resiliency. In this respect, the coastal realm is a major global playing field, wherein bottom-up protectionism can be elusive and laws and actions are piece-meal. Conservation may better be achieved by ecosystemic, regional, adaptive, and integrative strategies.

9.4.1 Sustainability and the environmental debt

An acknowledged feature of the modern world is that human demands for goods and services have exceeded the ability of ecosystems to produce them. That is, humans are living on the inherited richness of natural capital, whereas use of only the interest (net production) is sustainable. If ecosystem persistence, vigor, complexity, and resilience are to be maintained, a large-scale environmental debt (or "global environmental deficit," Bormann 1990) must be accounted for. In some cases this debt has been accumulating for centuries, even millennia, compounded in recent times by increasing intensity, extent, and duration of multiple impacts at all spatial and temporal scales (Fig. 9.4). Restoration is a costly attempt to repay that debt.

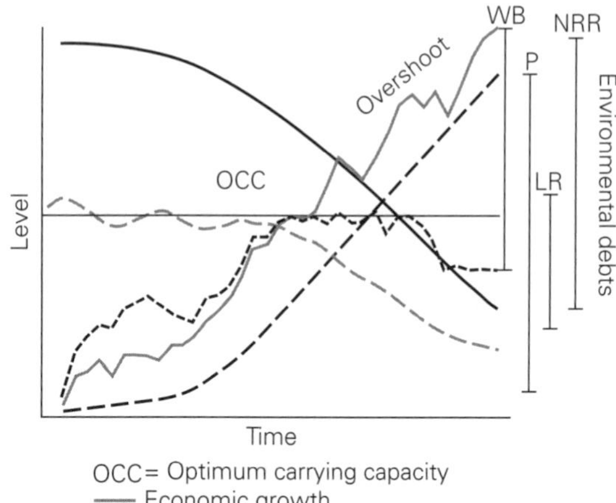

Fig. 9.4 The environmental debt is a function of increasing social uses and decreasing natural resources, observed as diverging trends over time. The optimum carrying capacity for the environment occurs at the intersection of economic growth measured as gross national product, and index of wellbeing measured as economic welfare, i.e., quality of life (Barrett & Odum 2000). An environmental debt is accrued when the optimum carrying capacity is exceeded (overshoot) by economic growth and/or pollution and resources no longer rebound, but decline. Repayment of the debt (or deficit) requires ecosystem readjustment or high capital investment in subsidies.

The coastal realm suffers from chronic human disturbance that involves addressing pollution, freshwater flow regimes, depleted fisheries, and greenhouse gases together at regional scales in order to repay the environmental debt. At present, most economies treat environmental functions as externalities, and resources as so-called "services" and investment opportunities, as if resources belonged only to the present generation. The emerging field of ecological economics, in contrast, attempts to bridge natural sciences, economics, and resource uses. In the present world of shortages and intergenerational inequity, ecological economics seeks collaborative solutions toward sustainability of the natural environment in the spirit of UNCED's *Rio Declaration*. Ecological economics recognizes that ecosystem services and the roles that plants and animals play do matter, and that local people, politicians, ecologists, economists, and managers are all stakeholders in the services that nature provides. Thus, to achieve sustainability, ecosystem goods and services need to be incorporated into economic accounting. Thus, integrating ecological and economic research requires that ecological systems be viewed as sets of processes rather than a collection of individual resources. Placing conservation in economic terms can relate directly to what people do in their daily lives, despite problems of disassembling "worth" from "value," but must also focus on ecosystem behavior, diversity, discontinuities, and the limits of ecosystem resilience.

9.4.2 Adaptive ecosystem management

Jackson et al. (2001) have convincingly drawn from history to show that coastal systems have undergone major structural and functional changes, "raising the possibility that many more marine ecosystems may be vulnerable to collapse in the near future." Because changes began centuries ago, recent observations provide only a starting point for stopping the hemorrhaging of ecosystems or extinction of species. Our case studies exemplify that conservation must include natural history, ecology, and people in a common paradigm for management. Attaining improved

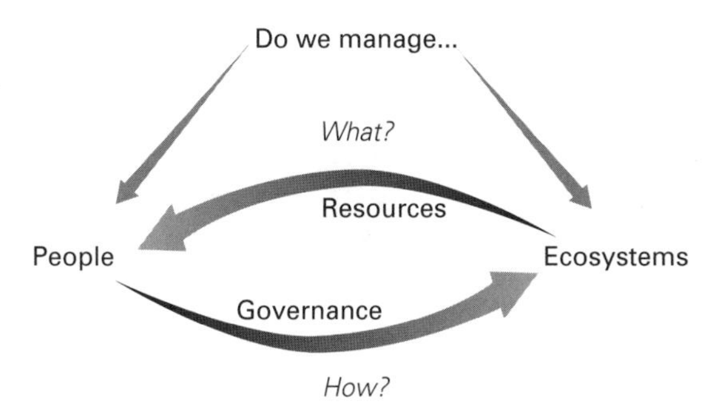

Fig. 9.5 Ecosystem management is about interactions and feedbacks among people and environment. "What" concerns managing resources sustainably. Inevitably, resource exploitation affects ecosystems, meaning that some ecosystem adjustment is required. "How" concerns effective governance that maintains ecosystem resiliency and resource sustainability.

ecological health of Chesapeake Bay requires consideration of whole watershed processes in the context of history. Conservation of marine mammals in the Bering Sea involves long-term dynamics of sea ice, climate, oceanography, and traditional human activities. Maintaining species and healthy environments in the Bahamas requires comprehensive protection of habitats that extend from nearshore to whole ocean basins, while taking account of the nation's need for economic development. Thus, a major conservation task is to relate environmental issues and their potential solutions to the spatial and temporal scales of environmental and social processes – that is, adaptive ecosystem management.

Adaptive ecosystem management is the antithesis of currently fragmented approaches. As Ludwig et al. (1993) observe for fisheries: "The complexity of the underlying biological and physical systems precludes a reductionist approach to management." However, ecosystem management is difficult to apply and has many interpretations (Box 9.8). Two approaches are possible: those tilted toward managing ecological systems themselves and those directed toward managing humans. The question is: what is being managed, the environment or people? The discussion is circular, as managing human activities inevitably affects ecosystems and *vice versa* (Fig. 9.5): that is, fisheries result in biotic community changes, and human activities in the coastal realm (filling wetlands, altering watersheds) clearly affect fisheries.

Ecosystem management is intended to integrate levels of government, economic sectors, and societies on the basis of sound ecosystem science, public support, and

Box 9.8 Sample definitions of ecosystem management

- ". . . regulating internal ecosystem structure and function, plus inputs and outputs, to achieve socially desirable conditions" (Agee & Johnson 1988)
- ". . . the careful and skillful use of ecological, economic, and managerial principles in managing ecosystems to produce, restore, or sustain ecosystem integrity and desired conditions, uses, products, values, and services over the long term" (Overbay 1992)
- ". . . a strategy or plan to manage ecosystems for all associated organisms, as opposed to a strategy or plan for managing individual species" (Forest Ecosystem Management Team 1993)
- ". . . integrating scientific knowledge of ecological relationships within a complex sociopolitical and values framework toward the general goal of protecting native ecosystem integrity over the long term" (Grumbine 1994)
- ". . . integration of ecological, economic, and social principles to manage biological and physical systems in a manner that safeguards the ecological sustainability, natural diversity, and productivity of the landscape" (Wood 1994)
- ". . . a method for sustaining or restoring natural systems and their functions and values. It is goal driven, and it is based on a collaboratively developed vision of desired future conditions that integrates ecological, economic, and social factors" (IEMTF 1995)
- ". . . is holistic, incorporating all the elements of the ecosystem, biological and physical, and their interrelations. *Sustainability* is at the core of ecosystem management, its essential element and precondition" (Haeuber & Franklin 1996)
- ". . . driven by explicit goals, protocols, and practices, and made adaptable by monitoring and research based on our best understanding of the ecological interactions and processes necessary to sustain ecosystem composition, structure, and function" (Christensen et al. 1996)
- ". . . is a paradigm shift from individual species to ecosystems . . . and from humans independent to humans as integral parts of ecosystems" (Sherman & Duda 1999)
- "The goals of ecosystem management are to maintain ecosystem health and sustainability" (EPAP 1999)
- ". . . a collaborative process that strives to reconcile the promotion of economic opportunities and livable communities with the conservation of ecological integrity and biodiversity" (Keystone Center 1996)

dedicated political will. Integration occurs over a continuum of activities, grading from a sector approach, through a communication forum, to coordination and synchronization, harmonization, and ultimately toward acceptable goals. When holistic goals and objectives are achieved in a context of sound information, segregated mechanisms can become incorporated into a strategic plan, with adaptive and precautionary forms of management. Flexibility is critical: ironically, it is often the case that when resources are sustained, success can be so great that management systems can become unresponsive to changing conditions (Holling & Meffe 1996), creating conditions for resource collapse, which can be broken only when a flexible, adaptive strategy is adopted.

9.4.3 A regional coastal-realm conservation model

During the early 1970s, UNESCO's Man and the Biosphere Programme (MAB) emerged as an offshoot of the International Biological Programme. Within MAB, a new concept for protected areas emerged – the "biosphere reserve." This concept specifically includes humans, research, monitoring, and adaptive strategies as parts of a regional conservation design. The concept requires that conservation problems not be addressed parochially nor for protection alone, but within a regional setting of greater depth and complexity. Three roles have been iterated by Batisse (1990): (i) a conservation role, providing protection of genetic resources, species, and ecosystems; (ii) a logistic role, providing research and monitoring within an international scientific program; and (iii) a development role, searching for sustainable resource use. These roles are served by a design that includes a core, a buffer, and a transition zone. The core may consist of any area for which a reasonable management and research program may exist or can be devised – a species, a protected area (existing or proposed), a fishery, a portion of a watershed where restoration is taking place, a research site, or an area of special esthetic or recreational value. To avoid core areas from becoming isolated patches within a land–seascape of disturbance, a buffer surrounding the core protects supporting processes. Beyond the buffer, a transition zone extends to a regional scale to ensure species dispersal and biogeographic, biophysical, or biogeochemical connectivities. The objective is to incorporate biogeographic regions into the management paradigm (Fig. 9.6) in which several core areas serve as a protected-area network.

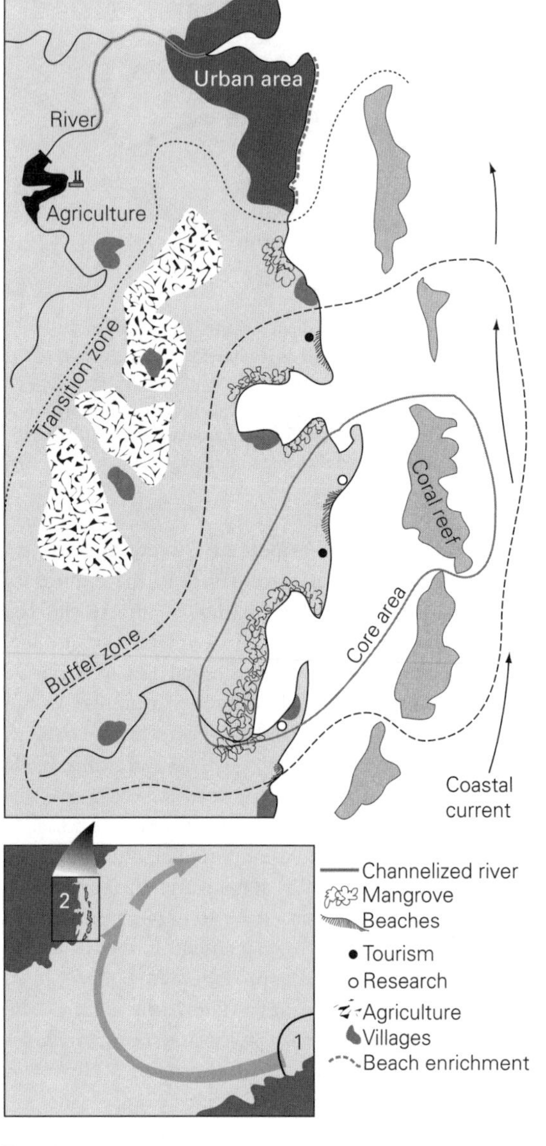

Fig. 9.6 An idealized ecosystem-management model for conservation in protected areas, embedded in a regional setting. In this case, a tropical reef–seagrass–mangrove complex is the "core" value to be protected; human activities are passive, restricted to ecotourism, education, research, and monitoring. Around the core is a "buffer zone" that helps maintain ecological processes while supporting compatible human activities, reducing pollution, and assuring living resource sustainability. A "transition zone" may accommodate farming, aquaculture, and light industry, while preserving environmental quality and variability. Account is also taken of influences of adjacent watersheds and the regional seashed (lower diagram). Several similar areas within the regions would comprise an interconnected network, accommodating species dispersal.

The biosphere reserve concept is best seen as a future design model, which a few large, region-scale marine parks and reserves have already approached by means of zonation (e.g., Australia's Great Barrier Reef Marine Park and some U.S. marine sanctuaries). However, influences from land and watersheds are rarely sufficiently incorporated. Nevertheless, many, if not most, scientific and governance tools for regional, coastal-realm conservation based on this model are on hand, lacking only social acceptance and political will to be put into practice.

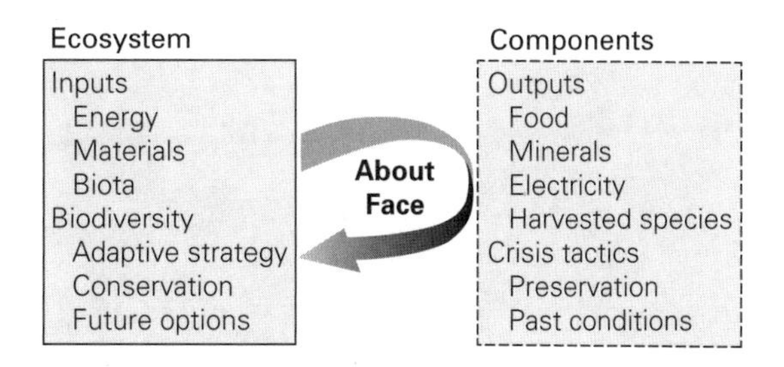

Fig. 9.7 Most resource managers, including conservationists, direct efforts toward ecosystem outputs, ignoring inputs, while attempting crisis management. An "about face" turns efforts toward maintaining inputs that provide ecosystem resiliency, and toward strategies of adaptive management for biodiversity conservation. Concept adapted from Odum (1987).

9.5 Conclusion

Evidence increasingly shows that human uses of the coastal realm are intensifying, conflicts and litigation are increasing, natural resources and environmental quality are decreasing, and costs of remedial action are spiraling. Furthermore, most coastal policies perpetuate the assumption that the coastal realm has an almost limitless capacity to absorb pollution and alteration and to sustain resource extraction. Many conventions, laws, and agreements that call attention to the coastal-realm environment and need for conservation have yet to be fulfilled. These facts testify to a need for a new alliance among society, science, and politics to alter current trends and affect successful conservation outcomes. Most important is the need for a holistic, unparochial, future-oriented, socially acceptable, scientific approach that accords with the dynamics of coastal-realm ecosystems.

Coastal-realm conservation science is in its infancy as a coherent body of knowledge. Conjecture and uncertainty plague solutions to conservation issues. Therefore, coastal-realm conservation may be perceived as the conduct of an experiment on ecosystem health, stress, and response. Present conservation practice, for the most part, remains concerned with managing primary and secondary issues separately. And while science seeks to explain, and conservation seeks to protect, society overwhelmingly makes choices on individual issues in a day-to-day decision-making process. A "tyranny of small decisions" made every day by individuals, agencies, and institutions adds up to an outcome no individual desires or intends (W. Odum 1982). Therefore, an "about face" is needed to redirect public support and policy attention toward inputs, adaptive strategies, and future options (Fig. 9.7).

The prospect for the future of the coastal realm and its resources is analogous to a "perfect storm," in which ecosystem shifts, emergent phenomena, sudden and unpredictable events, and human societal pressures synergize in unexpected ways toward maximum impact. Ironically, as human wellbeing has increased, resource issues have been exacerbated, placing immense pressure on the environments responsible for human wellbeing itself. Breaking this cycle is key to conservation. Inevitably, the human condition within the coastal realm will depend in large part on how well conservation succeeds by means of well-devised policies, strategies, and tactics. Human–environmental wellbeing will also depend on a new ethic. In Meeker's (1980) words: "An environmental ethic requires that human behavior be modified to agree with the ecology of the world, not that the world be rearranged to suit human desires." Emergent social and ecological imperatives have the potential to lead conservation into a new framework of governance, where everyone is a stakeholder in conserving and sustaining the values of the coastal realm – this most populous, diverse, and productive portion of Earth.

Citations and suggested readings

The references provided below offer only a "window" into the massive literature on coastal-realm science, policy, and management. They have been chosen from among the large volume of works pertinent to the subject matter of this book, and are presented for each chapter in two parts: citations in text and suggested readings. Citations include all that are mentioned in the text; suggested reading includes materials from which information has been derived, but which are not mentioned by name. The distinction is that citations refer to specific material, whereas suggested readings are representative of a larger body of works on the subject. A more complete listing is given on the Blackwell Publishing website: http://www.blackwellpublishing.com/ray/references.pdf.

We also recommend use of the internet for expanding the reader's information base; for example where the text refers to "on-line" websites. These website addresses are not given here due to continual changes of addresses in many cases. Website search mechanisms are available that can easily locate these sites. In many cases, care must be taken in the use of websites to verify the reliability of the material presented.

Preface

Holligan, P.M. & Reiners, W.A. (1992) Predicting the responses of the coastal zone to global change. *Advances in Ecological Research* **22**, 211–255.

Moore, J.A. (1993) *Science as a Way of Knowing. The Foundation of Modern Biology*. Harvard University Press, Cambridge, Mass.

Pernetta, J.C. & Milliman, J.D., eds. (1995) *Land–Ocean Interactions in the Coastal Zone: Implementation Plan*. Global Change Report **33**. International Geosphere–Biosphere Programme of ICSU, Stockholm.

Ray, C. (1970) Ecology, law, and the "marine revolution." *Biological Conservation* **3**(1), 7–17.

Chapter 1 Conservation issues

Citations

Aguirre, A.A. (1998) Fibropapillomas in marine turtles: A workshop at the 18th annual symposium on biology and conservation of sea turtles. *Marine Turtle Newsletter* **82**, 10–12.

Burton, R. (2003) Spain studies problems of *Prestige* oil spill. *Frontiers in Ecology and the Environment* **1**(1), 50.

Calmano, W. & Förstner, U. eds. (1996) *Sediments and Toxic Substances*. Springer-Verlag, Berlin.

Carlton, J.T. (1985) Transoceanic and interoceanic dispersal of coastal marine organisms: The biology of ballast water. *Oceanography and Marine Biology Annual Review* **23**, 313–71.

Carlton, J.T. (1993) Neoextinctions of marine invertebrates. *American Zoologist* **33**, 499–509.

Cervino, J.M., Goreau, T.J., Hayes, R.L. et al. (1998) Coral disease. *Science* **280**, 499–500.

Clark, R.B. (1997) *Marine Pollution*. Fourth edition. Clarendon Press, Oxford.

Diaz, R.J. & Rosenberg, R. (1995) Marine benthic hypoxia: A review of its ecological effects and the behavioral responses of benthic macrofauna. *Oceanography and Marine Biology Annual Review* **33**, 245–303.

Duxbury, A.C. & Duxbury, A.B. (1997) *An Introduction to the World's Oceans*. Fifth edition. Wm. C. Brown, Dubuque, Iowa.

FAO (1995) *Review of the State of World Fishery Resources: Marine Fisheries*. Food and Agriculture Organization, Rome.

FAO (2001) Online: http://www.fao.org/fi/default.asp.

Ferber, D. (2001) Keeping the Stygian waters at bay. *Science* **291**, 968–73.

Geiser, D.M., Taylor, J.W., Richie, K.B. & Smith, G.W. (1998) Cause of sea fan death in the West Indies. *Nature* **394**, 137–8.

Kaplan, R.D. (1994) The coming anarchy. *Atlantic Monthly* **273**(2), 44–75.

Harvell, C.D., Kim, K., Burkholder, J.M. et al. (1999) Emerging marine diseases – Climate links and anthropogenic factors. *Science* **285**, 1505–10.

Holt, S. (1969) The food resources of the ocean. *Scientific American* **221**(3), 2–15.

Irwin, R.J., VanMouwerik, M., Stevens, L., Seese, M.D. & Basham, W. (1997) *Environmental Contaminants Encyclopedia*. National Park Service, Water Resources Division, Fort Collins, Colo. Distributed within the federal government as an electronic document.

Jennings, S., Kaiser, M.J. & Reynolds, J.D. (2001) *Marine Fisheries Ecology*. Blackwell Science, Oxford.

Kenney, R.D., Payne, P.M., Heinemann, D.W. & Winn, H.E. (1996) Shifts in northwest shelf cetacean distributions relative to trends in Gulf of Maine/Georges Bank finfish abundance. In

The Northeast Shelf Ecosystem (eds. K. Sherman, N.A. Jaworski & T.J. Smayda), pp. 169–96. Blackwell Science, Oxford.

Kline, D.L. (1998) *Endocrine Disruption in Fish.* Kluwer Academic Publishers, Boston, Mass.

Laist, D. (1996) Impacts of marine debris: Entanglement of marine life in marine debris including a comprehensive list of species with entanglement and ingestion records. In *Marine Debris: Sources, Impacts, and Solutions* (eds. J.M. Coe & D.B. Rogers), pp. 99–139. Springer-Verlag, New York.

Littler, M.M. & Littler, D.S. (1995) Impact of CLOD pathogen on Pacific coral reefs. *Science* **267**, 1356–60.

Mangone, G.J. (1991) *Concise Marine Almanac.* Second edition. Van Nostrand Reinhold Co., New York.

Marine Mammal Commission (1995) *Annual Report to Congress, 1994.* U.S. Superintendent of Documents, Washington, D.C.

Montevecchi, B. & Kerry, M. (2001) Lessons from the Galapagos: International community can act now to prevent future oil spills. *The Gazette* (Montreal), February 14, 2001. Final edition. Southam, Inc.

Nagelkerken, I., Buchan, K., Smith, G.W. et al. (1997) Widespread disease in Caribbean sea fans: II. Patterns of infection and tissue loss. *Marine Ecology Progress Series* **160**, 255–63.

NAS (1995) *Understanding Marine Biodiversity: A Research Agenda for the Nation.* National Academy of Sciences. National Academy Press, Washington, D.C.

Norse, E., ed. (1993) *Global Marine Biodiversity.* Island Press, Washington, D.C.

NRC (1999) *Hormonally Active Agents in the Environment.* National Research Council. National Academy Press, Washington, D.C.

NRC (2002) *Oil in the Sea III: Inputs, Fates, and Effects.* Committee on Oil in the Sea, National Research Council. Prepublication version nationalacademies.org/publications. National Academy Press, Washington, D.C.

Patterson, K.L., Porter, J.W., Ritchie, K.B. et al. (2002) The etiology of white pox, a lethal disease of the Caribbean elkhorn coral, *Acropora palmata. Proceedings of the National Academy of Sciences* **99**(13), 8725–30.

Pauly, D., Christensen, V., Foese, R. & Palomares, M.L.L. (2000) Fishing down aquatic food webs. *American Scientist* **88**, 46–51.

Pollard, D.A. & Hutchings, P.A. (1990a) A review of exotic marine organisms introduced to the Australian region. I. Fishes. *Asian Fisheries Science* **3**, 205–21.

Pollard, D.A. & Hutchings, P.A. (1990b) A review of exotic marine organisms introduced to the Australian region. II. Invertebrates and algae. *Asian Fisheries Science* **3**, 223–50.

Popper, A.N., DeFerrari, H.A., Dolphin, W.F. et al. (2000) *Marine Mammals and Low-Frequency Sound: Progress since 1994.* National Academy Press, Washington, D.C.

Raloff, J. (1998) European crab leaps to Pacific prominence. *Science News* **153**, 373.

Reimer, D.C. & Lipscomb, T.P. (1998) Malignant seminoma with metastasis and herpesvirus infection in a free-living sea otter (*Enhydra lutris*). *Journal of Zoo and Wildlife Medicine* **29**, 35–9.

Resendes, A.R., Juan-Sallés, C., Almeria, S., Majó, N., Domingo, M. & Dubey, J.P. (2002) Hepatic sarcocystosis in a striped dolphin (*Stenella coeruleoalba*) from the Spanish Mediterranean coast. *Journal of Parasitology* **88**, 206–9.

Schneider, D.C. & Heinemann, D.W. (1996) The state of marine bird populations from Cape Hatteras to the Gulf of Maine. In *The Northeast Shelf Ecosystem* (eds. K. Sherman, N.A. Jaworski & T.J. Smayda), pp. 197–216. Blackwell Science, Oxford.

Sherman, K., Green, J., Solow, A. et al. (1996) Zooplankton prey field variability during collapse and recovery of pelagic fish in the northeast shelf ecosystem. In *The Northeast Shelf Ecosystem* (eds. K. Sherman, N.A. Jaworski & T.J. Smayda), pp. 217–36. Blackwell Science, Oxford.

Sindermann, C.J. (1996) *Ocean Pollution. Effects on Living Resources and Humans.* CRC Press, Boca Raton, Fla.

Steneck, R.S. & Carlton, J.T. (2001) Human alterations of marine communities. In *Marine Community Ecology* (eds. M.D. Bertness, S.D. Gaines & M.E. Hay), pp. 445–68. Sinauer Associates, Inc., Sunderland, Mass.

Stephens, J.S., Jr., Hose, J.E. & Love, M.S. (1988) Fish assemblages as indicators of environmental changes in nearshore environments. In *Marine Organisms as Indicators* (eds. D.F. Soule & G.S. Kleppel), pp. 91–105. Springer-Verlag, New York.

Upton, H.F. (1992) Biodiversity and conservation of the marine environment. *Fisheries* **17**(3), 20–5.

Walker, H.J. (1990) The coastal zone. In *The Earth as Transformed by Human Action. Local and Regional Changes in the Biosphere over the Past 300 Years* (eds. B.L. Turner, II, W.C. Clark, R.W. Kates et al.), pp. 271–94. Cambridge University Press, Cambridge.

Watling, L. & Norse, E.A. (1998) Disturbance of the seabed by mobile fishing gear: A comparison to forest clearcutting. *Conservation Biology* **12**(6), 1180–97.

WCMC (1992) *Sample Statistics on Protected Areas.* Report for the IV World Parks Congress. World Conservation Monitoring Centre, Cambridge.

Whitfield, P., Gardner, T., Vives, S.P. et al. (2002) Biological invasion of the Indo-Pacific lionfish (*Pterois volitans*) along the Atlantic coast of North America. *Marine Ecology Progress Series* **235**, 289–97.

Wilson, B., Grellier, K., Hammond, P.S., Brown, G. & Thompson, P.M. (2000) Changing occurrence of epidermal lesions in wild bottlenose dolphins. *Marine Ecology Progress Series* **205**, 283–90.

Woodham, A. (1997) Snow goose population explosion may threaten Arctic ecosystems. *Marine Pollution Bulletin* **34**(12), 935.

World Almanac (1999) *World Almanac 2000.* Primedia Ref., Inc., Mahwah, N.J.

World Almanac Books (1998) *The World Almanac and Book of Facts 1998.* K-III Reference Corporation, Mahwah, N.J.

World Commission on Dams (2000) *Dams and Development. A New Framework for Decision-Making.* Earthscan Publications Ltd., London.

WRI, UNEP, UNDP, WB (1998) *World Resources 1998–1999. A Guide to the Global Environment.* A report by the World Resources Institute in collaboration with the United Nations Environment Programme, the United Nations Development

Programme, and the World Bank. Oxford University Press, New York.

Suggested readings

Anderson, D.M. & Garison, D.J., eds. (1997) The ecology and oceanography of harmful algal blooms. *Limnology and Oceanography* **42**(5), 1009–22.

Bengtsson, B.-E., Nellbring, S., Hill, C. & Kessler, E., eds. (1999) Special issue: Reproductive disturbances in Baltic Sea fish: An international perspective. *Ambio* **XXVII**(1).

de Boer, J., Wester, P.G., Klamer, H.J.C., Lewis, W.E. & Boon, J.P. (1998) Do flame retardants threaten ocean life? *Nature* **394**, 28–9.

Bourgeois, T.M., Godfrey, D.G. & Bailey, M.J. (1998) Race on for deepwater acreage, 3500 meter depth capacity. *Offshore International Edition* **58**(10), 40–1, 152.

Cameron, P. & von Westernhagen, H. (1997) Malformation rates in embryos of North Sea fishes in 1991 and 1992. *Marine Pollution Bulletin* **34**(2), 129–34.

Casey, J.M. & Myers, R. (1998) Near extinction of a large, widely distributed fish. *Science* **281**, 690–2.

Caswell, H., Brault, S., Read, A.J. & Smith, T.D. (1998) Harbor porpoise and fisheries: An uncertainty analysis of incidental mortality. *Ecological Applications* **8**, 1226–38.

Champ, M.A. & Seligman, P.I.F., eds. (1996) *Organotin*. Chapman & Hall, London.

Cloern, J.E. (2001) *Our Evolving Model of the Coastal Eutrophication Problem*. Marine Ecology Progress Series **210**, 223–53.

Coe, J.M. & Rogers, D.B., eds. (1996) *Marine Debris: Sources, Impacts, and Solutions*. Springer-Verlag, New York.

Cosper, E.M., Bricelj, V.M. & Carpenter, E.J., eds. (1989) *Novel Phytoplankton Blooms: Courses and Impacts of Recurrent Brown Tides and Other Unusual Blooms*. Springer-Verlag, Berlin.

Crews, D., Gergeron, J.M. & McLachlan, J.A. (1995) The role of estrogen in turtle sex determination and the effect of PCBs. *Environmental Health Perspectives* **103**(suppl. 7), 73–7.

Crompton, T.R. (1997) *Toxicants in the Aqueous Ecosystem*. John Wiley & Sons, Chichester.

Daszok, P., Cunningham, A.A. & Hyatt, A.D. (2000) Emerging infectious diseases of wildlife – Threats to biodiversity and human health. *Science* **287**, 443–9.

De Mora, S.J., ed. (1996) *Tributyltin: Case Study of an Environmental Contaminant*. Cambridge University Press, Cambridge.

Domingo, M., Ferrer, L., Pumarola, M. et al. (1990) Morbillivirus in dolphins. *Nature* **348**, 21.

Gauthier, J.M., Pelletier, É., Brochu, C., Moore, S., Metcalfe, C.D. & Béland, P. (1998) Environmental contaminants in tissues of a neonate St. Lawrence beluga whale (*Delphinapterus leucas*). *Marine Pollution Bulletin* **36**(1), 102–8.

Hales, S., Weinstein, P., Souares, Y. & Woodward, A. (1999) El Niño and the dynamics of vectorborne disease transmission. *Environmental Health Perspectives* **107**(2), 99–102.

Hallegraeff, G.M. (1993) A review of harmful algal blooms and their apparent global increase. Phycological Reviews **13**. *Phycologia* **32**(2), 79–99.

Heywood, V.H. & Watson, R.T. (1995) *Global Biodiversity Assessment*. UNEP/Cambridge University Press, Cambridge.

Hilton-Taylor, C., compiler. (2000) *IUCN Red List of Threatened Species*. IUCN, Gland, Switzerland and Cambridge, UK.

Holsbeek, L., Siebert, U. & Joiris, C.R. (1998) Heavy metals in dolphins stranded on the French Atlantic Coast. *Science of the Total Environment* **217**(3), 241–9.

Kleivane, L. & Skaare, J.U. (1998) Organochlorine contaminants in northeast Atlantic minke whales (*Balaenoptera acutorostrata*). *Environmental Pollution* **101**(2), 231–9.

Law, R.J., Blake, S.J., Jones, B.R. & Rogan, E. (1998) Organotin compounds in liver tissue of harbour porpoises (*Phocoena phocoena*) and grey seals (*Halichoerus grypus*) from the coastal waters of England and Wales. *Marine Pollution Bulletin* **36**(3), 241–7.

Lewis, J.D. (1997) Abundance, distribution and partial mortality of the massive coral *Siderastrea siderea* on degrading coral reefs at Barbados, West Indies. *Marine Pollution Bulletin* **34**(8), 622–7.

Longwell, C., Chang, A.S., Herbert, A., Hughes, J.B. & Perry, D. (1992) Pollution and developmental abnormalities of Atlantic fishes. *Environmental Biology of Fishes* **35**, 1–21.

Maclean, J.L. (1989) Indo-Pacific red tides, 1985–1988. *Marine Pollution Bulletin* **20**(7), 304–10.

Malakoff, D. (1997) Extinction on the high seas. *Science* **277**, 486–8.

Mlot, C. (1998) News: The rise in toxic tides. *Science* **280**, 2053.

MPB (1997) News: California abalone under threat. *Marine Pollution Bulletin* **34**(11), 857.

MPB (1997) News: Mangroves under threat. *Marine Pollution Bulletin* **34**(9), 683.

MPB (1998) News: Illegal caviar threatens sturgeon survival. *Marine Pollution Bulletin* **36**(12), 254–5.

Nixon, S. (1998) Enriching the sea to death. *Scientific American* **9**(30), 48–53.

NRC (1985) *Oil in the Sea. Inputs, Fates, and Effects*. Steering Committee for the Petroleum in the Marine Environment Update, Board on Ocean Science and Policy. National Research Council. National Academy Press, Washington, D.C.

Osterhaus, A., Groen, J., Niesters, H. et al. (1997) Morbillivirus in monk seal mass mortality. *Nature* **388**, 838–9.

OTA (1987) *Wastes in Marine Environments*. Office of Technology Assessment, OTA-O-334, U.S. Congress. U.S. Government Printing Office, Washington, D.C.

Parsons, E.C.M. & Chan, H.M. (1998) Organochlorines in Indo-Pacific hump-backed dolphins (*Sousa chinensis*) and finless porpoises (*Neophocaena phocaenoides*) from Hong Kong. In *The Marine Biology of the South China Sea* **III** (ed. B. Morton), pp. 423–37. Hong Kong University Press, Hong Kong.

Parsons, T.R. & Lalli, C.M. (2003) Jellyfish population explosions: Revising a hypothesis of possible causes. *Le mer* **40**(3), 111–21.

Patnaik, P. (1992) *A Comprehensive Guide to the Hazardous Properties of Chemical Substances*. Van Nostrand Reinhold Co., New York.

Phillips, D.J.H. & Tanabe, S. (1989) Aquatic pollution in the Far East. *Marine Pollution Bulletin* **20**(7), 297–303.

Piatt, J.F. & van Pelt, T.I. (1997) Mass-mortality of guillemots (*Uria aalge*) in the Gulf of Alaska in 1993. *Marine Pollution Bulletin* **34**(8), 656–62.

Rabalais, N.N., Turner, R.E. & Wiseman, W.J., Jr. (1992) Distribution and characteristics of hypoxia on the Louisiana shelf in 1990 and 1991. In *Nutrient Enhanced Coastal Ocean Productivity (NECOP) Workshop Proceedings*, pp. 15–20. NOAA Coastal Ocean Program, TAMU-SG-92-109. Washington, D.C.

Raloff, J. (1995) How paper mill wastes may imperil fish. *Science News* **148**, 295.

Rasmussen, E. (1977) The wasting disease of eelgrass (*Zostera marina*) and its effects on environmental factors and fauna. In *Seagrass Ecosystems* (eds. C.P. McRoy & C. Helfferich), pp. 1–51. Marcel Dekker, Inc., New York.

Revenga, C., Murray, S., Abramovitz, J. & Hammond, A. (1998) *Watersheds of the World*. World Resources Institute and Worldwatch, Washington, D.C.

Rivinus, E.F. & Youssef, E.M. (1992) *Spencer Baird of the Smithsonian*. Smithsonian Institute Press, Washington, D.C.

Rosenberg, D.M., McCully, P. & Pringle, C.M. (2000) Global-scale environmental effects of hydrological alterations: Introduction. *BioScience* (special issue devoted to hydrological interactions) **50**(9), 746–51.

Sam, A.K., Ahamed, M.M.O., El Khangi, F.A., El Nigumi, Y.O. & Holm, E. (1998) Radioactivity levels in the Red Sea coastal environment of Sudan. *Marine Pollution Bulletin* **36**(1), 19–26.

Sanders, J.E. (1989) PCB pollution in the upper Hudson River. In *Contaminated Marine Sediments – Assessment and Remediation*, pp. 365–400. Committee on Contaminated Sediments (Marine Board, National Research Council). National Academy Press, Washington, D.C.

Shriadah, M.M.A. (1998) Impacts of an oil spill on the marine environment of the United Arab Emirates along the Gulf of Oman. *Marine Pollution Bulletin* **36**(11), 876–9.

Sindermann, C.J. (1990) *Principal Diseases of Marine Fish and Shellfish* **I, II**. Academic Press, San Diego, Calif.

Sydeman, W.J. & Jarman, W.M. (1998) Trace metals in seabirds, Steller sea lion, and forage fish and zooplankton from central California. *Marine Pollution Bulletin* **36**(10), 828–32.

Williams, S.J., Dodd, K. & Gohn, K.K. (1990) *Coasts in Crisis*. U.S. Geological Survey, Circular 1075. U.S. Government Printing Office, Washington, D.C.

Chapter 2 Mechanisms

Citations

Adams, A.B., ed. (1962) *First World Conference on National Parks*. U.S. Department of the Interior, Washington, D.C.

Archer, J.H., Connors, D.L., Laurence, K., Columbia, S.Ch. & Bowen, R. (1994) *The Public Trust Doctrine and the Management of America's Coasts*. University of Massachusetts Press, Amherst, Mass.

Borgese, E.M., Ginsbury, N. & Morgan, J.R., eds. (1994) *Ocean Yearbook* **11**. The University of Chicago Press, Chicago, Ill.

Coe, J.E. (1996) Introduction. In *Marine Debris: Sources, Impacts, and Solutions* (eds. J.M. Coe & D.B. Rogers), pp. xxxi–xxxv. Springer-Verlag, New York.

Crystal, D., ed. (1998) *The Cambridge Factfinder*. Third edition. Cambridge University Press, Cambridge.

Dutton, I. & Hotta, K. (1995) Introduction of coastal management. In *Coastal Management in the Asia-Pacific: Issues and Approaches* (eds. K. Hotta & I.M. Dutton), pp. 3–18. Japan International Marine Science and Technology Federation, Tokyo.

Encyclopædia Britannica (1999–2000) Online. Britannica.com Inc.

FAO (2001) Online: http://www.fao.org/fi/default.asp.

Fiorino, D.J. (1995) *Making Environmental Policy*. University of California Press, Berkeley, Calif.

Fox, J.R. (1992) *Dictionary of International and Comparative Law*. Oceana Publications, Inc., Dobbs Ferry, N.Y.

Gamboa, M.J. (1973) *A Dictionary of International Law and Diplomacy*. Central Lawbook Publishing Company, Inc., Quezon City, Philippines and Oceana Publications, Inc., Dobbs Ferry, N.Y.

Gleick, P.H. (2000) *The World's Water 2000–2001*. The Biennial Report on Freshwater Resources. Island Press, Washington, D.C.

Hinrichsen, D. (1998) *Coastal Waters of the World. Trends, Threats, and Strategies*. Island Press, Washington, D.C.

IMF (2001) International Monetary Fund. Online: http://www.imf.org/external/np/sec.decdo/contents.htm.

IUCN (1981) *Principles, Criteria and Guidelines for the Selection, Establishment and Management of Mediterranean Marine and Coastal Protected Areas*. International Union for the Conservation of Nature and Natural Resources, Doc. UNEP/IG.2, Gland, Switzerland. (Modified from UNEP/IG/20/3, prepared by IUCN for the Intergovernmental Meeting on Mediterranean Specially Protected Areas, October 1980, Athens, Greece.)

IUCN (1995) *The Law of the Sea: Priorities and Responsibilities in Implementing the Convention*. Part I. L.A. Kimball: *United Nations Convention on Law of the Sea: A Framework for Marine Conservation*. Part II. D.M. Johnson, P.M. Saunders & P. Payayo: *Conservation of the Marine Environment: A Marine Conservation and Development Report*. IUCN, Gland, Switzerland.

IUCN, UNEP, WWF (1980) *World Conservation Strategy*. IUCN, UNEP, WWF, Gland, Switzerland.

IUCN, UNEP, WWF (1991) *Caring for the Earth: A Strategy for Sustainable Living*. IUCN, Gland, Switzerland.

Jefferson, T. (1787) Letter to George Washington. Displayed at Monticello, Charlottesville, Virginia.

Ketchum, B.K., ed. (1972) *The Waters' Edge: Critical Problems of the Coastal Zone*. The MIT Press, Cambridge, Mass.

Mangone, G.J. (1986) *Concise Marine Almanac*. Van Nostrand Reinhold Co., New York.

Mangone, G.J. (1988) *Marine Policy for America*. Second edition. Taylor & Francis, New York.

Marine Mammal Commission (1994, 1997, 2000, 2001) *Annual Report to Congress*. U.S. Superintendent of Documents, Washington, D.C.

Morison, S.E. (1958) *Strategy and Compromise*. Little, Brown, & Co., Boston, Mass.

NOAA (2001) Online: http://www.edc.uri.edu/lme.

Ray, G.C. (1976) Critical marine habitats. Definition, description, criteria and guidelines for identification and management. In *Proceedings of an International Conference on Marine Parks and Reserves*, Tokyo, Japan, May 12–14, 1975, pp. 15–59. IUCN Publication New Series **37**. IUCN, Morges, Switzerland.

SBSTTA (1998) *The Jakarta Mandate on the Conservation and Sustainable Use of Marine and Coastal Biological Diversity*. Conference of the Parties, Secretariat of the Convention on Biological Diversity for the Convention on Biological Diversity Roster of Experts on Marine and Coastal Biological Diversity. First version. Draft for comment. Subsidiary Body on Scientific Technical and Technological Advice, October.

UN (1992) *Earth Summit. Agenda 21*. The United Nations Programme of Action from Rio. United Nations, New York.

UN (2000) Online. http://www.unep.org.

UNEP (1997) *Global Environment Outlook*. Oxford University Press, New York.

UNEP (2000) Regional Seas Programme. Online: http://www.unep.org.

University of Virginia School of Law (2001) Online: http://www.law.virginia.edu.

Wilder, R.J. (1998) *Listening to the Sea*. University of Pittsburgh Press, Pittsburgh, Pa.

WRI, IUCN, UNEP (1992) *Global Biodiversity Strategy: Guidelines for Action to Save, Study, and Use Earth's Biotic Wealth Sustainably and Equitably*. World Resources Institute, Washington, D.C.

Suggested readings

Beatley, T., Brower, D.J. & Schwab, A.K. (1994) *An Introduction to Coastal Zone Management*. Island Press, Washington, D.C.

Block, B.A., Dewar, H., Blackwell, S.B. et al. (2001) Migratory movements, depth preferences, and thermal biology of Atlantic bluefin tuna. *Science* **193**, 1310–14.

Bormann, F.H. (1966) The need for a federal system of natural areas for scientific research. *BioScience* **16**(9), 585–6.

Buck, S.J. (1996a) *Understanding Environmental Administration and Law*. Second edition. Island Press, Washington, D.C.

Buck, S.J. (1996b) *Comparative Marine Policy. Perspectives from Europe, Scandinavia, Canada, and the United States*. Island Press, Washington, D.C.

Buck, S.J. (1998) *The Global Commons*. Island Press, Washington, D.C.

Busha, T.S. (1986) The IMO Conventions. In *Ocean Yearbook* **6** (eds. E.M. Borgese & N. Ginsbury), pp. 160–70. The University of Chicago Press, Chicago, IU.

Elliott, L. (1998) *The Global Politics of the Environment*. New York University Press, New York.

Furtado, J.X. (1999) International law and the dispute over the Spratly Islands: Whither the UNCLOS. *Contemporary Southeast Asia* **21**(2), 386–404.

Glowka, L., Burhenne-Guilmin, F., Synge, H., McNeely, J.A. & Gundling, L. (1994) *A Guide to the Convention on Biological Diversity*. Environmental Policy and Law Paper **30**, IUCN Environmental Law Centre. IUCN Biodiversity Program, A contribution to the Global Biodiversity Strategy. IUCN, Gland, Switzerland.

Glowka, L. & de Klemm, C. (1996) International instruments, processes, organizations and non-indegenous species introductions: Is a protocol to the Convention on Biological Diversity necessary?. *Environmental Policy and Law* **26**(6), 389–405.

Grove, R.H. (1992) Origins of Western environmentalism. Strategies to reserve nature arose a newly colonized tropical lands were exploited in the 17th and 18th centuries. *Scientific American* **60**(7), 42–7.

Hazleton, W.A. (1993) United Nations. In *Encyclopedia America* **27**. Grolier Incorp., Danbury, Conn.

Hedley, C. & IGIFL (2000–1) Internet Guide to International Fisheries Law. Online: http://www.oceanlaw.net/texts/index2.htm.

Hershman, M. (1999) Seaport development and coastal management programs: A national overview. *Coastal Management* **27**, 271–90.

Hobart, W.L., ed. (1996) *Baird's Legacy: The History and Accomplishments of NOAA's National Marine Fisheries Service, 1871–1996*. NOAA Technical Mem. NMFS-F/Spo-18. U.S. Department of Commerce, National Marine Fisheries Service, Washington, D.C.

IUCN (1972) *Second World Conference on National Parks*. Yellowstone and Grand Teton National Parks, USA, 18–27 September. International Union for the Conservation of Nature and Natural Resources, Morges, Switzerland.

IUCN (2000) *2000 IUCN Red List of Threatened Species*. Compiled by Craig Hilton-Taylor, the IUCN Species Survival Commission. IUCN, Gland, Switzerland.

Kildow, J. (1997) The roots and context of the coastal zone movement. *Coastal Management* **25**, 231–63.

Marsh, G.P. (1864) *Man and Nature*. Reprinted in 1965, ed. D. Lowenthal. Belknap Press of Harvard University Press, Cambridge, Mass.

Ray, C. (1964) Inshore marine conservation. In *First World Conference on National Parks* (ed. A.B. Adams), pp. 77–87. National Park Service, Washington, D.C.

Reed, D. (1997) The environmental legacy of Bretton Woods: The World Bank. In *Global Governance. Drawing Insights from the Environmental Experience* (ed. O.R. Young), pp. 227–45. The MIT Press, Cambridge, Mass.

Rothwell, D.R. (1990) The Antarctic Treaty system: Resource development, environmental protection or disintegration? *Arctic* **43**(3), 284–91.

Salm, R.V., Clark, J.R. & Siirula, E. (2000) *Marine and Coastal Protected Areas: A Guide for Planners and Managers*. Third edition. IUCN, Washington, D.C.

Shah, M.J. (1986) Maritime law and the developing countries: Attitudes and trends. In *Ocean Yearbook* **6** (eds. E.M. Borgese, N. Ginsbury & J.R. Morgan), pp. 107–38. The University of Chicago Press, Chicago, Ill.

Sitarz, D., ed. (1994) *Agenda 21: The Earth Strategy to Save Our Planet*. Earthpress, Boulder, Colo.

UNEP (2002) Legal agreements relating to the marine environment. Online: http://www.unep.ch/seas/main/hlegal.html.

Young, O.R. (1994) *International Governance: Protecting the Environment in a Stateless Society*. Cornell University Press, Ithaca, N.Y.

Young, O.R. (1997) Rights, rules, and resources in world affairs. In *Global Governance. Drawing Insights from the Environmental Experience* (ed. O.R. Young), pp. 1–23. The MIT Press, Cambridge, Mass.

Chapter 3 The coastal-realm ecosystem

Citations

Bakun, A. (1996) *Patterns in the Ocean: Ocean Processes and Marine Population Dynamics*. California Sea Grant College System, National Oceanographic and Atmospheric Administration, La Jolla, Calif.

Birkeland, C., ed. (1997) *Life and Death of Coral Reefs*. Chapman & Hall, New York.

Carter, R.W.G. (1988) *Coastal Environments*. Academic Press, London.

Emery, K.O. (1969) The continental shelves. *Scientific American* **221**(3), 107–22.

Galloway, J.N., Howarth, R.W., Michaels, A.F., Nixon, S.W., Prospero, J.M. & Dentener, F.J. (1995) Nitrogen and phosphorus budgets of the north Atlantic ocean and its watershed. *Biogeochemistry* **35**, 3–25.

Galloway, W.E. (1983) *Terrigenous Clastic Depositional Systems*. Springer-Verlag, Heidelberg.

Garstang, M., Ellery, W.N., McCarthy, T.S. et al. (1998) The contribution of aerosol- and water-borne nutrients to the functioning of the Okavango Delta ecosystem, Botswana. *South African Journal of Science* **94**, 223–9.

Gross, M.G. & Gross, E. (1996) *Oceanography: A View of Earth*. Prentice Hall, Upper Saddle River, N.J.

Hayden, B.P., Dueser, R.D., Callahan, J.T. & Shugart, H.H. (1991) Long-term research at the Virginia Coast Reserve. *BioScience* **41**(5), 310–18.

Hayden, B.P., Ray, G.C. & Dolan, R. (1984) Classification of coastal and marine environments. *Environmental Conservation* **11**(3), 199–207.

Holligan, P.M. & Reiners, W.A. (1992) Predicting the responses of the coastal zone to global change. *Advances in Ecological Research* **22**, 211–55.

Hopkinson, C.S. & Schubauer, J.F. (1984) Static and dynamic aspects of nitrogen cycling in the salt marsh graminoid *Spartina alterniflora*. *Ecology* **65**(3), 961–9.

Inman, D.L. & Brush, B.M. (1973) The coastal challenge. *Science* **181**, 20–32.

Inman, D.L. & Nordstrom, C.E. (1971) Tectonic and morphologic classification of coasts. *Journal of Geology* **79**(1), 1–21.

Jackson, J.B.C. (1991) Adaptation and diversity of reef corals. *BioScience* **41**(7), 475–82.

Levin, S.A. (1992) The problem of pattern and scale in ecology. *Ecology* **73**(6), 1943–67.

Mann, K.H. & Lazier, J.R.N. (1991) *Dynamics of Marine Ecosystems. Biological–Physical Interactions in the Oceans*. Blackwell Scientific Publications, Boston, Mass.

Milliman, J.D. and Farnsworth, K.L. (in preparation) *River Runoff, Erosion and Delivery to the Coastal Ocean: A Global Analysis*. Oxford University Press, Oxford.

Nelson, J.S. (1994) *Fishes of the World*. 3rd edition, John Wiley & Sons, New York.

NRC (1992) *Coastal Meteorology: A Review of the State of the Science*. Panel on Coastal Meteorology, National Research Council. National Academy Press, Washington, D.C.

O'Neill, R.V., DeAngelis, D.L., Waide, J.B. & Allen, T.F.H. (1986) *A Hierarchical Concept of Ecosystems*. Monographs in Population Biology **23**. Princeton University Press, Princeton, N.J.

Orians, G.H. (1975) Diversity, stability, and maturity in natural ecosystems. In *Unifying Concepts in Ecology* (eds. W.H. van Dobben & R.H. Lowe-McConnell), pp. 139–50. Dr. W. Junk b.v., The Hague.

Pernetta, J.C. & Milliman, J.D., eds. (1995) *Land–Ocean Interactions in the Coastal Zone: Implementation Plan*. Global Change Report **33**. International Geosphere–Biosphere Programme of ICSU, Stockholm.

Prigogine, I. (1980) *From Being to Becoming*. Time and complexity in the Physical Sciences. W.H. Freeman and Co., New York.

Prospero, J.M., Barrett, K., Church, T., Pentener, F., Duce, R.A., Galloway, J.M., Levy, H., Moody, J. & Quinn, P. (1996) Atmospheric deposition of nutrients to the North Atlantic Basin. *Biochemistry* **35**, 27–73.

Prospero, P.M. (1999) Long-term measurements of the transport of African mineral dust to the southeastern United States: Implications for regional air quality. *Journal of Geophysical Research* **104**, 15917–27.

Ray, G.C. & Gregg, W.P., Jr. (1991) Establishing biosphere reserves for coastal barrier ecosystems. *BioScience* **41**(5), 301–9.

Ray, G.C. & Hayden, B.P. (1992) Coastal zone ecotones. In *Landscape Boundaries: Consequences for Biodiversity and Ecological Flows* (eds. F. di Castri & A.J. Hansen), pp. 403–20. Springer-Verlag, New York.

Ray, G.C. & McCormick-Ray, M.G. (1989) Coastal and Marine Biosphere Reserves. In *Proceedings of the Symposium on Biosphere Reserves* (eds., W.P. Gregg, Jr., S.L. Krugman & J.D. Wood, Jr.), pp. 68–78. The 4th World Wilderness Congress, *Worldwide Conservation*, held at Estes Park, Colorado, September 14–17, 1987. U.S. Department of the Interior, Atlanta, Ga.

Ray, G.C. & McCormick-Ray, M.G. (1992) Functional coastal-marine biodiversity. In *Transactions of the 57th Wildlife & Natural Resources Conference (1992)*, pp. 384–97. Wildlife Management Institute, Washington, D.C.

Rice, D.W. (1998) *Marine Mammals of the World: Systematics and Distribution*. Special Publication **4**. Society for Marine Mammalogy, Lawrence, Kans.

Roy, P.S. (1984) New South Wales estuaries: Their origin and evolution. In *Coastal Geomorphology in Australia* (ed. B.G. Thom), pp. 99–121. Academic Press, Sydney.

Stride, A.H., ed. (1982) *Offshore Tidal Sands*. Chapman & Hall, London.

Swap, R., Garstang, M., Greco, S., Talbot, R. & Kallberg, P. (1992) Saharan dust in the Amazon Basin. *Tellus* **44B**, 133–49.

Valiela, I. (1984) *Marine Ecological Processes*. Springer-Verlag, New York.

Weinberg, G.M. (1975, 2001) *An Introduction to General Systems Thinking*. John Wiley & Sons, New York. Reprinted, *Silver Anniversary Edition*, Dorset House Publishing Co., Inc., New York.

Suggested readings

Carter, R.W.G. & Woodroffe, C.D., eds. (1997) *Coastal Evolution*. Cambridge University Press, Cambridge.

De Angelis, D.L., Post, W.M. & Travis, C.C. (1986) *Positive Feedback in Natural Systems*. Biomathematics **15**. Springer-Verlag, Berlin.

Dolan, R. & Lins, H. (1987) Beaches and barrier islands. *Scientific American* **257**(1), 68–77.

Garstang, M. & Fitzjarrald, D.R. (1999) *Observations of Surface to Atmosphere Interactions in the Tropics*. Oxford University Press, New York.

Hanson, R.B., Ducklow, H.W. & Field, J.G., eds. (2000) *The Changing Ocean Carbon Cycle*. International Geosphere–Biosphere Programme Book Series. Cambridge University Press, Cambridge.

Griffin, D.W., Kellogg, C.A., Garrison, V.H. & Shinn, E.H. (2002) The global transport of dust. *American Science* **90**, 228–35.

Hayden, B.P. (1998) Ecosystem feedbacks on climate at the landscape scale. *Philosophical Transactions of the Royal Society of London B* **353**, 5–18.

Helfield, J.M. & Naiman, R.J. (2001) Effects of salmon-derived nitrogen on riparian forest growth and implications for stream productivity. *Ecology* **82**(9), 2403–9.

Howarth, R.W., Billen, G., Swaney, D. et al. (1996) Regional nitrogen budgets and riverine N and P fluxes for the drainages to the North-Atlantic Ocean – Natural and human influences. *Biogeochemistry* **35**(1), 75–139.

Inman, D.L. (1994) Types of coastal zones: Similarities and differences. In *Environmental Science in the Coastal Zone: Issues for Further Research*, pp. 67–84. National Academy Press, Washington, D.C.

Kjerfve, B., ed. (1988) *Hydrodynamics of Estuaries*. Vol. I. CRC Press, Inc., Boca Raton, Fla.

Kolber, Z.S., Plumley, F.G., Lang, A.S. et al. (2001) Contribution of aerobic photoheterotrophic bacteria to the carbon cycle in the ocean. *Science* **292**, 2492–5.

Lotspeich, F.B. (1980) Watersheds as the basic ecosystem: This conceptual framework provides a basis for a natural classification system. *Water Resources Bulletin* **16**(4), 581–6.

Mann, K.H. (1982) *Ecology of Coastal Waters. A Systems Approach*. Studies in Ecology **8**. University of California Press, Berkeley, Calif.

Milliman, J.D. (1992) Management of the coastal zone: Impact of onshore activities on the coastal environment. In *Use and Misuse of the Seafloor* (eds. H.J. Hsü & J. Thiebe), pp. 213–27. John Wiley & Sons, Chichester.

Milliman, J.D. & Meade, R.H. (1983) Worldwide delivery of river sediment to the oceans. *Journal of Geology* **91**(1), 1–20.

Nicolis, G. (1991) Non linear dynamics, self-organization and biological complexity. In *Perspectives on Biological Complexity* (eds. O.T. Solbrig & G. Nicolis), pp. 7–49. IUBS Monograph Series **6**. International Union of Biological Sciences, Paris.

Odum, H.T. (1983) *Systems Ecology: An Introduction*. John Wiley & Sons, New York.

Odum, H.T. (1988) Self-organization, transformity, and information. *Science* **242**, 1132–9.

Odum, W.E., Odum, E.P. & Odum, H.T. (1995) Nature's pulsing paradigm. *Estuaries* **18**(4), 547–55.

O'Neill, R.V. (1980) Perspectives in hierarchy and scale. In *Perspectives in Ecological Theory* (eds. J. Roughgarden, R.M. May & S.A. Levin), pp. 140–56. Princeton University Press, Princeton, N.J.

O'Neill, R.V., Johnson, A.R. & King, A.W. (1989) A hierarchical framework for the analysis of scale. *Landscape Ecology* **3**(3/4), 193–205.

Parsons, T.R., Takahashi, M. & Hargrave, B. (1977) *Biological Oceanographic Processes*. Second edition. Pergamon Press, Oxford.

Pethick, J.S. (1996) The geomorphology of mudflats. In *Estuarine Shores: Evolution, Environment and Human Alterations* (eds. K.F. Nordstrom & C.T. Roman), pp. 185–212. John Wiley & Sons, Chichester.

Pilson, M.E.Q. (1998) *An Introduction to the Chemistry of the Sea*. Prentice Hall, Englewood Cliffs, N.J.

Pimm, S.L. (1991) *The Balance of Nature*. Ecological Issues in the Conservation of Species and Communities. University of Chicago Press, Chicago.

Pirazzoli, P.A. (1997) Tectonic shorelines. In *Coastal Evolution* (eds. R.W.G. Carter & C.D. Woodroffe), pp. 451–76. Cambridge University Press, Cambridge.

Postina, H. & Zijlstra, J.J., eds. (1988) *Continental Shelves*. Ecosystems of the World **27**, Elsevier, Amsterdam.

Powell, T.M. (1989) Physical and biological scales of variability in lakes, estuaries, and the coastal ocean. In *Perspectives in Ecological Theory* (eds. J. Roughgarden, R.M. May & S.A. Levin), pp. 157–76. Princeton University Press, Princeton, N.J.

Pritchard, D. (1967) What is an estuary: Physical viewpoint. In *Estuaries* **83** (ed. G.H. Lauff), pp. 3–5. American Association for the Advancement of Science, Washington, D.C.

Ray, G.C. (2000) Estuarine ecosystems. In *Encyclopedia of Biodiversity* **1** (ed. S.A. Levin), pp. 579–91. Academic Press, San Diego, Calif.

Reichle, D.E., O'Neill, R.V. & Harris, W.F. (1975) Principles of energy and material exchange in ecosystems. In *Unifying Concepts in Ecology* (ed. W.H. van Dobben & R.H. Lowe-McConnell), pp. 27–43. Dr. W. Junk b.v., The Hague.

Ryan, J.C. (2001) The Caribbean gets dusted. *BioScience* **51**(5), 334–8.

Smetacek, V.S. (1986) Impact of freshwater discharge on production and transfer of materials in the marine environment. In *The Role of Freshwater Outflow in Coastal Marine Ecosystems* (ed. S. Skreslet), pp. 85–106. NATO ASI Series **G7**. Springer-Verlag, Berlin.

Steele, J.H. (1989) Discussion: Scale and coupling in ecological systems. In *Perspectives in Ecological Theory* (eds. J. Roughgarden, R.M. May & S.A. Levin), pp. 177–80. Princeton University Press, Princeton, N.J.

Stigebrandt, A. (1988) Dynamic control by topography in estuaries. In *Hydrodynamics of Estuaries*. **I**. *Estuarine Physics* (ed. B. Kjerfve), pp. 17–26. CRC Press, Boca Raton, Fla.

Suter, J.R. (1997) Deltaic coasts. In *Coastal Evolution* (eds. R.W.G. Carter & C.D. Woodroffe), pp. 87–120. Cambridge University Press, Cambridge.

Swap, R., Ulanski, S., Cobbett, M. & Garstang, M. (1996) Temporal and spatial characteristics of Saharan dust outbreaks. *Journal of Geophysical Research* **101**, 4205–20.

Valiela, I., Costa, J., Foreman, K., Teal, J.M., Howes, B. & Aubrey, D. (1990) Transport of groundwater-borne nutrients from watersheds and their effects on coastal waters. *Biogeochemistry* **10**, 177–97.

Van Dyne, G.M. (1995) Ecosystems, systems ecology, and systems ecologists. In *Complex Ecology. The Part–Whole Relation in Ecosystems* (eds. B.C. Patten & S.E. Jorgensen), pp. 1–27. Prentice Hall, Englewood Cliffs, N.J.

Wright, L.D. (1978) River deltas. In *Coastal Sedimentary Environments* (ed. R.A. Davis, Jr.), pp. 5–68. Springer-Verlag, New York.

Wright, L.D. (1995) *Morpho-dynamics of Inner Continental Shelves*. CRC Press, Boca Raton, Fla.

Zeitzschel, B. (1980) Sediment–water interactions in nutrient dynamics. In *Marine Benthic Dynamics* (eds. K.R. Tenore & B.C. Coull), pp. 195–218. University of South Carolina Press, Columbia, SC.

Chapter 4 Natural history of coastal-marine organisms

Citations

Able, K.W. & Fahay, M.P. (1998) *The First Year in the Life of Estuarine Fishes in the Middle Atlantic Bight*. Rutgers University Press, New Brunswick, N.J.

Bartholomew, G.A. (1970) A model for the evolution of pinniped polygyny. *Evolution* **24**, 546–59.

Bartholomew, G.A. (1986) The role of natural history in contemporary biology. *BioScience* **36**(5), 324–9.

Buchsbaum, R., Buchsbaum, M., Pearse, J. & Pearse, V. (1987) *Animals Without Backbones*. The University of Chicago Press, Chicago, IU.

Cushing, D.H. (1973) The natural regulation of fish populations. In *Sea Fisheries Research* (ed. F.R. Harden Jones), pp. 399–411. John Wiley & Sons, New York.

DeAngelis, D.L., Post, W.M. & Travis, C.C. (1986) *Positive Feedback in Natural Systems*. Springer-Verlag, Berlin.

Denny, M.W. (1988) *Biology and the Mechanics of the Wave-Swept Environment*. Princeton University Press, Princeton, N.J.

Denny, M. & Wethey, D. (2001) Physical processes that generate patterns in marine communities. In *Marine Community Ecology* (eds. M.D. Bertness, S.D. Gaines & M.E. Hay), pp. 3–37. Sinauer Associates, Inc., Sunderland, Mass.

Dobzhandsky, T. (1973) *American Biology Teacher* **35**, 125–9.

Hall, S.J., Raffaelli, D. & Thrush, S.F. (1994) Patchiness and disturbance in shallow water benthic assemblages. In *Aquatic Ecology: Scale, Pattern, and Process* (eds. P.S. Giller, A.G. Hildrew & D.G. Raffaelli), pp. 333–75. Blackwell Science, Oxford.

Hayden, B.P. & Dolan, R. (1979) Barrier islands, lagoons, and marshes. *Journal of Sedimentary Petrology* **49**(4), 1061–72.

Hayden, B.P., Ray, G.C. & Dolan, R. (1984) Classification of coastal and marine environments. *Environmental Conservation* **11**(3), 199–207.

Horn, H.S. & Rubenstein, D.I. (1984) Behavioural adaptation and life history. In *Behavioural Ecology. An Evolutionary Approach* (eds. J.R. Krebs & N.B. Davies), pp. 279–98. Second edition. Sinauer Associates, Inc., Sunderland, Mass.

Huston, M.A. (1994) *Biological Diversity*. Cambridge University Press, Cambridge.

Hutchinson, G.E. (1965) *The Ecological Theater and the Evolutionary Play*. Yale University Press, New Haven, Conn.

Kidwell, S.M. & Jablonski, D. (1983) Taphonomic feedback: Ecological consequences of shell accumulation. In *Biotic Interactions in Recent and Fossil Benthic Communities* (ed. M.J.S. Teresz & P.L. McCall), pp. 195–250. Plenum Press, New York.

Margulis, L. & Schwartz, K.V. (1998) *Five Kingdoms: An Illustrated Guide to the Phyla of Life on Earth*. W.H. Freeman & Co., New York.

May, R.M. (1988) How many species are there on Earth? *Science* **241**, 1441–9.

Munson, B.R., Young, D.F. & Okiishi, T.H. (1994) *Fundamentals of Fluid Mechanics*. Second edition. John Wiley & Sons, New York.

Nelson, J.S. (1984) *Fishes of the World*. John Wiley & Sons, New York.

NRC (1995) *Understanding Marine Biodiversity*. Committee on Biological Diversity in Marine Systems, National Research Council, Washington, D.C.

Odum, W.E. & Heald, E.J. (1975) The detritus-based food web of an estuarine mangrove community. In *Estuarine Research* **I** (ed. L.E. Cronin), pp. 265–87. Academic Press, New York.

Petraitis, P.S. & Lantham, R.E. (1999) The importance of scale in testing the origins of alternative community states. *Ecology* **80**, 429–42.

Pianka, E.R. (1970) *r* and *K* selection. *The American Naturalist* **104**, 592–7.

Raffaelli, D.G. & Hawkins, S. (1996) *Intertidal Ecology*. Chapman & Hall, London.

Ray, G.C. (1981) The role of large organisms. In *Analysis of Marine Ecosystems* (ed. A.R. Longhurst), pp. 397–413. Academic Press, New York.

Ray, G.C. (1991) Coastal-zone biodiversity patterns. *BioScience* **41**(7), 490–8.

Ray, G.C. (1997) Do the metapopulation dynamics of estuarine fishes influence the stability of shelf ecosystems? *Bulletin of Marine Science* **60**(3), 1040–9.

Ray, G.C., Hayden, B.P., McCormick-Ray, M.G. and Smith, T.M. (1997) Land-seascape diversity of the U.S. east coast coastal zone with particular reference to estuaries. In *Marine Biodiversity: Causes and Consequences* (eds. R.F.G. Ormond, J.D. Gage, & M.V. Angel), pp. 337–71. Cambridge University Press, Cambridge.

Rhoads, D.C. (1974) Organism–sediment relations on the muddy sea floor. *Oceanography and Marine Biology Annual Review* **12**, 263–300.

Robins, C.R., Bailey, R.M., Bond, C.E. et al. (1991) *Common and Scientific Names of Fishes from the United States and Canada*. Fifth edition. Special Publication **20**. American Fisheries Society, Bethesda, Md.

Silliman, B.R. & Bartness, M.D. (2002) A trophic cascade regulates salt marsh primary production. *Proceedings of the National Academy of Sciences* **99**(16), 10500–5.

Skov, H., Durinck, J., Danielsen, F. & Bloch, D. (1995) Co-occurrence of cetaceans and seabirds in the northeast Atlantic. *Journal of Biogeography* **22**, 71–88.

Vogel, S. (1983) *Life in Moving Fluids*. Princeton University Press, Princeton, N.J.

Vogel, S. (1994) *Life in Moving Fluids*. Second edition. Princeton University Press, Princeton, N.J.

Winslow, F. (1882) Methods and results. In *Report of the oyster beds of the James River, Va. and of Tangier and Pocomoke sounds, Maryland and Virginia*. U.S. Government Printing Office, Washington, D.C.

Suggested readings

Bertness, M.D., Gaines, S.D. & Hay, M.E., eds. (2001) *Marine Community Ecology*. Sinauer Associates, Inc., Sunderland, Mass.

Briggs, J.C. (1974) *Marine Zoogeography*. McGraw-Hill, New York.

Brown, J., Colling, A., Park, D. et al. (1994) *Waves, Tides, and Shallow-Water Processes*. Prepared by an Open University course team. Pergamon and The Open University, Walton Hall, Milton Keynes, UK.

Caughley, G. & Gunn, A. (1996) *Conservation Biology in Theory and Practice*. Blackwell Science, Oxford.

Chapman, A.R.O. (1979) *Biology of Seaweeds*. University Park Press, Baltimore, Md.

Cowardin, L.M., Carter, V., Golet, F.C. & LaRoe, E.T. (1979) *Classification of the Wetlands and Deepwater Habitats of the United States*. U.S. Department of the Interior, Fish and Wildlife Service, Office of Biological Services, Washington, D.C.

Cowen, R.K., Lwiza, K.M.M., Sponaugle, S. et al. (2000) Connectivity of marine populations: Open or closed? *Science* **287**, 857–9.

Denny, M. (1995) Survival in the surf zone. *American Scientist* **83**, 166–73.

Denny, M.W. (1993) *Air and Water. The Biology and Physics of Life's Media*. Princeton University Press, Princeton, N.J.

Dudgeon, S.R. & Buss, L.W. (1996) Growing with the flow: On the maintenance and malleability of colony form in the hydroid *Hydractinia*. *The American Naturalist* **147**(5), 667–91.

Duffy, J.E. & Hay, M.E. (2001) The ecology and evolution of marine consumer–prey interactions. In *Marine Community Ecology* (eds. M.D. Bertness, S.D. Gaines & M.E. Hay), pp. 131–57. Sinauer Associates, Inc., Sunderland, Mass.

Dyer, K.R. (1988) Tidally generated estuarine mixing processes. In *Hydrodynamics of Estuaries*. **I**. *Estuarine Physics* (ed. B. Kjerfve), pp. 41–57. CRC Press, Boca Raton, Fla.

Ekman, S. (1953) *Zoogeography of the Sea*. Sidgwick and Jackson Ltd, London.

Estes, J.A. (1979) Exploitation of marine mammals: *r*-selection of K-strategists? *Journal of the Fisheries Research Board of Canada* **36**, 1009–17.

Estes, J.A. & Palmisano, J.F. (1974) Sea otters: Their role in structuring nearshore communities. *Science* **185**, 1058–60.

Farnsworth, E.J. & Ellison, A.M. (1996) Scale-dependent spatial and temporal variability in biogeography of mangrove root epibiont communities. *Ecological Monographs* **66**(1), 45–66.

Forman, R.T.T. & Godron, M. (1986) *Landscape Ecology*. John Wiley & Sons, New York.

Garman, G.C. & Macko, S.A. (1998) Contribution of marine-derived organic matter to an Atlantic coast, freshwater, tidal stream by anadromous clupeid fishes. *Journal of the North American Benthological Society* **17**(3), 277–85.

Genin, A., Dayton, P.K., Lonsdale, P.F. & Spiess, F.N. (1986) Corals on seamount peaks provide evidence of current acceleration over deep-sea topography. *Nature* **322**, 59–61.

Grassle, J.F. (1991) Deep-sea benthic biodiversity. *BioScience* **41**(7), 464–9.

Grassle, J.F. & Maciolek, N.J. (1992) Deep-sea species richness: Regional and local diversity estimates from quantitative bottom samples. *The American Naturalist* **139**, 313–41.

Gray, J.S., Poore, G.C.B., Ugland, K.I., Wilson, R.S., Olsgard, F. & Johannessen, O. (1999) Coastal and deep-sea benthic diversities compared. *Marine Ecology Progress* series 159, 97–103.

Grossman, D.H., Bourgeron, P., Busch, W.-D.N. et al. (1999) Principles for ecological classification. In *Ecological Stewardship: A Common Reference for Ecosystem Management* (eds. N.C. Johnson, A.J. Malk, W.T. Sexton & R. Szaro), pp. 353–93. Elsevier Science, Oxford.

Hanski, I. & Gyllenberg, M. (1993) Two general metapopulation models and the core-satellite species hypothesis. *The American Naturalist* **142**(1), 17–41.

Hanski, I., Moilanen, A. & Gyllenberg, M. (1996) Minimum viable metapopulation size. *The American Naturalist* **147**(4), 527–41.

Hayden, B.P. & Dolan, R. (1976) Coastal marine fauna and marine climates of the Americas. *Journal of Biogeography* **3**, 71–81.

Hedgpeth, J.W. (1957) Marine biogeography. *Geological Society of America Memoir* **67**(1), 359–82.

Hengeveld, R. (1990) *Dynamic Biogeography*. Cambridge University Press, Cambridge.

Heppell, S.S., Crowder, L.B. & Menzel, T.R. (1999) Life table analysis of long-lived marine species with implications for conservation and management. In *Life in the Slow Lane* (ed. J.A. Musick), pp. 137–48. Symposium **23**. American Fisheries Society, Bethesda, Md.

Jones, C.G., Lawton, J.H. & Shachak, M. (1997) Positive and negative effects of organisms as physical ecosystem engineers. *Ecology* **78**(7), 1946–57.

Lawton, J.H. (1994) What do species do in ecosystems? *Oikos* **71**, 367–74.

Layman, C.A. (2000) Fish assemblage structure of the shallow ocean surf-zone on the Eastern shore of Virginia barrier islands. *Estuarine, Coastal and Shelf Science* **51**, 201–13.

Layman, C.A., Smith, D.E. & Herod, J.D. (2000) Seasonally varying importance of abiotic and biotic factors in marsh-pond fish communities. *Marine Ecology Progress Series* **207**, 155–69.

Lenihan, H.S. (1999) Physical–biological coupling on oyster reefs: How habitat structure influences individual performance. *Ecological Monographs* **69**(3), 251–75.

Lewis, R. (1997) *Dispersion in Estuaries and Coastal Waters*. John Wiley & Sons, Chichester.

Mackintosh, N.A. (1970) Whales and krill in the twentieth century. In *Antarctic Ecology* **1** (ed. M.V. Holdgate), pp. 195–212. Academic Press, London.

Magurran, A.E. (1987) *Ecological Diversity and Its Measurement*. Princeton University Press, Princeton, N.J.

Martinez, N.D. (1994) Scale-dependent constraints on food-web structure. *The American Naturalist* **144**(6), 935–53.

Massel, S.R. (1999) *Fluid Mechanics for Marine Ecologists*. Springer-Verlag, Berlin.

Maxwell, J.R., Edwards, C.J., Jensen, M.E. et al. (1995) *A Hierarchical Framework of Aquatic Ecological Units in North America (Nearctic Zone)*. General Technical Report, NC-176, 72. U.S. Department of Agriculture, Forest Service, N. Central Forest Experimental Station.

May, R.M. & Southwood, T.R.E. (1990) Introduction. In *Living in a Patchy Environment* (eds. B. Shorrocks & I.R. Swingland), pp. 1–22. Oxford University Press, Oxford.

Miya, M. & Nishida, M. (1997) Speciation in the open ocean. *Nature* **389**, 803–4.

Moore, M.N. (1981) Elemental accumulation in organisms and food chains. In *Marine Ecosystems* (ed. A.R. Longhurst), pp. 535–69. Academic Press, London.

Morgan, S.G. (2001) The larval ecology of marine communities. In *Marine Community Ecology* (eds. M.D. Bertness, S.D. Gaines & M.E. Hay), pp. 159–81. Sinauer Associates, Inc., Sunderland, Mass.

Morris, D.W. & Knight, T.W. (1996) Can consumer–resource dynamics explain patterns of guild assembly? *The American Naturalist* **147**(4), 558–75.

Myers, A.A. (1997) Biogeographic barriers and the development of marine biodiversity. *Estuarine, Coastal and Shelf Science* **44**, 241–8.

Osborn, T. (1995) Copepod behavior and turbulence: An evolutionary response? In *Ecology of Fjords and Coastal Waters* (eds. H.R. Skjoldal, C. Hopkins, K.E. Erikstad & H.P. Leinaas), pp. 27–32. Elsevier Science, Amsterdam.

Osman, R.W. (1977) The establishment and development of a marine epifaunal community. *Ecological Monographs* **47**, 37–63.

Paine, R.T. (1974) Intertidal community structure: Experimental studies on the relationship between a dominant competitor and its principal predator. *Oecologia* **15**, 93–120.

Palmer, J.D. (1995) *The Biological Rhythms and Clocks of Intertidal Animals*. Oxford University Press, New York.

Parrish, J.K. & Edelstein-Keshet, L. (1999) Complexity, pattern, and evolutionary trade-offs in animal aggregation. *Science* **284**, 99–101.

Peterson, C.H. (1991) Intertidal zonation of marine invertebrates in sand and mud. *American Scientist* **41**, 236–49.

Polis, G.A. & Hurd, S.D. (1996) Linking marine and terrestrial food webs: Allochthonous input from the ocean supports high secondary productivity on small islands and coastal land communities. *The American Naturalist* **147**(3), 396–423.

Polis, G.A. & Strong, D.R. (1996) Food web complexity and community dynamics. *The American Naturalist* **147**(5), 813–46.

Polis, G.A. & Winemiller, K.O. (1996) *Food Webs: Integration of Patterns and Dynamics*. Chapman & Hall, New York.

Polis, G.A., Sears, A.L.W., Huxel, G.R. et al. (2000) When is a trophic cascade a trophic cascade? *Trends in Ecology and Evolution* **15**(11), 473–5.

Prosser, C.L. (1986) *Adaptational Biology*. John Wiley & Sons, New York.

Ray, G.C. (1988) Ecological diversity in coastal zones and oceans. In *BioDiversity* (eds. E.O. Wilson & F. Peter), pp. 36–50. National Academy Press, Washington, D.C.

Ray, G.C. & Grassle, J.F. (1991) Marine biological diversity. *BioScience* **41**(7), 453–63.

Ricklefs, R.E. (1987) Community diversity: Relative roles of local and regional processes. *Science* **235**, 167–72.

Risser, P. (1990) Landscape pattern and its effects on energy and nutrient distribution. In *Changing Landscape: An Ecological Perspective* (eds. I.S. Zonneveld & R.T.T. Forman), pp. 45–56. Springer-Verlag, New York.

Rohr, J.J., Fish, F.E. & Gilpatrick, J.W., Jr. (2002) Maximum swim speeds of captive and free-ranging delphinids: Critical analysis of extraordinary performance. *Marine Mammal Science* **18**(1), 1–19.

Rosemond, A.D., Pringle, C.M., Ramírez, A. & Paul, M.J. (2001) A test of top-down and bottom-up control in a detritus-based food web. *Ecology* **82**(8), 2279–93.

Schoenly, K. & Cohen, J.E. (1990) Temporal variation in food web structure: 16 empirical cases. *Ecological Monographs* **61**(3), 267–98.

Shapiro, D.Y., Hensley, D.A. & Appledorn, R.S. (1988) Pelagic spawning and egg transport in coral-reef fishes: A skeptical overview. *Environmental Biology of Fishes* **22**(1), 3–14.

Shorrocks, B. & Swingland, I.R., eds. (1990) *Living in a Patchy Environment*. Oxford University Press, Oxford.

Sieburth, J.M. (1979) *Sea Microbes*. Oxford University Press, New York.

Simkiss, K. & Wilbur, K.M. (1989) *Biomineralization. Cell Biology and Mineral Deposition*. Academic Press, San Diego, Calif.

Swearer, S.E., Caselle, J.E., Lea, D.W. & Warner, R.R. (1999) Larval retention and recruitment in an island population of a coral-reef fish. *Nature* **402**, 799–802.

Takahashi, J.S. & Zatz, M. (1982) Regulation of circadian rhythmicity. *Science* **217**, 1104–12.

Underwood, A.J. (1989) The analysis of stress in natural populations. *Biological Journal of the Linnean Society* **37**, 51–78.

Wilson, E.O. & Peter, F., eds. (1988) *BioDiversity*. National Academy Press, Washington, D.C.

Winemiller, K.O. (1990) Spatial and temporal variation in tropical fish trophic networks. *Ecological Monographs* **60**, 331–67.

Winemiller, K.O. (1992) Life-history strategies and the effectiveness of natural selection. *Oikos* **63**(2), 318–27.

Winemiller, K.O. (1995a) Why do most fish produce so many tiny offspring? *The American Naturalist* **142**(4), 585–603.
Winemiller, K.O. (1995b) Fish ecology. In *Encyclopedia of Environmental Biology* **2**, pp. 49–65. Academic Press, San Diego, Calif.
Winemiller, K.O. & Rose, K.A. (1992) Patterns of life-history diversification in North American fishes: Implications for population regulation. *Fisheries and Oceans* **49**(10), 2196–218.
Winston, J.E., Gregory, M.R. & Stevens, L.M. (1996) Encrusters, epibionts, and other biota associated with pelagic plastics: A review of biogeographical, environmental, and conservation issues. In *Marine Debris: Sources, Impacts, and Solutions* (eds. J.M. Coe & D.B. Rogers), pp. 81–97. Springer-Verlag, New York.
Wood, R. (1990) Reef-building sponges. *American Scientist* **78**, 224–34.

Chapter 5 Chesapeake bay: estuarine alteration and restoration

Citations

Bahr, L.M., Jr. (1974) Aspects of the structure and function of the intertidal oyster reef community in Georgia. Dissertation submitted to the Graduate Faculty of the University of Georgia. Athens, Ga.
Bahr, L.M., Jr. & Lanier, W.P. (1981) *The Ecology of Intertidal Oyster Reefs of the South Atlantic Coast: A Community Profile.* U.S. Fisheries and Wildlife Service, Office of Biological Services, FWS/OBS-81/15. Dept of Interior, Washington, D.C.
Baird, D. & Ulanowicz, R.E. (1989) The seasonal dynamics of the Chesapeake Bay ecosystem. *Ecological Monographs* **59**(4), 329–64.
Barnard, T. & Mason, P. (1990) *Compensatory Mitigation within the Tidal Wetlands of Virginia.* Technical Report 90-7. Virginia Institute of Marine Science, College of William and Mary, Gloucester Point, Va.
Barnard, T. & Priest, W. (1993) Nekton utilization of an anthropogenic brackish water wetland bank. In *Proceedings of the Annual Meeting of the Society of Wetland Scientists*, May 30–June 3, 1993. University of Alberta, Edmonton, Canada.
Bearman, G.G., ed. (1989) *Waves, Tides and Shallow-Water Processes.* The Open University, Walton Hall, Milton Keynes, U.K.
Boicourt, W.C., Kuzmic, M. & Hopkins, T.S. (1999) The inland sea: Circulation of Chesapeake Bay and the Northern Adriatic. In *Ecosystems at the Land–Sea Margin: Drainage Basin to Coastal Sea.* (eds. T.C. Malone, A. Malej, L.W. Harding, Jr. et al.), pp. 81–129. Coastal and Estuarine Studies **55**. American Geophysical Union, Washington, D.C.
Boynton, W.R., Garber, J.H., Summers, R. & Kemp, W.M. (1995) Inputs, transformations and transport of nitrogen and phosphorus in Chesapeake Bay and selected tributaries. *Estuaries* **18**(1B), 285–314.
Brush, G.S. (1986) Estuarine biostratigraphic monitoring. *Journal of the Washington Academy of Sciences* **76**(3), 146–60.
CBP (1999) Chesapeake Bay Program's 1999 Executive Council Meeting. Online: http://www.chesapeakebay.net/1999exec.htm.
CBP (2000, 2001, 2002) Online: http://www.chesapeakebay.net/tsc.htm.
CBP Toxic Subcommittee (1992) *Chesapeake Bay Basin Comprehensive List of Toxic Substances.* U.S. Environmental Protection Agency for the Chesapeake Bay Program. April. U.S. Gov. Print. #312-014/40145.
Cooper, S.R. & Brush, G.S. (1993) A 2500 year history of anoxia and eutrophication in Chesapeake Bay. *Estuaries* **16**(3B), 617–26.
Correll, D.L. (1978) Estuarine productivity. *BioScience* **28**(10), 646–50.
Cross, F.A., Willis, J.N., Hardy, L.H., Jones, N.Y. & Lewis, J.M. (1975) Role of juvenile fish in cycling of Mn, Fe, Cu, and Zn in a coastal-plain estuary. In *Estuarine Research* **I** (ed. L.E. Cronin), pp. 45–63. Academic Press, New York.
Custer, J.F. (1986) Prehistoric use of the Chesapeake Estuary: A diachronic perspective. *Journal of the Washington Academy of Sciences* **76**(3), 161–72.
Dame, R., Childers, D. & Koepfler (1992) A geohydrologic continuum theory for the spatial and temporal evolution of marsh-estuarine ecosystems. *Netherlands Journal of Sea Research* **30**, 63–72.
Dame, R.F. (1996) *Ecology of Marine Bivalves. An Ecosystem Approach.* CRC Press, Boca Raton, Fla.
Davies, T.T. (1972) Effect of environmental gradients in the Rappahannock River estuary on the molluscan fauna. In *Environmental Framework of Coastal Plain Estuaries*, (ed. B.W. Nelson), pp. 263–90. Memoir **133**. Geological Society of America, Boulder, Colo.
DeAlteris, J.T. (1988) The geomorphic development of wreck shoal, a subtidal oyster reef of the James River, Virginia. *Estuaries* **11**(4), 240–9.
EPA (1983) *Chesapeake Bay: A Profile of Environmental Change.* EPA Region 3. U.S. Environmental Protection Agency, Philadelphia, Pa.
Erwin, R.M. (2002) Integrated management of waterbirds: Beyond the conventional. *Waterbirds* **25** (special publication 2), 5–12.
Erwin, R.M., Haramis, G.M., Krementz, D.G. & Funderburk, S.L. (1993) Resource protection for waterbirds in Chesapeake Bay. *Environmental Management* **17**, 613–19.
Funderburk, S.L., Mihursky, J.A., Jordan, S.J. & Riley, D., eds. (1991) *Habitat Requirements for Chesapeake Bay Living Resources.* Second edition. Chesapeake Research Consortium, Inc., Solomons, Md.
Galtsoff, P.S. (1964) The American oyster *Crassostrea virginica* Gmelin. *Fishery Bulletin* **64**, 1–480.
Gross, M.G. & Gross, E. (1996) *Oceanography: A View of Earth.* Prentice Hall, Upper Saddle River, N.J.
Harding, L.W., Jr., Degobbis, D. & Precali, R. (1999) Production and fate of phytoplankton: Annual cycles and interannual variability. In *Ecosystems at the Land–Sea Margin: Drainage Basin to Coastal Sea* (eds. T.C. Malone, A. Malej, L.W. Harding, Jr. et al.), pp. 131–72. Coastal and Estuarine Studies **55**. American Geophysical Union, Washington, D.C.

Hayes, M.O. & Sexton, W.J. (1989) IGC Field Trip T371: Modern clastic depositional environment, South Carolina. In *Coastal and Marine Geology of the United States, 28th International Geological Congress*, pp. T371 1–70. American Geophysical Union, Washington, D.C.

Homer, M.L. & Mihursky, J.A. (1991) Spot. In *Habitat Requirements for Chesapeake Bay Living Resources* (eds. S.L. Funderburk, J.A. Mihursky, S.J. Jordan & D. Riley), pp. 11, 1–19. Second edition. Chesapeake Research Consortium, Inc., Solomons, Md.

Horwitz, M.J. (1977) *The Transformation of American Law, 1780–1860*. Harvard University Press, Cambridge, Mass.

Karlsen, A.W., Cronin, T.M., Ishman, S.E. et al. (2000) Historical trends in Chesapeake Bay dissolved oxygen based on benthic foraminifera from sediment cores. *Estuaries* **23**(4), 488–508.

Kennedy, V.S., ed. (1984) *The Estuary as a Filter*. Academic Press, Orlando, Fla.

Kennedy, V.S. (1989) The Chesapeake Bay oyster fishery: Traditional management practices. In *Marine Invertebrate Fisheries: Their Assessment and Management* (ed. J.F. Caddy), pp. 455–77. John Wiley & Sons, New York.

Kennedy, V.S. & Breisch, L.L. (1983) Sixteen decades of political management of the oyster fishery in Maryland's Chesapeake Bay. *Journal of Environmental Management* **16**, 153–71.

Kennedy, V.S., Newell, R.I.E. & Eble, A.F., eds. (1996) *The Eastern Oyster* Crassostrea Virginica. Maryland Sea Grant Book, College Park, Md.

Kline, D.L. (1998) *Endocrine Disruption in Fish*. Kluwer Academic Publishers Boston, Mass.

Kull, T.K. & Hopkins, H.T. (1987) Virginia water supply and use. In *National Water Summary*, US Geological Survey Water Supply Paper #2350, State Summaries, pp. 505–14.

Kuo, A.Y. & Neilson, B.J. (1987) Hypoxia and salinity in Virginia estuaries. *Estuaries* **10**(4), 277–83.

Largier, J.L. (1993) Estuarine fronts: How important are they? *Estuaries* **16**(1), 1–11.

Magnien, R.E., Summers, R.M. & Sellner, K.G. (1992) External nutrient sources, internal nutrient pools, and phytoplankton production in Chesapeake Bay. *Estuaries* **15**(4), 497–516.

McCormick-Ray, M.G. (1998) Oyster Reefs in 1878 Seascape Pattern – Winslow Revisited. *Estuaries* **21**(4B), 784–800.

Miller, H.M. (1986) Transforming a "splendid and delightsome land": colonists and ecological change in the 17th and 18th century Chesapeake. *Journal of the Washington Academy of Sciences* **76**(3), 173–88.

Möbius, K. (1883) The oyster. *Report of the Commissioner*, part VIII, Appendix H. U.S. Commission of Fish and Fisheries, Spencer F. Baird, Commissioner. U.S. Government Printing Office, Washington, D.C.

Nichols, M.M. (1972) Sediments of the James River estuary. In *Environmental Framework of Coastal Plain Estuaries* (ed. B.W. Nelson), pp. 169–212. Memoir **133**. Geological Society of America, Boulder, Colo.

Nichols, M.M. (1994) Response of estuaries to storms in the Chesapeake Bay region; Summary. In *Changes in Fluxes in Estuaries: Implications from Science to Management* (eds. K.R. Dyer & R.J. Orth), pp. 67–70. Olsen & Olsen, Fredensborg, Denmark.

Odum, W.E., Smith, T.J., III, Hover, J.K. & McIvor, C.C. (1984) *The Ecology of Tidal Freshwater Marshes of the United States East Coast: A Community Profile*. FWS/OBS-83/17. U.S. Fish and Wildlife Service, Washington, D.C.

Officer, C.B., Biggs, R.B., Taft, J.L. et al. (1984a) Chesapeake Bay anoxia: Origin, development, and significance. *Science* **223**, 22–7.

Officer, C.B., Lynch, D.R., Setlock, G.H. & Helz, G.R. (1984b) Recent sedimentation rate in Chesapeake Bay. In *The Estuary as a Filter* (ed. V.S. Kennedy), pp. 131–57. Academic Press, Orlando, Fla.

Perry, J. & Anderson, J. (1998) *An Evaluation of Southern Virginia Bottomland Hardwood Swamps*. U.S. EPA Wetlands Program Grant Final Report to Region III. Philadelphia, Pa.

Perry, J.E. & Atkinson, R.A. (1997) Plant diversity along a salinity gradient for four marshes of the York River, Virginia. *Castanea* **62**(2), 112–18.

Perry, M. & Deller, A. (1995) Waterfowl population trends in the Chesapeake Bay area. In *Toward a Sustainable Coastal Watershed: The Chesapeake Experiment* (eds. P. Hills & S. Nelson), pp. 490–504. Proceedings of a Conference, June 1–3, 1994, Norfolk, Va. Publication **149**. Chesapeake Research Consortium, Edgewater, Md.

Perry, J.E., Barnard, T., Bradshaw, I. et al. (2001) Created tidal salt marshes in the Chesapeake Bay. *J. Coastal Research* S.I. **27**, 170–92.

Ray, G.C., Hayden, B.P., McCormick-Ray, M.G. & Smith, T.M. (1997) Land-seascape diversity of the U.S. East coast coastal zone with particular reference to estuaries. In *Marine Biodiversity: Causes and Consequences* (eds. R.F.G. Ormond, J.D. Gage & M.V. Angel), pp. 337–71. Cambridge University Press, Cambridge.

Richardson, D.L. (1992) *Hydrogeology and Analysis of the Ground-Water-Flow System of the Eastern Shore, Virginia*. Open-file Report 91-490. U.S. Geological Survey, Richmond, Va.

Rilling, G.R. & Houde, E. (1999) Regional and temporal variability in distribution and abundance of bay anchovy (*Anchoa mitchilli*) eggs, larvae, and adult biomass in the Chesapeake Bay. *Estuaries* **22**(4), 1096–109.

Stevenson, J.C. & Kearney, M.S. (1996) Shoreline dynamics on the windward and leeward shores of a large temperate estuary. In *Estuarine Shores* (eds. K.F. Nordstrom & C.T. Roman), pp. 233–59. John Wiley & Sons, Chichester.

Tiner, R. (1984) *Wetlands of the United States: Current Status and Recent Trends*. National Wetlands Inventory, Fish and Wildlife Service, U.S. Department of the Interior, Washington, D.C.

Uncles, R.J., Stephens, J.A. & Barton, M.L. (1991) The nature of near-bed currents in the upper reaches of an estuary. In *Estuaries and Coasts: Spatial and Temporal Intercomparisons* (eds. M. Elliott & J.P. Ducrotoy), pp. 43–7. ECS 19 Symposium. Olsen & Olsen, Fredensborg, Denmark.

VMRC (1996) *Mitigation Guidelines*. Draft. Virginia Marine Resources Commission, Newport News, Va.

Wass, M.L. & Wright, T.D. (1969) *Coastal Wetlands of Virginia*. Interim report. Special Report in Applied Marine Science and Ocean Engineering **10**. Virginia Institute of Marine Science, Gloucester Point, Va.

Winslow, F. (1882) Methods and results. In *Report of the oyster beds of the James River, Va. and of Tangier and Pocomoke sounds, Maryland and Virginia.* U.S. Coast and Geological Survey, Appendix No. 11. U.S. Government Printing Office, Washington, D.C.

Suggested readings

Breitburg, D.I., Coen, L.D., Luckenbach, M.W., Mann, R., Posey, M. & Wesson, J.A. (2000) Oyster reef restoration: Convergence of harvest and conservation strategies. *Journal of Shellfish Research* **19**(1), 371–7.

Brinson, M.M., Christian, R.R. & Blum, L.K. (1995) Multiple states in the sea-level induced transition from terrestrial forest to estuary. *Estuaries* **18**(4), 648–59.

Carter, H.H. & Pritchard, D. (1988) Oceanography of Chesapeake Bay. In *Hydrodynamics of Estuaries.* **II**. *Estuarine Case Studies.* (ed. B. Kjerfve), pp. 1–16. CRC Press, Boca Raton, Fla.

Chesapeake–Potomac Study Commission (1948) *Report on Fish and Shellfish in the Chesapeake Bay and Potomac River, with Recommendations for their Future Management.* Press of the Daily Record, Baltimore, Md.

Cronin, L.E. (1986) Fisheries and resource stress in the 19th century. *Journal of the Washington Academy of Sciences* **76**(3), 188–98.

Curtin, P.D., Brush, S. & Fisher, G.W., eds. (2001) *Discovering the Chesapeake Bay. The History of an Ecosystem.* Johns Hopkins University Press, Baltimore, Md.

Galtsoff, P.S., Chipman, Jr, W.A., Engle, J.B. & Calderwood, H.N. (1947) *Ecological and Physiological Studies of the Effect of Sulfate Pulp Mill Wastes on Oysters in the York River, Virginia. Fishery Bulletin* **43**, Fishery Bulletin of the Fish and Wildlife Service **51**. U.S. Government Printing Office, Washington, D.C.

Gerstell, R. (1998) *American Shad. A Three-Hundred Year History.* Pennsylvania State University Press, University Park, Pa.

Gottschalk, L.C. (1945) Effects of soil erosion on navigation in upper Chesapeake Bay. *Geographical Review* **35**(2), 219–38.

Malone, T.C., Malej, A., Harding, L.W., Jr. et al., eds. (1999) *Ecosystems at the Land–sea Margin: Drainage Basin to Coastal Sea*, pp. 131–72. Coastal and Estuarine Studies **55**. American Geophysical Union, Washington, D.C.

Hargis, W.J., Jr. & Haven, D.S. (1995) *The Precarious State of the Chesapeake Public Oyster Resource.* Contribution **1965**. Virginia Institute of Marine Science, Gloucester Point, Va.

Haven, D.S., Hargis, W.J., Jr. & Kendall, P.C. (1978) *The Oyster Industry of Virginia: Its Status, Problems and Promise.* Special Report **168**. Virginia Institute of Marine Science, Gloucester Point, Va.

Haven, D.S. & Morales-Alamo, R. (1972) Biodeposition as a factor in sedimentation of fine suspended solids in estuaries. *Geological Society of America Memoir* **133**, 121–30.

Hobbie, J.E., ed. (2000) *Estuarine Science, A Synthetic Approach to Research and Practice.* Island Press, Washington, D.C.

JHU (1893) *Maryland: Its Resources, Industries, and Institutions.* Johns Hopkins University Press, Baltimore, Md.

Kennedy, V. (1991) Eastern oyster *Crassostrea virginica.* In *Habitat Requirements for Chesapeake Bay Living Resources* (eds. S.L. Funderburk, J.A. Mihursky, S.J. Jordan & D. Riley), pp. 3:1–20. Second edition. Chesapeake Research Consortium, Inc., Solomons, Md.

Kennedy, V.S. (1996) The ecological role of the eastern oyster, *Crassostrea virginica*, with remarks on disease. *Journal of Shellfish Research* **15**(1), 177–83.

Kjerfve, B., ed. (1988) *Hydrodynamics of Estuaries.* **I**; II. *Estuarine Physics.* CRC Press, Boca Raton, Fla.

Lauff, G.H., ed. (1967) *Estuaries* **83**. American Association for the Advancement of Science, Washington, D.C.

Mann, R. (2000) Restoring the oyster reef communities in the Chesapeake Bay. A commentary. *Journal of Shellfish Research* **19**(1), 335–99.

McCormick-Ray, J. (1995) A watershed perspective in Chesapeake Bay management. *Proceedings of a Conference: 1994 Chesapeake Bay Conference Toward a Sustainable Coastal Watershed: The Chesapeake Experiment*, June 1–3 in Norfolk, Va, pp. 235–47. Chesapeake Research Consortium, Solomons, Md.

McCormick-Ray, J. (2001) Estuarine biodiversity is multi-ordered: The oyster example. In *Conservation of Biological Diversity: A Key to the Restoration of the Chesapeake Bay Ecosystem and Beyond. Proceedings, 1998* (ed. G.D. Therres), pp. 214–24. Maryland Department of Natural Resources, Annapolis, Md.

Mountford, K. (2000) History of dredge reveals deeper need to understand bay's bottom line. *Bay Journal* **10**(5), 8–10.

Newell, R.I.E. (1988) Ecological changes in Chesapeake Bay: Are they the result of overharvesting the American oyster, *Crassostrea virginica*? In *Understanding the Estuary: Advances in Chesapeake Bay Research.* Proceedings of a conference, March 29–31, 1988, Baltimore, Md. Chesapeake Research Consortium **129**. CBP/TRS 24/88.

NRC (1992) *Restoration of Aquatic Ecosystems: Science, Technology, and Public Policy.* Water Science and Technology Board, Committee on Restoration of Aquatic Ecosystem, National Research Council. National Academy Press, Washington, D.C.

Pritchard, D. (1967) Observation of circulation in coastal plain estuaries. In *Estuaries* **83** (ed. G.H. Lauff), pp. 37–44. American Association for the Advancement of Science, Washington, D.C.

Pritchard, D.W. & Schubel, J.R. (2001) Human influences on the physical characteristics of the Chesapeake Bay. In *Discovering the Chesapeake Bay. The History of an Ecosystem* (eds. P.D. Curtin, S. Brush, & G.W. Fisher), pp. 60–82. Johns Hopkins University Press, Baltimore, Md.

Rothschild, B.J., Ault, J.S., Goulletquer, P. & Héral, M. (1994) Decline of the Chesapeake Bay oyster population: A century of habitat destruction and overfishing. *Marine Ecology Progress Series* **111**, 29–39.

Roundtree, H.C. & Davidson, T.E. (1997) *Eastern Shore Indians of Virginia and Maryland.* University Press of Virginia, Charlottesville, Va.

Schubel, J. (1986) *Life and Death of the Chesapeake Bay.* Sea Grant College, Publication UM-SG-86-01. University of Maryland, College Park, Md.

Silberhorn, G.M. (1999) *Common Plants of the Mid-Atlantic Coast: A Field Guide*. Revised edition. Johns Hopkins University Press, Baltimore, Md.

Stevenson, C.H. (1892) The oyster industry of Maryland. *Bulletin of the United States Fish Commission* **12**, 203–97.

Ulanowicz, R.E. & Tuttle, J.H. (1992) The trophic consequences of oyster stock rehabilitation in Chesapeake Bay. *Estuaries* **15**(3), 298–306.

Ward, L.G., Rosen, P.S., Neal, W.J. et al. (1989) *Living with Chesapeake Bay and Virginia's Ocean Shores*. Duke University Press, Durham, N.C.

Wennersten, J.R. (1981) *The Oyster Wars of Virginia*. Tidewater Publ., Centreville, Md.

Wharton, J. (1957) *The Bounty of the Chesapeake. Fishing in Colonial Virginia*. Virginia 350th Anniversary Celebration Corp., Williamsburg, Va.

Chapter 6 Bering Sea: marine mammals in a regional sea

Citations

Bickham, J.W., Patton, J.C. & Loughlin, T.R. (1996) High variability for control-region sequences in a marine mammal: Implications for conservation and biogeography of Steller sea lions (*Eumetopias jubatus*). *Journal of Mammalogy* **77**(1), 95–108.

Fay, F.H. (1982) *Ecology and Biology of the Pacific Walrus*, Odobenus rosmarus divergens *Illiger*. *American Fauna* **74**. U.S. Fish and Wildlife Service Washington, D.C.

Fay, F.H., Eberhardt, L.L., Kelly, B.P. et al. (1997) Status of the Pacific walrus population, 1950–1989. *Marine Mammal Science* **13**(4), 537–65.

Fay, F.H., Kelly, B.P. & Sease, J.L. (1989) Managing the exploitation of Pacific walruses: A tragedy of delayed response and poor communication. *Marine Mammal Science* **5**(1), 1–16.

Fay, F.H., Ray, G.C. & Kibal'chich, A.A. (1984) Time and location of mating and associated behavior of the Pacific walrus, *Odobenus rosmarus divergens* Illiger. In *Soviet–American Cooperative Research on Marine Mammals*. **1**. *Pinnipeds* (eds. F.H. Fay & G.A. Fedoseev), pp. 89–99. NOAA Technical Report NMFS **12**. NOAA, Washington, D.C.

Fitzhugh, W.W. & Kaplan, S.A. (1982) *Inua: Spirit World of the Bering Sea Eskimo*. Smithsonian Institution Press, Washington, D.C.

Gorbics, C.S., Garlich-Miller, J.L. & Schliebe, S.L. (1998) *Alaska Marine Mammal Stock Assessments 1998: Sea Otters, Polar Bear and Walrus*. Draft. Marine Mammal Management, U.S. Fish and Wildlife Service, Anchorage, Alaska.

Hopkins, D.M. (1967) *The Bering Land Bridge* (ed. D.M. Hopkins). Stanford University Press, Stanford, Calif.

Loughlin, T.R. (1997) Using the phylogeographic method to identify Steller sea lion stocks. In *Molecular Genetics of Marine Mammals* (eds. A. Dizon, S.J. Chivers & W.F. Perrin), pp. 159–71. Special Publication **3**. Society for Marine Mammalogy, Lawrence, Kans.

Loughlin, T.R. & Merrick, R.L. (1989) Comparison of commercial harvest of walleye pollock and northern sea lion abundance in the Bering Sea and Gulf of Alaska. In *Proceedings of the International Symposium on the Biology and Management of Walleye Pollock*, Alaska Sea Grant **89-1**, pp. 679–700. University of Alaska, Fairbanks, Alaska.

Loughlin, T.R. & Ohtani, K., eds. (1999) *Dynamics of the Bering Sea*. Sea Grant, AK-SG-**99-03**. University of Alaska Sea Grant Press, Fairbanks, Alaska.

Loughlin, T.R., Sukhanova, I.N., Sinclair, E.H. & Ferrero, R.C. (1999) Summary of biology and ecosystem dynamics in the Bering Sea. In *Dynamics of the Bering Sea* (eds. T.R. Loughlin & K. Ohtani), pp. 387–407. University of Alaska Sea Grant Press, Fairbanks, Alaska.

NOAA (1988) *Bering, Chukchi, and Beaufort Seas. Coastal and Ocean Zones, Strategic Assessment: Data Atlas*. U.S. Department of Commerce. U.S. Government Printing Office, Washington, D.C.

NRC (1996) *The Bering Sea Ecosystem*. Committee on the Bering Sea Ecosystem, Polar Research Board, National Research Council. National Academy Press, Washington, D.C.

Ray, G.C. & Curtsinger, B. (1979) Learning the ways of the walrus. *National Geographic* **156**(4), 564–80.

Ray, G.C. & Hayden, B.P. (1993) Marine biogeographic provinces of the Bering, Chukchi, and Beaufort Seas. In *Large Marine Ecosystems: Stress, Mitigation, and Sustainability* (eds. K. Sherman, L.M. Alexander & B.D. Gold), pp. 175–84. American Association for the Advancement of Science, Washington, D.C.

Ray, G.C. & Hufford, G.L. (1989) Relationships among Beringian marine mammals and sea ice. *Rapport Procés-verbal Réunion Conseil International Exploration de la Mer* **188**, 22–39.

Ray, G.C. & Wartzok, D. (1974) *BESMEX: Bering Sea Marine Mammal Experiment*. NASA TM X-62399. NASA Ames Research Center, Woffetl Field, CA.

Springer, A.M. (1992) A review: Walleye pollock in the north Pacific – How much difference do they really make? *Fisheries Oceanography* **1**(1), 80–96.

Springer, A.M., McRoy, C.P. & Flint, M.V. (1996) The Bering Sea green belt: Shelf-edge processes and ecosystem production. *Fisheries Oceanography* **5**(3/4), 205–23.

Suggested readings

Alverson, D.L. (1992) A review of commercial fisheries and the Steller sea lion (*Eumetopias jubatus*): The conflict arena. *Reviews in Aquatic Sciences* **6**, 203–56.

Anderson, P.J. & Piatt, J.F. (1999) Community reorganization in the Gulf of Alaska following ocean climate regime shift. *Marine Ecology Progress Series* **189**, 117–23.

Antonson, J.M. & Hanable, Wm.S. (1992) *Alaska's Heritage*. Second edition, vols. 1 and 2. Alaska Historical Commission Studies in History **133**. Alaska Historical Commission, Fairbanks, Alaska.

Bailey, K.M., Quinnell, T.J., Benzen, P. and Brant, W.S. (2000) Population structures and dynamics of walleye pollock *Theragra chalcogramme*. In *Advances in Marine Biology* (eds. A.J. Southward, P.A. Tyler, & G.M. Young), pp. 179–255. Academic Press, San Diego.

Barlow, J., Swartz, S.L., Eagle, T.C. & Wade, P.R. (1995) *U.S. Marine Mammal Stock Assessments: Guidelines for Preparation, Background, and a Summary of the 1995 Assessments*. U.S. Department of Commerce, NOAA Technical Memorandum NMFS-OPR-6, 1–73.

Becker, P.R. (2000) Concentration of chlorinated hydrocarbons and heavy metals in Alaska Arctic marine mammals. *Marine Pollution Bulletin* **40**(10), 819–29.

Benson, A.J., and Trites, A.W. (2002) Ecological effects of regime shifts in the Bering Sea and eastern North Pacific Ocean. *Fish and Fisheries* **3**, 95–113.

Bockstoce, J.R. & Botkin, D.B. (1982) The harvest of Pacific walruses by the pelagic whaling industry. *Arctic and Alpine Research* **14**, 183–8.

Burns, J.J., Shapiro, L.H. & Fay, F.H. (1981) Ice as marine mammal habitat in the Bering Sea. In *The Eastern Bering Sea Shelf: Oceanography and Resources* **2** (eds. D.W. Hood & J.A. Calder), pp. 781–97. University of Washington Press, Seattle, Wash.

Dickerson, L. (1999) Monitoring the Pacific walrus harvest in Alaska. In *Proceedings of a Workshop Concerning Walrus Monitoring* (eds. J. Garlich-Miller & C. Pungowiyi), pp. 15–20. Fish and Wildlife Service, Alaska.

Estes, J.A. & Gilbert, J.R. (1978) Evaluation of an aerial survey of Pacific walruses (*Odobenus rosmarus divergens*). *Journal of the Fisheries Research Board of Canada* **35**, 1130–40.

Fay, F.H. & Ray, C. (1968) Influence of climate on the distribution of walruses, *Odobenus rosmarus* (Linnaeus). I. Evidence from thermoregulatory behavior. *Zoologica* **53**(1), 1–14 + plates I–IV.

Francis, R.C., Aydin, K., Merrick, R. & Bollens, S. (1999) Modeling and management of the Bering Sea ecosystem. In *Dynamics of the Bering Sea* (eds. T.R. Loughlin & K. Ohtani), pp. 409–33. University of Alaska Sea Grant Press, Fairbanks, Alaska.

Francis, R.C. & Hare, S.R. (1994) Decadal scale regime shifts in the large marine ecosystem of the North Pacific: A case for historical science. *Fisheries Oceanography* **3**, 279–91.

Francis, R.C., Hare, S.R., Hollowed, A.B. & Wooster, W.S. (1998) Effects of interdecadal climate variability on the ocean ecosystems of the N.E. Pacific. *Fisheries Oceanography* **7**, 1–21.

Hare, S.R. & Mantua, N.J. (2000) Empirical evidence for North Pacific regime shifts in 1977 and 1989. *Progress in Oceanography* **47**(2–4), 103–46.

Hopkins, D.M. (1996) Introduction: The concept of Beringia. In *American Beginnings. The Prehistory and Palaeoecology of Beringia* (ed. F.H. West), pp. xvii–xxi. The University of Chicago Press, Chicago, IU.

Johnson, K.R. & Nelson, C.H. (1984) Side-scan sonar assessment of gray whale feeding in the Bering Sea. *Science* **225**, 1150–2.

Krupnik, I. (1993) *Arctic Adaptations: Native Whalers and Reindeer Herders of Northern Eurasia*. University Press of New England, Dartmouth, Hanover, N.H.

Loughlin, T.R. & York, A.E. (2000) An accounting of the sources of Steller sea lion, *Emmetopias jubatas*, mortality. *Marine Fisheries Review* **62**(4), 40–5.

Mantua, N.J., Hare, S.R., Zhang, Y. et al. (1997) A Pacific interdecadal climate oscillation with impacts on salmon production. *Bulletin of the American Meteorological Society* **78**(3), 1069–79.

Miller, E.H. (1975) Walrus ethology. I: The social role of tusks and applications of multidimensional scaling. *Canadian Journal of Zoology* **53**, 590–613.

Napp, J.M. & Hunt, G.L., Jr. (2001) Anomalous conditions in the south-eastern Bering Sea 1977: Linkages among climate, weather, ocean, and biology. *Fisheries Oceanography* **10**(1), 61–8.

Nelson, C.H. & Johnson, K.R. (1987) Whales and walruses as tillers of the sea floor. *Scientific American* **256**(2), 112–17.

Pitcher, K.W. (1981) Prey of the Steller sea lion, *Eumetopias jubatus*, in the Gulf of Alaska. *Fisheries Bulletin* **79**, 467–72.

Ray, C. & Fay, F.H. (1968) Influence of climate on the distribution of walruses, *Odobenus rosmarus* (Linnaeus). II. Evidence from physiological characteristics. *Zoologica* **53**(1), 19–32.

Ray, G.C. & Watkins, W.A. (1975) Social functions of underwater sounds in the walrus *Odobenus rosmarus*. *Rapp. P.-v. Reun. Cons. Int. Explor. Mer.* **169**, 524–6.

Seagars, D.L. & Garlich-Miller, J. (2001) Organochlorine compounds and aliphatic hydrocarbons in Pacific walrus blubber. *Marine Pollution Bulletin* **43**(1–6), 122–31.

Simpson, J.J., Hufford, G.L., Fleming, M.D. & Ashton, J.B. (in press) Long-term climate patterns in Alaska surface temperature and precipitation and their biological consequences. *Journal of Hydrology*.

Taylor, B.L. & Wade, P.R. (2000) "Best" abundance estimates and best management: Why they are not the same. In *Quantitative Methods for Conservation Biology* (eds. S. Ferson & M. Burgman), pp. 96–108. Springer-Verlag, New York.

Trites, A.W., Christensen, V. & Pauly, D. (1997) Competition between fisheries and marine mammals for prey and primary production in the Pacific Ocean. *Journal of Northwest Atlantic Fishery Science* **22**, 173–87.

Udevitz, M.S., Gilbert, J.R. & Fedoseev, G.A. (2001) Comparison of methods used to estimate numbers of walruses on sea ice. *Marine Mammal Science* **17**(3), 601–16.

Wade, P.R. & DeMaster, D.P. (1999) Determining the optimum interval for abundance surveys. In *Marine Mammal Survey and Assessment Methods* (eds. G.W. Garner, S.C. Amstrup, J.L. Laake et al.), pp. 53–66. A.A. Balkema, Rotterdam, The Netherlands.

Chapter 7 The Bahamas: tropical-oceanic island nation

Citations

BEST (1997) *Commonwealth of The Bahamas: National Biodiversity Strategy and Action Plan*. Prepared by Task Force: The Bahamas National Biodiversity Strategy and Action Plan for The Bahamas Environment, Science and Technology Commission (BEST), under contract from the United Nations Environment Programme. Nassau, The Bahamas.

Birkeland, C., ed. (1997) *Life and Death of Coral Reefs*. Chapman & Hall, New York.

Bjorndal, K.A., ed. (1995) *Biology and Conservation of Sea Turtles*. Revised edition. Smithsonian Institution Press, Washington, D.C.

Carr, A. (1956/1979) *The Windward Road*. Re-issue 1979. University Press of Florida, Tallahassee, Fla.

Colin, P.L. (1975) *The Neon Gobies: The Comparative Biology of the Gobies of the Genus* Gobiosoma *Subgenus* Elacatinus *(Pisces, Gobiidae) in the Tropical Western North Atlantic Ocean*. T.F. Publ., Neptune City, N.J.

Colin, P.L. (1992) Reproduction of the Nassau grouper, *Epinephelus striatus* (Pisces: Serranidae) and its relationship to environmental conditions. *Environmental Biology of Fishes* **34**, 357–77.

Correll, D.S. & Correll, H.B. (1982) *Flora of the Bahamian Archipelago*. J. Cramer, Vaduz, Germany.

Crowder, L.B., Lyman, S.J., Figueira, W.F. & Priddy, J. (2000) Source–sink population dynamics and the problem of siting marine reserves. *Bulletin of Marine Science* **66**, 799–820.

Cunningham, R.L. (1997) The biological impacts of 1492. In *The Indigenous People of the Caribbean* (ed. S.M. Wilson), pp. 31–5. University Press of Florida, Gainesville, Fla.

Dietz, R.S., Holden, J.C. & Sproll, W.P. (1970) Geotectonic evolution and subsidence of the Bahama platform. *Geological Society of America Bulletin* **81**, 1915–28.

Keegan, W.F. (1997) *Bahamian Archeology*. Media Publishing, Nassau, Bahamas.

Lipcius, R.N., Stockhausen, W.T. & Eggleston, D.B. (2001) Marine reserves and Caribbean spiny lobster: Empirical evaluation and theoretical metapopulation recruitment dynamics. *Marine and Freshwater Research* **52**, 1589–98.

Lipcius, R.N., Stockhausen, W.T., Eggleston, D.B. et al. (1997) Hydrodynamic decoupling of recruitment, habitat quality and adult abundance in the Caribbean spiny lobster: Source–sink dynamics? *Marine and Freshwater Research* **48**, 807–15.

Lutz, P.L. & Musick, J.A., eds. (1997) *The Biology of Sea Turtles*. CRC Press, New York.

Polunin, N.V.C. & Roberts, C.M., eds. (1996) *Reef Fisheries*. Chapman & Hall, London.

Robins, C.R. (1971) Distributional patterns of fishes from coastal and shelf waters of the tropical western Atlantic. In *Symposium on Investigations and Resources of the Caribbean Sea and Adjacent Regions*. FAO, Rome.

Sadovy, Y. & Eklund, A.-M. (1999) *Synopsis of Biological Data on the Nassau Grouper,* Epinephelus striatus *(Bloch, 1792), and the Jewfish,* E. itajara *(Lichtenstein, 1822)*. NOAA Technical Report NMFS 146, A Technical Report of the *Fishery Bulletin*, FAO Fisheries Synopsis **157**. U.S. Department of Commerce, Seattle, Wash.

Sale, P.F., ed. (1991) *The Ecology of Fishes on Coral Reefs*. Academic Press, San Diego, Calif.

Sale, P.F., ed. (2002) *Advances in the Ecology of Fishes on Coral Reefs*. 2nd edition, Academic Press, San Diego, Calif.

Schlager, W. & Ginsburg, R.N. (1981) Bahama carbonate platforms – The deep and the past. *Marine Geology* **44**, 1–24.

Sealey, N.E. (1994) *Bahamian Landscapes: An Introduction to the Geography of the Bahamas*. Second edition. Media Publishing, Nassau, Bahamas.

Sealey, N.E., ed. (2001) *Caribbean Certificate Atlas*. Third edition. Macmillan Education Ltd., Oxford.

Spalding, M.D., Ravilious, C. & Green, E.P. (2001) *World Atlas of Coral Reefs*. Prepared at the UNEP World Conservation Monitoring Centre. University of California Press, Berkeley, Calif.

Stockhausen, W.T. & Lipcius, R.N. (2001) Single large or several small marine reserves for the Caribbean spiny lobster? *Marine and Freshwater Research*, **52**, 1605–14.

Stockhausen, W.T., Lipcius, R.N. & Hickey, B.H. (2000) Joint effects of larval dispersal, population regulation, marine reserve design, and exploitation on production and recruitment in the Caribbean spiny lobster. *Bulletin of Marine Science* **66**, 957–90.

Wallace, A.R. (1880) *Island Life*. Re-published, with an Introduction by H.J. Birk. Prometheus Books, Great Minds Series. Amherst, New York.

Suggested readings

Auyong, J. (1995) Tourism and conservation. In *Coastal Management in the Asia-Pacific Region: Issues and Approaches* (eds. K. Hotta & I.M. Dutton), pp. 95–115. Japan International Marine Science and Technology Federation, Tokyo.

Barnett, F.G. (1996) Shipping and marine debris in the wider Caribbean: Answering a difficult challenge. In *Marine Debris: Sources, Impacts, and Solutions* (eds. J.M. Coe & D.B. Rogers), pp. 219–27. Springer-Verlag, New York.

Birkeland, C. (1996) Why some species are especially influential on coral-reef communities and others are not. *Galaxea* **13**, 77–84.

Bjorndal, K.A., Bolten, A.B. & Chaloupka, M.Y. (2000) Green turtle somatic growth model: Evidence for density dependence. *Ecological Applications* **10**(1), 269–82.

Bjorndal, K.A., Wetherall, J.A., Bolten, A.B. & Mortimer, J.A. (1999) Twenty-six years of green turtle nesting at Tortuguero, Costa Rica: An encouraging trend. *Conservation Biology* **13**(1), 126–34.

Bourrouilh-Le Jan, F.G. (1998) The role of high-energy events (hurricanes and/or tsunamis) in the sedimentation, diagenesis and karst initiation of tropical shallow water carbonate platforms and atolls. *Sedimentary Geology* **118**(1–4), 3–36.

Broecker, W.S. & Takahashi, T. (1966) Calcium carbonate precipitation on the Bahama banks. *Journal of Geophysical Research* **71**(6), 1575–662.

Coleman, F.C., Koenig, C.C. & Collins, L.A. (1996) Reproductive styles of shallow-water groupers (Pisces: Serranidae) in the eastern Gulf of Mexico and the consequences of fishing spawning aggregations. *Environmental Biology of Fishes* **47**, 129–41.

Coleman, F.C., Koenig, C.C., Huntsman, G.R. et al. (2000) Long-lived reef fishes: The grouper–snapper complex. *Fisheries* **25**(3), 14–21.

Colin, P.L., Laroche, W.A. & Brothers, E.B. (1997) Ingress and settlement in the Nassau grouper, *Epinephelus striatus* (Pisces: Serranidae), with relationship to spawning occurrence. *Bulletin of Marine Science* **60**(3), 656–67.

Dahlgren, C.P. (1999) The biology, ecology, and conservation of the Nassau grouper *Epinephelus striatus* in the Bahamas. *Bahamas Journal of Science* **7**(1), 6–12.

Dahlgren, C.P. & Eggleston, D.B. (in press) Ecological processes underlying ontogenetic habitat shifts in a coral reef fish. *Ecology*.

Doherty, P.J. & Williams, D.M. (1988) The replenishment of coral reef fish populations. *Annual Review of Oceanography and Marine Biology* **26**, 487–551.

Domeier, M.L. & Colin, P.L. (1997) Tropical reef-fish spawning aggregations defined and reviewed. *Bulletin of Marine Science* **60**(3), 698–726.

Eggleston, D.B. (1995) Recruitment in Nassau grouper, *Epinephelus striatus*: post-settlement abundance, microhabitat features, and ontogenetic habitat shifts. *Marine Ecology Progress Series* **124**, 9–22.

Eggleston, D.B., Lipcius, R.N. & Grover, J.J. (1997) Predator and shelter size effects on coral reef fish and spiny lobster prey. *Marine Ecology Progress Series* **149**, 43–59.

Frantz, D. (2000) Gaps in sea laws shield pollution by cruise lines. Sovereign islands. *New York Times* **CXLVIII** (51 391), 1.

Gould, S.J. (1993) Cerion – The evolving snail. *Bahamas Journal of Science* **1**(1), 10–15. (First published in *Natural History*, April 1983.)

Gruber, S.H. & Parks, W. (2002) Mega-resort development on Bimini: Sound economics or environmental disaster? *Bahamas Journal of Science* **9**(2), 2–18.

Hammerton, J.L. (2001) Casuarinas in the Bahamas: A clear and present danger. *Bahamas Journal of Science* **9**(1), 2–14.

Huntsman, G.R., Potts, J., Mays, R.W. & Vaughan, D. (1999) Groupers (Serranidae, Epinephelinae): Endangered apex predators of reef communities. In *Life in the Slow Lane* (ed. A.J. Musick), pp. 217–31. Symposium **23**. American Fisheries Society, Washington, D.C.

Johannes, R.E. (1978) Reproductive strategies of coastal fishes in the tropics. *Environmental Biology of Fishes* **3**(1), 65–84.

Jones, G.P., Millcich, M.J., Emslie, M.J. & Lunow, C. (1999) Self-recruitment in a coral reef fish population. *Nature* **402**, 802–4.

Knowlton, N. & Jackson, J.B.C. (2001) The ecology of coral reefs. In *Marine Community Ecology* (eds. M.D. Bertness, S.D. Gaines & M.E. Hay), pp. 395–422. Sinauer Associates, Inc., Sunderland, Mass.

Lahanas, P.N., Bjorndal, K.A., Bolten, A.B. et al. (1998) Genetic composition of a green turtle (*Chelonia mydas*) feeding ground population: Evidence for multiple origins. *Marine Biology* **130**, 345–52.

Layman, C.A. & Silliman, B.R. (2002) Preliminary survey and diet analysis of juvenile fishes of an estuarine creche on Andros Island, Bahamas. *Bulletin of Marine Science* **70**(1), 199–210.

Lugo, A.E., Rogers, C.S. & Nixon, S.W. (2000) Hurricanes, coral reefs and rainforests: Resistance, ruin and recovery in the Caribbean. *Ambio* **29**(2), 108–14.

Macintyre, I.G., Reid, R.P. & Steneck, R.S. (1996) Growth history of stromatolites in a Holocene fringing reef, Stocking Island, Bahamas. *Journal of Sedimentary Research* **66**(1), 231–42.

Nagelkerken, I., van der Velde, G., Gorissen, M.W. et al. (2000) Importance of mangroves, seagrass beds, and the shallow coral reef as a nursery for important reef fishes, using a visual census technique. *Estuarine, Coastal and Shelf Science* **51**, 31–44.

Newell, N.D., Purdy, E.G. & Imbrie, J. (1960) Bahamas oölitic sand. *Journal of Geology* **68**(5), 481–97.

Peggs, D. (1955) *A Short History of The Bahamas*. The Deans Peggs Research Fund, Nassau, Bahamas.

Ray, G.C. (1998) Bahamian protected areas. Part 1: How it all began. *Bahamas Journal of Science* **6**(1), 2–11.

Reid, R.P., Macintyre, I.G., Browne, K.M. et al. (1995) Modern marine stromatolites in the Exuma Cays, Bahamas: Uncommonly common. *Facies* **33**, 1–18.

Reid, R.P., Macintyre, I.G. & Steneck, R.S. (1999) A microbialite/alga ridge fringing reef complex, Highbourne Cay, Bahamas. *Atoll Research Bulletin* **465**, 1–18.

Sadovy, Y. & Colin, P.L. (1995) Sexual development and sexuality in the Nassau grouper. *Journal of Fish Biology* **46**, 961–76.

Sale, P.F., ed. (2002) *Coral Reef Fishes: Dynamics and Diversity in a Complex Ecosystem*. Academic Press, San Diego, Calif.

Shaklee, R.V. (1996) Tropical cyclone frequency in the Bahamas 1900–1994. *Bahamas Journal of Science* **3**(2), 23–9.

Sluka, R., Chiappone, M., Sullivan, K.M. & Wright, R. (1996) *Habitat and Life in the Exuma Cays, The Bahamas: The Status of Groupers and Coral Reefs in the Northern Cays*. The Nature Conservancy, Florida and Caribbean Marine Conservation Science Center, University of Miami, Coral Gables, Fla.

Smith, C.L. (1972) A spawning aggregation of Nassau grouper, *Epinephelus striatus* (Bloch). *Transactions of the American Fisheries Society* **2**, 257–61.

Smith, S.V. (1978) Coral-reef area and the contributions of reefs to processes and resources of the world's oceans. *Nature* **273**, 225–6.

Tilmant, J.T., Curry, R.W., Jones, R. et al. (1994) Hurricane Andrew's effects on marine resources. The small underwater impact contrasts sharply with the destruction in mangrove and upland-forest communities. *BioScience* **44**(4), 230–7.

Turner, M.S.A. (1999) *Constitutional and Administrative Law in the Bahamas*. Media Publishing, Nassau, Bahamas.

Wilson, S.M. (1997) Introduction to the study of the indigenous people of the Caribbean. In *The Indigenous People of the Caribbean* (ed. S.M. Wilson), pp. 1–8. University Press of Florida, Gainesville, Fla.

Chapter 8 Coastal-realm change

Citations

Alongi, D.M. (1998) *Coastal Ecosystem Processes*. CRC Press, Boca Raton, Fla.

Angel, M.V. (1991) Biodiversity in the oceans. *Ocean Challenge* **2**, 28–36.

Bakun, A. (1996) *Patterns in the Ocean: Ocean Processes and Marine Population Dynamics*. California Sea Grant College System, National Oceanographic and Atmospheric Administration, La Jolla, Calif.

Birkeland, C. (1982) Terrestrial runoff as a cause of outbreaks of *Acanthaster planci* (Echinodermata: Asteroidea). *Marine Biology* **69**, 175–85.

Birkeland, C. (1989) The Faustian traits of the crown-of-thorns. *American Scientist* **77**, 154–63.

Caley, M.J., Carr, M.H., Hixon, M.A. et al. (1996) Recruitment and the local dynamics of open marine populations. *Annual Review of Ecology and Systematics* **27**, 477–500.

Campbell, D.G., Lowell, K.S. & Lightbourn, M. (1991) The effect of introduced hutias (*Geocapromys ingrahami*) on the woody vegetation of Little Wax Cay, Bahamas. *Conservation Biology* **5**(4), 536–41.

Carlton, J.T. (1992) Introduced marine and estuarine mollusks of North America: An end-of-the–20th-century perspective. *Journal of Shellfish Research* **11**(2), 489–505.

Carlton, J.T. (1993) Neoextinctions of marine invertebrates. *American Zoologist* **33**, 499–509.

Carlton, J.T. (1999) The scale and ecological consequences of biological invasions in the world's oceans. In *Invasive Species and Biodiversity Management* (eds. O.T. Sandlund, P.J. Schei & Å. Viken), pp. 195–212. Kluwer Academic Publishers, Dordrecht.

Carlton, J.T. (2000) Global change and biological invasions in the oceans. In *Invasive Species in a Changing World* (eds. H.A. Mooney & R.J. Hobbs), pp. 31–53. Island Press, Washington, D.C.

Carlton, J.T. & Geller, J.B. (1993) Ecological roulette: The global transport of nonindigenous marine organisms. *Science* **261**, 78–82.

Carter, R.W.G. (1988) *Coastal Environments*. Academic Press, London.

Ehrlich, P.R. (1986) Which animal will invade? In *Ecology of Biological Invasions of North America and Hawaii* (eds. H.A. Mooney & J.A. Drake), pp. 79–95. Springer-Verlag, New York.

Elton, C.S. (1958) *The Ecology of Invasions of Animals and Plants*. Methuen & Company, Ltd., London, and John Wiley & Sons, New York.

Farrington, J.W. (1989) Bioaccumulation of hydrophobic organic pollutant compounds. In *Ecotoxicology: Problems and Approaches* (eds. S.A. Levin, M.A. Harwell, J.R. Kelly & K.D. Kimball), pp. 279–313. Springer-Verlag, New York.

Folsom, B.R. & Wood, J.M. (1986) Predictions for the mobility of elements in the estuarine environment. In *Biogeochemical Processes at the Land–Sea Boundary* (eds. P. Lasserre & J.M. Martin), pp. 201–14. Elsevier Science, Amsterdam.

Fowler, S.W. (1982) Biological transfer and transport processes. In *Pollutant Transfer and Transport in the Sea* (ed. G. Kullenberg), pp. 1–65. CRC Press, Boca Raton, Fla.

Frijedl, G. & Wüest, A. (2002) Disrupting biogeochemical cycles – Consequences of damming. *Aquatic Sciences* **64**(1), 55–65.

Gold, J.R. & Richardson, L.R. (1999) Population structure of two species targeted for marine stock enhancement in the Gulf of Mexico. *Bulletin of the National Research Institute for Aquaculture* Suppl. **1**, 79–87.

Goldberg, E.D. (1998) Book reviews: Tributyltin. *Marine Pollution Bulletin* **34**(4), 280.

Golubev, G. (1996) Caspian and Aral Seas: Present state and management problems. Abstract. In *Our Coastal Seas. What is their Future?* (eds. A. Brooks, W. Bell & J. Greer), p. 77. Maryland Sea Grant Publication, College Park, Md.

Gray, J.S. (1993) Eutrophication in the sea. In *Marine Eutrophication and Population Dynamics* (eds. I. Ferrari, V.U. Cechherelli & R. Rossi), pp. 3–15. Olsen & Olsen, Fredensborg, Denmark.

Guo, X., Ford, S.E. & Zhang, F. (1999) Molluscan aquaculture in China. *Journal of Shellfish Research* **18**(1), 19–31.

Hayes, M. (1985) Beach erosion. In *Coasts. Coastal Resources Management: Development Case Studies*, (ed. J.R. Clark), pp. 67–200. Coastal Publication **3**, Renewable Resources Information Series. Research Planning Institute, Inc., Columbia, S. Ca.

Henderson, S. (1998) Sea otter decline. *Marine Pollution Bulletin* **36**(8), 565.

Hilty, J. & Merenlender, A. (2000) Faunal indicator taxa selection for monitoring ecosystem health. *Biological Conservation* **92**, 185–97.

Hixon, M.A. (1998) Population dynamics of coral-reef fishes: Controversial concepts and hypotheses. *Australian Journal of Ecology* **23**, 192–201.

Hixon, M.A. & Brostoff, W.N. (1996) Succession and herbivory: Effects of differential fish grazing on Hawaiian coral-reef algae. *Ecological Monographs* **66**(1), 67–90.

Hope, B., Scatolini, S., Titus, E. & Coter, J. (1997) Distribution patterns of polychlorinated biphenyl congeners in water, sediment and biota from Midway Atoll (North Pacific Ocean). *Marine Pollution Bulletin* **34**(7), 548–63.

Hughes, T.P. (1994a) Catastrophes, phase shifts, and large-scale degradation of a Caribbean coral reef. *Science* **265**, 1547–51.

Hughes, T.P. (1994b) Coral reef catastrophe: Response. *Science* **266**, 1932–3.

Irwin, R.J., VanMouwerik, M., Stevens, L. et al. (1997) *Environmental Contaminants Encyclopedia*. National Park Service, Water Resources Division, Fort Collins, Colo. Distributed within the federal government as an electronic document.

Jackson, J.B.C. (1994) Community unity? *Science* **264**, 1412–13.

Jennings, S., Kaiser, M.J. & Reynolds, J.D. (2001) *Marine Fisheries Ecology*. Blackwell Science, Oxford.

Kan-Atireklap, S., Tanabe, S.S. & Sanguansin, J. (1997) Contamination by butyltin compounds in sediments from Thailand. *Marine Pollution Bulletin* **34**(11), 894–9.

Kelly, J.R. & Harwell, M.A. (1989) Indicators of ecosystem response and recovery. In *Ecotoxicology: Problems and Approaches* (eds. S.A. Levin, M.A. Harwell, J.R. Kelly & K.D. Kimball), pp. 9–35. Springer-Verlag, New York.

Kennish, M.J. (1998) *Pollution Impacts on the Marine Biotic Communities*. CRC Press, Boca Raton, Fla.

Knowlton, N. (1992) Thresholds and multiple states in coral reef community dynamics. *American Zoologist* **32**, 674–82.

Kohane, M.J. & Parsons, P.A. (1988) Domestication: Evolutionary change under stress. In *Evolutionary Biology* **23** (eds. M.K. Hecht & B. Wallace), pp. 31–48. Plenum Press, New York.

Lachner, E.A., Robins, C.R. & Courtenay, W.R., Jr. (1970) Exotic fishes and other aquatic organisms introduced into North America. *Smithsonian Contributions to Zoology* **59**, 1–29.

Lalli, C.M. & Parsons, T.R. (1997) *Biological Oceanography. An Introduction*. Second edition. The Open University, Oxford.

Langton, R.W. & Auster, P.J. (1999) Marine fishery and habitat interactions: To what extent are fisheries and habitat interdependent? *Fisheries* **24**(6), 14–21.

Larsen, P.F., Huggett, R.J. & Unger, M.A. (1998) Assessment of organotin in waters of selected Gulf of Maine estuaries. *Marine Pollution Bulletin* **34**(10), 802–4.

Law, R.J., Blake, S.J., Jones, B.R. & Rogan, E. (1998) Organotin compounds in liver tissue of harbour porpoises (*Phocoena phocoena*) and grey seals (*Halichoerus grypus*) from the coastal waters of England and Wales. *Marine Pollution Bulletin* **36**(3), 241–7.

Lessios, H.A., Robertson, D.R. & Cubit, J.D. (1984) Spread of *Diadema* mass mortality through the Caribbean. *Science* **226**, 335–7.

Mayr, E. (1991) *One Long Argument*. Harvard University Press, Cambridge, Mass.

Meyer, J.L. & Schultz, E.T. (1985) Migrating haemulid fishes as a source of nutrients and organic matter on coral reefs. *Limnology and Oceanography* **30**(1), 146–56.

Milliman, J.D. (1992) Management of the coastal zone: Impact of onshore activities on the coastal environment. In *Use and Misuse of the Seafloor* (eds. K.J. Hsü & J. Thiede), pp. 213–27. John Wiley & Sons, Chichester.

Mlot, C. (1997) The rise in toxic tides. *Science News* **152**, 202–4.

Morgan, E., Murphy, J. & Lyons, R. (1998) Imposex in *Nucella lapillus* from TBT contamination in South and Southwest Wales: A continuing problem around ports. *Marine Pollution Bulletin* **36**(10), 840–3.

MPB (1998) News: Mercury and other toxins contaminate fish. *Marine Pollution Bulletin* **36**(12), 937–8.

Nichols, F.H., Cloern, J.E., Luoma, S.N. & Peterson, D.H. (1986) The modification of an estuary. *Science* **231**, 567–73.

Nichols, M.M. (1988) Consequences of dredging. In *Hydrodynamics of Estuaries*. **II**. *Estuarine Case Studies* (ed. B. Kjerfve), pp. 89–99. CRC Press, Boca Raton, Fla.

NRC (1990) *Managing Coastal Erosion*. Committee on Coastal Erosion Zone Management, National Research Council. National Academy Press, Washington, D.C.

NRC (1995) *Understanding Marine Biodiversity*. Committee on Biological Diversity in Marine Systems, National Research Council, Washington, D.C.

NRC (1999) *Sustaining Marine Fisheries*. National Research Council. National Academy Press, Washington, D.C.

NRC (2002) *Oil in the Sea III: Inputs, Fates, and Effects*. Committee on Oil in the Sea, National Research Council. Prepublication version online: www.nationalacademies.org/publications. National Academy Press, Washington, D.C.

Nriagu, J.O. (1993) Legacy of mercury pollution. *Nature* **363**, 589.

Ogden, J.C. (1988) The influence of adjacent systems on the structure and function of coral reefs. *Proceedings of the Sixth International Coral Reef Symposium* **1**, 123–9.

Ogden, J.C. (1994) Coral reef catastrophe. *Science* **266**, 1931.

Ogden, J.C. (1997) Ecosystem interactions in the tropical coastal seascape. In *Life and Death of Coral Reefs* (ed. C. Birkeland), pp. 288–97. Chapman & Hall, London.

Mrosovsky, V. (1990) *Rheostasis. The Physiology of Change*. Oxford University Press, New York.

O'Neill, R.V. (2001) Is it time to bury the ecosystem concept? (with full military honors, of course!). *Ecology* **82**(12), 3275–84.

Patterson, R.G., Burby, R.J. & Nelson, A.C. (1991) Sewering the coast: Bane or blessing to marine water quality. *Coastal Management* **19**, 239–52.

Pauly, D. & Christensen, V. (1995) Primary production required to sustain global fisheries. *Nature* **374**, 255–7.

Pauly, D., Christensen, V., Dalsgaad, J. et al. (1998) Fishing down marine food webs. *Science* **279**, 860–3.

Pauly, D., Christensen, V., Foese, R. & Palomares, M.L.L. (2000) Fishing down aquatic food webs. *American Scientist* **88**, 46–51.

Peterson, C.H. & Estes, J.A. (2001) Conservation and management of marine communities. In *Marine Community Ecology* (eds. M.D. Bertness, S.D. Gaines & M.E. Hay), pp. 469–507. Sinauer Associates, Inc., Sunderland, Mass.

Pilkey, O.H. & Dixon, K.L. (1996) *The Corps and the Shore*. Island Press, Washington, D.C.

Pimentel, D. & Levitan, L. (1986) Pesticides. *BioScience* **36**(2), 86–91.

Randall, J.E. (1974) The effect of fishes on coral reefs. *Proceedings of the Second International Coral Reef Symposium* **1**, 159–66.

Rapport, D., Costanza, R., Epstein, P.R. et al., eds. (1998) *Ecosystem Health*. Blackwell Science, Boston, Mass.

Ruiz, G.M., Rawlings, T.K., Dobbs, F.C. et al. (2000) Global spread of microorganisms by ships. *Science* **408**, 49.

Schubel, J.R. & Hirschberg, D.J. (1982) The Chang Jiang (Yangtze) estuary: Establishing its place in the community of estuaries. In *Estuarine Comparisons* (ed. V.S. Kennedy), pp. 649–66. Academic Press, New York.

Selye, H. (1976) *The Stress of Life*. Revised edition. McGraw-Hill, New York.

Sepúlveda, M.S., Ochoa-Acuña, H. & Sundlof, S.F. (1997) Heavy metal concentrations in Juan Fernández fur seals (*Arctocephalus philippii*). *Marine Pollution Bulletin* **34**(8), 663–5.

Shapiro, H.A. (1988) The landfilled coast of Japan's Inland Sea. In *Ocean Yearbook* **7** (eds. E.M. Borgese, N. Ginsbury & J.R. Morgan), pp. 294–316. University of Chicago Press, Chicago, Ill.

Sissenwine, M.P. & Cohen, E.B. (1991) Resource productivity and fisheries management of the Northeast Shelf Ecosystem. In *Food Chains, Yields, Models and Management of Large Marine Ecosystems* (eds. K. Sherman, L.M. Alexander & B.D. Gold), pp. 107–21. Westview Press, Boulder, Colo.

Smodlaka, N., Malone, T.C., Malej, A. & Harding, L.W., Jr. (1999) Introduction. In *Ecosystems at the Land–Sea Margin: Drainage Basin to Coastal Sea* (eds. T.C. Malone, A. Malej, L.W. Harding, Jr. et al.) pp. 1–6. American Geophysical Union, Washington, D.C.

Steele, J.H. (1991) Marine functional diversity. *BioScience* **41**(7), 470–4.

Steele, J.H. & Schumacher, M. (2000) Ecosystem structure before fishing. *Fisheries Research* **44**, 201–5.

Stevenson, C.H. (1892) The oyster industry of Maryland. *Bulletin of the United States Fish Commission* **12**, 203–97.

Tanabe, S. (1999) Butyltin contamination in marine mammals – A review. *Marine Pollution Bulletin* **39**(1–12), 62–72.

Tang, Q. (1993) Effects of long-term physical and biological perturbations on the contemporary biomass yields of the Yellow Sea ecosystem. In *Large Marine Ecosystems. Stress, Mitigation, and Sustainability* (eds. K. Sherman, L.M. Alexander & B.D. Gold), pp. 79–93. American Association for the Advancement of Science, Washington, D.C.

Tilman, D. & Kareiva, P., eds. (1997) *Spatial Ecology. The Role of Space in Population Dynamics and Interspecific Interactions.* Princeton University Press, Princeton, N.J.

UNEP (1982) *GESAMP: The Health of the Oceans.* Joint Group of Experts on the Scientific Aspects of Marine Pollution. United Nations Environment Programme, UNEP Regional Seas Reports and Studies **16**.

Walker, H.J. (1990) The coastal zone. In *The Earth as Transformed by Human Action. Local and Regional Changes in the Biosphere over the Past 300 Years* (eds. B.L. Turner, II, W.C. Clark, R.W. Kates et al.), pp. 271–94. Cambridge University Press, Cambridge.

Weinberg, G.M. (1975) *An Introduction to General Systems Thinking.* John Wiley & Sons, New York.

Whitfield, P., Gardner, T., Vives, S.P. et al. (2002) Biological invasion of the Indo-Pacific lionfish (*Pterois volitans*) along the Atlantic coast of North America. *Marine Ecology Progress Series* **235**, 289–97.

Wiggerts, H. (1981) Coastal management in the Netherlands. In *Comparative Marine Policy*, pp. 75–81. Praeger Special Studies, Center for Ocean Management Studies. University of Rhode Island, New York.

Zaitsev, Y. & Mamaev, V. (1997) *Marine Biological Diversity in the Black Sea. A Study of Change and Decline.* United Nations Development Programme, New York.

Suggested readings

Bahr, L.M. Jr. & Lanier, W.P. (1981) *The Ecology of Intertidal Oyster Reefs of the South Atlantic Coast: A Community Profile.* U.S. Fish and Wildlife Service, Office of Biological Services, FWS/OBS-81/15. Department of the Interior, Washington, D.C.

Barrett, G.W., van Dyne, G.M. & Odum, E.P. (1976) Stress ecology. *BioScience* **26**(3), 192–4.

Baumard, P., Budzinski, H., Garrigues, P., Sorbe, J.C., Burgeot, T. & Bellocq, J. (1998) Concentrations of PAHs (Polycyclic Aromatic Hydrocarbons) in various marine organisms in relation to those in sediments and to trophic level. *Marine Pollution Bulletin* **36**(12), 951–60.

Botsford, L.W., Castilla, J.C. & Peterson, C.H. (1997) The management of fisheries and marine ecosystems. *Science* **277**, 509–14.

Caddy, J.F. (1993) Toward a comparative evaluation of human impacts on fishery ecosystems of enclosed and semi-enclosed seas. *Reviews in Fisheries Science* **1**(1), 57–95.

Capuzzo, J.M. (1996) The bioaccumulation and biological effects of lipophilic organic contaminants. In *The Eastern Oyster Crassostrea Virginica* (eds. V.S. Kennedy, R.I.E. Newell & A.F. Eble), pp. 539–57. Maryland Sea Grant book, College Park, Md.

Carriker, M.R. (1992) Introductions and transfers of molluscs: Risk considerations and implications. *Journal of Shellfish Research* **11**(2), 507–10.

Courtney, W.R., Jr. & Robins, C.R. (1989) Fish introductions: Good management, mismanagement, or no management. *Aquatic Science* **1**(1), 159–72.

Darnell, R.M. (1992) Ecological history, catastrophism, and human impact on the Mississippi/Alabama continental shelf and associated waters: A review. *Gulf Research Reports* **8**(4), 375–86.

Dynesius, M. & Nilsson, C. (1994) Fragmentation and flow regulation of river systems in the northern third of the world. *Science* **266**, 753–62.

Farley, A. (1992) Mass mortalities and infectious lethal disease in bivalve molluscs and associations with geographic transfers of populations. In *Dispersal of Living Organisms into Aquatic Ecosystems* (eds. A. Rosenfield & R. Mann), pp. 139–54. Maryland Sea Grant, College Park, Md.

Hall, J.A., Frid, C.L.J. & Gill, M.E. (1997) The response of estuarine fish and benthos to an increasing discharge of sewage effluent. *Marine Pollution Bulletin* **34**(7), 527–35.

Hall, S.J. (1999) *The Effects of Fishing on Marine Ecosystems and Communities.* Blackwell Science, Oxford.

Hallerman, E.M. & Kapuscinski, A.R. (1990) Transgenic fish and public policy: Regulatory concerns. *Fisheries* **15**(1), 12–20.

Hilborn, R. (1990) Marine biota. In *The Earth as Transformed by Human Action. Local and Regional Changes in the Biosphere over the Past 300 Years* (eds. B.L. Turner, II, W.C. Clark, R.W. Kates et al.), pp. 371–85. Cambridge University Press, Cambridge.

Holmes, B. (1994) Biologists sort the lessons of fisheries collapse. *Science* **264**, 1252–3.

Jousson, O., Pawlowski, J., Zaninetti, L. et al. (2000) Invasive alga reaches California. *Nature* **408**, 157–8.

Levin, S.A., Harwell, M.A., Kelly, J.R. & Kimball, K.D., eds. (1989) *Ecotoxicology: Problems and Approaches.* Springer-Verlag, New York.

May, R.M., Beddington, J.R., Clark, C.W. et al. (1979) Management of multispecies fisheries. *Science* **20**(4403), 267–77.

McLachlan, J.A. & Korach, K.S. (1995) Symposium on estrogens in the environment, III. *Environmental Health Perspectives* **103**(suppl. 7), 3–4.

Moore, M.N. (1981) Elemental accumulation in organisms and food chains. In *Marine Ecosystems* (ed. A.R. Longhurst), pp. 525–64. Academic Press, London.

Munford, J.G. & Baxter, J.M. (1991) Conservation and aquaculture. *In Aquaculture and the Environment* (eds. N. De Pauw & J. Joyce), pp. 279–98. Special Publication **16**. European Aquaculture Society, Gent, Belgium.

Nee, S. & May, R.M. (1997) Extinction and the loss of evolutionary history. *Science* **278**, 692–4.

NRC (1992) *Marine Aquaculture. Opportunities for Growth.* Committee on Assessment of Technology and Opportunities for Marine Aquaculture in the United States, National Research Council. National Academy Press, Washington, D.C.

NRC (1999) *Hormonally Active Agents in the Environment.* National Research Council. National Academy Press, Washington, D.C.

Páez-Osuna, F., Guerrero-Galván, S.R. & Ruiz-Fernández, A.C. (1998) The environmental impact of shrimp aquaculture and the coastal pollution in Mexico. *Marine Pollution Bulletin* **36**(1), 65–75.

Pauly, D. (1995) Anecdotes and the shifting baseline syndrome of fisheries. *Trends in Ecology and Evolution* **10**(10), 430.

Pullin, R.S.V. (1994) Exotic species and genetically modified organisms in aquaculture and enhanced fisheries: ICLARM's position. *NAGA, International Center for Living Aquatic Resources Management Q* **17**(4), 19–24.

Ramakrishnan, P.S. & Vitousek, P.M. (1989) Ecosystem-level processes and the consequences of biological invasions. In *Biological Invasions: A Global Perspective* (eds. J.A. Drake, H.A. Mooney, F. di Castri et al.), pp. 281–300. John Wiley & Sons, Chichester.

Ray, G.C. (1996) Coastal-marine discontinuities and synergisms: Implications of biodiversity and conservation. *Biodiversity and Conservation* Special Issue **5**, 1095–108.

Ray, G.C., Hayden, B.P., Bulger, A.J., Jr. & McCormick-Ray, M.G. (1992) Effects of global warming on the biodiversity of coastal-marine zones. In *Global Warming and Biological Diversity* (eds. R.L. Peters & T.E. Lovejoy), pp. 91–104. Yale University Press, New Haven, Conn.

Reiners, W.A. (1983) Disturbance and basic properties of ecosystem energetics. In *Disturbance and Ecosystems* (eds. H.A. Mooney & M. Godron), pp. 83–98. Springer-Verlag, New York.

Richards, W.J. & Bohnsack, A.A. (1992) The Caribbean Sea. A large marine ecosystem in stress. In *Large Marine Ecosystems. Patterns, Processes and Yields* (eds. K. Sherman, L.M. Alexander & B.D. Gold), pp. 44–53. American Association for the Advancement of Science, Washington, D.C.

Rusnak, G.A. (1967) Rates of sediment accumulation in modern estuaries. In *Estuaries* **83** (ed. G.H. Lauff), pp. 180–4. American Association for the Advancement of Science, Washington, D.C.

Scheffer, M., Carpenter, S., Foley, J.A. et al. (2001) Catastrophic shifts in ecosystems. *Nature* **413**, 591–6.

Sheppard, C. (1995) The shifting baseline syndrome. *Marine Pollution Bulletin* **30**(12), 766–7.

Simenstad, C.A., Estes, J.A. & Kenyon, K.W. (1978) Aleuts, sea otters, and alternate stable-state communities. *Science* **200**, 403–11.

Smith, R.I.L. (1988) Destruction of Antarctic terrestrial ecosystems by a rapidly increasing fur seal population. *Biological Conservation* **45**(1), 55–72.

Soule, D.F. & Kleppel, G.S., eds. (1988) *Marine Organisms as Indicators*. Springer-Verlag, New York.

Stachowicz, J.J., Whitlatch, R.B. & Osman, R.W. (1999) Species diversity and invasion resistance in a marine ecosystem. *Science* **286**, 1577–9.

Steele, J.H. (1996) Regime shifts in fisheries management. *Fisheries Research* **25**, 19–23.

Steneck, R.S. & Carlton, J.T. (2001) Human alterations of marine communities. In *Marine Community Ecology* (eds. M.D. Bertness, S.D. Gaines & M.E. Hay), pp. 445–68. Sinauer Associates, Inc., Sunderland, Mass.

Vitousek, P.M., Mooney, H.A., Lubchenco, J. & Melillo, J.M. (1997) Human domination of Earth's ecosystems. *Science* **277**, 494–9.

Vörösmarty, C.J. & Sahagian, D. (2000) Anthropogenic disturbance of the terrestrial water cycle. *BioScience* **50**(9), 753–65.

Webster, P.J. & Palmer, T.N. (1997) The past and the future of El Niño. *Nature* **390**, 562–4.

World Resources Institute (2000) UNEP Regional Seas Programme, 2000. Online: http://www.wri.org/.

Young, G.J., Dooge, J.C.I. & Rodda, J.C. (1994) *Global Water Resource Issues*. Cambridge University Press, Cambridge.

Zaitsev, Y. & Mamaev, V. (1997) *Marine Biological Diversity in the Black Sea: A Study of Change and Decline*. United Nations Development Program, New York.

Zaletaev, V.S. (1996) Modern regression of the Aral Sea: Desertification of former sea beds and problems of preventing ecological catastrophes. Abstract. In *Our Coastal Seas. What is their Future?* (eds. A. Brooks, W. Bell & J. Greer), p. 95. Maryland Sea Grant Publication, College Park, Md.

Chapter 9 Synthesis

Citations

Adams, A.B., ed. (1962) *First World Conference on National Parks*. U.S. Department of the Interior, Washington, D.C.

Agee, J. and Johnson, D. (1988) *Ecosystem Management for Parks and Wilderness*. University of Washington Press, Seattle.

Anderson, P.J. & Piatt, J.F. (1999) Community reorganization in the Gulf of Alaska following ocean climate regime shift. *Marine Ecology Progress Series* **189**, 117–23.

Barrett, G.W., & Odum, E.P. (2001) The twenty-first century: The world at carrying capacity. *BioScience* **50**(4), 363–8.

Batisse, M. (1990) Development and implementation of the biosphere reserve concept and its applicability to coastal regions. *Environmental Conservation* **17**, 111–16.

Beatley, T. (1994) *Ethical Land Use*. Johns Hopkins University Press, Baltimore, Md.

Belsky, M.H. (1984) Environmental policy law in the 1980s: Shifting back the burden of proof. *Ecology Law Quarterly* **12**(1), 1–88.

Belsky, M.H. (1989) The ecosystem model mandate for a comprehensive United States ocean policy and law of the sea. *San Diego Law Review* **26**(3), 417–95.

Björklund, M.I. (1974) Achievements in marine conservation, I. Marine parks. *Environmental Conservation* **1**(3), 205–23.

Bormann, F.H. (1990) Viewpoint. The global environment deficit. *BioScience* **40**(2), 74.

Bormann, F.H. (2000) On respect for nature. *Northern Rockies Conservation Cooperative News* **13**, 4–5.

Carr, M.H., Neigel, J.E., Estes, J.A. et al. (2003) Comparing marine and terrestrial ecosystems: implications for the design of coastal marine reserves. *Ecological Applications*, **13**(1), 590–1.

Carson, R. (1962) *Silent Spring*. Houghton Mifflin Company, Boston, Mass.

Carson, R.L. (1951) *The Sea Around Us*. Oxford University Press, New York.

Christensen, N.L., Bartuska, A.M., Brown, J.H. et al. (1996) The report of the Ecological Society of America Committee on the scientific basis for ecosystem management. *Ecological Applications* **6**(3), 665–91.

CMSER (1969) *Our Nation and the Sea. A Plan for National Action.* Commission on Marine Science, Engineering and Resources. Superintendent of Documents, U.S. Government Printing Office, Washington, D.C.

Cousteau, J.-Y. & Dumas, F. (1953) *The Silent World.* Harper and Brothers Publishers, New York.

Daly, H.E. (1993) Sustainable growth: An impossibility theorem. In *Valuing the Earth* (eds. H. Daly & K.N. Townsend), pp. 1–25. The MIT Press, Cambridge, Mass.

Daly, H.E. & Cobb, J.B., Jr. (1989) *For the Common Good: Redirecting the Economy toward Community, the Environment, and a Sustainable Future.* Beacon Press, Boston, Mass.

Dasmann, R.F. (1968) *A Different Kind of Country.* The MacMillan Company, New York and Collier-MacMillan Limited, London.

Dasmann, R.F. (1971) *No Further Retreat.* The MacMillan Company, New York and Collier-MacMillan Limited, London.

Dayton, P.K. (1998) Reversal of the burden of proof in fisheries management. *Science* **279**, 821–2.

Dobbs, D. (2000) *The Great Gulf: Fishermen, Scientists, and the Struggle to Revive the World's Greatest Fishery.* Island Press, Washington, D.C.

EPAP (1999) *Ecosystem-Based Fisher Management. A Report to Congress by the Ecosystem Principles Advisory Panel.* U.S. Department of Commerce, National Marine Fisheries Service.

Elliott, L. (1998) *The Global Politics of the Environment.* New York University Press, New York.

Estes, J.A., Tinker, M.T., Williams, T.M. & Doak, D.F. (1998) Killer whale predation on sea otters linking coastal with oceanic ecosystems. *Science* **282**, 473–6.

Forest Ecosystem Management Team (FEMAT) (1993) *Forest Ecosystem Management: An Ecological, Economic, and Social Assessment.* Joint Publication of the United States Department of Agriculture, Forest Service; United States Department of Commerce, National Oceanic and Atmospheric Administration and National Marine Fisheries Service; United States Department of Interior, Bureau of Land Management, Fish and Wildlife Service, and National Park Service; and United States Environmental Protection Agency, Washington, D.C.

Fiorino, D.J. (1995) *Making Environmental Policy.* University of California Press, Berkeley, Calif.

Folke, C., Jansson, A., Larsson, J. & Costanza, R. (1997) Ecosystem appropriation by cities. *Ambio* **25**(3), 167–72.

Franklin, J.F. (1993) Preserving biodiversity: Species, ecosystems, or landscape? *Ecological Applications* **3**, 202–5.

Fujiwara, M. & Caswell, H. (2001) Demography of the endangered North Atlantic right whale. *Nature* **414**, 537–41.

Gerber, L.R., DeMaster, D.P. & Roberts, S.P. (2000) Measuring success in conservation. *American Scientist* **88**(4), 316–24.

Goodland, R. & Daly, H. (1996) Environmental sustainability: Universal and non-negotiable. *Ecological Applications* **6**(4), 1002–17.

Gottgens, J.F., Perry, J.E., Fortney, R.H. et al. (2001) The Parana–Paraguay Hidrovia: Protecting the Pantanal with lessons from the past. *BioScience* **51**(4), 559–70.

Grumbine, R.E. (1994) *Environmental Policy and Biodiversity.* Island Press, Washington, D.C.

Harrison, D. (1994) Tourism, capitalism and development in less developed countries. In *Capitalism and Development* (ed. L. Sklair), pp. 232–57. Routledge, London.

Haeuber, R. & Franklin, J. (1996) Forum: Perspectives on ecosystem management. *Ecological Applications* **6**(3), 692–3.

Hixon, M.A., Boersina, P.D., Hunter, M.L. Jr. et al. (2001) Ocean at risk: Research priorities in marine conservation biology. In *Conservation Biology: Research Priorities for the Next Decade* (eds. M.E. Soul & G.H. Orvans). Island Press, Washington, D.C.

Holling, C.S. & Meffe, G.K. (1996) Command and control and the pathology of natural resources management. *Conservation Biology* **10**(2), 328–37.

IEMTF (1995) *The Ecosystem Approach: Healthy Ecosystems and Sustainable Economics.* Interagency Ecosystem Management Task Force. National Technical Information Service **I**: PB95-265583. U.S. Department of Commerce, Springfield, Va.

IUCN (1976) *Proceedings of an International Conference on Marine Parks and Reserves*, Tokyo, Japan, 1975. International Union for the Conservation of Nature and Natural Resources Publication New Series **37**. IUCN, Morges, Switzerland.

IUCN (1978) Marine programme report. *IUCN Bulletin*, New Series, **9**(1–2), 1–16.

Jackson, J.B.C., Kirby, M.X., Berger, W.H. et al. (2001) Historical overfishing and the recent collapse of coastal ecosystems. *Science* **293**, 629–38.

Kareiva, P. (2001) When on whale matters. *Nature* **414**, 493–4.

Kelleher, G., Bleakly, C. & Wells, S., eds. (1995) *A Global Representative System of Marine Protected Areas* **1–4**. Great Barrier Reef Marine Park Authority, the World Bank, and the World Conservation Union, Washington, D.C.

Ketchum, B.K., ed. (1972) *The Waters' Edge: Critical Problems of the Coastal Zone.* The MIT Press, Cambridge, Mass.

Keystone Center (1996) *The Keystone National Policy Dialogue on Ecosystem Management.* Final Report, the Keystone Center, Colorado.

Kullenberg, G.E.B. (1999) The ocean challenge. *Natural Resources Forum* **23**, 99–103.

Kurlansky, M. (1997) *Cod: A Biography of the Fish that Changed the World.* Penguin Books, New York.

Larkin, P.A. (1977) An epitaph for the concept of maximum sustainable yield. *Transactions of the American Fisheries Society* **106**(1), 1–11.

Leopold, A. (1950) *A Sand County Almanac and Sketches Here and There.* Oxford University Press, New York.

Lovelock, J.E. (1979) *Gaia: A New Look at Life on Earth.* Oxford University Press, Oxford.

Ludwig, D., Hilborn, R. & Walters, C. (1993) Uncertainty, resource exploitation, and conservation: Lessons from history. *Science* **260**, 17 & 36.

Meeker, J.W. (1980) *The Comedy of Survival. In Search of an Environmental Ethic.* Guild of Tutors Press, Los Angeles, Calif.

Miller, M. (1995) *The Third World in Global Environmental Politics.* Lynne Reinner, Boulder, Colo.

Moore, J.A. (1984) Science as a way of knowing. *American Zoologist* **24**(2), 421–31.

Marsh, G.P. (1864) *Man and Nature: Or Physical Geography as Modified by Human Action.* Republished 1965 by Belknap Press of Harvard University Press, Cambridge, Mass.

Mayr, E. (1997) *This is Biology.* Belknap Press of Harvard University Press, Cambridge, Mass.

NAS (1996) *Upstream: Salmon and Society in the Pacific Northwest.* National Research Council Committee on Protection and Management of Pacific Northwest Anadromous Salmonids. National Academy Press, Washington, D.C.

Norse, E.A., Rosenbaum, K.L., Wilcove, D.S. (1986) *Conserving Biological Diversity in Our National Forests.* Wilderness Society, Washington, D.C.

NRC (2000) *Marine Protected Areas: Tools for Sustaining Ocean Ecosystems.* National Research Council. National Academy Press, Washington, D.C.

Odum, E.P. (1989) *Ecology and Our Endangered Life-Support Systems.* Sinauer Associates, Inc., Sunderland, Mass.

Odum, W. (1982) Tyranny of small decisions. *BioScience* **32**(9), 728–9.

Osborn, F. (1948) *Our Plundered Planet.* Little, Brown, & Co., New York.

Osborn, F. (1953) *The Limits of the Earth.* Little, Brown, & Co., New York.

Overbay, J.C. (1992) Ecosystem management. In *Taking an Ecological Approach to Management.* United States Department of Agriculture Forest Service Publication Wo-Wsa-3.

Pauly, D., Christensen, V., Guénette, S. et al. (2002) Towards sustainability in world fisheries. *Nature* **418**, 689–95.

Peterson, C.H. & Estes, J.A. (2001) Conservation and management of marine communities. In *Marine Community Ecology* (eds. M.D. Bertness, S.D. Gaines & M.E. Hay) pp. 469–507. Sinauer Associates, Inc. Sunderland, Massachusetts.

Prescott-Allen, R. (2001) *The Wellbeing of Nations.* Island Press, Washington, D.C.

Ray, C. (1970) Ecology, law, and the "marine revolution." *Biological Conservation* **3**(1), 7–17.

Ray, G.C. (1976) Critical marine habitats. Definition, description, criteria and guidelines for identification and management. In *Proceedings of an International Conference on Marine Parks and Reserves*, Tokyo, Japan, 1975, pp. 15–59. IUCN Publication New Series **37**. IUCN, Morges, Switzerland.

Ray, G.C. (1998) Ecological diversity in coastal zones and oceans. In *Biodiversity* (eds. F.O. Wilson & F. Riter), pp. 36–50. National Academy Press, Washington, D.C.

Ray, G.C. (1999) Coastal-marine protected areas: Agonies of choice. *Aquatic Conservation: Marine and Freshwater Ecosystems* **9**, 607–14.

Roberts, C.M. (1997) Ecological advice for the global fisheries crisis. *Trends in Ecology and Evolution* **12**(1), 35–8.

Rolls, K., Brownell, R.L. & Ballon, J. (1983) Genetic diversity in California sea otters: Theoretical considerations and management implications. *Biological Conservation* **25**, 209–32.

Rykiel, E.J., Jr. (2001a) What is, what might be, and what ought to be. *BioScience* **51**(6), 423.

Rykiel, E.J., Jr. (2001b) Scientific objectivity, value systems, and policymaking. *BioScience* **51**(6), 433–6.

Schevill, W.E., Ray, G.C. & Norris, K.S., eds. (1974) *The Whale Problem.* Harvard University Press, Cambridge, Mass.

Sherman, K. & Duda, A.M. (1999) An ecosystem approach to global assessment and management of coastal waters. *Marine Ecology Progress Series* **190**, 271–87.

Soulé, M.L. (1985) What is conservation biology? *BioScience* **35**(11), 727–34.

Soulé, M.L. (1986) *Conservation Biology: The Science of Scarcity and Diversity.* Sinauer Associates, Inc., Sunderland, Mass.

Steele, J.H. (1985) A comparison of terrestrial and marine ecological systems. *Nature* **313**, 355–8.

UNCED (1993) *Earth Summit. Agenda 21.* The United Nations Programme of Action from Rio. United Nations, New York.

Weber, M.L. (2002) *From Abundance to Scarcity: A History of U.S. Marine Fisheries Policy.* Island Press, Washington, D.C.

Wilson, E.O. & Peter, F., eds. (1988) *BioDiversity.* National Academy Press, Washington, D.C.

Wood, C.A. (1994) Ecosystem management: Achieving the new land ethic. *Renewable Natural Resources Journal* **12**, 6–12.

WWF (1976) The seas must live. In *World Wildlife Fund: Conservation Programme 1977/78*, pp. 6–17. World Wildlife Fund, Morges, Switzerland.

Suggested readings

Allison, G.W., Lubchenco, J. & Carr, M.A. (1998) Marine reserves are necessary but not sufficient for marine conservation. *Ecological Applications* Suppl. **8**(1), S79–S92.

Boulding, K.E. (1991) What do we want to sustain? Environmentalism and human evaluations. In *Ecological Economics. The Science and Management of Sustainability* (ed. R. Costanza), pp. 22–31. Columbia University Press, New York.

Brack, D. (1996) *International Trade and the Montreal Protocol.* The Brookings Institute, Washington, D.C.

Burbridge, P. (2001) Sustainability and human use of coastal systems. *Land-Ocean Interactions in the Coastal Zone (LOICZ) Newsletter* **21**, 3–5.

Costanza, R. (1996) Ecological economics: Reintegrating the study of humans and nature. *Ecological Applications* **6**(4), 978–90.

Costanza, R., d'Arge, R., de Groot, R. et al. (1997) The value of the world's ecosystem services and natural capital. *Nature* **387**, 253–60.

Daily, G., Dasgupta, P., Bolin, B. et al. (1998) Food production, population growth, and the environment. *Science* **281**, 1291–2.

Daily, G.C., Söderqvist, T., Aniyar, S. et al. (2000) The value of nature and the nature of value. *Science* **289**, 395–6.

Daily, G.C. & Walker, B.H. (2000) Seeking the great transition. *Nature* **403**, 243–5.

Farrow, R.S. (1990) *Managing the Outer Continental Shelf Lands.* Taylor & Francis, New York.

Finn, D.P. *Managing the Ocean Resources of the United States: The Role of the Federal Marine Sanctuaries Program.* Springer-Verlag, Berlin.

Folke, C., Holling, C.S. & Perrings, C. (1996) Biological diversity, ecosystems, and the human scale. *Ecological Applications* **6**(4), 1018–24.

Folke, C., Kautsky, N. & Troell, M. (1994) The costs of eutrophication from salmon farming: Implications for policy. *Journal of Environmental Management* **40**, 173–82.

Ford, J. & Martinez, D. (2000) Invited feature: Traditional ecological knowledge, ecosystem science, and environmental management. *Ecological Applications* **10**(5), 1249–50.

Foster, K.R. Vecchia, P. & Repacholi, M.H. (2000) Science and the precautionary principle. *Science* **288**, 979–81.

Galbraith, J. (1958) How much should a country consume? In *Perspectives on Conservation* (ed. H. Jarrett). Johns Hopkins University Press, Baltimore, Md.

Gunderson, L.H., Holling, C.S. & Light, S.S., eds. (1995) *Barriers and Bridges to the Renewal of Ecosystems and Institutions*. Columbia University Press, New York.

Hempel, L.C. (1995) *Environmental Governance: The Global Challenge*. Island Press, Washington, D.C.

Holling, C.S., ed. (1978) *Adaptive Environmental Assessment and Management*. John Wiley & Sons, Chichester.

Inman, D.L. & Brush, B.M. (1973) The coastal challenge. Fragile ribbons which border our land require more understanding, new technology, and resolute planning. *Science* **181**, 20–31.

Johannes, R.E. (1982) Traditional conservation methods and protected marine areas in Oceania. *Ambio* **11**(5), 258–61.

Johannes, R.E. (1988) The case for data-less marine resource management: Examples from tropical nearshore fisheries. *Trends in Ecology and Evolution* **13**(6), 243–6.

Levy, J.P. (1999) A democratic approach to ocean governance. *Natural Resources Forum* **23**, 115–21.

Litfin, K.T. (1998) The greening of sovereignty. In *The Greening of Sovereignty in World Politics* (ed. K.T. Litfin), pp. 1–27. The MIT Press, Cambridge, Mass.

Livingston, R.J. (2001) *Eutrophication Processes in Coastal Systems*. CRC Press, Boca Raton, Fla.

Maltby, E. (1997) Ecosystem management: The concept and the strategy. *World Conservation* **3**, 3–24.

Margules, C.R. & Pressey, R.L. (2000) Systematic conservation planning. *Nature* **405**, 243–53.

McCaffrey, S.C. (1993) Water, politics, and international law. In *Water in Crisis. A Guide to the World's Fresh Water Resources* (ed. P.H. Gleick), pp. 92–104. Oxford University Press, New York.

McNeely, J.A. (1999) Strange bed fellows: Why science and policy don't mesh and what can be done about it. In *The Living Planet in Crisis: Biodiversity Science and Policy* (eds. J. Cracraft & F.T. Grifo), pp. 275–86. Columbia University Press, New York.

Miller, M.L. (1993) The rise of coastal and marine tourism. *Ocean and Coastal Management* **20**, 181–99.

Mittermeier, R.A., Myers, N., Thomsen, J.B. et al. (1998) Biodiversity hotspots and major tropical wilderness areas: Approaches to setting conservation priorities. *Conservation Biology* **12**(3), 516–20.

Murray, S.N., Ambrose, R.F., Bohnsack, J.A. et al. (1999) No-take reserve networks: Sustaining fishery populations and marine ecosystems. *Fisheries* **24**(11), 11–25.

Myers, N. (1996) Two key challenges for biodiversity: Discontinuities and synergisms. *Biodiversity and Conservation* Special Issue **5**, 1025–34.

Myers, N., Mittermeier, R.A., Mittermeier, C.G. et al. (2000) Biodiversity hotspots for conservation priorities. *Nature* **403**, 853–8.

NAS (1990) *Managing Troubled Waters. The Role of Marine Environmental Monitoring*. Committee on a Systems Assessment of Marine Environmental Monitoring, National Academy of Sciences. National Academy Press, Washington, D.C.

Naylor, R.L., Goldburg, R.J., Mooney, H. et al. (1998) Nature's subsidies to shrimp and salmon farming. *Science* **282**, 883–4.

NRC (1995) *Understanding Marine Biodiversity: A Research Agenda for the Nation*. Oceans Studies Board. National Academy Press, Washington, D.C.

O'Neill, R.V. (2001) Is it time to bury the ecosystem concept? (with full military honors, of course!). *Ecology* **82**(12), 3275–84.

Odum, E.P. (1989) *Ecology and our Endangered Life-Support Systems*. Sinauer Associates, Inc., Sunderland, Mass.

Olsen, S. & Hale, L. (1994) Coasts: The ethical dimension. *People and Planet* **3**(1), 29–31.

Olson, D.M. & Dinerstein, E. (1998) The Global 2000: A representation approach to conserving the earth's most biologically valuable ecoregions. *Conservation Biology* **12**(3), 502–15.

Parsons, D.J., Swetnam, T.W. & Christensen, N.L. (1999) Invited feature: Uses and limitations of historical variability concepts in managing ecosystems. *Ecological Applications* **9**(4), 1177–8.

Parsons, T.R. (1991) Editorial: Impact of fish harvesting on ocean ecology. *Marine Pollution Bulletin* **22**(5), 217.

Parsons, T.R. & Seki, H. (1995) A historical perspective of biological studies in the ocean. *Aquatic Living Resources* **8**, 113–22.

Pimm, S.L. (1997) The value of everything. *Nature* **387**, 231–2.

Porter, G. & Brown, J.W. (1991) *Global Environmental Politics*. Westview Press, Boulder, Colo.

Pressey, R.L., Humphries, C.J., Margules, C.R. et al. (1993) Beyond opportunism: Key principles for systematic reserve selection. *Trends in Ecology and Evolution* **8**(4), 124–8.

Pringle, C.M. (2001) Hydrologic connectivity and the management of biological reserves: A global perspective. *Ecological Applications* **11**(4), 981–98.

Ray, G.C. (1999) Petruchio's Paradox: The oyster or the pearl. In *The Living Plantation in Crisis* (eds. J. Cracraft and F.T. Gritos), pp. 128–36. Columbia University Press, New York.

Ray, G.C. & McCormick-Ray, M.G. (1992) *Marine and Estuarine Protected Areas: A Strategy for a National Representative System within Australian Coastal and Marine Environments*. Consultancy report for the Australian National Parks and Wildlife Service, Canberra, Australia.

Rhodes, R.A.W. (1997) *Understanding Governance*. Open University Press, Buckingham.

Roberts, C.M., McClean, C.J., Veron, J.E.N. et al. (2002) Marine biodiversity hotspots and conservation priorities for tropical reefs. *Science* **295**, 1280–4.

Roberts, C.M. & Polunin, N.V.C. (1991) Are marine reserves effective in management of reef fisheries? *Review Fish Biology and Fisheries* **1**, 65–91.

Ruggiero, R., Wilson, B., Gummer, J. et al. (2000) *Trade, Investment and Environment.* The Brookings Institute, Washington, D.C.

Russ, G.C. & Acala, A.C. (1996) Marine reserves: Rates and patterns of recovery and decline of large predatory fish. *Ecological Applications* **6**(3), 947–61.

Russ, G.C. & Acala, A.C. (1998a) Natural fishing experiments in marine reserves (1983–1993): Community and trophic responses. *Coral Reefs* **17**, 383–97.

Russ, G.C. & Acala, A.C. (1998b) Natural fishing experiments in marine reserves (1983–1993): Roles of life history and fishing intensity in family responses. *Coral Reefs* **17**, 399–416.

Safina, C. (1998) The world's imperiled fish. *Scientific American* **9**(3), 58–63.

Salm, R. (1986) Coral reefs and tourist carrying capacity: The Indian ocean experience. *UNEP: Industry and Environment* **9**(1), 11–14.

Salvat, B. (1976) Guidelines for the planning and management of marine parks and reserves. In *Proceedings of an International Conference on Marine Parks and Reserves*, Tokyo, Japan, 1975, pp. 75–90. IUCN Publication New Series **37**. IUCN, Morges, Switzerland.

Sampson, G.P. & Bradnee, W., eds. (2000) *Trade, Environment, and the Millennium.* The Brookings Institute, Washington, D.C.

Schubel, J.R. (1996) Integrated coastal management (ICM) and the new governance: Who's on first? Abstract. In *Our Coastal Seas. What is their Future?* (eds. A. Brooks, W. Bell & J. Greer), p. 68. Maryland Sea Grant Publication, College Park, Md.

Sharp, G.D. (2000) The past, present and future of fisheries oceanography: Refashioning a responsible fisheries science. In *Fisheries Oceanography. An Integrative Approach to Fisheries Ecology and Management* (eds. P.J. Harrison & T.R. Parsons), pp. 207–62. Blackwell Science, Oxford.

Sherman, K. & Duda, A.M. (1999) Large marine ecosystems: An emerging paradigm for fishery sustainability. *Fisheries* **24**(12), 15–26.

Snape, W.J., III, ed. (1995) *Biodiversity and the Law.* Island Press, Washington, D.C.

Solbrig, O.T. (1991) *From Genes to Ecosystems: A Research Agenda for Biodiversity.* Report of an IUBS–UNESCO Workshop, Harvard Forest, Mass., June 1991. International Union of Biological Sciences, Paris.

Steele, J.H. & Schumacher, M. (2000) Ecosystem structure before fishing. *Fisheries Research* **44**, 201–5.

Stokke, O.S. (1999) Regimes as governance systems. In *Global Governance. Drawing Insight from the Environmental Experience* (ed. O.R. Young), pp. 27–63. The MIT Press, Cambridge, Mass.

Thorne-Miller, B. & Catena, J. (1991) *The Living Ocean: Understanding and Protecting Marine Biodiversity.* Island Press, Washington, D.C.

Van Dyke, J.M., Zaelke, D. & Hewison, G., eds. (1993) *Freedom for the Seas in the 21st Century: Ocean Governance and Environmental Harmony.* Island Press, Washington, D.C.

Wade, B.A. (1996) The challenges of ship-generated garbage in the Caribbean. In *Marine Debris: Sources, Impacts, and Solutions* (eds. J.M. Coe & D.B. Rogers), pp. 229–37. Springer–Verlag, New York.

Wallerstein, I. (1994) Development: Lodestar or illusion? In *Capitalism and Development* (ed. L. Sklair), pp. 3–20. Routledge, London.

Ward, T.J., Vanderklift, M.A., Nicholls, A.O. & Kenchington, R.A. (1999) Selecting marine reserves using habitats and species assemblages as surrogates for biological diversity. *Ecological Applications* **9**(2), 691–8.

Watson, R. & Pauly, D. (2001) Systematic distortions in world fisheries catch trends. *Nature* **414**, 534–6.

WCED (1987) *Our Common Future.* World Commission on Environment and Development. Oxford University Press, Oxford.

Yeager, P.C. (1991) *The Limits of Law: The Public Regulation of Private Pollution.* Cambridge University Press, Cambridge.

Index

NOTES:

- This index is organized hierarchically: that is, "sperm whale" is listed as "whale, sperm" together with all other whales; etc.
- Names of oceans, seas, and place names occur often in the text and are listed only with reference to particular topics.
- Species are listed sparingly, and include only those that are considered as best examples of specific topics. Most species are listed by their common names, followed by scientific names in parentheses; if more than one species fall under a single genus, the listing is, for example, Crab, king (*Paralithodes* spp.).
- Page numbers in *italics* refer to figures and/boxes; those in **bold** refer to tables.

A

B

C

D

Q

R

S

T

U

QH
541.5
C65
C5916
2003

Coastal-marine
conservation.

$69.95

DATE			

QH
541.5
C65

CARROLL COMMUNITY COLLEGE LMTS

Coastal – marine conservation.

00000009598616

WITHDRAWN

Library & Media Ctr.
Carroll Community College
1601 Washington Rd.
Westminster, MD 21157

JUN 18 2004

BAKER & TAYLOR